Laser Direct Energy Deposition 3D Printing of Superalloys

Xiaoqi Chen
South China University of Technology (SCUT)
Guangzhou, China

Ze Chai
Tongji University
Shanghai, China

Xiaoqiang Zhang
Jiangsu University of Science and Technology
Zhenjiang, China

CRC Press
Taylor & Francis Group
Boca Raton London New York

CRC Press is an imprint of the
Taylor & Francis Group, an **informa** business
A SCIENCE PUBLISHERS BOOK

First edition published 2026
by CRC Press
2385 NW Executive Center Drive, Suite 320, Boca Raton FL 33431

and by CRC Press
4 Park Square, Milton Park, Abingdon, Oxon, OX14 4RN

CRC Press is an imprint of Taylor & Francis Group, LLC

Library of Congress Cataloging-in-Publication Data (applied for)

ISBN: 978-1-032-52349-1 (hbk)
ISBN: 978-1-032-52355-2 (pbk)
ISBN: 978-1-003-40632-7 (ebk)

DOI: 10.1201/9781003406327

Typeset in Times New Roman
by Prime Publishing Services

Preface

The convergence of additive manufacturing and nickel-based superalloys has catalyzed a paradigm shift in high-value component manufacturing, particularly for aerospace turbines, energy systems, critical maintenance and repair applications. *Laser Direct Energy Deposition 3D Printing of Superalloys* consolidates a decade of the authors' translational research systematically addressing fundamental challenges in process modelling and optimization, microstructure control, and mechanical performance enhancement for industrial-scale deployment.

This book begins with a comprehensive introduction to additive manufacturing technologies covering their fundamental principles, development, classifications, applications, and emerging trends (Chapter 1).

Chapter 2 provides an overview of additively manufactured solution- and precipitation-strengthened superalloys and their applications, with a focus on their unique microstructural features and resulting anisotropic mechanical properties.

Chapter 3 presents a physics-informed neural network methodology that incorporates heat transfer physics as constraints within the loss function, treating the thermal diffusion coefficient as a learnable parameter. This physics-guided hybrid methodology enables accurate 3D temperature field predictions for Inconel 718 multi-layer deposition, boasting superior interpretability and computational efficiency.

Chapter 4 establishes mathematical correlations between deposition geometry, molten pool temperatures, and process parameters to develop quantitative optimization strategies for Z-axis lift height, laser power, and scanning speed—effectively addressing uneven deposition-layer thickness issues in Inconel 718.

Chapter 5 develops comprehensive decision-making models, comprising a data-driven XGBoost model, a reverse inference model and a horizontal overlap step matching model, to enable efficient parameter determination in laser direct energy deposition.

Chapter 6 introduces a novel gradient-laser-power deposition strategy that breaks coarse long-chain Laves phases, refines matrix grains, alleviates micro-segregation, and enhances mechanical properties of Inconel 718 thin walls.

Chapter 7 further elucidates the influence of gradient-laser-power parameters on the thermal accumulation, cooling rates, macroscopic features, microstructural characteristics, and mechanical properties for the manufactured Inconel 718 thin walls.

Chapter 8 examines post-heat-treatment microstructural evolution and mechanical performance of both constant-laser-power and gradient-laser-power deposited Inconel 718 thin walls, subjected to direct aging, solution treatment plus aging, or homogenization plus solution treatment plus aging.

Chapter 9 introduces ultrasonic vibration-assisted deposition for microstructure control and mechanical performance enhancement of Inconel 718, leveraging on multi-physics simulations to reveal the role of ultrasound in melt pool flow and heat transfer.

Chapter 10 systematically reviews cracking mechanisms, susceptibility metrics, and suppression strategies for laser direct energy deposition of γ'-rich nickel-based superalloys.

Chapter 11 explores columnar dendritic structures, phase constituents, and cracking mechanisms in laser direct energy deposition of Inconel 738LC—a difficult-to-weld superalloy with high Al/Ti content.

Chapter 12 employs multiscale advanced characterization techniques to establish quantitative correlations between process parameters and liquation/solid-state cracking, providing guidelines for defect-minimization in Inconel 738LC deposition.

Finally, Chapter 13 presents bimetallic layered heterostructures via alternating IN738 and IN718 deposition, designed to suppress cracking and enhance mechanical performance in laser direct energy deposition of difficult-to-weld superalloys.

This book provides materials and manufacturing engineers with methods and techniques for implementing robust laser-directed energy deposition (L-DED) processes. It equips graduate students in these fields with a rigorous understanding of process-structure-property relationships through experimentally validated models, while offering academics cutting-edge methodologies to advance research paradigms. Complex concepts are elucidated with minimal prerequisites, ensuring accessibility without sacrificing technical depth.

The authors would like to express sincere gratitude to all team members for their valuable contributions to this book. These include Dr. Ze Chai, former postdoctoral fellow, for his role in co-supervising postgraduate research projects; former PhD students Xiaoqiang Zhang, Luming Xu, and Bo Peng; and former master's students Jibing Xie and Anqi Wang. Their dedication to cutting-edge research has been instrumental in bringing this book project to fruition. Additionally, the authors wish to extend their appreciation to the editors at CRC Press for their professional guidance throughout the book's preparation process.

Xiaoqi Chen
Shien-Ming Wu School of Intelligent Engineering
South China University of Technology
Guangzhou, China

Ze Chai
School of Aerospace Engineering and Applied Mechanics
Tongji University
Shanghai, China

Xiaoqiang Zhang
School of Materials Science and Engineering
Jiangsu University of Science and Technology
Zhenjiang, China

Contents

Chapter 1
Overview of Additive Manufacturing

1.1 Additive manufacturing technology

Additive manufacturing (AM), developed in the 1980s, is also known as rapid prototyping [1]. It is defined by the American Society for Testing and Materials (ASTM) as "the process of joining materials to make objects from 3D model data, typically layer by layer, in contrast to subtractive manufacturing methods; synonyms: 3D printing, additive fabrication, additive processes, additive techniques, additive layer manufacturing, layer manufacturing, and freeform fabrication" [2]. According to ASTM, the scope of additive manufacturing technologies is divided into several categories (Table 1.1): binder jetting, direct energy deposition, material extrusion, material jetting, powder bed fusion, sheet lamination, and vat photopolymerization [2, 3].

Binder Jetting: The process of depositing liquid binder onto a powder bed. It can be used with materials such as gypsum, sand, glass, metals, etc.

DED (Directed Energy Deposition): Metal is fed as powder or wire feedstock into an energy source (such as an electron beam or laser beam) mounted on a multi-axis robotic arm. The material is melted layer by layer onto a substrate. It is used with metals like Titanium and Cobalt-chrome.

Material Extrusion: Material is deposited onto a substrate from an extrusion machine. Typically, thermoplastic filaments are melted through a heating mechanism and extruded through a nozzle. However, the same process can be used for viscous materials like concrete, clay, organic tissue, and even food.

Material Jetting: Specialized print heads, similar to piezoelectric print heads in 2D inkjet printers, jet liquid material onto a substrate. In most cases, the material is a photosensitive plastic resin (also known as photopolymer) that is then cured with ultraviolet (UV) light.

Powder Bed Fusion: This is a process where an energy source (such as a laser or electron beam) is directed at a powder bed, heating individual particles until they melt together. This technology is commonly associated with metals like titanium and plastics like Nylon.

Sheet Lamination: In this process, sheets are fused together and shapes are selectively etched out in each layer. The final object is then removed from the binding sheets. This rare 3D printing technique is currently used not only with paper but also with metals and plastics.

Vat Photopolymerization: Large vat of photopolymer resin is exposed to an energy source, such as a laser beam or digital light projector, to selectively cure the material layer by layer. This process is often associated with thermosetting plastics.

Each category encompasses several different processes, but they all share the principle of selective modeling for layers. Parts manufactured using additive manufacturing technologies exhibit some degree of anisotropy due to the presence of layers. Anisotropy can be reduced by selecting appropriate orientations during the part manufacturing process. In terms of materials, various

DOI: 10.1201/9781003406327-1

Table 1.1. ASTM classification (2012) [2].

Process categories	Technology	Material
Binder Jetting	3D printing Ink-jetting S-Print M-Print	Metal Polymer Ceramic
Direct Energy Deposition	Direct Metal Deposition Laser Deposition Laser Consolidation Electron Beam Direct Melting	Metal Powder And wire
Material extrusion	Fused Deposition Modeling	Polymer
Material Jetting	Polyjet Ink-jetting Thermojet	Photopolymer Wax
Powder bed fusion	Selective Laser Sintering Selective Laser Melting Electron Beam Melting	Metal Polymer Ceramic
Sheet lamination	Ultrasonic Consolidation Laminated Object Manufacture	
Vat photopolymerization	Stereolithography Digital Light Processing	

polymers, ceramics, metals, and composite materials can be used for AM. The choice of materials depends on the type of AM process employed [4].

The first application of additive manufacturing was in the field of rapid prototyping, followed by the mold-making industry. These application areas continue to be developed, but the growing performance of additive manufacturing technologies has expanded their use in direct production. Industries such as aerospace, which require highly complex aircraft components, have recognized the potential of additive manufacturing technology and are investing in research to improve its reliability and applicability [4, 5]. In the medical field, highly personalized applications such as orthodontics, prosthetics, orthotics, implants, and substitute organs, produced using AM processes, have reached a certain level of maturity and utilization [6].

Additive manufacturing and traditional manufacturing face trade-offs in terms of production capabilities. Additive manufacturing has the potential to reduce waste, shorten delivery times and costs, and enable the design of products with complex parts. Additionally, AM allows for the construction of assembled objects by reducing expenses associated with documentation and production planning [13]. Furthermore, by eliminating the need for tools, part designs can achieve the same specifications as those manufactured through traditional processes but with less material. Additive manufacturing technologies can quickly respond to the market by producing spare parts on-demand, reducing or eliminating the need for inventory.

This chapter will provide an overview of the development of 3D printing, from its origins to contemporary manifestations, while also providing information on how different 3D printing processes work.

1.2 The development of AM and vat photopolymerization

Like many other technologies, additive manufacturing was conceived by several independent inventors around the same time. Hideo Kodama of Nagoya Municipal Industrial Research Institute

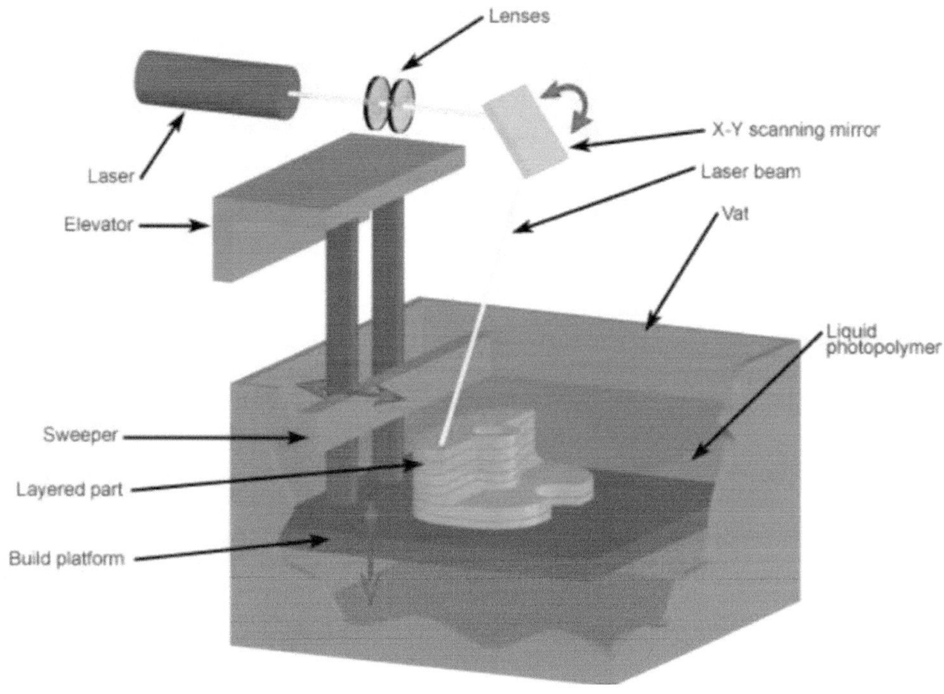

Figure 1.1. Working principle of SL.

developed two 3D printing technologies in 1981 [8]. Another version of the technology was patented in 1984 by Alain Le Méhauté, Olivier de Witte, and Jean Claude André of the French General Electric and CILAS companies [9]. Just three weeks prior, an inventor from the United States, Chuck Hull, filed his own patent for a 3D printing technology known as Stereolithography (SL) [10], which is a form of photopolymerization (Figure 1.1). Chuck Hull became widely recognized as the inventor of this technology once Compagnie Générale d'Électricité and CILAS relinquished their patent, despite the significant work done by his predecessors. With his SL technology, Hull established 3D Systems in Valencia, California, in 1986. The company's first printer, the SLA-1 [11], was commercialized in 1987. It relied on selectively curing a photosensitive resin point by point and layer by layer using ultraviolet lasers, recreating physical objects based on the design of CAD files. The company also introduced the Stereolithography (STL) file format [12], which is still widely used as a file type for 3D printing. Photopolymerization has been pursued and continuously developed by numerous companies since its inception. In 2001, EnvisionTEC, headquartered in Germany, introduced the first machine in their Perfactory production line, showcasing a unique form of photopolymerization technology called Digital Light Processing (DLP).

While SL relies on precise laser curing of photosensitive resin point by point, Envision TEC's DLP technology projects light from a UV projector onto the resin vat, curing the entire layer at once, enabling fine details and fast printing speeds [13].

In recent years, during the 2010s, this technology has evolved to utilize light sources other than DLP projectors, similar to the ones people might purchase from electronics stores. For example, the use of LED projectors has further significantly reduced the cost of this technology.

Currently, most available SL and DLP technologies can generate intricate objects very quickly. However, since these parts are typically translucent and made of photopolymers, their color tends to degrade over time when exposed to sunlight. Photopolymers are thermosetting polymers, which means that once they are solidified, these plastics cannot be melted and reshaped. Despite the development of various photopolymers for 3D printing, the parts may not always be suitable for functional applications as they tend to fracture rather than yield under stress.

Vat photopolymerization is known for its ability to produce high-resolution and detailed parts with smooth surface finishes. It is commonly used in applications where precision and fine details are crucial, such as jewelry, dental models, and prototypes in various industries.

1.3 Metallic additive manufacturing

1.3.1 Directed energy deposition

Although 3D printing with plastics may become highly valuable for many industries, aerospace and defense manufacturers are particularly interested in the development of metal 3D printing technologies, especially Directed Energy Deposition (DED) and powder bed fusion processes.

DED, also known as blown powder additive manufacturing or laser cladding, which involves introducing metal powder into a heat source, such as a laser, and melting the metal particles together during deposition (Figure 1.2). In 1988, Frank Arcella from Westinghouse Electric Corporation filed the first patent for powder bed metal 3D printing technology [15], and later in 1997, he developed DED technology at Johns Hopkins University and commercialized it through his company, Aeromet [16]. Around the same time, Optomec in New Mexico began commercializing similar technology developed at Sandia National Laboratories. While Aeromet closed its operations in 2005, Optomec continued producing its laser-engineered net shaping (LENS) metal 3D printers [17]. Due to the ability to directly feed metal powder into the heat source, typically attached to a 4-axis or 5-axis arm, DED systems are not limited to 3D printing on flat substrates. Instead, metals can be printed onto curved surfaces, such as existing metal structures. For this reason, laser cladding is commonly used for repairing damaged components, particularly in the aerospace industry. DED machines may also not be limited in terms of printing volume. Companies like Sciaky have developed extremely large-scale systems to 3D print massive metal parts close to their final geometries before further machining (Figure 1.3). Some companies have even started researching the capability of using DED to manufacture entire aircraft fuselages, although the results of this work have not been publicly disclosed.

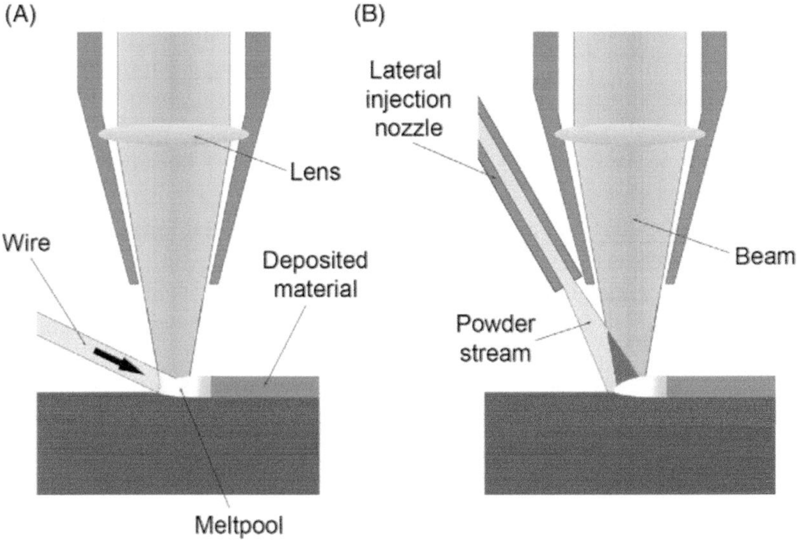

Figure 1.2. In directed energy deposition, a metal feedstock is introduced to an energy source in the form of a wire (A) or as a powder (B) [14].

Figure 1.3. Directed energy deposition (DED) processes involve creating a nearnet shape object, that is, a part will need to be further processed (above) to reach the actual desired shape (below). Though rough, DED is quick and capable of producing largescale objects [14].

1.3.2 Power bed fusion

Unlike DED systems, powder bed machines are housed in an inert gas chamber, where a high-power energy source (usually a laser) selectively melts metal particles layer by layer, similar to the SLS process for plastics. In fact, EOS is one of the global leaders in SLS printers and was the first to commercialize the powder bed metal 3D printing process in 1995, known as Direct Metal Laser Sintering [16].

Around the same time, the Fraunhofer Institute for Laser Technology in Aachen, Germany, developed a similar powder bed technology called Selective Laser Melting (SLM) [18]. The main difference between Fraunhofer's technology and EOS's technology is that the metal particles are fully melted together rather than just sintered, resulting in denser and stronger parts that do not require post-printing heat treatment. The commercial forms of SLM come from a company now known as SLM Solutions and another company called Realizer. Electron Beam Melting (EBM) is a specialized category of SLM technology that relies on an electron beam instead of a laser, resulting in faster build times. This technology was developed by Chalmers University of Technology in Gothenburg, Sweden, and later commercialized by Arcam AB [19].

Powder bed fusion technology may be better suited for producing intricate parts and can be used for small batch production when the machine is large enough. While parts produced using powder bed melting 3D printing require extensive post-processing to remove support structures and improve surface finish, they do not require the same level of machining as parts produced using DED (Figure 1.4) [14]. Although the Xline 2000R from Concept Laser is the largest powder bed fusion 3D printer in the world, with a build volume of up to 2.6 feet × 1.3 feet × 1.6 feet (0.8 meters × 0.4 meters × 0.5 meters) [20], it is still not as large-scale as some DED machines. Sciaky's DED technology, known as Electron Beam Additive Manufacturing (EBAM), can be used to produce parts with dimensions up to 19 feet × 4 feet × 4 feet (5.79 meters × 1.22 meters × 1.22 meters) [21].

Figure 1.4. Parts made with powder bed fusion are capable of much finer detail than directed energy deposition, but may be limited in size and will still require postprocessing to obtain this final result [14].

1.3.3 Selective Laser Melting (SLM)

The Selective Laser Melting (SLM) process was jointly developed by Dr. M. Fockele and Dr. D. Schwarze from F&S Stereolithographietechnik GmbH, along with Dr. W. Meiners, Dr. K. Wissenbach, and Dr. G. Andres from Fraunhofer ILT. It is used for producing metal components from metal powder. SLM is a powder bed fusion process that utilizes a high-intensity laser as the energy source to selectively melt the powder layer by layer based on computer-aided design (CAD) data.

The SLM process involves a series of steps from CAD data preparation to the removal of the fabricated components from the build platform. Before the 3D CAD data is uploaded to the SLM machine, the STL (Standard Tessellation Language) file must be processed through software to provide support structures for any overhanging features and generate the sliced data (2D cross-sections) for laser scanning each layer. The building process starts by laying a thin layer of metal powder on the build platform in the build chamber. Once the powder is spread, the laser selectively fuses the chosen areas based on the processed data using a high-energy density laser. After the laser scan is completed, the build platform is lowered, and the next layer of powder is deposited on top, followed by laser scanning of the new layer. This process is repeated until the component is completed. In the SLM process, due to the high temperatures required for melting, the parts are manufactured in a controlled oxygen atmosphere using inert process gases (argon or nitrogen) to prevent oxidation and other issues that may affect the mechanical properties of the finished parts [22].

In SLM, laser power, scanning speed, hatch distance, and layer thickness are common process parameters for optimizing the process. These parameters affect the volumetric energy density available for heating and melting the powder, mechanical properties, and surface roughness of the produced parts. During the heating and melting process, the heat capacity and latent heat are largely material-dependent and proportional to the mass being melted.

Low laser power, high scanning speed, and large layer thickness often result in insufficient energy, leading to the phenomenon of balling due to poor wetting of the previous layer on the melt pool [23]. The balling phenomenon significantly affects the quality of the formed part. Additionally, inadequate hatch distance between adjacent melt tracks often leads to regular porosity in manufactured parts because the adjacent molten lines do not fully fuse together. The laser scanning strategy used for manufacturing each layer affects the thermal gradient within the part, thus influencing the performance of the finished part [24–27]. The primary scanning patterns include parallel stripe arrays, spirals, contour lines from the part perimeter to the center, and zigzag trajectories [24–25, 27]. Specifying certain process variables, such as the scan spacing between adjacent laser passes, is necessary when choosing any scanning strategy [22, 24]. There needs to be a certain degree of overlap between adjacent melt zones to ensure complete material densification. It is evident that recoating process parameters, such as recoater type, recoater speed, powder dosage, and layer thickness, impact the density and performance of each layer and the finished part [22, 28–30].

The SLM process parameters can be used to improve surface quality [31–34], although they are generally not as good as traditional manufacturing methods. Therefore, various post-processing techniques have been adopted, including sandblasting, machining, etching, electro-polishing, and plasma spraying. However, these techniques are not always applicable to complex components. The alloys currently available for this process include, stainless steel, cobalt-chromium (Co-Cr) alloys, nickel-based alloys, aluminum (Al-Si-Mg) alloys, and titanium (Ti6Al4V) alloy. Martensitic precipitation hardening steels, which combine martensite and aging, exhibit high strength, toughness, good weldability, and dimensional stability during the aging heat treatment process. They are primarily used in the aircraft and aerospace industries, where excellent mechanical properties, weldability, and superior machinability are essential for tool applications [35]. Stainless steel 316L is one of the most extensively studied materials in SLM due to its wide applications in marine, biomedical devices, and fuel cells [36–39]. Cobalt-chromium-molybdenum-based high-temperature alloys offer excellent mechanical properties, corrosion resistance, and temperature resistance. Over the years, cobalt-chromium-molybdenum (CoCrMo) has been used in biomedical, dental restorations, and orthopedic implants (such as joint replacements and fracture fixation applications) due to being the hardest known biocompatible alloy with good corrosion resistance, high tensile and fatigue properties [36, 40–43]. Nickel-based superalloys (chromium-nickel-iron alloys, chromium-nickel-iron-cobalt alloys, etc.) are widely used in aerospace turbine engines, high-speed body components, high-temperature bolts and fasteners, and nuclear engineering, among other fields, due to their excellent balance of creep, damage tolerance, tensile properties, corrosion resistance, and oxidation resistance [44]. Aluminum alloys are of primary interest for lightweight applications in the aerospace and automotive industries. AlSi10Mg is an age-hardenable casting alloy with good mechanical properties. It has good castability and weldability due to its composition being close to the eutectic Al-Si. The addition of Mg enables age hardening, improving strength through the sequential precipitation of Mg2Si. For these reasons, AlSi10Mg is a good candidate for SLM. The process parameters, support structures, microstructure, mechanical properties, and surface roughness of this alloy have been extensively studied [45–49]. Tensile strength of parts produced by SLM using titanium alloys (usually Ti6Al4V) can surpass those produced by traditional manufacturing processes. Many properties of Ti6Al4V are limited by the microstructure (α and β phases) controlled during solidification. It has been shown that the performance of the Ti6Al4V alloy is significantly influenced by the solidification process. A significant amount of research has been conducted on the fabrication of Ti6Al4V alloy parts using SLM, making it an ideal material for aerospace, dental, and facial applications [50–53]. Additionally, titanium exhibits excellent corrosion resistance in physiological environments.

SLM technology offers the possibility to create complex architectures inspired by biological systems. The lightweight internal and external structures of the powerripper developed by Festo [54] take into account the forces acting on the component and can only be produced in this form through metal laser melting. The powerripper is modeled based on the complex kinematics of bird beak breaking.

1.3.4 Electron Beam Melting (EBM)

In the Electron Beam Melting (EBM) process, a high-power electron beam of approximately 4 kilowatts is used as the energy source instead of a laser, which can only be used for conductive metals. This process is also referred to as Selective Electron Beam Melting (SEBM) or Electron Beam Additive Manufacturing (EBAM). In fact, the energy per unit volume (energy density) is higher than that of laser equipment, and the electron beam is controlled by electromagnetic coils, allowing for increased melting capacity, resulting in higher productivity compared to SLM machines. Arcam, based in Sweden, is the only company that has developed and commercialized EBM machines [55]. The first patent was filed in 1992 [56], describing a process for producing 3D components by layer-wise melting of conductive powder using an electron beam. However, it was not until 2002 that the first commercial EBM system was developed by Arcam. Currently, several commercial EBM machines are available, and Arcam has also developed and commercialized multi-beam machines. So far, over 150 EBM systems have been installed in universities, research centers, and industries worldwide. In addition to Arcam, there are also research centers and universities contributing to the development of evidence-based medical technologies for specific applications [57]. Due to its higher energy density compared to SLM technology, EBM usually allows for processing higher powder layer thicknesses (even exceeding 100 μm). However, when grown at high production rates, the surface quality of the parts is poorer, and post-processing is a drawback [58].

Due to the thickness of powder layers and the typical grain size, the powder distribution in EBM ranges from 45 to 150 μm, which is larger than the particle size distribution used in SLM. Karlsson et al. [59] reported a study that evaluated the effect of using smaller particle size powder (25–45 μm): material properties were not affected, so there might be an opportunity to improve the surface smoothness of parts even with reduced productivity when using smaller particle size powder. In line with this, compared to SLM, the EBM process typically results in lower resolution and higher surface roughness. So far, the surface roughness of EBM parts has been around 30–50 μm in terms of Ra [60]. Similar to roughness, there have been efforts in the literature to improve the geometrical accuracy of EBM parts, as described by Smith et al. [61]. Like SLM, the powder can be recycled in the EBM process, and there are some studies on the recyclability of powder in EBM. For example, Tang et al. [62] and Petrovic et al. [63] evaluated the impact of powder recycling time on Ti6Al4V, and it appears that many builds using recycled powder can be completed without significant changes in an inert atmosphere in the SLM chamber.

One fundamental difference between SLM and EBM powder bed processes is that the EBM chamber operates in a vacuum atmosphere (approximately 1×10^{-5} millibar), reducing any contamination picked up during the process. This low contamination level is particularly useful when dealing with reactive materials such as titanium alloys or gamma titanium aluminide intermetallics. Another important difference is that the electron beam, due to its high scanning rate, allows for general preheating of the powder before melting (depending on the processing material, even up to 1100°C), which serves as a focal point for reducing thermal stresses in the growing components [57]. This results in stress-relieved parts, limiting the risk of crack formation and reducing the production of materials with poor toughness or even brittleness, such as gamma titanium aluminide intermetallics [64–67]. In terms of the industrial market for EBM, two main areas can be highlighted, as also mentioned on the Arcam website [55]: aerospace and orthopedic implants.

In the field of orthopedic implants, EBM has been used since 2007 to produce components such as acetabular cups, knee joints, maxillofacial plates, hip joints, mandibular replacements, etc., which have obtained CE certification and approval from the U.S. Food and Drug Administration (FDA) in 2010 [68–72]. Currently, since April 2014, over 40,000 titanium alloy acetabular cups produced by EBM have been implanted, featuring porous surfaces for improved bone integration, accounting for approximately 2% of the total production of acetabular cups [57]. The two most commonly used metals for EBM production of orthopedic implants are Ti6Al4V and CoCr, and there have been good evaluations of these two materials in the literature by Sing et al. [68] and Murr et al. [73].

1.4 Other technologies of additive manufacturing

1.4.1 Fused deposition modeling

In the late 1980s and early 1990s, several other companies started introducing new non-SL technologies to the rapidly growing 3D printing market. One of the most important companies was Stratasys, which invented a material extrusion process called Fused Deposition Modeling

(FDM), introduced in 1992 [74]. FDM is an extrusion-based process where thermoplastic materials (primarily in the form of filament) are heated to their melting point and deposited layer by layer onto a build platform. Unlike thermosetting plastics, thermoplastics can be melted and cooled multiple times. Due to the heat extrusion process associated with FDM, the technology is known for producing functional and robust parts that are more durable than those manufactured using SL. In fact, Stratasys is one of the few FDM manufacturers capable of 3D printing with aerospace-grade thermoplastics such as polyetherimide (PEI). For this reason, FDM parts can be found in performance-critical applications such as aircraft or spacecraft. At the same time, FDM parts may have lower resolution and slower print speeds compared to SL. By using soluble support materials to 3D print support structures that hold up the parts during printing, complex objects and moving parts can be created. Once the object is complete, the support material is dissolved away, enabling the ability to 3D print fully assembled components in a single print job. Another unique aspect of FDM is that the low-cost components required to build a simple plastic extrusion 3D printer eventually made it possible to construct affordable open-source 3D printers like the RepRaps, which stands for "replicating rapid prototyper." The low cost of the printers and the openness of the RepRap project, founded by Professor Adrian Bowyer of the University of Bath, led to an explosion of desktop 3D printers [75]. This includes the now infamous MakerBot, which became the poster child for the recent boom in the 3D printing industry in the early 2010s. MakerBot was acquired by Stratasys in 2013, but prior to that, the brand had already brought 3D printing technology to the attention of the world [76]. Around the same time Bowyer and his students were developing open-source FDM 3D printers, Hod Lipson at Cornell University also developed an open-source 3D printer called Fab@Home. Lipson and his students demonstrated various applications of material extrusion, including 3D printing food and other viscous materials. While Lipson continued his research in 3D printing, Fab@Home played a crucial role in showcasing the versatility of using various materials with this open-source system [77].

1.4.2 Sheet lamination

In 1991, a company called Helisys pioneered the introduction of sheet lamination technology to the market. The company's laminated object manufacturing process fused together layers of material and used digitally guided laser cutting to form the desired objects. While Helisys eventually ceased operations in 2000, other companies began using proprietary sheet lamination technologies since then to achieve new manufacturing purposes.

In 2003, Conor MacCormack and Fintan MacCormack developed a unique sheet lamination technology that relied on standard office paper to 3D print objects. In this process, a standard inkjet printer would print a shape onto a sheet of paper, which was then adhered to another sheet of paper. A tungsten carbide blade would then trace the shape on the stacked sheets. This process would continue until the object was complete and had to be peeled away from the paper to be revealed. This technology was commercialized by the MacCormack brothers' company, Mcor, and later advanced by the company to produce full-color 3D printing. By printing CMYK colors onto the paper, Mcor's paper-based printing process could produce vivid objects most suited for marketing purposes and visual prototypes [78].

As a partial subtractive process, sheet lamination typically cannot achieve the same geometric complexity as other 3D printing processes. Due to the limited access to the internal parts of an object, it is also not possible to remove excess material from within the object.

1.4.3 Selective Laser Sintering (SLS)

Selective Laser Sintering (SLS) is a powder bed fusion technology that was developed at the University of Texas in Austin with funding from the Defense Advanced Research Projects Agency (DARPA) before being commercialized by DTM Corporation. SLS was introduced to the market in 1992 and uses a laser to selectively fuse plastic powder, typically nylon, into complete 3D objects.

DTM was eventually acquired by its biggest competitor, 3D Systems, in 2001 [79]. It's worth noting that in the early 1990s, a German company called Electro Optical Systems (EOS) entered the field and commercialized the SL system in 1990 before introducing the SLS system in 1994. EOS later became one of the major manufacturers of SLS systems worldwide [80].

What sets SLS apart is that the powder bed in an SLS printing system can serve as built-in support for the part being sintered, unlike SL and FDM, which require separate support structures for 3D printing. This allows for the printing of highly complex geometries, including interlocking and moving parts such as gears and hollow objects. While the strength of the parts produced through SLS may be comparable to FDM parts, the cost depends on the materials used, and SLS is generally more expensive. For this reason, SLS machines are typically used in industrial environments, while FDM may be used by consumers, prosumers (or "makers"), and industrial settings.

1.4.4 Material jetting

Another technology developed in the 1990s was wax 3D printing, invented by Solidscape, later known as Sanders Prototype Inc., and commercialized in 1993 [81]. What sets this technology apart is that the objects it produces can be immediately cast into high-precision metal parts with high burnout, laying the foundation for 3D printing as a secondary process for creating end-use parts. In 1996, 3D Systems introduced their own wax 3D printing technology, which relied on material jetting. This would later serve as the basis for the company's MultiJet Printing (MJP) technology, which relies on piezoelectric inkjet printheads, similar to traditional 2D printers, to deposit photopolymerizable plastic resin or casting wax with high resolution. The ink is then cured with UV light, solidifying each layer until completion [82].

In 2000, an Israeli company called Objet entered the field with its own material jetting technology called PolyJet, surpassing 3D Systems in this area. Objet eventually merged with Stratasys in 2013, forming Stratasys Ltd., integrating global FDM and PolyJet technologies [83]. Stratasys has surpassed 3D Systems and holds a leading market share in 3D printing globally [81]. The strength of material jetting technologies like MJP and PolyJet lies in their ability to adjust material properties based on various physical characteristics. With machines like Stratasys' J750, it becomes possible to 3D print different physical properties in a single object, enabling multi-color, multi-material printing in practice.

1.4.5 Binder jetting

While 3D Systems may not have completely dominated the field of material jetting, it has taken a leading position compared to another color 3D printing technology under 3D Systems—ColorJet Printing (CJP). This binder jetting process was invented by MIT and commercialized by Z Corporation in 1996, also utilizing piezoelectric inkjet printheads. However, this technology does not deposit photosensitive ink but instead deposits liquid binder and colored ink onto a layer of gypsum powder, resulting in full-color sandstone-like prints [84].

When Z Corporation's full-color 3D printers entered the market at a price of tens of thousands of dollars, they were much cheaper than other industrial systems [85]. In the 2010s, the emergence of low-cost RepRaps also sparked the imagination of consumers with full-color page binder jetting, allowing them to create realistic 3D prints of people captured by 3D scanning. Since the initial media attention to these 3D printed selfies, consumer fascination with page binder jetting has somewhat diminished. However, the industrial applications of binder jetting are still extensive. In 1999, Extrude Hone AM, now known as ExOne, began licensing binder jetting technology from MIT to develop its own 3D printing metal parts technology. By using fine metal powder instead of gypsum powder, ExOne's technology was obtained. These metal parts are then placed in a furnace for further sintering into dense metal components. ExOne also applied the same method to 3D print large-scale sand molds, which are cast into metal parts, as well as ceramics and glass [86]. Similarly, a German company called Voxeljet has licensed MIT's technology for large-scale sand 3D printing [87], while the American pharmaceutical company Aprecia has licensed binder jetting technology to develop a new method for 3D printing medications. These FDA-approved drugs can dissolve immediately in a patient's mouth, making them ideal for patients with swallowing issues [88].

1.5 Applications of additive manufacturing

1.5.1 Manufacturing and damage repair of high-end components in the manufacturing field

For example, Directed Energy Deposition with Laser (DED-L) is commonly used in the manufacturing and damage repair of high-end components. The basic principle of DED-L involves the delivery of metal powder through a gas stream and its deposition by a laser processing head. Under the effect of a high-energy laser beam, the metal powder is melted along a predetermined scanning path. After rapid cooling and solidification, a deposited layer is formed. By layer-by-layer deposition, a three-dimensional model is ultimately created [89]. The main features of DED-L technology are as follows: (1) The formable size is only limited by the range of motion mechanisms, allowing for efficient production of large-sized components; (2) Different types of materials can be mixed during deposition, resulting in compositional gradient materials; (3) It enables rapid repair of damaged parts.

In recent years, in order to fully integrate the advantages of PBF and DED, hybrid manufacturing processes combining multiple materials and techniques have emerged. As shown in Figure 1.5, the National Aeronautics and Space Administration (NASA) used PBF-L technology to additively manufacture a GRCop-42 copper alloy combustion chamber. Using this as a base, they employed DED-L technology to additively manufacture a large-scale integral channel nozzle using NASA's self-developed Fe-Ni-based high-temperature alloy NASA-HR-1 [90], thereby achieving hybrid manufacturing with dual materials and dual processes.

Currently, additive manufacturing technologies can achieve good mechanical properties for several main metal materials, including stainless steel (304 [91], 316L, 17-4 PH [92–95]),

Figure 1.5. Coupled BP-DED nozzle with L-PBF chamber development. (a) GRCop-42 chamber with bimetallic joint prepared for BP-DED, (b) BP-DED process of coupled manufacturing demonstrator, (c) Completed coupled BP-DED/L-PBF bimetallic demonstrator, (d) 2K-lbf coupled hardware for hot-fire testing [90].

nickel-based alloys (Inconel 718, Inconel 625, Hastelloy X, Invar 36) [96], titanium alloys (Ti-6Al-4V [97, 98]), and lightweight alloys (AlSi10Mg [99]), among others. However, many commonly used materials in industrial applications have not been extensively studied in relation to additive manufacturing, or they possess poor printability, making them challenging to apply in additive manufacturing, thus severely restricting the widespread application of additive manufacturing.

In the field of high-end equipment manufacturing such as aviation, shipbuilding, tanks, and power generation, gas turbines often serve as expensive core power components. The processing of gas turbines involves complex processes, significant material loss, and extremely high technical difficulties, making them the crown jewels of the manufacturing industry. Typically, gas turbine blades operate in harsh environments characterized by high temperature, high speed, strong vibrations, and severe corrosion, resulting in occurrences of damage and failure. Figure 1.6 shows

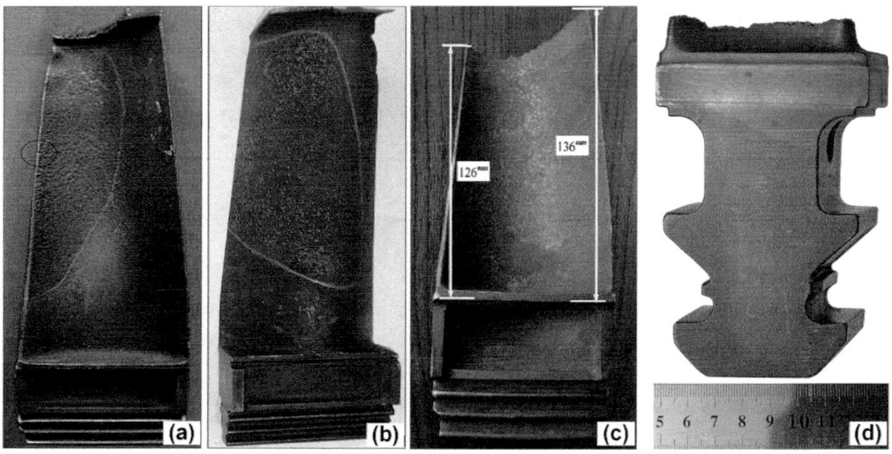

Figure 1.6. Damaged Inconel 738 blades, (a) pitting on front concave, (b) pitting on convex (c) fracture at the top [100], (d) fracture on the root [101].

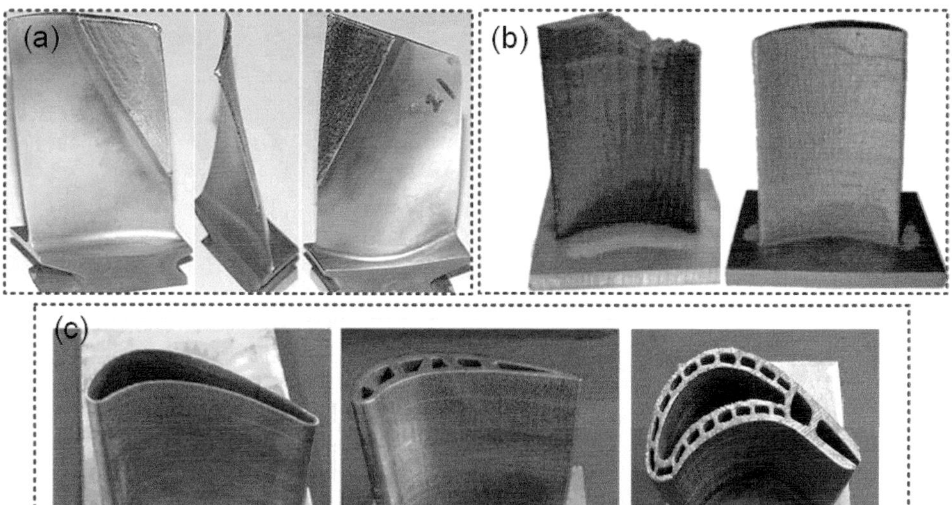

Figure 1.7. Additive Manufacture and repair of blade by DED-L, (a) Inconel 718 alloy [104], (b) Inconel 625 alloy [105], (c) 316L stainless steel [106, 107].

examples of damaged Inconel 738 blades, including pitting on the blade sides and fractures at the top or root [100, 101]. To save costs, such damaged blades require extensive repair and remanufacturing [102, 103]. DED-L technology happens to be one of the key technologies for high-quality repair. As shown in Figure 1.7 Qi Huan et al. [104] repaired Inconel 718 alloy blades, Wilson et al. [105] performed blade additive manufacturing and repair using Inconel 625 alloy, and a team led by Academician Lu Bingheng from Xi'an Jiaotong University conducted research on turbine blade additive manufacturing using 316L stainless steel [106, 107].

1.5.2 Visualization

Traditionally, 3D printing has been known as a rapid prototyping technology used for visually displaying pre-production designs. As a prototyping technique, 3D printing can be a faster and more accurate method compared to manual design fabrication. These models can be essential for saving costs associated with mass production. Before investing in expensive tools and molds required for manufacturing large quantities of goods, 3D printed prototypes can be used to visualize critical design elements. The various processes mentioned above are suitable for different prototyping applications. For example, SL (stereolithography) and DLP (digital light processing) are suitable for fine-detail features, while FDM (fused deposition modeling) is used for mechanical testing. Although SL and DLP parts may be fragile, they can reflect the details that will be incorporated into the final product. On the other hand, FDM may not achieve the same resolution, but it can produce parts using the same material that will be used for mass production. CJP (color jet printing) can be used for visualizing the color and aesthetic qualities of a product; however, PolyJet can reflect actual material properties, such as the flexibility of rubber or the transparency of glass.

Prototyping and Product Development: Additive manufacturing enables rapid prototyping and iteration, allowing designers and engineers to quickly create physical prototypes for testing and validation. It accelerates the product development cycle, reduces costs, and facilitates design optimization.

Manufacturing and Production: Additive manufacturing is increasingly being used for small-batch and customized production. It offers the flexibility to produce complex geometries and customized products on-demand, eliminating the need for costly tooling or molds. Industries such as aerospace, automotive, consumer goods, and healthcare are adopting additive manufacturing for low-volume production.

Healthcare and Medical Applications: Additive manufacturing has transformed the medical field by enabling personalized healthcare solutions. It is used for the production of patient-specific implants, prosthetics, surgical tools, and anatomical models. Additive manufacturing also plays a crucial role in tissue engineering, biofabrication, and drug delivery systems.

Automotive Industry: Additive manufacturing is increasingly integrated into automotive manufacturing processes. It allows for the production of lightweight and complex components, such as engine parts, brackets, exhaust systems, and customized interior components. Additive manufacturing in automotive applications enables design optimization, weight reduction, and improved performance.

Architecture and Construction: Additive manufacturing is being explored for architectural and construction purposes. Large-scale 3D printers can create building components, such as walls, facades, and structural elements, with intricate designs and reduced material waste. It has the potential to revolutionize construction techniques, enable architectural innovation, and enhance sustainability.

Consumer Goods and Design: Additive manufacturing provides opportunities for customization and personalized design in the consumer goods industry. It enables the production of unique jewelry, fashion accessories, home decor, and consumer electronics with intricate designs and personalized features.

Education and Research: Additive manufacturing is increasingly integrated into educational institutions and research laboratories. It allows students and researchers to explore design possibilities, learn about advanced manufacturing processes, and innovate in various fields.

These are just a few examples of the application areas of additive manufacturing. As the technology continues to advance and evolve, new applications and industries are likely to adopt additive manufacturing for enhanced design capabilities, production efficiency, and customization.

1.5.3 Medical and healthcare field

Additive manufacturing, also known as 3D printing, has made significant contributions to the medical and healthcare field. Here are some key applications of additive manufacturing in this industry.

Customized Implants and Prosthetics: Additive manufacturing enables the production of patient-specific implants and prosthetics tailored to an individual's anatomy. It allows for precise customization, better fit, and improved patient comfort. Examples include cranial implants, hip and knee implants, dental implants, and prosthetic limbs.

Surgical Instruments and Tools: Additive manufacturing allows for the production of complex and specialized surgical instruments and tools. Surgeons can have access to customized instruments designed for specific procedures, enhancing surgical precision and efficiency.

Anatomical Models and Surgical Planning: Additive manufacturing facilitates the creation of accurate anatomical models from medical imaging data. These models help surgeons visualize complex structures, plan surgeries, and simulate procedures, leading to better surgical outcomes and reduced operating time.

Tissue and Organ Fabrication: Additive manufacturing is being explored for the bioprinting of tissues and organs. Researchers are working on developing techniques to print living cells, biomaterials, and bioinks to create functional tissues and even organs for transplantation. While still in the experimental stage, this technology holds the potential to revolutionize regenerative medicine.

Prosthetic and Orthotic Devices: Additive manufacturing allows for the production of cost-effective and customized prosthetic and orthotic devices. It offers rapid prototyping and production capabilities, making these assistive devices more accessible and affordable.

Pharmaceutical Applications: Additive manufacturing is used in pharmaceutical research and development, including the production of personalized medicine, drug delivery systems, and dosage forms. 3D printing allows for precise control over drug release profiles and the creation of complex drug formulations.

Surgical Guides and Implant Templates: Additive manufacturing enables the production of surgical guides and implant templates that assist surgeons in accurate and minimally invasive procedures. These guides help ensure precise implant placement, reduce surgical complications, and enhance patient outcomes.

Medical Education and Training: Additive manufacturing provides medical professionals and students with realistic anatomical models for training and education purposes. It allows for hands-on learning, surgical simulation, and the practice of complex procedures in a risk-free environment.

The application of additive manufacturing in the medical and healthcare field continues to advance, offering innovative solutions for personalized patient care, improved surgical outcomes, and advancements in regenerative medicine.

1.5.4 Architecture and construction industries

Additive manufacturing, also known as 3D printing, is increasingly being utilized in the architecture and construction industries.

Rapid Prototyping and Concept Modeling: Additive manufacturing allows architects and designers to quickly create physical models and prototypes of their designs, facilitating better visualization, communication, and understanding of the architectural concepts.

Complex Geometries and Customization: Additive manufacturing enables the production of intricate and complex architectural elements, structures, and components that would be challenging or costly to manufacture using traditional methods. It allows for the realization of unique, custom-designed features and intricate details in building designs.

Construction Components and Elements: Additive manufacturing can produce construction components and elements, such as façade panels, walls, columns, and beams, using various materials like concrete, polymers, and composites. 3D printing these elements offers design freedom, reduces material waste, and can speed up the construction process.

Sustainable Construction: Additive manufacturing allows for more sustainable construction practices by reducing material waste and energy consumption. It enables the precise deposition of materials only where needed, minimizing waste generation. Additionally, the ability to reuse and recycle materials in the additive manufacturing process contributes to sustainable construction practices.

On-Site Construction and Repair: Portable and mobile additive manufacturing systems can be deployed to construction sites for on-demand production of construction components and repair parts. This can save time and costs associated with sourcing and transporting materials and components.

Prototyping and Testing of Building Systems: Additive manufacturing can be used to prototype and test building systems and innovative construction techniques before full-scale implementation. This helps identify design flaws, optimize performance, and ensure the viability of new construction approaches.

Architectural Installations and Artistic Creations: Additive manufacturing allows architects and artists to create large-scale architectural installations, sculptures, and artistic creations with intricate designs and details. It offers new possibilities for artistic expression and the integration of art into architectural projects.

Disaster Relief and Affordable Housing: Additive manufacturing has the potential to address housing challenges in disaster-stricken areas and regions with housing shortages. It offers rapid and cost-effective construction solutions, enabling the production of affordable and resilient housing units.

1.5.5 Education and research

Additive manufacturing, also known as 3D printing, has had a significant impact on education and research across various disciplines.

Hands-on Learning: Additive manufacturing provides students with a hands-on learning experience. It allows them to design and fabricate physical objects, enabling a deeper understanding of concepts in fields such as engineering, architecture, design, and science.

Prototyping and Design Iteration: Additive manufacturing allows students and researchers to rapidly prototype and iterate their designs. They can bring their ideas to life and test their concepts in a tangible and visual form, facilitating design refinement and innovation.

STEM Education: Additive manufacturing plays a crucial role in STEM (Science, Technology, Engineering, and Mathematics) education. It helps students apply theoretical concepts in practical settings, fostering critical thinking, problem-solving, and creativity.

Research Tool: Additive manufacturing serves as a valuable research tool in various disciplines. Researchers can use it to fabricate custom equipment, specialized components, and prototypes for experiments, enabling them to explore new ideas, validate hypotheses, and conduct innovative studies.

Material Science and Engineering: Additive manufacturing allows researchers to investigate and develop new materials and material properties. It facilitates the exploration of advanced materials, composites, and alloys, as well as their processing techniques and applications.

Biomedical Research: Additive manufacturing has revolutionized biomedical research. Researchers can create patient-specific anatomical models, develop tissue engineering scaffolds, and fabricate biomedical devices and implants. It aids in studying human anatomy, surgical planning, and advancements in regenerative medicine.

Archaeology and Paleontology: Additive manufacturing enables the replication of artifacts, fossils, and archaeological findings with great precision. Researchers can create accurate replicas for study, conservation, exhibition, and educational purposes, while preserving the original artifacts.

Collaborative Research and Data Sharing: Additive manufacturing facilitates collaboration among researchers by enabling the sharing of digital designs and 3D printing files. Researchers can collaborate on projects, share expertise, and replicate experiments, enhancing knowledge exchange and interdisciplinary research.

Educational Models and Teaching Aids: Additive manufacturing allows for the creation of educational models, teaching aids, and visual aids. It helps explain complex concepts, demonstrate scientific principles, and enhance classroom engagement and comprehension.

Entrepreneurship and Innovation: Additive manufacturing fosters an entrepreneurial and innovative mindset among students and researchers. It provides opportunities for developing prototypes, starting businesses, and exploring novel applications and solutions across various industries.

The use of additive manufacturing in education and research continues to expand, providing new avenues for learning, experimentation, and breakthrough discoveries. It equips students and researchers with practical skills, promotes interdisciplinary collaboration, and drives innovation in multiple fields of study.

1.6 Future technologies of additive manufacturing

The future of additive manufacturing, also known as 3D printing, holds tremendous potential for various advancements and innovations. Here are some key areas that could shape the future of additive manufacturing:

1.6.1 Increased Materials Diversity

The increased materials diversity in additive manufacturing refers to the expansion of available materials beyond the traditional plastics, metals, and ceramics. Here are some advancements and possibilities in this area:

Advanced Composites: Additive manufacturing is exploring the use of advanced composite materials, such as carbon fiber-reinforced polymers (CFRP) and glass fiber composites. These materials offer high strength-to-weight ratios and can be used in applications where lightweight and strong parts are required, such as aerospace, automotive, and sporting goods.

Bioactive Materials: Researchers are developing 3D-printable biomaterials that are compatible with living tissues and can be used in biomedical applications. These materials can be tailored to promote cell growth, tissue regeneration, and implant integration, opening up new possibilities in areas like personalized medicine and prosthetics.

Conductive Inks and Electronics: Additive manufacturing techniques are being developed to print conductive inks, allowing the integration of electronics directly into 3D-printed objects. This enables the creation of functional circuits, sensors, and antennas within complex geometries, expanding the potential for smart devices and Internet of Things (IoT) applications.

Food-Based Materials: The field of food 3D printing is emerging, allowing the creation of edible and customized food products. By using edible materials, such as chocolate, dough, or purees, 3D printers can produce intricately designed food items or even personalized nutrition based on individual needs.

Sustainable and Recyclable Materials: In line with growing environmental concerns, additive manufacturing is exploring sustainable materials, including bio-based polymers, recycled plastics, and biodegradable materials. These materials aim to reduce waste, promote circular economy practices, and minimize the environmental impact of 3D printing.

Hybrid Materials: Researchers are combining different materials to create hybrid structures with unique properties. For example, combining metals and ceramics can produce parts with enhanced strength and thermal properties. These hybrid materials offer new possibilities for functional and customized applications.

Smart Materials: Additive manufacturing is enabling the integration of smart materials that can respond to external stimuli, such as temperature, light, or electrical signals. These materials, known as shape memory alloys, shape memory polymers, or stimuli-responsive polymers, have applications in fields like aerospace, robotics, and medicine.

The expansion of materials diversity in additive manufacturing opens up new possibilities for various industries and applications. As research and development in this area continue, we can expect to see even more innovative and specialized materials that cater to specific needs and requirements in different fields.

1.6.2 Multi-material and multi-property printing

Multi-material and multi-property printing in additive manufacturing refers to the ability to simultaneously print objects using different materials with distinct properties. This capability allows for the creation of complex structures and functional parts with varying mechanical, electrical, thermal, or chemical characteristics. Here are some key aspects of multi-material and multi-property printing:

Simultaneous Printing: Additive manufacturing technologies are being developed to print objects using multiple materials in a single printing process. This can involve either combining different materials in a single extrusion system or using multiple print heads to deposit different materials simultaneously. By seamlessly integrating different materials, complex geometries and structures can be created.

Gradient Materials: With multi-material printing, it is possible to create objects with graded material compositions or properties. This means that the material properties gradually change within the printed part, allowing for smooth transitions between different mechanical, chemical, or thermal characteristics. Graded materials enable the production of parts with optimized performance and functionality.

Functionally Graded Materials (FGM): FGMs involve printing objects with varying material compositions and properties in specific regions. This technique allows for the design of parts with tailored characteristics at different locations. For example, a part can have regions with high strength, flexibility, or conductivity precisely where they are needed.

Embedded Components: Multi-material printing enables the integration of embedded components or structures within a printed object. This involves printing around or within pre-existing components, such as electronics, sensors, or reinforcement elements. Embedding components directly into the printed object reduces the need for assembly and improves the overall functionality and integrity of the part.

Material Mixing and Blending: Additive manufacturing techniques are being developed to mix or blend different materials on a microscale. This allows for the creation of composite materials with tailored properties by controlling the distribution and alignment of different materials within the printed object. Material mixing techniques enhance the strength, durability, and functionality of the printed parts.

Sequential Printing: Some additive manufacturing systems allow for sequential printing of different materials or property regions. This involves printing one material or property region first and then switching to another material or property region in subsequent layers. Sequential printing allows for the creation of parts with distinct sections or functionalities.

Applications of multi-material and multi-property printing span various industries, including aerospace, automotive, electronics, healthcare, and consumer goods. This capability enables the production of complex and customized objects with diverse functionality and performance, leading to enhanced product designs, improved efficiency, and expanded possibilities for innovation.

1.6.3 High-speed printing

High-speed printing in additive manufacturing refers to the ability to rapidly fabricate objects with additive processes, significantly reducing the overall printing time. This advancement in speed has several benefits, including increased productivity, reduced production costs, and improved scalability. Here are some key aspects of high-speed printing in additive manufacturing:

Enhanced Printing Systems: High-speed printing requires advancements in printing systems and technologies. This includes faster and more precise motion control systems, improved extrusion or deposition mechanisms, and optimized printing paths. These enhancements enable faster and more efficient material deposition, resulting in reduced printing times.

Rapid Layering Techniques: Traditional layer-by-layer printing can be time-consuming, especially for large and complex objects. High-speed printing explores innovative layering techniques that allow for the simultaneous deposition of multiple layers or larger layers in a single pass. This reduces the overall printing time while maintaining printing accuracy and quality.

Advanced Material Formulations: The development of materials specifically designed for high-speed printing is crucial. These materials should have optimized flow properties, faster curing or solidification times, and improved interlayer adhesion. High-speed printing often requires materials that can be rapidly deposited and solidified without sacrificing part quality or mechanical properties.

Parallel Printing: High-speed printing can involve the simultaneous printing of multiple parts or objects using an array of print heads or multiple printing platforms. Parallel printing significantly increases production throughput as multiple parts are fabricated simultaneously. This approach is particularly useful for small-sized objects or objects with simple geometries.

Integration of Automation: Automation plays a vital role in high-speed printing. It involves the integration of robotic systems, advanced material handling, and automated post-processing steps. Automation streamlines the entire printing process, reduces human intervention, and minimizes idle time, leading to increased printing speed and overall productivity.

Continuous Printing: Continuous printing techniques aim to eliminate the need for layer-wise printing, enabling a continuous flow of material deposition. This can involve the use of continuous extrusion processes or continuous liquid resin printing. Continuous printing allows for faster fabrication times as the printing process is not interrupted between layers.

High-speed printing is particularly beneficial for industries that require rapid prototyping, small-batch production, or on-demand manufacturing. It enables the efficient production of functional prototypes, customized products, and spare parts, reducing lead times and increasing flexibility in manufacturing operations. As high-speed printing technologies continue to advance, additive manufacturing is becoming more competitive with traditional manufacturing methods in terms of production speed and efficiency.

1.6.4 Continuous and large-scale printing

Continuous and large-scale printing in additive manufacturing refers to the ability to fabricate objects without the need for layer-wise printing and the ability to produce objects on a large scale. This approach offers several advantages, including increased production efficiency, reduced fabrication time, and the ability to create objects of significant size. Here are key aspects of continuous and large-scale printing in additive manufacturing:

Continuous Printing Processes: Continuous printing techniques aim to eliminate the layer-by-layer approach by enabling a continuous flow of material deposition. This can be achieved through various methods such as continuous extrusion processes, continuous liquid resin printing, or other innovative approaches. Continuous printing allows for faster fabrication times as the printing process is not interrupted between layers, resulting in a seamless and uninterrupted production process.

Robust Printing Systems: Large-scale printing requires robust and reliable printing systems capable of handling the increased volume of materials and extended printing durations. This includes reinforced gantry systems, stable platforms, and extrusion or deposition mechanisms that can handle high material flow rates and prolonged operation. The printing systems should also have advanced monitoring and control systems to ensure print quality and consistency throughout the process.

Material Handling and Supply Systems: Continuous and large-scale printing requires efficient material handling and supply systems. This includes automated material feeding, integrated material storage and delivery systems, and optimized material flow pathways. These systems ensure a steady and uninterrupted supply of materials during the printing process, minimizing downtime and maximizing productivity.

Scalable Printing Platforms: To achieve large-scale printing, the printing platforms need to be scalable to accommodate larger object sizes. This may involve expanding the build volume of the printers or developing modular systems that allow for the seamless integration of multiple printing units to work in parallel. Scalable printing platforms enable the fabrication of objects of various sizes, from small prototypes to full-scale industrial components.

Post-Processing Integration: Large-scale printing often requires post-processing steps such as curing, cooling, or surface finishing. Integration of these post-processing steps within the printing system or in close proximity to it minimizes handling and transportation times, further optimizing the overall production workflow.

Real-Time Monitoring and Quality Control: Continuous and large-scale printing can benefit from advanced real-time monitoring and quality control systems. These systems utilize sensors, cameras, and software algorithms to monitor the printing process, detect defects or anomalies, and make necessary adjustments in real-time. Real-time monitoring ensures the quality and integrity of the printed objects throughout the continuous and large-scale fabrication process.

Continuous and large-scale printing has the potential to revolutionize manufacturing industries by enabling the efficient production of large objects, architectural structures, and even entire buildings. It offers significant advantages in terms of speed, cost-effectiveness, and design freedom, opening up new possibilities for applications in construction, aerospace, automotive, and other industries that require the fabrication of large and complex components.

1.6.5 Advanced design and simulation tools

Advanced design and simulation tools play a crucial role in additive manufacturing by facilitating the optimization of designs, predicting the behavior of printed parts, and ensuring their structural integrity. Here are some key aspects of advanced design and simulation tools in additive manufacturing:

Design Optimization: Advanced design tools allow engineers and designers to optimize the geometry, topology, and lattice structures of objects to maximize their performance while minimizing material usage and weight. These tools use generative design algorithms, parametric modeling, and optimization algorithms to explore a vast design space and identify the most efficient and functional designs for additive manufacturing.

Finite Element Analysis (FEA): FEA is widely used in additive manufacturing to simulate and analyze the mechanical behavior and structural integrity of printed parts. FEA software can predict stress distribution, deformation, and failure points, helping to identify potential issues and optimize the design to enhance strength and durability. It enables virtual testing of designs, reducing the need for physical prototyping and iterative testing.

Thermal Analysis: Thermal simulation tools are crucial for additive manufacturing processes that involve melting or sintering of materials. These tools simulate the heat transfer, temperature distribution, and thermal stresses during the printing process, allowing designers to optimize the printing parameters, such as laser power, scan speed, and cooling strategies, to ensure proper material melting, minimal residual stresses, and dimensional accuracy.

Microstructure Simulation: Advanced simulation tools can predict the microstructure and material properties of printed parts based on the printing parameters and material characteristics. By simulating the solidification, phase transformation, and grain growth during the printing process, engineers can optimize the printing parameters to achieve desired material properties, such as strength, hardness, and thermal conductivity.

Support Structure Optimization: Additive manufacturing often requires the use of support structures to prevent deformation and maintain the stability of overhanging or complex geometries. Advanced design tools can automatically generate and optimize support structures to minimize material usage, reduce post-processing efforts, and improve the surface finish of printed parts.

Process Simulation: Process simulation tools simulate the entire additive manufacturing process, including material deposition, heat transfer, and part distortion. These tools help identify potential issues such as warping, residual stresses, and build failures, allowing engineers to optimize the printing process parameters, such as scan strategy, layer thickness, and build orientation, to achieve accurate and high-quality prints.

Multi-scale Modeling: Additive manufacturing involves multiple length scales, from microscale features to macroscopic part dimensions. Multi-scale modeling tools enable the integration of different modeling techniques and scales to capture the behavior of printed parts accurately. This includes the coupling of macroscopic structural analysis with mesoscale material modeling and microscale analysis of specific features like lattices or surface roughness.

By leveraging these advanced design and simulation tools, engineers and designers can optimize designs, predict performance, reduce material waste, and minimize the need for physical prototyping and iterative testing. This leads to faster development cycles, improved part quality, and increased confidence in the performance and reliability of additive manufactured components.

1.6.6 Integration of AI and machine learning

The integration of AI (Artificial Intelligence) and machine learning techniques in additive manufacturing has the potential to revolutionize the industry by enabling more efficient processes, improved part quality, and advanced capabilities. Here are some ways in which AI and machine learning are being integrated into additive manufacturing:

Process Optimization: AI algorithms can analyze large volumes of data collected during the additive manufacturing process, including sensor data, process parameters, and material characteristics. By using machine learning techniques, AI systems can identify patterns, correlations, and optimal process settings for achieving desired outcomes, such as minimizing defects, reducing build time, or improving material utilization. This enables automated process optimization and adaptive control, leading to more reliable and efficient manufacturing processes.

Defect Detection and Quality Control: AI-powered vision systems can analyze images or scan data of printed parts to detect defects, such as surface irregularities, porosity, or dimensional errors. Machine learning algorithms can be trained on a dataset of defect images to classify and identify various types of defects automatically. Real-time monitoring and analysis of the printing process can enable early detection of anomalies and prompt adjustments, ensuring consistent quality and reducing post-processing efforts.

Design Assistance and Generative Design: AI algorithms can assist engineers in generating optimized designs for additive manufacturing. By analyzing design constraints, performance requirements, and material characteristics, AI systems can generate design suggestions or automatically generate complex geometries and lattice structures that are optimized for additive manufacturing processes. This helps to unlock the full potential of additive manufacturing by creating designs that are difficult to achieve through traditional manufacturing methods.

Material Development and Selection: AI can support the development and selection of materials for additive manufacturing. By analyzing material databases, experimental data, and simulation results, machine learning algorithms can identify material properties, predict material behavior under specific process conditions, and assist in material selection for specific applications. This enables the exploration of new materials and the optimization of material properties to meet the requirements of additive manufacturing processes and printed parts.

Predictive Maintenance: AI algorithms can analyze sensor data from additive manufacturing machines to monitor their condition and predict maintenance needs. By monitoring parameters such as temperature, vibration, or energy consumption, machine learning models can identify patterns indicative of machine degradation or potential failures. This enables proactive maintenance scheduling, minimizing downtime and maximizing machine utilization.

Supply Chain Optimization: AI techniques can be applied to optimize the additive manufacturing supply chain by analyzing factors such as material availability, production schedules, and demand fluctuations. Machine learning algorithms can predict demand patterns, optimize inventory management, and enable dynamic production planning, reducing lead times and improving responsiveness to customer needs.

Overall, the integration of AI and machine learning in additive manufacturing has the potential to enhance productivity, quality, and innovation. These technologies enable process optimization, defect detection, design assistance, material development, predictive maintenance, and supply chain optimization, driving the advancement and widespread adoption of additive manufacturing in various industries.

1.6.7 Bioprinting and tissue engineering

Bioprinting is a technique that uses 3D printing technology to create structures composed of living cells, bioinks, and biomaterials. Instead of using traditional inks or materials, bioprinters deposit living cells, growth factors, and supporting biomaterials layer by layer to create complex living tissues and organs.

The process of bioprinting typically involves three main components: bioink, bioprinter, and a computer-aided design (CAD) model. Bioinks serve as the "ink" for bioprinters and are composed of living cells encapsulated in a biocompatible material. These bioinks are carefully formulated to provide the necessary environment for cell survival and function.

Bioprinters are specialized 3D printers that are capable of handling the delicate nature of living cells. They deposit the bioink layer by layer according to the CAD model, creating complex tissue structures. There are different bioprinting techniques available, such as extrusion-based, inkjet-based, and laser-based bioprinting, each with its own advantages and limitations.

Tissue engineering is a multidisciplinary field that combines principles of biology, engineering, and medicine to create functional tissues and organs in the laboratory. It involves the use of scaffolds, cells, and biochemical factors to regenerate or repair damaged tissues and organs.

In tissue engineering, a scaffold serves as a temporary support structure that guides cell growth and tissue formation. These scaffolds can be produced using additive manufacturing techniques, such as 3D printing, to create intricate structures with defined architectures and mechanical properties. The cells are then seeded onto or into the scaffold, where they attach, proliferate, and differentiate to form functional tissue.

The biochemical factors used in tissue engineering, such as growth factors and signaling molecules, are often incorporated into the scaffold or delivered to the cells to regulate their behavior. These factors can influence cell proliferation, differentiation, and tissue development.

The combination of bioprinting and tissue engineering has the potential to revolutionize regenerative medicine and personalized healthcare. Some of the key applications include:

Organ transplantation: Bioprinting holds the promise of creating patient-specific organs, such as kidneys or livers, that can be transplanted without the need for immunosuppression.

Drug testing and development: Bioprinted tissues can be used as models for drug testing, allowing researchers to study the effects of drugs on human tissues in a controlled environment, reducing the reliance on animal testing.

Disease modeling: Bioprinting can be used to create disease models, enabling researchers to study the mechanisms of various diseases, screen potential therapeutics, and develop personalized treatment strategies.

Tissue repair and regeneration: Bioprinted scaffolds can be used to facilitate the repair and regeneration of damaged tissues, such as bone, cartilage, or skin, by providing structural support and promoting cell growth.

It's important to note that while bioprinting and tissue engineering have shown great promise, there are still numerous technical and regulatory challenges that need to be overcome before these technologies can be widely implemented in clinical settings. However, ongoing research and advancements in the field continue to push the boundaries of what is possible in the realm of additive manufacturing for healthcare applications.

1.6.8 On-demand manufacturing and decentralization

On-demand manufacturing and the decentralization of additive manufacturing are two interconnected concepts that have the potential to transform the manufacturing industry. Let's explore these ideas further:

On-demand Manufacturing: On-demand manufacturing, also known as just-in-time manufacturing, refers to the production of goods in response to customer orders or demand rather than producing items in bulk and storing them in inventory. This approach minimizes the need for large-scale production, storage costs, and the risks associated with unsold inventory.

Additive manufacturing, such as 3D printing, is well-suited for on-demand manufacturing because it enables the production of customized, complex, and low-volume parts without the need for expensive tooling or molds. With 3D printing, the digital design can be directly translated into a physical product, allowing for quick and efficient production.

On-demand manufacturing offers several advantages, including:

Cost reduction: By producing items only when there is a demand, companies can minimize inventory costs associated with storage, maintenance, and potential waste.

Customization and personalization: Additive manufacturing enables the production of highly customized products tailored to individual customer needs and preferences.

Reduced lead times: On-demand manufacturing allows for faster turnaround times, as products can be manufactured and delivered more quickly compared to traditional manufacturing methods.

Supply chain optimization: By producing items closer to the point of demand, companies can reduce transportation costs, minimize logistical complexities, and achieve a more sustainable supply chain.

Decentralization of Additive Manufacturing: The decentralization of additive manufacturing refers to the shift from centralized manufacturing facilities to a distributed network of smaller manufacturing centers or even localized production setups. This shift is made possible by advancements in additive manufacturing technology and the availability of affordable 3D printers.

The decentralization of additive manufacturing offers several benefits:

Accessibility: Smaller-scale additive manufacturing setups can be established in various locations, making the technology more accessible to businesses and individuals. This can empower entrepreneurs, startups, and local manufacturers to produce goods without relying on large-scale centralized facilities.

Localized production: Decentralization enables localized production, reducing the need for long-distance shipping and transportation. This localization can help in reducing carbon emissions, promoting sustainability, and supporting local economies.

Flexibility and agility: Decentralized additive manufacturing allows for greater flexibility and agility in responding to market demands. Local manufacturers can quickly adapt their production capabilities to cater to changing customer needs and emerging trends.

Distributed innovation: Decentralized manufacturing setups encourage localized innovation and experimentation. Small-scale manufacturers can iterate and improve their products more rapidly, driving innovation at a grassroots level.

Challenges and Considerations: While on-demand manufacturing and the decentralization of additive manufacturing offer numerous advantages, there are some challenges and considerations to keep in mind:

Quality control: Maintaining consistent quality across distributed manufacturing setups can be a challenge. Ensuring that products meet the required standards may require standardized processes, quality control measures, and effective communication.

Intellectual property protection: With decentralized manufacturing, there is an increased risk of intellectual property infringement. Safeguarding designs and protecting intellectual property becomes crucial when manufacturing is distributed.

Infrastructure requirements: Establishing a distributed network of additive manufacturing centers requires adequate infrastructure, including access to 3D printers, materials, and skilled technicians. Investments in equipment and training may be necessary.

Regulatory considerations: Regulations related to product safety, quality control, and intellectual property may vary across different regions. Navigating these regulatory landscapes in a decentralized manufacturing environment can be complex.

Overall, on-demand manufacturing and the decentralization of additive manufacturing have the potential to transform traditional manufacturing processes by offering greater flexibility, customization, and efficiency. As technology continues to advance and the ecosystem evolves, these concepts are likely to play an increasingly important role in shaping the future of manufacturing.

1.6.9 Collaboration and open innovation

Collaboration and open innovation are key drivers in advancing additive manufacturing technologies. By fostering collaboration and embracing open innovation principles, the additive manufacturing community can accelerate the development and adoption of new materials, techniques, and applications. Let's explore how collaboration and open innovation contribute to the progress of additive manufacturing:

Knowledge sharing: Collaboration allows researchers, engineers, and industry experts to share knowledge, experiences, and best practices. This sharing of information helps the additive manufacturing community collectively learn and overcome challenges more effectively.

Cross-disciplinary collaboration: Additive manufacturing is a multidisciplinary field that intersects with various domains such as materials science, engineering, biology, and medicine. Collaboration encourages experts from different fields to come together, exchange ideas, and contribute their expertise. This cross-disciplinary collaboration often leads to innovative solutions and breakthroughs.

Standards development: Collaboration is vital for establishing industry standards in additive manufacturing. Standardization ensures interoperability, quality control, and safety across different additive manufacturing systems and materials. Collaborative efforts can lead to the development of standardized testing methods, material properties, and design guidelines, fostering the growth of the industry.

Open-source initiatives: Open-source platforms and communities have played a significant role in advancing additive manufacturing. Open-source initiatives allow for the sharing of software, hardware designs, and knowledge freely. This encourages innovation, lowers barriers to entry, and facilitates collaboration among individuals and organizations.

Crowdsourcing and user feedback: Open innovation principles can be applied to additive manufacturing through crowdsourcing and involving users in the innovation process. By gathering feedback and ideas from a diverse range of individuals and organizations, additive manufacturing technologies can be refined and improved based on real-world needs and perspectives.

Collaborative research and development projects: Collaborative research and development projects bring together academia, industry, and government organizations to work towards common goals in additive manufacturing. These projects enable shared resources, funding, and expertise, leading to accelerated progress and technology transfer.

Consortia and partnerships: Consortia and industry partnerships bring together stakeholders, including companies, research institutions, and government bodies, to collaborate on additive manufacturing initiatives. These collaborations pool resources, expertise, and funding to address technological, commercial, and regulatory challenges collectively.

Education and training: Collaboration plays a crucial role in education and training initiatives in additive manufacturing. Partnerships between academic institutions, industry, and training organizations facilitate the development of curricula, workshops, and certification programs to upskill the workforce and promote knowledge transfer.

By fostering collaboration and embracing open innovation principles, additive manufacturing can benefit from the collective expertise, insights, and resources of a diverse community. This collaborative approach can accelerate the pace of innovation, address challenges more effectively, and promote the widespread adoption of additive manufacturing technologies in various industries.

1.7 Summary

Additive manufacturing technology has begun to demonstrate significant potential and advantages in many industrial sectors. It provides an efficient solution for producing small-batch customized products with complex geometries. The materials used in additive manufacturing offer comparable, if not superior, performance to those obtained through traditional manufacturing methods.

While there are still challenges to address, such as material limitations, scalability for high-volume manufacturing, and regulatory considerations, the prospects of additive manufacturing are significant. Ongoing research, technological advancements, and collaboration within the additive manufacturing community continue to unlock new opportunities and drive its widespread adoption across industries.

References

[1] Kruth, J.-P., Leu, M. C. and Nakagawa, T. (1998). Progress in additive manufacturing and rapid prototyping. CIRP Ann.-Manuf. Technol. 47(2): 525–540.

[2] ASTM F2792-12a. (2012). Standard Terminology for Additive Manufacturing Technologies (Withdrawn 2015). ASTM International, pp. 1–3, West Conshohocken, PA.

[3] Petrovic, V., Gonzalez, J. V. H., Ferrando, O. J., Gordillo, J. D., Puchades, J. R. B. and Griñan, L. P. (2011). Additive layered manufacturing: Sectors of industrial application shown through case studies. Int. J. Prod. Res. 49(4): 1061–1079.

[4] Guo, N. and Leu, M. C. (2013). Additive manufacturing: Technology, applications and research needs. Frontiers Mech. Eng. 8(3): 215–243, Sep. 2013.

[5] Lyons, B. (2012). Additive manufacturing in aerospace: Examples and research outlook. Bridge, Linking Eng. Soc. 42(1): 13–19.

[6] Sandström, C. C. (2015). Adopting 3D printing for manufacturing—Evidence from the hearing aid industry. Technol. Forecasting Social Change 102: 160–168.

[7] Wohlers Report 2014—3D Printing and Additive Manufacturing State of the Industry, Annual Worldwide Progress Report. Wohlers Associates, Cary, NC, USA, 2014.

[8] Kodama, H. (1981). A scheme for three-dimensional display by automatic fabrication of threedimensional model. J. IEICE 64: 1981–4.

[9] Jean-Claude, A. (1986). Disdpositif pour realiser un modele de piece industrielle. National De La Propriete Industrielle.

[10] Hull, C. W. (1984). Apparatus for Production of Three-Dimensional Objects by Stereolithography. US Patent Office.

[11] 3D Systems'First 3D Printer named Historic Mechanical Engineering Landmark by ASME, Press Releases, Latest News and Other Related Resources, 2016. https://www.asme.org/ about-asme/news/press-releases/3d-systems-first-3d-printer-named-historic-mechani (accessed 03.01.17).

[12] Stereo Lithography Interface Specification, 3D Systems, Inc., October 1989.

[13] Molitch-Hou, M. (2016). From Photopolymers to Composites, EnvisionTEC CEO Discusses the Company's 3D Printing Tech. ENGINEERING.com. http://www.engineering.com/3 DPrinting/3DPrintingArticles/ArticleID/13585/ From-Photopolymers-to-Composites-EnvisionTEC-CEO-Discusses-the-Companys-3D-Printing-Tech.aspx (accessed 04.01.17).

[14] Molitch-Hou, M. (2018). Overview of additive manufacturing process[M]. Additive Manufacturing. Butterworth-Heinemann, 2018: 1–38.

[15] Arcella, F. (1989). Casting shapes, US patent 4,818,562 filed March 24th, 1988, published April 4th.

[16] Shellabear, M. and Nyrhilä, O. (2004). DMLS – Development History and the State of the Art. Presented at LANE 2004 Conference; 2004, September 21–24, Erlangen, Germany.
[17] Molitch-Hou, M. (2016). Optomec Introduces Hybrid Metal 3D Printing at Half the Cost. ENGINEERING.com. http://www.engineering.com/3DPrinting/3DPrintingArticles/ArticleID/13251/Optomec-Introduces-Hybrid-Metal-3D-Printing-at-Half-the-Cost.aspx (accessed 04.01.17).
[18] Meiners, W. (1998). Shaped body especially prototype or replacement part production, German patent 19649865 filed December 2nd, 1996, published February 12th, 1998.
[19] History, Arcam, A. B. (n.d.). http://www.arcam.com/company/about-arcam/history/ (accessed 04.01.17).
[20] Sher, D. (2015). World's Largest Laser Metal 3D Printer Finds 1st Customer, 3D Printing Industry. https://3dprintingindustry.com/news/first-x-line-2000-r-metal-3d-printernetherlandspurchased-blok-group-54933/ (accessed 05.01.17).
[21] ELECTRON BEAM ADDITIVE MANUFACTURING (EBAM™), Sciaky. (n.d.). http:// www.sciaky.com/additive-manufacturing/electron-beam-additive-manufacturing-technology (accessed 05.01.17).
[22] Kurzynowski, T., Chlebus, E., Kuźnicka, B. and Reiner, J. (2012). Parameters in selective laser melting for processing metallic powders. Proc. SPIE 8239: 823914, Feb. 2012, doi: 10.1117/12.907292.
[23] Li, R., Liu, J., Shi, Y., Wang, L. and Jiang, W. (2011). Balling behavior of stainless steel and nickel powder during selective laser melting process. Int. J. Adv. Manuf. Technol. 59: 1025–1035, Apr. 2011.
[24] Van Elsen, M. M. (2007). Complexity of Selective Laser Melting: A New Optimization Approach. Ph.D. Dissertation, Department of Mechanical Engineering, Katholieke Univ. Leuven, Leuven, Belgium.
[25] Niebling, F., Otto, A. and Geiger, M. (2002). Analyzing the DMLS-process by a macroscopic FE-model. pp. 384–391. In Proc. 13th Solid Freeform Fabrication Symp.
[26] Kruth, J.-P., Froyen, L., Van Vaerenbergh, J., Mercelis, P., Rombouts, M. and Lauwers, B. (2014). Selective laser melting of iron-based powder. J. Mater. Process. Technol. 149: 616–622, Jun. 2014.
[27] Hauser, C., Childs, T. H. C., Taylor, C. M., Badrossamay, M., Akhtar, S., Wright, C. S. et al. (2003). Direct selective laser sintering of tool steel powders to high density. Part A: Effects of laser beam width and scan strategy. pp. 644–655. In Proc. 14th Solid Freeform Fabrication Symp.
[28] Lott, P., Schleifenbaum, H., Meiners, W., Wissenbach, K., Hinke, C. and Bültmann, J. (2011). Design of an optical system for the *in situ* process monitoring of selective laser melting (SLM). Phys. Procedia 12: 683–690.
[29] Van der Schueren, B. and Kruth, J.-P. (1995). Powder deposition in selective metal powder sintering. Rapid Prototyping J. 1(3): 23–31.
[30] van Elsen, M., Al-Bender, F. and Kruth, J.-P. (2008). Application of dimensional analysis to selective laser melting. Rapid Prototyping J. 14(1): 15–22.
[31] Calignano, F., Manfredi, D., Ambrosio, E. P., Iuliano, L. and Fino, P. (2013). Influence of process parameters on surface roughness of aluminum parts produced by DMLS. Int. J. Adv. Manuf. Technol. 67(9): 2743–2751.
[32] Kruth, J.-P., Badrossamay, M., Yasa, E., Deckers, J., Thijs, L. and Van Humbeeck, J. (2010). Part and material properties in selective laser melting of metals. pp. 1–12. In Proc. 16th Int. Symp. Electromach.
[33] Alrbaey, K., Wimpenny, D., Tosi, R., Manning, W. and Moroz, A. (2014). On optimization of surface roughness of selective laser melted stainless steel parts: A statistical study. J. Mater. Eng. Perform. 23(6): 2139–2148.
[34] Yadroitsev, I. and Smurov, I. (2011). Surface morphology in selective laser melting of metal powders. Phys. Procedia 12: 264–270.
[35] ASM Handbook: Properties and Selection: Irons, Steels, and High-Performance Alloys, vol. 1. Materials Park, OH, USA: ASM International, 1990, pp. 1303–1408.
[36] Miranda, G., Faria, S., Bartolomeu, F., Pinto, E., Madeira, S., Mateus, A. et al. (2016). Predictive models for physical and mechanical properties of 316L stainless steel produced by selective laser melting. Mater. Sci. Eng. A 657: 43–56, Mar. 2016.
[37] Dewidar, M. M., Khalil, K. A. and Lim, J. K. (2007). Processing and mechanical properties of porous 316L stainless steel for biomedical applications. Trans. Nonferrous Met. Soc. China 17(3): 468–473.
[38] Wang, H., Sweikart, M. A. and Turner, J. A. (2003). Stainless steel as bipolar plate material for polymer electrolyte membrane fuel cells. J. Power Sour. 115(2): 243–251.
[39] Xu, F.-L., Duan, J.-Z., Lin, C.-G. and Hou, B.-R. (2015). Influence of marine aerobic biofilms on corrosion of 316L stainless steel. J. Iron Steel Res. Int. 22(8): 715–720.
[40] Wataha, J. C. and Schmalz, G. (2009). Dental alloys. pp. 221–254. In Biocompatibility of Dental Materials. Berlin, Germany: Springer.
[41] Hedberg, S. Y., Qian, B., Shen, Z., Virtanen, S. and Wallinder, I. O. (2014). *In vitro* biocompatibility of CoCrMo dental alloys fabricated by selective laser melting. Dental Mater. 30(5): 525–534.
[42] Takaichi, A., Nakamoto, T., Joko, N., Nomura, N., Tsutsumi, Y., Migita, S. et al. (2013). Microstructures and mechanical properties of Co–29Cr–6Mo alloy fabricated by selective laser melting process for dental applications. J. Mech. Behavior Biomed. Mater. 21: 67–76, May 2013.
[43] Barucca, G., Santecchia, E., Majni, G., Girardin, E., Bassoli, E., Denti, L. et al. (2015). Structural characterization of biomedical Co–Cr–Mo components produced by direct metal laser sintering. Mater. Sci. Eng. C 48: 263–269, Mar. 2015.

[44] Chlebus, E., Gruber, K., Kuxnicka, B., Kurzac, J. and Kurzynowski, T. (2015). Effect of heat treatment on the microstructure and mechanical properties of Inconel 718 processed by selective laser melting. Mater. Sci. Eng. A 639: 647–655, Jul. 2015.

[45] Manfredi, D., Calignano, F., Krishnan, M., Canali, R., Ambrosio, E. P., Biamino, S. et al. (2014). Additive manufacturing of Al alloys and aluminium matrix composites (AMCs). pp. 3–34. In Light Metal Alloys Applications, W. A. Monteiro, Ed. Rijeka, Croatia: InTech.

[46] Calignano, F. (2014). Design optimization of supports for overhanging structures in aluminum and titanium alloys by selective laser melting. Mater. Des. 64: 203–213, Dec. 2014.

[47] Olakanmi, E. O. (2013). Selective laser sintering/melting (SLS/SLM) of pure Al, Al–Mg, and Al–Si powders: Effect of processing conditions and powder properties. J. Mater. Process. Technol. 213: 1387–1405, Aug. 2013.

[48] Brandl, E., Heckenberger, U., Holzinger, V. and Buchbinder, D. (2012). Additive manufactured AlSi10Mg samples using selective laser melting (SLM): Microstructure, high cycle fatigue, and fracture behavior. Mater. Des. 34: 159–169, Feb. 2012.

[49] Siddique, S., Imran, M., Wycisk, E., Emmelmann, C. and Walther, F. (2015). Influence of process-induced microstructure and imperfections on mechanical properties of AlSi12 processed by selective laser melting. J. Mater. Process. Technol. 221: 205–213, Jul. 2015.

[50] Zhou, Y., Wen, S. F., Song, B., Zhou, X., Teng, Q., Wei, Q. S. et al. (2016). A novel titanium alloy manufactured by selective laser melting: Microstructure, high temperature oxidation resistance. Mater. Des. 89: 1199–1204, Jan. 2016.

[51] Pattanayak, D. K., Fukuda, A., Matsushita, T., Takemoto, M., Fujibayashi, S., Sasaki, K. et al. (2011). Bioactive Ti metal analogous to human cancellous bone: Fabrication by selective laser melting and chemical treatments. Acta Biomater. 7(3): 1398–1406.

[52] Vilaro, T., Colin, C. and Bartout, J. D. (2011). As-fabricated and heat-treated microstructures of the Ti–6Al–4V alloy processed by selective laser melting. Metall. Mater. Trans. A 42: 3190–3199, Oct. 2011.

[53] Bertol, L. S., Kindlein, W. Jr., da Silva, F. P. and Aumund-Kopp, C. (2010). Medical design: Direct metal laser sintering of Ti–6Al–4V. Mater. Des. 31(8): 3982–3988.

[54] Festo Group, [Online]. Available: http://www.festo.com.

[55] Arcam AB, [Online]. Available: http://www.arcam.com/.

[56] Larson, R. (1994). Method and device for producing three-dimensional bodies. WO Patent 1994026446 A1.

[57] Wang, J. and Tang, H. (2016). Review on metals additively manufactured by SEBM. Mater. Technol., Adv. Perform. Mater. 31(2): 86–89.

[58] Guo, N. and Leu, M. C. (2013). Additive manufacturing: Technology, applications and research needs. Frontiers Mech. Eng. 8(3): 215–243.

[59] Karlsson, J., Snis, A., Engqvist, H. and Lausmaa, J. (2013). Characterization and comparison of materials produced by electron beam melting (EBM) of two different Ti–6Al–4V powder fractions. J. Mater. Process. Technol. 213(12): 2109–2118.

[60] Algardh, J. K., Horn, T., West, H., Aman, R., Snis, A., Engqvist, H. et al. (2016). Thickness dependency of mechanical properties for thin-walled titanium parts manufactured by electron beam melting (EBM). Additive Manuf. 12: 45–50, Oct. 2016.

[61] Smith, C. J., Derguti, F., Nava, E. H., Thomas, M., Tammas-Williams, S., Gulizia, S. et al. (2016). Dimensional accuracy of electron beam melting (EBM) additive manufacture with regard to weight optimized truss structures. J. Mater. Process. Technol. 229: 128–138, Mar. 2016.

[62] Tang, H. P., Qian, M., Liu, N., Zhang, X. Z., Yang, G. Y. and Wang, J. (2015). Effect of powder reuse times on additive manufacturing of Ti-6Al-4V by selective electron beam melting. JOM 67(3): 555–563.

[63] Petrovic, V. and Niñerola, R. (2015). Powder recyclability in electron beam melting for aeronautical use. Aircraft Eng. Aerosp. Technol. Int. J. 87(2): 147–155.

[64] Tang, H. P., Yang, G. Y., Jia, W. P., He, W. W., Lu, S. L. and Qian, M. (2015). Additive manufacturing of a high niobium containing titanium aluminide alloy by selective electron beam melting. Mater. Sci. Eng. A 636: 103–107, Jun. 2015.

[65] Biamino, S., Penna, A., Ackelid, U., Sabbadini, S., Tassa, O., Fino, P. et al. (2011). Electron beam melting of Ti–48Al–2Cr–2Nb alloy: Microstructure and mechanical properties investigation. Intermetallics 19(6): 776–787.

[66] Murr, L. E., Gaytan, S. M., Ceylan, A., Martinez, E., Martinez, J. L., Hernandez, D. H. et al. (2010). Characterization of titanium aluminide alloy components fabricated by additive manufacturing using electron beam melting. Acta Mater. 58(5): 1887–1894.

[67] Schwerdtfeger, J. and Körrner, C. (2014). Selective electron beam melting of Ti–48Al–2Nb2Cr: Microstructure and aluminium loss. Intermetallics 49: 29–35, Jun. 2014.

[68] Sing, S. L., An, J., Yeong, W. Y. and Wiria, F. E. (2016). Laser and electron-beam powderbed additive manufacturing of metallic implants: A review on processes, materials and designs. J. Orthopaedic Res. 34(3): 369–385.

[69] Cronskär, M., Bäckström, M. and Rännar, L. E. (2013). Production of customized hip stem prostheses—A comparison between conventional machining and electron beam melting (EBM). Rapid Prototyping J. 19(5): 365–372.

[70] Mazzoli, A., Germani, M. and Raffaeli, R. (2009). Direct fabrication through electron beam melting technology of custom cranial implants designed in a PHANToM-based haptic environment. Mater. Design 30(8): 3186–3192.

[71] Jardini, A. L., Larosa, M. A., de Carvalho Zavaglia, C. A., Bernardes, L. F., Lambert, C. S., Kharmandayan, P. et al. (2014). Customised titanium implant fabricated in additive manufacturing for craniomaxillofacial surgery. Virtual Phys. Prototyping 9(2): 115–125.

[72] Jardini, A. L., Larosa, M. A., Maciel Filho, R., de Carvalho Zavaglia, C. A., Bernardes, L. F., Lambert, C. S. et al. (2014). Cranial reconstruction: 3D biomodel and custom-built implant created using additive manufacturing. J. Cranio-Maxillofacial Surgery 42(8): 1877–1884.

[73] Murr, L. E., Gaytan, S. M., Martinez, E., Medina, F. and Wicker, R. B. (2012). Next generation orthopaedic implants by additive manufacturing using electron beam melting. Int. J. Biomater., pp. 1–14, Jun. 2012.

[74] Crump, S. S. (1992). Apparatus and method for creating three-dimensional objects. US Patent Office.

[75] Jones, R., Haufe, P., Sells, E., Iravani, P., Olliver, V., Palmer, C. et al. (2011). Reprap – the replicating rapid prototyper. Robotica 29(1): 177–91.

[76] Stratasys to Acquire MakerBot, Merging Two Global 3D Printing Industry Leaders, Stratasys to Acquire MakerBot, Merging Two Global 3D Printing Industry Leaders | Business Wire, 2013. http://www.businesswire.com/news/home/20130619006431/en/StratasysAcquire-MakerBot-Merging-Global-3D-Printing (accessed 03.01.17).

[77] Malone, E. and Lipson, H. (2006). Fab@Home: The Personal Desktop Fabricator Kit. In: Proceedings of the 17th Solid Freeform Fabrication Symposium; 2006, August 2006, Austin, TX.

[78] Wolf, M. (2013). Article in Forbes, 3D Printing With Paper At Your Local Office Supply Store? Yep, If Mcor Has Its Way. Forbes; 2013, March.

[79] Lou, A. and Grosvenor, C. (2012). Selective Laser Sintering, Birth of an Industry. The University of Texas; 2012. (retrieved 22.03.13).

[80] Turning innovative ideas into a success story: with Additive Manufacturing from prototypes to serial production, Your expert in Additive Manufacturing since 25 years. https:// www.eos.info/about_eos/history (accessed 04.01.17).

[81] Wohlers, T., Caffrey, T., Campbell, I.. Wohlers Report. 2016. Fort Collins, CO: Wohlers Associates; 2016.

[82] Chua, C. K. and Leong, K. F. (2015). 3D Printing and Additive Manufacturing: Principles and Applications. 4th ed. Hackensack, NJ: World Scientific.

[83] Stratasys and Objet Complete Merger. BusinessWire; 2012, December 3.

[84] Grimm, T. (2004). User's Guide to Rapid Prototyping. SME; 2004. ISBN 978-0-87263-697-2, p. 163, (retrieved 31.10.2008).

[85] Molitch-Hou, M. (2016). Ultimaker's John Kawola Talks Dual Extrusion, Automation and the Ultimaker 3 3D Printer. ENGINEERING.Com; 2016. http://www.engineering.com/3DPrint ing/3DPrintingArticles/ArticleID/13435/Ultimakers-John-Kawola-Talks-Dual-ExtrusionAutomation-and-the-Ultimaker-3-3D-Printer.aspx (accessed 04.01.17).

[86] History of ExOne, History | ExOne. http://www.exone.com/about-exone/history (accessed 04.01.17).

[87] Z Corporation Nonexclusive Patent License SublicenseAndAgreement With Voxeljet. SEC. gov; 2014. https://www.sec.gov/Archives/edgar/data/1582581/000104746913009124/ a2216674zex-10_4.htm (accessed 04.01.17).

[88] Our Story, Our Story | Aprecia. (n.d.). https://www.aprecia.com/about-us/our-story.php (accessed 04.01.17).

[89] An Overview of Direct Laser Deposition for Additive Manufacturing Part I: Transport Phenomena, Modeling and Diagnostics [J]. Additive Manufacturing, 2015, 8: S2214860415000317.

[90] Gradl, P. R., Protz, C., Fikes, J., Ellis, D., Evans, L., Clark, A. et al. (2020). Lightweight Thrust Chamber Assemblies using Multi-Alloy Additive Manufacturing and Composite Overwrap; proceedings of the AIAA Propulsion and Energy 2020 Forum, F, 2020 [C].

[91] Wang, Z., Palmer, T. A. and Beese, A. M. (2016). Effect of processing parameters on microstructure and tensile properties of austenitic stainless steel 304L made by directed energy deposition additive manufacturing [J]. Acta Materialia 110: 226–35.

[92] Ali, U., Esmaeilizadeh, R., Ahmed, F., Sarker, D., Muhammad, W., Keshavarzkermani, A. et al. (2019). Identification and characterization of spatter particles and their effect on surface roughness, density and mechanical response of 17-4 PH stainless steel laser powder-bed fusion parts [J]. Materials Science and Engineering A 756: 98–107.

[93] Caballero, A., Ding, J., Ganguly, S. and Williams, S. (2019). Wire + Arc Additive Manufacture of 17-4 PH stainless steel: Effect of different processing conditions on microstructure, hardness, and tensile strength [J]. Journal of Materials Processing Technology 268: 54–62.

[94] Mahmoudi, M., Elwany, A., Yadollahi, A., Thompson, S. M., Bian, L. and Shamsaei, N. (2017). Mechanical properties and microstructural characterization of selective laser melted 17-4 PH stainless steel [J]. Rapid Prototyping Journal 23(2): 280–94.

[95] Sun, Y,, Hebert, R. J. and Aindow, M. (2018). Non-metallic inclusions in 17-4PH stainless steel parts produced by selective laser melting [J]. Materials and Design 140: 153–62.

[96] DebRoy, T., Wei, H. L., Zuback, J. S., Mukherjee, T., Elmer, J. W., Milewski, J. O. et al. (2018). Additive manufacturing of metallic components – Process, structure and properties [J]. Progress in Materials Science 92: 112–224.

[97] Xu, W., Brandt, M., Sun, S., Elambasseril, J., Liu, Q., Latham, K. et al. (2015). Additive manufacturing of strong and ductile Ti-6Al-4V by selective laser melting via in situ martensite decomposition [J]. Acta Materialia 85: 74–84.

[98] Tan, X., Kok, Y., Tan, Y. J., Descoins, M., Mangelinck, D., Tor, S. B. et al. (2015). Graded microstructure and mechanical properties of additive manufactured Ti-6Al-4V via electron beam melting [J]. Acta Materialia 97: 1–16.

[99] Wu, J., Wang, X. Q., Wang, W., Attallah, M. M. and Loretto, M. H. (2016). Microstructure and strength of selectively laser melted AlSi10Mg [J]. Acta Materialia 117: 311–20.

[100] Poursaeidi, E., Aieneravaie, M. and Mohammadi, M. R. (2008). Failure analysis of a second stage blade in a gas turbine engine [J]. Engineering Failure Analysis 15(8): 1111–29.

[101] Salehnasab, B., Hajjari, E. and Mortazavi, S. A. (2017). Failure assessment of the first stage blade of a gas turbine engine [J]. Transactions of the Indian Institute of Metals.

[102] Chen, X. Q., Lin, W. J. and Ng, T. J. (2009). Neutral line-based Robust Profile Reconstruction for adaptive machining of turbine blade tip welds [J]. International Journal of Manufacturing Research 4(2): 236.

[103] Hodges, V. E. and Mo, J. P. (2018). Transitioning defence aerospace support solution to service commercial sector maintenance repair and overhaul [J]. International Journal of Production Research.

[104] Qi, H., Azer, M. and Singh, P. (2010). Adaptive toolpath deposition method for laser net shape manufacturing and repair of turbine compressor airfoils [J]. The International Journal of Advanced Manufacturing Technology 48(1): 121–31.

[105] Wilson, J. M., Piya, C., Shin, Y. C., Zhao, F. and Ramani, K. (2014). Remanufacturing of turbine blades by laser direct deposition with its energy and environmental impact analysis [J]. Journal of Cleaner Production 80(oct.1): 170–8.

[106] Lu, Z. L., Zhang, A. F., Tong, Z. Q., Yang, X. H., Li, D. C. and Lu, B. H. (2011). Fabricating the steam turbine blade by direct laser forming [J]. Materials and Manufacturing Processes 26(7): 879–85.

[107] Lu, Z. L., Li, D. C., Tong, Z. Q., Lu, Q. P., Traore, M. M., Zhang, A. F. et al. (2011). Investigation into the direct laser forming process of steam turbine blade [J]. Optics and Lasers in Engineering 49(9): 1101–10.

Chapter 2

Additive Manufactured Superalloys and their Applications

2.1 Introduction

Superalloys refer to a class of metallic materials primarily based on iron (Fe), cobalt (Co), and nickel (Ni), capable of maintaining excellent performance under high temperatures and certain stress conditions for prolonged periods. Generally, superalloys are defined as alloys that can maintain their mechanical properties, heat resistance, and chemical stability above 600°C [1]. They exhibit superior high-temperature strength, low creep rates, excellent oxidation resistance, heat resistance, corrosion resistance, good fatigue properties, and fracture toughness. They find extensive applications in fields such as aviation, aerospace, petroleum, and chemical industries, where exposure to high temperatures, high pressures, and corrosive environments is prevalent [2].

Ni-based superalloys represent the most widely employed class of superalloys in high-temperature components, with the majority of advanced engines comprising parts crafted from these alloys. In comparison to Fe-based and Co-based superalloys, Ni-based superalloys boast distinctive characteristics and advantages in terms of composition, microstructure, mechanical properties, and applications:

1) *Composition*: Ni-based superalloys primarily consist of Ni, chromium (Cr), aluminum (Al), and other elements, with Ni content typically exceeding 50%. Cr and Al are incorporated to enhance the alloy's heat and corrosion resistance [3].
2) *Microstructure*: The microstructure of Ni-based superalloys remains stable with minimal harmful phases. It typically comprises solid solution, transitional phases, and dispersed phases. The solid solution primarily consists of Ni-rich phases. Transitional phases remain stable at high temperatures, and dispersed phases serve as minute precipitates contributing to the alloy's strength and stability.
3) *Mechanical properties*: Operating at elevated temperatures, Ni-based superalloys exhibit outstanding high-temperature strength, resistance to oxidation and thermal erosion, and resistance to creep deformation. These alloys retain excellent mechanical properties in high-temperature environments, exhibiting minimal plastic deformation or fracture [4].
4) *Applications*: Ni-based superalloys find extensive applications in aerospace, energy, and chemical industries, among others, being utilized in the manufacturing of jet engines, gas turbines, aircraft turbine blades, high-temperature furnaces, and other high-temperature and high-pressure equipment. They play critical roles in these domains, maintaining stable performance in extremely high-temperature, high-pressure, and corrosive environments.

DOI: 10.1201/9781003406327-2

On the whole, compared to Fe-based and Co-based superalloys, Ni-based superalloys demonstrate superior performance and find broader applications in high-temperature environments, establishing themselves as one of the pivotal categories within high-temperature materials.

Traditional manufacturing of Ni-based superalloys typically involves conventional processes such as casting, forging, and heat treatment. These methods are suitable for mass production and relatively simple geometric shapes of components. Material shaping is achieved through casting molds or forging dies, followed by multiple steps including heating, plastic deformation, and cooling to obtain the final products. Traditional preparation methods offer advantages such as low cost and suitability for large-scale production. However, for Ni-based superalloys produced through casting, issues such as porosity and inclusions are often encountered, along with larger grain sizes and uneven microstructures. These drawbacks can adversely affect the performance of superalloy components, making them less conducive to long-term use under high temperature and stress conditions. For forging Ni-based superalloys, the manufacturing of complex geometric shapes is significantly constrained, and overall processing time is prolonged, resulting in low processing efficiency.

Laser additive manufacturing (LAM), a departure from traditional subtractive machining methods, involves an additive approach wherein material is incrementally deposited layer by layer using a computer-controlled laser beam as the energy source. This technique allows for rapid shaping of metal components without the need for molds. Components manufactured through LAM exhibit high density, superior performance, uniform and dense microstructures, and absence of macroscopic segregation [5]. Consequently, there has been a growing body of research and application in the realm of Ni-based superalloy laser additive manufacturing.

Traditional manufacturing of superalloys typically yields uniform microstructures, yet they may exhibit certain defects such as uneven grain sizes and grain orientation. However, compared to traditional directional solidification, laser additive manufacturing processes entail extremely high-temperature gradients and rapid solidification rates, resulting in manufactured superalloys with finer microstructures and more uniform chemical composition distributions. Moreover, by flexibly controlling processing modes and parameters during additive manufacturing (AM), grain orientation can be regulated. Directionally solidified columnar grains produced via LAM significantly reduce transverse grain boundaries perpendicular to the stress axis, thus yielding superior high-temperature creep resistance.

In terms of mechanical properties, traditionally manufactured superalloys generally demonstrate commendable mechanical performance, including high tensile strength and oxidation resistance. Nonetheless, due to factors such as deformation and residual stresses during manufacturing, some degree of non-uniformity may occasionally exist. Superalloys manufactured through additive methods typically exhibit excellent mechanical properties, with reports suggesting performance comparable to or even surpassing forged counterparts [6]. Due to the ability to achieve finer microstructures, additive-manufactured superalloys boast enhanced fatigue and creep resistance. Furthermore, additive manufacturing enables the realization of more intricate geometric shapes and internal structures, thereby enhancing component functionality and performance. Additive manufacturing offers unique advantages in small-batch production, complex geometric shapes, customization, and performance optimization.

2.2 Superalloy category in additive manufacturing

High-temperature alloys exhibit remarkable microstructural stability and outstanding tensile and yield strength, as well as excellent fatigue, creep resistance, and corrosion properties, making them widely utilized in industrial applications under service conditions at temperatures reaching 650°C and beyond. These alloys possess complex chemical compositions and can be categorized based on their matrix elements and strengthening mechanisms, as illustrated in Figure 2.1.

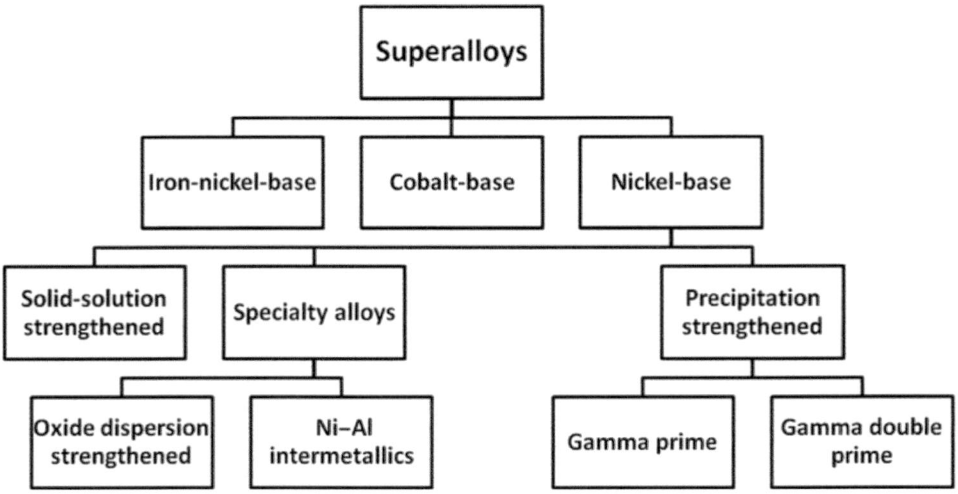

Figure 2.1. Classification of superalloys [7].

Ni-based superalloys typically consist of Ni as the primary element, alloyed with 30–40 wt% of a combination of at least eight other significant and minor alloying elements. They can be divided into two main categories based on their strengthening methods: solution strengthening and precipitation strengthening [8].

Face-centered cubic (FCC) grain formation is the primary source of strength in solution-strengthened superalloys. The FCC phase, referred to as γ, is established when a mixture of Ni and elements like Co, Fe, molybdenum (Mo), Al, and tungsten (W) is heated to the solid solution temperature, which generally ranges from approximately 1120°C to 1180°C for many superalloys. The differing atomic sizes of the added alloying elements induce distortions in the lattice of the matrix metal, enhancing its strength. Solution alloys are commonly used for components in aircraft and aerospace engine combustion chambers and casings.

A major factor in the high-temperature strength and creep resistance of precipitation-hardened superalloys is the precipitation of the FCC phase, denoted as γ' (Ni_3Al), which gives the alloys their strength. When Ni is mixed with elements such as Al and titanium (Ti) and heated to temperatures between 1170°C and 1190°C, this precipitate is created [5]. Precipitation-hardened alloys are typically employed in structural components such as turbine discs and blades in aerospace engines.

Figure 2.2 shows the compositions and main phase of Ni-based superalloys. In terms of compositions, Ni is the principal alloying element and the dominant component of the FCC γ matrix, which forms the foundational structure of many Ni-based superalloys. Al and Ti are crucial for enhancing the alloy's strength via precipitation hardening, as they enable the formation of a coherent γ' intermetallic phase ($Ni_3(Al, Ti)$) with a highly ordered $L1_2$ crystal structure. Ta, acting as a potential substitute for Al in the γ' phase, also strengthens this phase by promoting stability and increasing resistance to deformation. Nb is frequently incorporated in specific Ni-based superalloys, where it functions similarly to Al and Ti by precipitating the γ'' phase (Ni_3Nb) with a $D0_{22}$ structure, which also contributes to strengthening mechanisms. Elements like Al, Cr, Y, Ce, and Mo are commonly added to enhance oxidation and corrosion resistance.

Mo, Fe, Cr, V and Co are often included as solid-solution strengtheners, further improving the alloy's resistance to deformation. Small additions of other elements, including Cr, Ti, Ta, Mo, Hf, W, B and C are strategically used to control the grain structure, thereby enhancing mechanical properties at high temperatures. These elements tend to form stable borides and carbides, which localize at grain boundaries and contribute to structural stability. However, excessive amounts of refractory elements like W, Mo Nb and Re can promote the precipitation of hard intermetallic phases

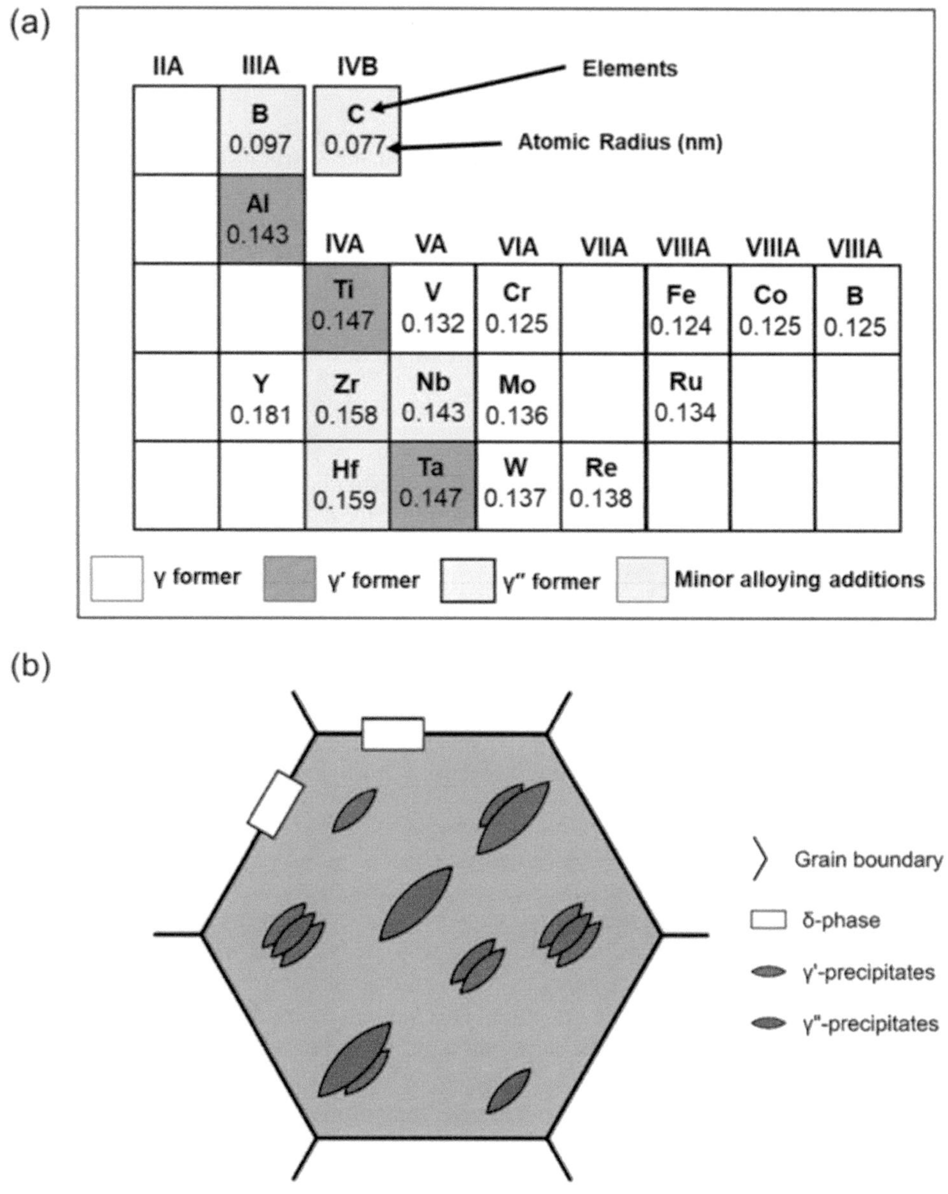

Figure 2.2. Compositions and main phases of Ni-based superalloys: (a) alloying elements in Ni-based superalloys and (b) schematic illustrating the typical phases in the microstructure of a typical Ni-based superalloys (Inconel 718) after aging [9, 10].

like the δ phase and Topologically Closed Packed (TCP) phases (e.g., Laves, μ and σ), which may affect alloy performance by introducing brittleness under certain conditions.

As shown in Figure 2(b), the microstructure of Ni-based superalloys is characterized by an FCC γ-matrix containing different solute elements. Depending on the specific composition, processing methods and post-heat treatments, precipitates such as the ordered intermetallic phases γ′ and γ″, as well as Laves phases, δ phases, and various carbides, may form within the matrix [11]. However, a high volume fraction of Laves and δ phases is generally deemed detrimental to the mechanical properties. This is due to their tendency to deplete the γ′ and γ″ forming elements from the γ-matrix and to act as sites for crack initiation under cyclic loading conditions.

This section will provide a detailed overview of several Ni-based superalloys commonly used in additive manufacturing.

2.2.1 Inconel 625

Inconel 625 is a typical Ni-based superalloy strengthened primarily through solid-solution strengthening by relatively high levels of the refractory metal components, Mo and Nb, in a Ni–Cr matrix [12]. Ni and Cr offer resistance to oxidizing environments, whereas Ni and Mo provide protection in non-oxidizing environments [13]. Mo plays a crucial role in preventing pitting and crevice corrosion. Nb serves to stabilize the alloy against sensitization during welding and exhibits excellent resistance to chloride stress-corrosion cracking. The primary phase of Inconel 625 is an austenitic Ni-based FCC γ phase. Additionally, carbide phases with reflections aligning with those of Nb-rich MC, Cr-rich $M_{23}C6$, and M_6C carbides and intermetallic phases such as γ'' and δ have been observed [14].

Inconel 625 is renowned for its exceptional corrosion resistance, mechanical properties, and high-temperature performance. It demonstrates outstanding resistance to a wide array of severe oxidation and corrosive environments, making it ideal for applications in severely harsh environments. Additionally, it boasts excellent processability, fabricability and weldability, with no sensitivity to post-weld cracking, further enhancing its appeal for various industrial uses. One of its notable attributes is its certification for manufacturing pressure vessels with wall temperatures ranging from –196°C to 450°C, underscoring its suitability for demanding applications. This superalloy finds extensive application across advanced industrial sectors, including aerospace engines, chemical equipment, nuclear industries, petrochemical, and marine industries owing to its combination of robust mechanical properties, excellent resistance to high-temperature corrosion, and versatility across a broad temperature spectrum from cryogenic conditions to ultra-high environments exceeding 1000°C [15].

2.2.2 Hastelloy-X

Hastelloy X (HX) is a Ni-based superalloy renowned for its exceptional strength and corrosion resistance. Its composition primarily consists of a Mo-rich γ matrix along with carbides (M_6C, $M_{23}C_6$), σ, and μ phases [16]. Hastelloy X is a solid-solution strengthened alloy featuring a single-phase FCC Ni-based austenitic structure. This structure is reinforced by alloying elements, with strength bolstered notably by the incorporation of Mo and Fe. Additionally, the addition of Cr enhances its resistance to corrosion.

Hastelloy X is renowned for its exceptional performance in demanding environments. Notably, it offers excellent strength and ductility at elevated temperatures, coupled with remarkable oxidation and corrosion resistance [17]. This alloy exhibits outstanding weldability and formability, making it highly suitable for various fabrication processes. Furthermore, it boasts exceptional weldability, adding to its versatility in manufacturing applications. Its ability to withstand stress-corrosion cracking is noteworthy, ensuring reliability in critical environments. Hastelloy X alloy finds extensive applications in various demanding environments characterized by high temperatures and corrosive atmospheres. Specifically, it is widely utilized in gas turbine engines and petrochemical reactors due to its exceptional durability and resistance to corrosion, as well as its ability to maintain good performance for extended periods at temperatures below 900°C [18]. Moreover, Hastelloy X is favored for its suitability in advanced engineering applications such as compact heat exchangers, heat shields, and gas turbine engine components, where extreme conditions prevail [19]. It can be used in gas turbine combustion chamber components, chemical reaction vessels, and high-temperature industrial furnaces. Additionally, it is essential for hot-end parts in the energy, petroleum,andaerospaceindustries,suchasfuelnozzlesandgasturbinecombustors,duetoitsexceptional high-temperature mechanical qualities, resistance to oxidation, and resistance to corrosion.

Due to its elevated work hardening capacity, processing intricate components from Hastelloy X presents challenges. Jinoop et al. [20] studied the hot deformation behaviors of Hastelloy X fabricated via laser directed energy deposition (LDED). The experiments revealed that the activation energy obtained is lower compared to analogous alloys in wrought conditions. This disparity primarily arises from the increased volume fraction of vacancies in LDED, which is regulated by the solidification rate. The reduced activation energy underscores the feasibility of hot-working LDED-fabricated Hastelloy X as a promising approach for constructing high-strength engineering components.

2.2.3 Inconel 718

Inconel 718 alloy stands as one of the most extensively utilized Ni-based superalloys to date, owing to its exceptional amalgamation of properties, rendering it widely applied in the realm of additive manufacturing. In the aviation sector, Inconel 718 alloy finds extensive application, with components such as turbine blades, shafts, supports, and fasteners in aircraft engines predominantly composed of this alloy. Inconel 718 exhibits high tensile strength, yield strength, creep resistance, and ductility within the temperature range of –253°C to 650°C. Additionally, it boasts commendable corrosion resistance, radiation resistance, thermal processing capabilities, and weldability, thereby establishing itself as a pivotal material in the aviation industry [21].

Inconel 718 alloy, primarily comprised of large proportions of Nb, Mo, and minor quantities of Ti and Al elements, is strengthened predominantly by precipitation hardening phases and the solid-solution hardening effect of refractory metal constituents, notably Nb and Mo, within a Ni-Cr based γ matrix [22]. The phase composition of Inconel 718 primarily encompasses the austenitic γ phase, which manifests as a FCC Ni–Fe–Cr solid solution. The primary strengthening phase comprises the $D0_{22}$-ordered body-centered tetragonal (BCT) γ'' (Ni_3Nb), while the secondary strengthening phase comprises the $L1_2$-ordered FCC γ' ($Ni_3(Al, Ti, Nb)$). This alloy exhibits commendable weldability attributed to the sluggish precipitation of primary strengthening phases within its structure [23].

Furthermore, Inconel 718 contains small quantities of other incoherent phases, such as the brittle intermetallic hexagonal Laves phase ($(Ni, Fe, Cr)_2(Nb, Mo, Ti)$), the $D0_a$-structural orthorhombic δ (Ni_3Nb), and the FCC MX (Nb,Ti)(C,N). These incoherent phases, while capable of refining grain size, can prove detrimental to mechanical performance when excessively coarse and acicular, as they offer diminished dispersion hardening [24]. Furthermore, they will prevent grain boundary sliding if they are located near grain boundaries, which will decrease ductility but perhaps increase creep resistance. Among these phases, the most severe in terms of performance degradation is the coarse and long-chain Laves phase, as it depletes the principal elements required for precipitation strengthening and facilitates easy crack initiation and propagation, thereby increasing susceptibility to cracking.

2.2.4 Inconel 738

Apart from the aforementioned easily machinable Ni-based superalloys like Inconel 625 and Inconel 718, non-weldable superalloys with outstanding high-temperature properties have also emerged as focal points of recent research endeavours in additive manufacturing. These alloys, such as Inconel 738, Inconel 939, CM247LC, and CMSX-4, exhibit exceptional resistance to creep-rupture at high temperatures because of a significant volume fraction of γ' phases that are directly related to the high Al + Ti concentrations. Nonetheless, their amenability to additive manufacturing is compromised due to (i) the broad solidification interval stemming from extensive alloying, (ii) the emergence of several phases including low-melting-point borides and γ/γ' eutectics, and (iii) the escalation of residual stresses and reduction in ductility attributed to rapid γ' precipitation during the additive manufacturing process [25].

Inconel 738 stands as a quintessential representative within the realm of non-weldable superalloys. This alloy, strengthened by the ordered and coherent $L1_2$ intermetallic $Ni_3(Al,Ti)$ γ' phase, embodies a precipitation-strengthened Ni-based superalloy. In addition to this, it benefits from solid solution strengthening derived from Cr, Mo, Co, and W, along with grain boundary strengthening by carbides and grain boundary γ'. With serviceable stability at temperatures soaring up to 980°C, this alloy owes its resilience to its substantial content of the γ' phase (approximately 40–43% in volume fraction), which plays a pivotal role in sustaining its exceptional high-temperature mechanical performance [26]. Inconel 738 offers outstanding comprehensive performances of high-temperature strength, fatigue performance, resistance to high-temperature creep, corrosion and oxidation, as well as excellent microstructural stability [27]. These properties with high reliability and longevity in harsh service conditions make it suitable for a wide range of applications, including static and rotating hot-end components in gas turbines and aircraft engines.

However, compared to Ni-based superalloys commonly used in additive manufacturing such as Inconel 718 and Inconel 625, the application of Inconel 738 is constrained by its poorer machinability, weldability, and higher susceptibility to cracking. Figure 2.3 illustrates the weldability performance of Ni-based superalloys as the Ti + Al content varies. Henderson et al. [28] investigated weldability of a broad spectrum of Ni-based superalloys and determined that alloys with an Al + Ti content of less than 2.5 wt% are more conducive to welding than those exceeding this threshold. Typically, when the combined Al and Ti content exceeds 4 wt%, an alloy is deemed non-weldable. Al and Ti are essential constituents of the primary precipitation-strengthening γ' phase for Inconel 738 and the Al + Ti content in Inconel 738 can reach 6.77 wt%, significantly surpassing the 4 wt% limit. This high content of Al + Ti leads to heightened susceptibility to cracking in the heat-affected zone, exacerbated by the more severe grain boundary liquation resulting from the nonequilibrium melting of lower-melting-point phases.

During the laser additive manufacturing process of Inconel 738, cracking can occur due to various mechanisms including liquation cracking, solidification cracking, ductility dip cracking, strain age cracking and others. The presence of $\gamma + \gamma'$ eutectic on grain boundaries is widely recognized as a microstructural condition conducive to cracking. The absence of liquid feeding in the final stages of solidification processes and elevated levels of thermal stress during additive manufacturing further exacerbate the propensity for cracking. Therefore, merely adjusting processing parameters simplistically cannot effectively eliminate the cracking of Inconel 738. Numerous researchers endeavour to suppress the cracking of Ni-based superalloys during laser additive manufacturing by optimizing processing parameters, preheating the substrate, and employing modified

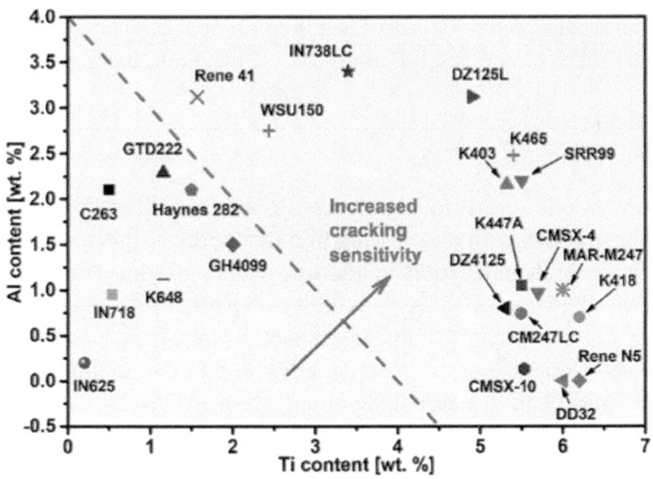

Figure 2.3. Weldability of Ni-based superalloys as a function of Al + Ti content [29].

laser-beam profiles [30]. Furthermore, researchers are actively exploring methods such as introducing nanoparticles and varying the design of superalloys to alter the constituents of materials.

2.3 Additive manufacturing superalloys: microstructure and mechanical properties

2.3.1 Microstructure characteristics

A high-energy laser creates a melt pool on the surface of the material to be deposited during the additive manufacturing process. After that, the laser quickly moves along the scanning path, creating a sharp temperature gradient from the top to the bottom of the melt pool as well as a rapid cooling rate. This produces a unique thermal environment that has a big impact on the deposited alloys' microstructure. The temperature gradient (G) and the solidification rate (R) are the two main solidification parameters that affect the microstructure. While their product, GR, affects the microstructural size, the ratio of G to R determines the microstructure's morphology.

In AM-fabricated samples, grains typically grow epitaxially from the preceding layer, resulting in coarse columnar grains spanning several layers. The pronounced texture observed in these samples may stem from the interaction between epitaxial and competitive growth, which is governed by the alignment of the grains' preferred orientation with the local thermal gradient. Moreover, micro-segregation during the solidification process is a critical factor, particularly in superalloys that contain substantial amounts of solid solution elements such as Cr and Mo in Hastelloy X, Nb and Mo in Inconel 718 and Inconel 625. This segregation leads to the development of heterogeneous substructures, such as cellular structures, and the formation of secondary phases. These secondary phases typically form in the interdendritic regions, contributing to the material's strong anisotropic mechanical properties.

Figure 2.4(a) depicts the microstructure of the as-deposited Hastelloy X alloy produced by laser directed energy deposition. The dendrite trunk constituted the γ matrix, while the interdendritic region comprised the Mo-rich γ phase and carbides. The as-deposited Hastelloy X exhibits prominent characteristics of epitaxial growth along the building direction, which is one of the typical microstructural features of additive manufacturing samples. As illustrated in Figure 2.4(a), the microstructure along building direction primarily consists of columnar dendrites, initiated by the primary dendrites' epitaxial development from the previously deposited layer's remelted area. The direction of growth is mostly determined by the direction of heat dissipation, and these previously deposited layers act as pre-nuclei for microstructural growth.

Electron backscatter diffraction (EBSD) results from Figure 2.4(b) illustrate the presence of coarse columnar grains ranging from approximately 31 μm to about 332 μm along the building direction. As shown in Figure 2.4(c), fine columnar cellular/dendritic structures can be distinctly observed in the longitudinal section with subtle secondary dendrite arms. However, the columnar dendrites reveal a cellular network in the transversal section. A typical micro-segregation phenomenon is demonstrated in Figure 2.4(d). EDS mapping and line scanning studies reveal that the cellular structure is enriched in Cr and Mo elements. Mo and Cr accumulated in the interdendritic areas as a result of segregation during solidification, ultimately creating this substructure. The precipitation of the second phase between dendrites was further aided by segregation.

Parimi et al. [32] examined the influence of laser scanning path and laser power on the texture of Inconel 718 thin-walled builds fabricated through direct laser fabrication (DLF). Figure 2.5 presents EBSD results of the longitudinal direction of the Inconel 718 thin-walled samples. Specifically, B1 represents a unidirectional deposition path, B2 corresponds to a bidirectional deposition path, and B3 indicates a bidirectional deposition path with increased laser power.

Fine grain zones at the layer interfaces between inclined columnar grains were a characteristic of the banded grain structure seen in both the B1 and B2 samples (Figure 2.5(a) and (b)). While the columnar grains primarily displayed <001> and <101> textures, the tiny grains revealed random

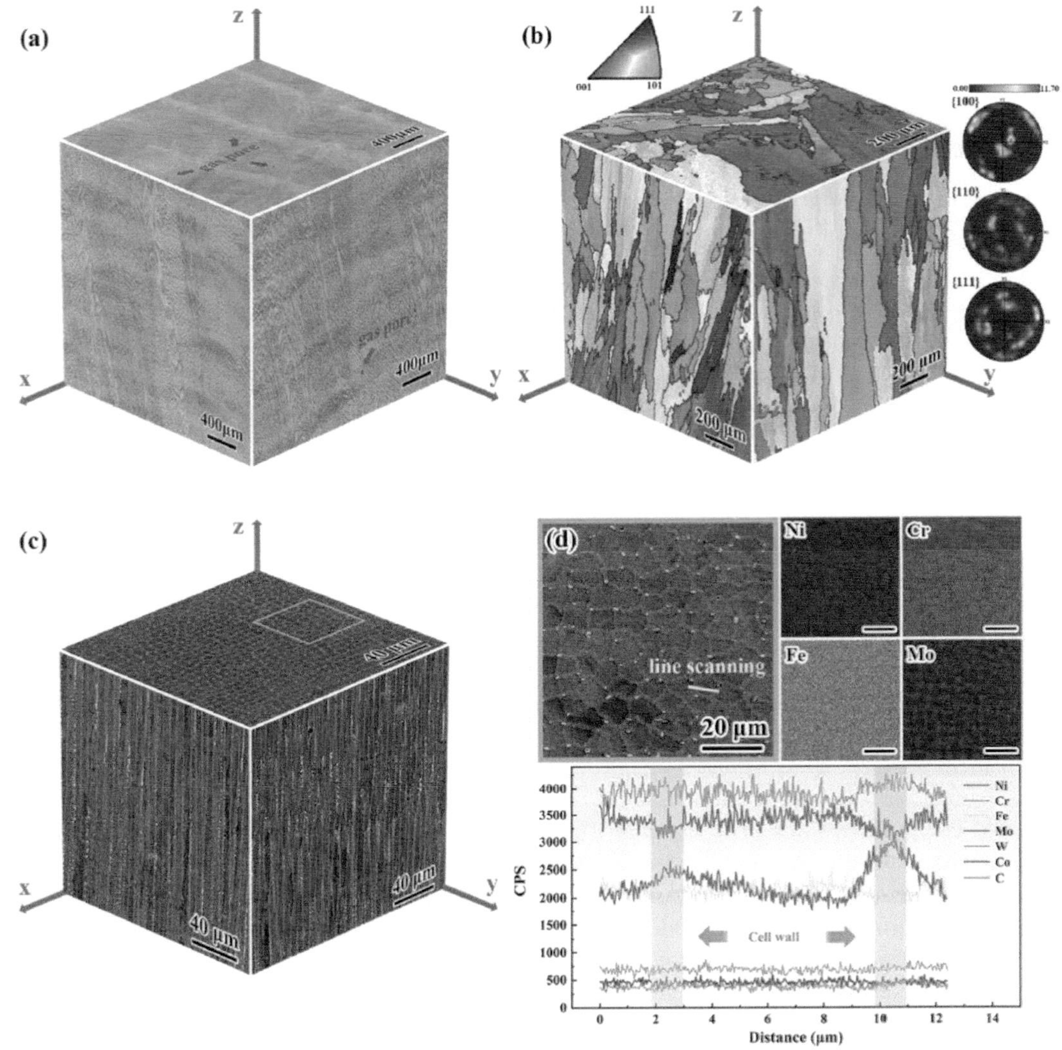

Figure 2.4. Microstructure of the as-deposited Hastelloy X alloy fabricated by laser directed energy deposition. (a) optical microscopy (OM); (b) EBSD inverse pole figure (IPF); (c) scanning electron microscopy-backscattered electron (SEM-BSE) image; (d) magnified SEM-BSE image from the transverse direction shown in (c) along with corresponding energy dispersive spectroscopy (EDS) results [31].

textures. Because of the high thermal conductivity of Ni-superalloys, which causes a rapid cooling process during DLF and prevents epitaxial development, this banded structure forms.

As illustrated in Figure 2.5(c), the B3 sample, deposited with higher laser power, exhibits fine columnar grains at the bottom regions. As deposition progresses, grains transition into coarser columnar structures that grow along the building direction. The <001> pole figures for B3 demonstrate a pronounced texture.

The microstructure has a preferred texture with grains growing epitaxially parallel to the <001>, which corresponds to the direction of the maximum thermal gradient. This alignment is typical for FCC alloys due to its crystallographic favourability. Because of the substrate's diminishing heat sink effect, higher laser powers are associated with lower cooling rates, which causes the melt pool temperature to rise steadily with build height. Consequently, when building height grows, the grain structure gets coarser. Furthermore, it is anticipated that the high melt pool temperature and

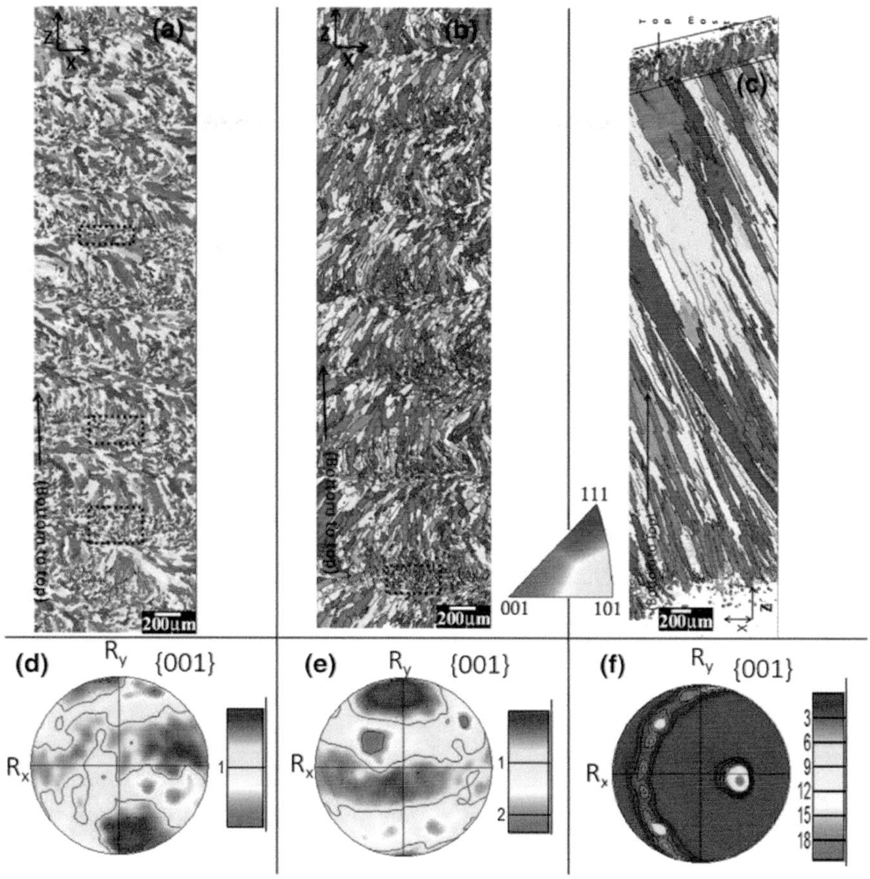

Figure 2.5. EBSD maps for samples (a) B1, (b) B2, and (c) B3, along with their corresponding <001> pole figures shown in (d), (e), and (f) [32].

continuous heat flux during bidirectional high-power deposition will result in the full re-melting of succeeding layers, encouraging the formation of epitaxial grains.

As one of the characteristics of additive manufacturing microstructures, micro-segregation is a significant challenge in LAM of Ni-based superalloys. Micro-segregation refers to the non-uniform distribution of alloying elements within the microstructure of the material. During the rapid solidification process inherent to LAM, different elements solidify at different rates, leading to compositional gradients within the dendritic structures. This segregation can result in the formation of secondary phases or precipitates, which may be detrimental to the material's mechanical properties. The presence of micro-segregation impacts the overall performance of Ni-based superalloys produced by LAM. Non-uniform distribution of elements can lead to localized areas of weakness, increasing susceptibility to cracking under thermal and mechanical stresses.

Inconel 718, a Ni-based superalloy, is widely studied in the context of laser additive manufacturing and shows pronounced micro-segregation characteristics. Because of their low solubility in the γ-matrix, elements like Nb, Mo, and Ti segregate into the interdendritic regions during the solidification of Inconel 718. The local thermodynamics of Inconel 718 are drastically changed by this segregation, which also affects the forces that propel the formation of different phases. Consequently, during solidification, phases like Laves, δ, NbC, and TiN form in the interdendritic areas. Additionally, the precipitation kinetics of the strengthening phases (γ'/γ'') in the γ-matrix are impacted by the segregation of these elements. The impact of micro-segregation on the performance of Inconel 718 is considerable. The as-built microstructure of additively manufactured

Inconel 718 typically consists of a supersaturated γ matrix, Laves phase, and a limited amount of strengthening γ″ and γ′ particles. The presence of brittle Laves phase in the interdendritic regions, due to the segregation of Nb, Mo, and Ti, can be detrimental to the mechanical properties of the material.

Zhang et al. [33] investigated the micro-segregation and formation of Laves phases in Inconel 718, comparing samples fabricated using selective laser melting (SLM) to those fabricated by laser-directed energy deposition (L-DED) and casting. Figure 2.6 presents the microstructure of Inconel 718 produced by casting and L-DED, alongside quantitative analyses of the Laves phase, Nb-rich regions, and the matrix under different conditions. Unlike the discrete Laves phases observed in cast samples, the L-DED samples exhibited long, chain-like Laves phases located in the interdendritic regions.

Notably, large Nb-rich regions surrounding the Laves phase are observed in both cast and L-DED-built Inconel 718, as shown in Figure 2.6(b) and (e). The volume fraction of these Nb-rich regions is approximately 33.0% in cast Inconel 718 and 14.0% in L-DED-built Inconel 718. EDS research shows that Nb concentration rises from the dendrite trunk into the interdendritic Laves phase. In cast Inconel 718, the average Nb concentrations in the Laves phase, Nb-rich area, and dendritic core are around 31.01 wt%, 6.80 wt%, and 1.23 wt%, respectively. For L-DED-built Inconel 718, Nb concentrations are 22.15 wt% in the Laves phase, 10.59 wt% in the Nb-rich region, and 1.62 wt% in the dendrite trunk. The significant difference in Nb concentration between the Nb-rich region and the γ-matrix leads to distinct corrosion morphologies and a clear interface between these regions in both cast and L-DED-built Inconel 718. However, in SLM-built Inconel 718, the average Nb concentrations in the Laves phase and dendritic trunk are 19.35 weight percent and 2.37 weight percent, respectively, and Nb-rich areas are hardly noticeable. SLM attains noticeably greater cooling rates than casting and L-DED. This rapid cooling enhances the dendritic growth rate and solute trapping during the solidification of the molten pool. Consequently, in SLM-built Inconel 718 more Nb atoms are kept in the matrix and less Nb atoms are released to produce inter-dendritic Laves phases. This results in reduced micro-segregation.

In this study, we emphasize that understanding and controlling micro-segregation are crucial for optimizing the mechanical performance and reliability of LAM-fabricated nickel-based superalloys.

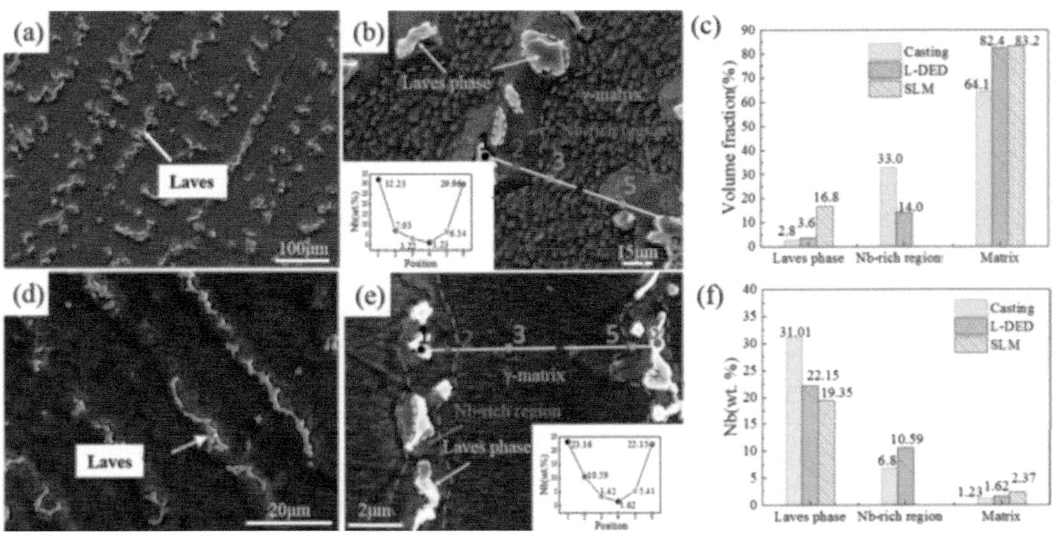

Figure 2.6. Microstructure of Inconel 718 produced by different fabrication methods: (a, b) casting; (d, e) L-DED; quantitative analysis of the Laves phase, Nb-rich regions, and the matrix in cast, L-DED-built, and SLM-built Inconel 718: (c) volume fraction; (f) Nb content [33].

Effective control of the solidification process is essential in minimizing micro-segregation and enhancing the performance of these high-temperature materials. Currently, many researchers are exploring various laser deposition techniques to mitigate micro-segregation. These efforts aim to improve the uniformity of element distribution within the alloy, thereby enhancing its structural integrity and performance under extreme conditions.

Chen et al. [34] investigated the effects of two different laser beam modes, flat top laser beam (FTLB) and gaussian distribution laser beam (GDLB), on micro-segregation and Laves phase distribution. The distinct energy distribution profiles of the FTLB, produced using a diode laser, and the GDLB, generated by a fibre laser, are illustrated schematically in Figure 2.7(a) and (b). Figure 2.7(c–d) and Figure 2.7(e–h) display the typical distribution of Laves phase in the transverse section of as-deposited FTLD-Inconel 718 and GDLD-Inconel 718 samples, respectively.

It is evident that the Laves phase distribution in both FTLD-Inconel 718 and GDLD-Inconel 718 samples varies with the deposition height. At the bottom, the Laves phase appears fine and in low fraction due to the significant cooling effect from the substrate, which results in high cooling rates. In the middle section, there is a coarse and high fraction of Laves phase formation due to increased thermal accumulation as the deposition progresses. At the top, the formation of the Laves phase is reduced again. The significant cooling at the base leads to the formation of finer Laves phases. As deposition continues, heat accumulation increases.

The interdendritic region in the heat-affected zone re-melts when the subsequent layer is heated and when secondary solidification takes place. Increased Laves phase development is the outcome of this process, which permits further dendritic growth and interdendritic micro-segregation.

Compared to the GDLD-Inconel 718 samples, the Laves phase distribution in FTLD-Inconel 718 samples is more discrete and uniform, with a lower overall content, indicating a reduced level

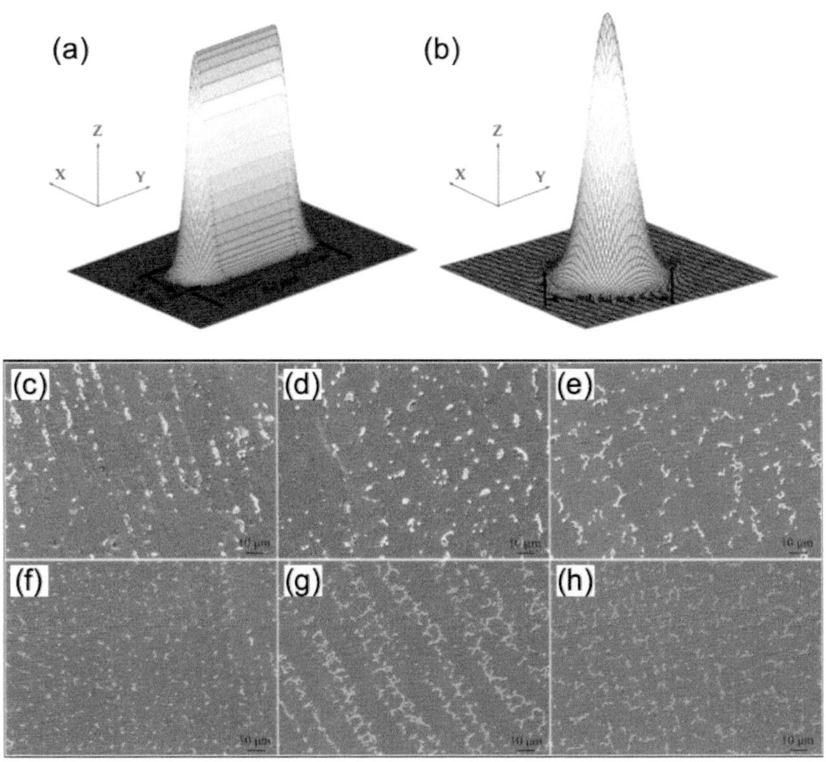

Figure 2.7. Schematic of laser energy distribution patterns and Laves phase distribution: (a) FTLB, (b) GDLB; (c–e) Laves phases in bottom, middle and top regions of FTLD-Inconel 718; (f–h) Laves phases morphologies in bottom, middle and top regions of GDLD-Inconel 718 [34].

of micro-segregation. Higher melt pool temperatures and a smaller ratio of horizontal to vertical thermal gradients at the solid-liquid interface are the outcomes of the GDLD method's extreme laser energy concentration at its core. The interdendritic micro-segregation and Laves phase development in the GDLD-Inconel 718 samples are greatly enhanced by these thermal properties of the GDLD-induced melt pool, which also results in a decreased redistribution coefficient for alloying elements. This comparison analysis shows that the FTLD mode outperforms the GDLD mode for laser additive manufacturing.

Xiao et al. [35] conducted a comprehensive study on the effects of both continuous-wave (CW) and quasi-continuous-wave (QCW) laser modes on micro-segregation and Laves phase formation during laser additive manufacturing. Figure 2.8 illustrates the morphologies of the Laves phase and Nb segregation under these two laser modes. Under the CW mode (Figure 2.8(a)), coarse and interconnected Laves phase particles are observed. In contrast, the QCW mode results in the formation of fine and discrete Laves phase particles throughout the entire sample (Figure 2.8(b)). According to the EDS analysis, Nb segregation is more pronounced in the CW sample compared to the QCW sample (Figure 2.8(c–f)). The average Nb concentration in the Laves phase particles of the CW sample is approximately 18.73 wt%, whereas it is reduced to 15.77 wt% in the QCW sample.

The QCW laser mode encourages the formation of fine, discrete Laves phase particles, decreases Nb segregation, and improves the synthesis of fine equiaxed dendrites. This is explained by the QCW mode's enhanced solute trapping, quicker dendritic growth rates, and higher cooling rates. Conversely, columnar dendrites, more severe Nb segregation, and continuously distributed long-chained Laves phases are typically produced by the CW laser mode.

Further, Xiao et al. [36] studied the influence of laser modes on the control of Nb-rich precipitates in laser additive-manufactured Inconel 718. As shown in Figure 2.9, the SEM-EDS mapping reveals the segregation patterns of samples fabricated under quasi-continuous-wave laser additive manufacturing with different processing parameters, specifically laser pulse frequencies. The analysis shows significant enrichment of Nb and Mo elements in the Laves phase within the interdendritic regions, while these elements are depleted in the dendrite core regions. The interdendritic eutectic regions surrounding the Laves phase exhibit notable micro-segregation of Nb and Mo, displaying a slight increase in brightness. It is observed that the dendritic growth mode is closely linked to the segregation pattern in the as-fabricated samples. Variations in laser pulse frequency lead to different thermal histories during solidification, which in turn affect the distribution and concentration of segregated elements. These microstructural characteristics play a crucial role in determining the mechanical properties and performance of the final alloy.

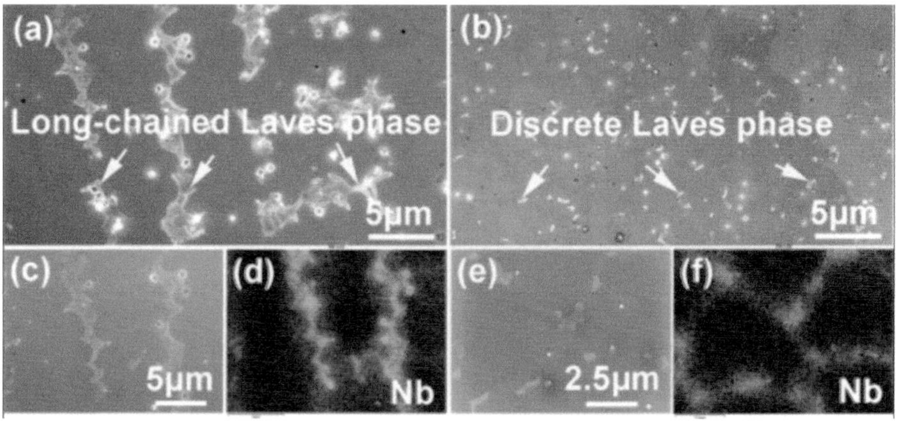

Figure 2.8. SEM images illustrating the morphology of Laves phase particles and the distribution of Nb elements in the interdendritic regions: (a, c, and d) CW sample and (b, e, and f) QCW sample [35].

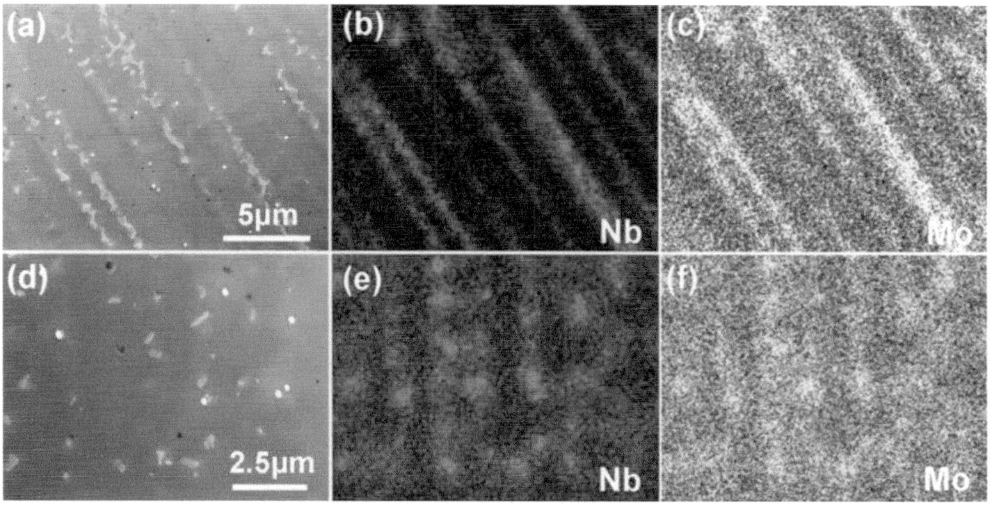

Figure 2.9. SEM images and corresponding EDS mappings of Nb and Mo for as-deposited Inconel 718 samples by quasi-continuous-wave laser additive manufacturing with different laser pulse frequencies. (a–c) 50 Hz; (e–h) 10 Hz [36].

2.3.2 Mechanical properties

Due to the complex thermal gradients and cooling rates inherent in additive manufacturing processes, the microstructural anisotropy, grain morphology, and crystal orientation of produced Ni-based superalloy components often exhibit significant differences. Consequently, these components typically display pronounced mechanical anisotropy.

It is well established that horizontally oriented AM parts generally possess higher mechanical strength compared to vertically oriented ones. Various studies have shown that when the load is parallel to the building direction, AM samples exhibit higher ductility, lower strength, and lower elastic modulus, including a reduced Young's modulus [37]. Conversely, when the load is perpendicular to the building direction, the samples demonstrate lower ductility, higher susceptibility to cracking, higher elastic modulus, and increased strength.

Factors influencing the anisotropy of tensile properties include grain morphology (e.g., columnar grain morphology), crystallographic texture (e.g., <100> fiber texture), precipitate arrangement, and lack-of-fusion defects [38]. The anisotropy of tensile properties are closely related to the preferred grain orientation and grain morphology. During laser additive manufacturing, the directional heat flux and significant thermal gradient along the building direction result in the directional growth of grains and the development of texture.

Ni-based superalloys produced through LAM exhibit significant variations in strength depending on the loading direction. When the load is applied parallel to the building direction, the alloy typically demonstrates lower strength. This phenomenon occurs because, in this direction, the grains grow along the building direction, forming columnar grains. These columnar grains have weaker grain boundary strengthening effects, resulting in reduced overall material strength. This reduction in strength is primarily due to the ease with which dislocation slip occurs within the columnar grains under tensile load.

Conversely, when the load is applied perpendicular to the building direction, the material's strength is significantly enhanced. The grain boundaries of the columnar grains prevent dislocations from moving in this direction, which increases the material's strength because the grain growth direction is perpendicular to the loading direction. This anisotropic behavior can be explained by the crystallographic orientation and the morphology of the columnar grains.

Zhao et al. [39] investigated the impact of homogenization heat treatment on the properties of Inconel 718 alloy deposited through laser powder bed fusion (LPBF) along both vertical and

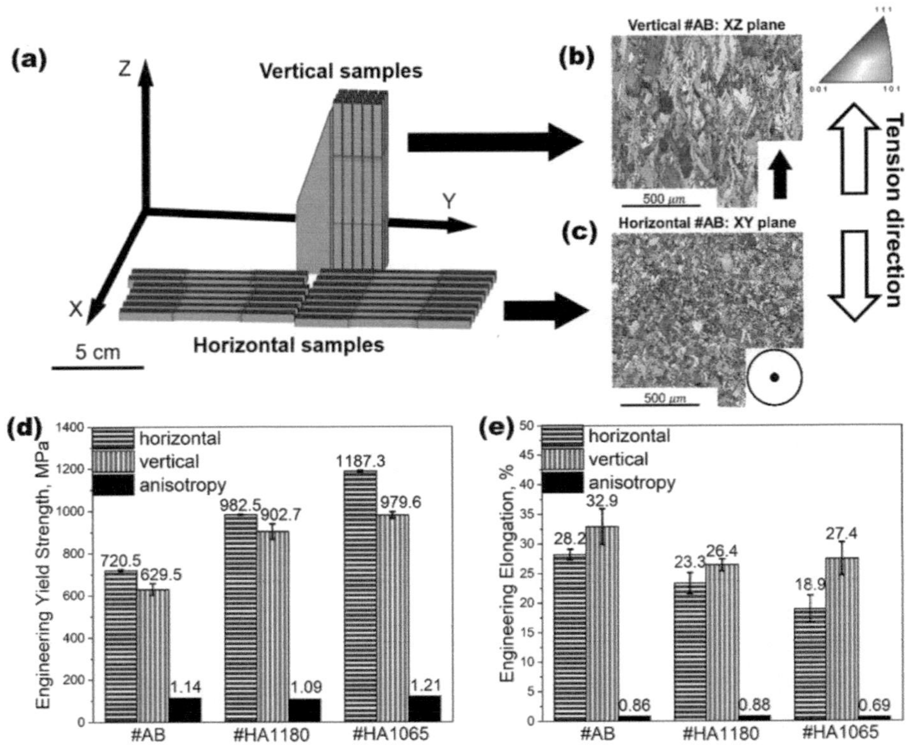

Figure 2.10. Illustration of the correlations between process, structure, and properties in LPBF Inconel 718 tensile samples: (a) vertically and horizontally printed tensile samples schematic diagram; (b) grain morphology and tensile direction for the vertical sample; (c) grain morphology and tensile direction for the horizontal sample; (d) engineering yield strength; (e) engineering elongation of horizontal and vertical tensile samples under different heat treatments [39].

horizontal directions. The method, structure, and property relationships in LPBF Inconel 718 tensile samples are shown in Figure 2.10. Although the ductility shows the opposite tendency, the data show that for as-deposited samples, the horizontal samples' yield strength is greater than the vertical samples'. Specifically, the vertical sample's XZ plane has a grain size of 61.4 μm, while the horizontal sample's XY plane has an average grain size of 29.5 μm. Because of the Hall-Petch effect, the horizontal sample exhibits a larger yield strength than the vertical sample (Figure 2.10(d)).

As illustrated in Figure 2.10(d) and (e), the ratios of the mechanical properties of horizontal and vertical tensile samples were used to assess the anisotropic qualities. In an ideal isotropic scenario, the ratio value is 1. According to the findings, even after homogenization heat treatment, the anisotropy in mechanical properties cannot be eliminated.

Figure 2.11 illustrates the mechanical analysis of various microstructures under mechanical load. Equiaxed dendrites exhibit uniform force distribution, while columnar dendrites display significant mechanical anisotropy. As shown in Figure 2.11(ii), columnar grains can withstand greater tensile/compressive stress along the direction of grain growth. Studies have indicated that columnar grains provide optimal high-temperature mechanical performance when the load is applied along the grain growth direction [40]. However, for applications involving multidirectional stress, this anisotropy can be detrimental.

When considering columnar dendrites with multiple grain growth directions, the ability of individual grains to bear external loads is further diminished. This is due to the conflicting orientations among different columnar dendrites, which increases the likelihood of premature failure at the midpoint of a columnar grain, as depicted in Figure 2.11(iii). This interaction among grains with varying growth directions compromises the structural integrity and load-bearing capacity of the material.

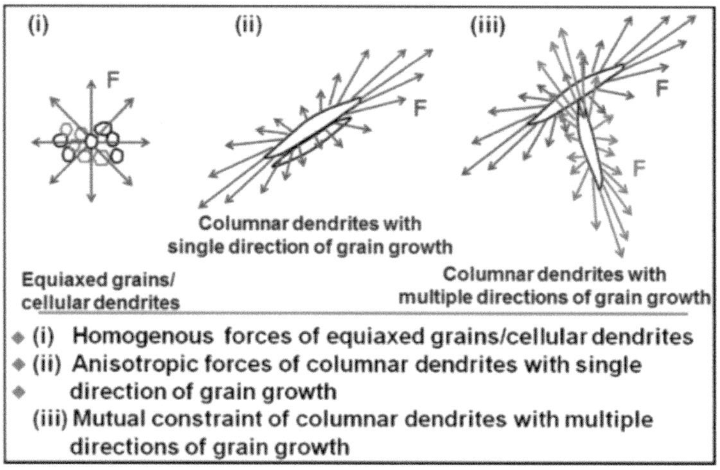

Figure 2.11. Mechanical analysis of various microstructures under mechanical load [41].

Ductility is influenced by several factors, including grain size, texture, precipitates, grain boundary conditions, and defects. In the context of laser additive manufacturing, specimens printed horizontally exhibit lower ductility, whereas those printed vertically demonstrate higher ductility. This phenomenon can be attributed to columnar grains in horizontally printed samples being oriented perpendicular to the tensile load direction, facilitating easier crack propagation along grain boundaries, and thereby reducing material ductility. Conversely, in vertically printed samples, columnar grains are aligned parallel to the tensile load direction, aiding crack propagation within the grains and thereby enhancing material ductility.

The anisotropy in ductility can be due to variations in grain boundary elongation, which results in various cracking mechanisms. Columnar grain boundaries serve as pathways where damage preferentially accumulates, ultimately causing failure. The cracking mechanisms are determined by the angles between the tensile load and the directions of columnar grain boundaries [42].

Schematic diagram of different fracture mode is shown in Figure 2.12. In horizontal samples, tensile loads are perpendicular to the columnar grain boundaries (Figure 2.12(b)), conforming to Mode I opening tension. This alignment facilitates easy crack opening and results in low ductility, as the long axis of the grain boundaries aligns well with Mode I crack openings under tension. Dislocation pile-ups encourage the development of microcracks along the columnar grain boundaries, which lessens the horizontal samples' ductility.

In vertical samples, the tensile loads are aligned parallel to the columnar grain boundaries, as illustrated in Figure 2.12(a). Consequently, only a few the grains' short axes are subjected to Mode I opening tension, making it more challenging for opening failure to occur. The crack tip openings are deflected when the columnar grains are parallel to the tensile stress. Friction stress is introduced on the crack surfaces between adjacent columnar grains as a result of this deflection, which persists as the crack widens [44]. The fatigue fracture opening and growth process is slowed down by this friction, suggesting increased ductility. Additionally, this alignment implies that grains that are more closely aligned with the tensile axis promote slip at lower applied stresses, increasing the material's ability to accommodate plastic deformation prior to fracture.

Moreover, the non-uniform distribution of grain sizes can significantly impact the material's ductility. Research indicates that during plastic deformation, the uneven distribution of grain size and morphology leads to stress concentration and variations in dislocation density between different grains, which in turn affects the material's strain hardening behavior and overall ductility. For instance, larger grains, due to their weaker strain hardening effect, are more prone to becoming pathways for crack propagation, thereby reducing the material's ductility [45].

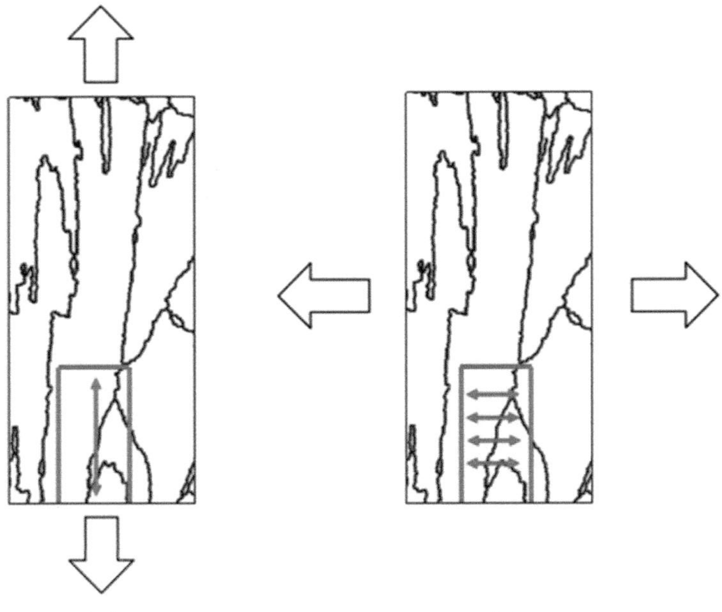

Figure 2.12. Schematic diagram of different fracture mode: (a) vertical tensile sample (tensile loads are parallel to the grain boundary) and (b) horizontal tensile sample (tensile loads are perpendicular to the grain boundary) [43].

2.4 Engineering application of additive manufactured superalloys

2.4.1 High performance complex geometric structure

Due to their exceptional resistance to high-temperature corrosion, dimensional stability, and outstanding mechanical strength—particularly in terms of tensile strength, creep resistance, and fatigue resistance—Ni-based superalloys are well-suited for high-temperature, high-stress environments that are exposed to severe corrosive conditions [46]. These superalloys maintain their remarkable strength at elevated temperatures through mechanisms including solid solution strengthening and precipitation hardening. Their properties make them ideal for applications where temperatures can exceed 1000°C, particularly in critical sectors such as aerospace, automotive, and energy industries that operate under high-temperature conditions [47]. Many Ni-based superalloys are currently utilized in aircraft components. For instance, Ni-based superalloys make up more than half of the mass of a sophisticated aviation engine [48]. It is reported that 18.2% of the revenue in the AM industry is received from the aerospace and aeronautical sectors [49].

Because of their great strength and large work hardening rate, Ni-based superalloys are difficult to machine despite having excellent mechanical properties. Complex geometries are difficult to fabricate using conventional manufacturing methods like labor-intensive casting and forging because of these factors, which also result in low material removal rates and higher tool wear. With the rapid advancement of aerospace components, there is a growing demand for intricate structures and lightweight designs, which poses substantial challenges to traditional manufacturing processes.

In the aerospace industry, LAM presents a variety of applications for Ni-based superalloys. It enables the production of high-performance engine components, including turbine blades, turbine discs and combustion chambers, which require complex geometries, high strength, and resistance to elevated temperatures.

LAM's ability to fabricate freeform designs makes it especially suitable for aerospace applications, offering significant design flexibility that traditional manufacturing methods cannot achieve. The key advantages of additive manufacturing include the ability to create complex internal structures and cooling channels, thus optimizing component performance and enabling higher service temperatures. This capability has sparked considerable interest among turbine manufacturers. LAM also provides significant benefits such as reduced costs and lead times, the introduction of novel materials, lightweight designs, and the consolidation of multiple components. These improvements enhance performance and risk management by integrating internal cooling features and minimizing traditional joining processes.

The Ariane Group has successfully developed the injector components for the Ariane 6 rocket utilizing LAM technology. As illustrated in Figure 2.13, the injector core was produced using LAM technology with Inconel 718. Typically, injector systems in rocket engines comprise hundreds of individually manufactured parts, which were traditionally joined through brazing and welding into a single injector head. The implementation of LAM significantly reduces both the production time and costs associated with this process.

As a key component of high temperature and high performance in rocket engine systems, regenerative cooling channel wall nozzles (CWN) are a widely used design in the propulsion sector for fabricating nozzle structures with internal coolant channels. The scale and complexity of CWN are challenging to design, with stringent requirements for materials, tight tolerances for manufacturing and assembly, thin-wall characteristics, manufacturing processes, and more, resulting in long lead times and expensive costs. DED technology is being tried in regen-cooled nozzles. DED can net shape individual components through coaxial powder feeding (as shown in Figure 2.14), and this fully integrated channel configuration close to the final shape can minimize the overall number of parts. DED has the potential to create whole CWN with integrated coolant channels in a single additive manufacturing. Through the complex and thin-walled properties of DED manufacturing, the manufacturing schedule is greatly reduced.

NASA has effectively utilized DED for the fabrication of large structures, particularly for rocket engine components such as large-scale CWN and powerhead components. Typical examples of large DED components at NASA are shown in Figure 2.15. As illustrated in Figure 2.15(a), the integral channel DED nozzle can achieve a diameter of up to 1.52 m and a height of 1.78 m, confirming the feasibility of producing large parts with integrated channel features via DED. When compared to nozzles that are normally built in full-scale designs, these nozzles demonstrate the possibility of reducing the number of parts.

Additional components fabricated using DED include various large-scale manifolds and complex parts such as the Inconel 718 powerhead half-shell and the DED JBK-75 nozzle. The use of AM can significantly minimize the extensive machining operations associated with traditional forging and casting methods, resulting in considerable cost and schedule savings due to shorter deposition times with DED.

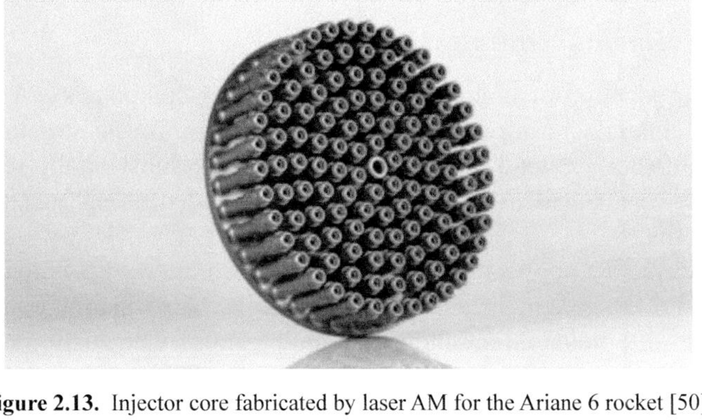

Figure 2.13. Injector core fabricated by laser AM for the Ariane 6 rocket [50].

Figure 2.14. Nozzle manifold fabrication by DED [51].

Figure 2.15. Large DED components at NASA (a) integral channel nozzle fabricated by DED form NASA, (b) Inconel 718 powerhead half shell and (c) DED JBK-75 nozzle [50].

2.4.2 Repair and remanufacturing

In addition to the production of high-performance complex components, AM is widely utilized for the repair and remanufacturing of parts. For high-value, intricate structures, advanced repair techniques can significantly extend their service life, thereby substantially reducing expenditures in the maintenance, repair, and overhaul (MRO) sector and offering substantial economic and environmental benefits.

Figure 2.16 shows AM applications in repair and restoration in different fields. AM is used in different parts and equipment in aviation, automobile, shipbuilding, railway, mining, mold, pharmaceutical and other fields. Many industrial equipment service conditions are severe, in high temperature, high pressure, overload, high impact, vibration, fatigue, wear, erosion and other harsh

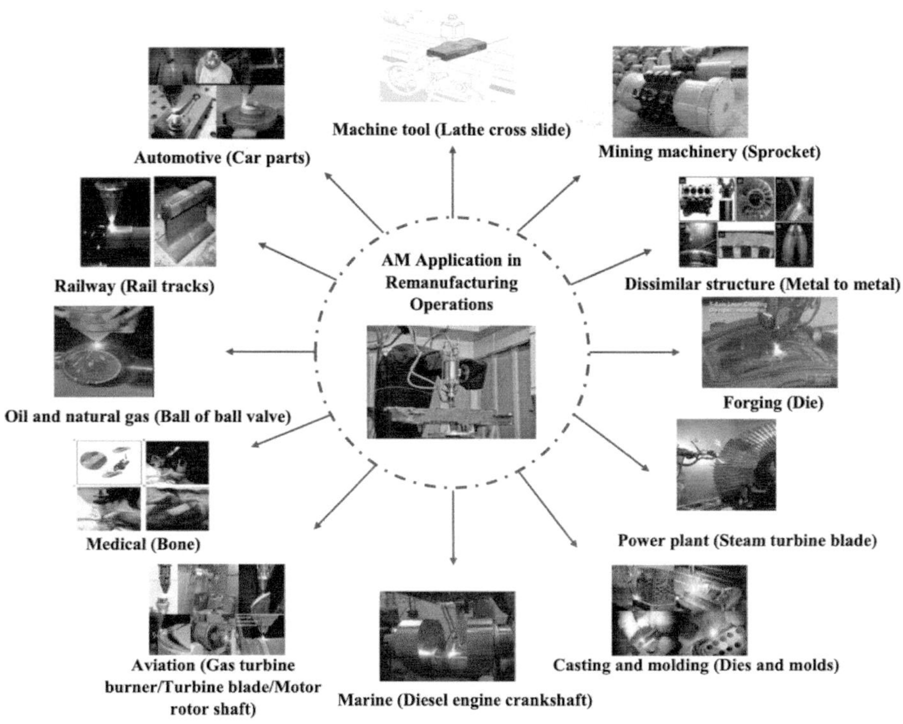

Figure 2.16. AM applications in repair and restoration in different fields [52].

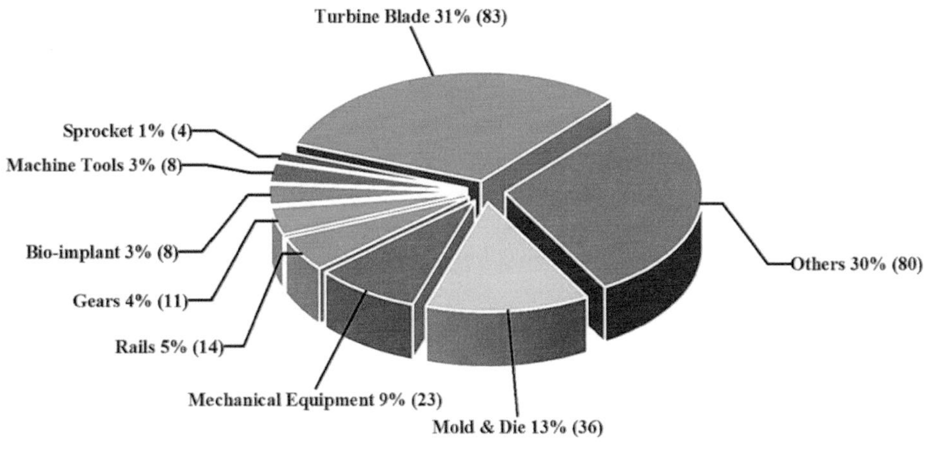

Figure 2.17. Representative remanufactured parts deployment by AM [52].

conditions of service, resulting in scratches, peeling, cracks and pits, partial fracture, corrosion, etc., damage and loss of function. AM can greatly reduce the abandonment rate of parts and equipment, and reduce the loss of parts and materials.

Figure 2.17 shows some typical parts remanufactured using the AM method. It can be seen that turbine blades, which are typical of the aviation sector, account for the highest proportion, up to 31%. In addition, mold & die and mechanical equipment accounted for 13%, and the 9%, respectively, listing the second and third. This confirms that aerospace continues to be the sector with the highest adoption of AM remanufacturing.

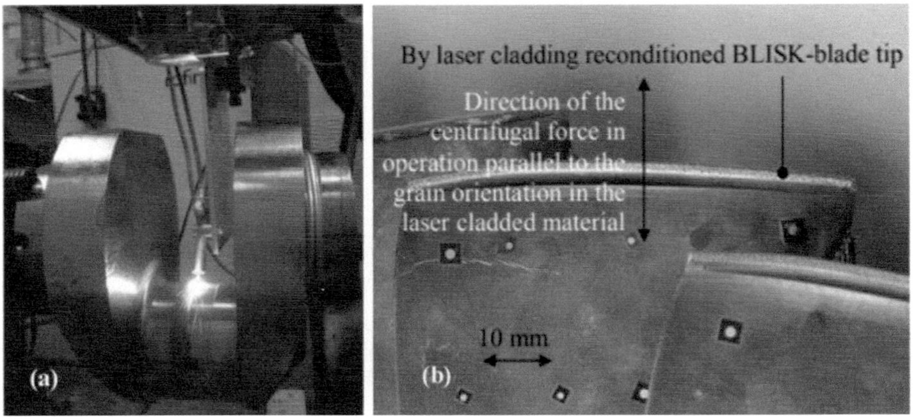

Figure 2.18. Laser additive manufacturing repair of (a) marine diesel engines and (b) aviation turbine blade tip [55].

As one of the most promising AM technologies for repair and restoration, DED technology operates independently of a powder bed, providing a high degree of geometrical freedom without the constraints of build size limitations. This enables the fabrication of very large components using this method. Different from the repair method of traditional welding techniques, DED has a smaller heat input, accompanied by a smaller geometric distortion and residual stress in the repaired parts. Furthermore, DED can produce parts with intricate geometries and enhanced dimensional accuracy, making it both advantageous and cost-effective.

Consequently, DED has become as a pivotal technology for automated repair and restoration of worn-out products, particularly for expensive components, returning them to as-new states for subsequent life cycles. DED offers extended flexibility for strong metallurgical bonds and has become widely employed in the repair of high-value Ni-based superalloy components in aerospace applications, such as turbine blades, airfoils and impellers.

Figure 2.18 illustrates two representative applications of laser additive manufacturing repair in marine and aerospace industries. The repair of marine diesel engine crankshafts demonstrated that the reconstructed structure exhibits strong metallurgical bonding, with a favourable grain structure between the substrate and the cladding material [53]. This repair process resulted in minimal dilution and distortion, as well as improvements in hardness, sufficient strength, increased fatigue resistance, and enhanced corrosion resistance. As depicted in Figure 2.18(b), Kelbassa et al. [54] utilized laser cladding to repair the leading edges of blisks using various aerospace material powders, including Inconel 718. The repaired components achieved the necessary static and dynamic mechanical properties following the application of suitable post-weld heat treatments.

Laser cladding techniques have been successfully used by Guo et al. [56] for gas turbine blade fusion repair. Their study concentrated on employing Inconel 713 powder and laser powder fusion to fix the Z notch on Inconel 713 turbine blades. The results indicated that gas turbine blades can be effectively repaired via laser cladding, with the repaired areas demonstrating good deposition, and strong bonding to the substrate without cracks, pores and other defects.

Figure 2.19 presents a case study of blisk repair. The T700 Blisk, constructed from AM355 steel, exhibits wear and erosion along its leading edge. To address this, a superalloy material was employed to repair the leading edge using the DED technique. Mechanical performance tests revealed that the repaired blisk demonstrated favorable metallurgical, tensile, and erosion resistance properties, allowing it to meet certification requirements. Furthermore, it successfully passed spin and low-cycle fatigue assessments.

Figure 2.19. DED repair process of T700 Blisk (a) DED repair process, (b) T700 blisk after leading-edge repair and (c) repaired blisk after finishing [50].

2.5 Summary

This chapter provides a summary of the content related to additive manufactured superalloys and their applications. It primarily discusses the following aspects: superalloy categories in additive manufacturing, microstructure and mechanical properties of additive manufacturing superalloys, and engineering applications of additively manufactured superalloys. The main conclusions are summarized as follows:

1) Ni-based superalloys are crucial in high-temperature applications due to their excellent properties. These alloys, typically used in environments above 650°C, are categorized into solution-strengthened and precipitation-strengthened types. Solution-strengthened superalloys, such as those with an FCC matrix, are strengthened by elements like Co, Fe, Mo, and W. Precipitation-strengthened alloys, like Inconel 718, gain strength from the γ' phase. Key superalloys in LAM include Inconel 625, Hastelloy X, Inconel 718, and Inconel 738, each offering unique properties for aerospace, industrial, and high-temperature applications. However, superalloys like non-weldable Inconel 738 face challenges in weldability and machinability due to high Ti and Al contents.
2) In LAM, high-energy lasers induce rapid cooling and thermal gradients that significantly influence the microstructure of Ni-based superalloys. Typical microstructural features of AM-fabricated superalloys include epitaxial growth along the build direction and pronounced grain orientation along the <001> direction. Severe micro-segregation during solidification, particularly in Hastelloy X and Inconel 718, leads to secondary phases that weaken mechanical properties. Altering the laser mode, such as using quasi-continuous-wave or flat-top laser beams, can reduce segregation and promote finer, equiaxed dendrites, improving alloy performance and reliability.
3) LAM of Ni-based superalloys results in significant mechanical anisotropy due to complex thermal gradients and cooling rates. This anisotropy is affected by grain morphology, crystallographic texture, precipitate arrangement, and defects. When tensile loads are applied parallel to the building direction, lower strength is observed due to the alignment of columnar grains facilitating dislocation movement. Conversely, loads applied perpendicularly enhance strength as grain boundaries obstruct dislocation motion. Additionally, grain size and orientation significantly influence ductility, with vertical samples exhibiting higher ductility compared to horizontal samples, where crack propagation along grain boundaries occurs.
4) Ni-based superalloys are essential for high-performance and high-temperature applications in aerospace, automotive, and energy sectors due to their exceptional strength, corrosion resistance, and high-temperature stability. However, traditional manufacturing methods struggle with their complex geometries. LAM offers unique advantages, enabling the production of intricate components with internal cooling channels while reducing costs. It is particularly effective for

fabricating parts like rocket injectors and turbine blades, and plays a vital role in repair and remanufacturing, enhancing the lifespan and performance of high-value components.

Despite the widespread adoption of LAM in industrial applications, several critical challenges remain unresolved, hindering its full potential. One of the key issues is the difficulty in controlling and eliminating defects, such as lack of fusion, inclusions, and porosities, which can significantly compromise the integrity of the final product. Additionally, the high-temperature mechanical behaviour of parts produced via LAM is often unpredictable, leading to potential fatigue performance degradation over time. Real-time defect detection, especially during the manufacturing process, is another challenge that limits quality assurance and process control. Moreover, LAM is still struggling with issues related to low production rates for high-volume manufacturing, making it less suitable for mass-production applications. The limited range of materials available for LAM, especially for high-performance superalloys, calls for the development of new, additive-friendly superalloys or customized materials to meet specific performance requirements. Post-processing challenges also remain, as many LAM-produced parts require extensive finishing to meet dimensional and surface quality standards.

Considering techno-economic considerations, part complexity, and surface finish quality requirements, future advancements are likely to focus on hybrid manufacturing approaches, combining additive and subtractive methods into a single system. This integrated approach simplifies part fabrication by reducing the number of production steps, and enables the manufacture of complex, high-performance components with internal structures that were previously difficult or impossible to achieve with traditional methods. This evolution will help improve both the efficiency and capability of LAM, potentially transforming its application in advanced industries.

References

[1] Akande, I. G., Oluwole, O. O., Fayomi, O. S. I. and Odunlami, O. A. (2021). Overview of mechanical, microstructural, oxidation properties and high-temperature applications of superalloys[J]. Materials Today: Proceedings 43: 2222–2231.

[2] GaoLe, Z., Yun, J., XiaoAn, H., Jia, H., JiWu, W. and Yun, W. (2020). High-temperature mechanical properties of nickel-based superalloys manufactured by additive manufacturing[J]. Materials Science and Technology 36(14): 1523–1533.

[3] Ganji, D. K. and Rajyalakshmi, G. (2019). Influence of alloying compositions on the properties of nickel-based superalloys: a review[J]. Recent Advances in Mechanical Engineering: Select Proceedings of NCAME 2020: 537–555.

[4] Shahwaz, M., Nath, P. and Sen, I. (2022). A critical review on the microstructure and mechanical properties correlation of additively manufactured nickel-based superalloys[J]. Journal of Alloys and Compounds 907: 164530.

[5] Jones, J., Whittaker, M., Buckingham, R., Johnston, R., Bache, M. and Clark, D. (2017). Microstructural characterisation of a nickel alloy processed via blown powder direct laser deposition (DLD)[J]. Materials & Design 117: 47–57.

[6] Zhao, Y., Guo, Q., Li, C., Yang, Z., Zhang, J., Huang, Y. et al. (2024). Achieving superior elevated temperature properties in additive manufactured nickel-based superalloys through unique alloy design[J]. Additive Manufacturing 88: 104273.

[7] Minet, K., Saharan, A., Loesser, A. and Raitanen, N. (2019). Superalloys, powders, process monitoring in additive manufacturing [M]. Additive Manufacturing for the Aerospace Industry. Elsevier, 2019: 163–185.

[8] Wei, Q., Xie, Y., Teng, Q., Shen, M., Sun, S. and Cai, C. (2022). Crack types, mechanisms, and suppression methods during high-energy beam additive manufacturing of nickel-based superalloys: a review[J]. Chinese Journal of Mechanical Engineering: Additive Manufacturing Frontiers 1(4): 100055.

[9] Theska, F., Stanojevic, A., Oberwinkler, B. and Primig, S. (2020). Microstructure-property relationships in directly aged Alloy 718 turbine disks[J]. Materials Science and Engineering: A 776: 138967.

[10] Sims, C. T., Stoloff, N. S. and Hagel, W. C. (1987). Superally II[M]. John Wiley & Sons. New York 8(9): 10.

[11] Radavich, J. F. (1989). The physical metallurgy of cast and wrought alloy 718[J]. Superalloy 718(33): 229–240.

[12] Hu, Y. L., Li, Y. L., Zhang, S. Y., Lin, X., Wang, Z. H. and Huang, W. D. (2020). Effect of solution temperature on static recrystallization and ductility of Inconel 625 superalloy fabricated by directed energy deposition[J]. Materials Science and Engineering: A 772: 138711.

[13] Paul, C. P., Ganesh, P., Mishra, S. K., Bhargava, P., Negi, J. and Nath, A. K. (2007). Investigating laser rapid manufacturing for Inconel-625 components[J]. Optics & Laser Technology 39(4): 800–805.

[14] Rombouts, M., Maes, G., Mertens, M. and Hendrix, W. (2012). Laser metal deposition of Inconel 625: Microstructure and mechanical properties[J]. Journal of Laser Applications 24(5).
[15] Hu, Y. L., Lin, X., Li, Y. L., Zhang, S. Y., Gao, X. H., Liu, F. G. et al. (2020). Plastic deformation behavior and dynamic recrystallization of Inconel 625 superalloy fabricated by directed energy deposition[J]. Materials & Design 186: 108359.
[16] Zhang, S., Liu, J., Lin, X., Huang, Y., Zhang, Y., Guo, P. et al. (2021). Microstructure and anodic electrochemical behavior of additive manufactured Hastelloy X alloy via directed energy deposition[J]. Additive Manufacturing 39: 101824.
[17] Jinoop, A. N., Pau, C. P., Denny, J., Nayak, S. K., Krishna, V. and Bindra, K. S. (2020). Laser Additive Manufacturing (LAM) of Hastelloy-X thin walls using directed energy deposition (DED): Parametric investigation and multi-objective analysis[J]. Lasers in Engineering (Old City Publishing) 46.
[18] Zhang, W., Zheng, Y., Liu, F., Wang, D., Liu, F., Huang, C. et al. (2021). Effect of solution temperature on the microstructure and mechanical properties of Hastelloy X superalloy fabricated by laser directed energy deposition[J]. Materials Science and Engineering: A 820: 141537.
[19] Jinoop, A. N., Paul, C. P. and Bindra, K. S. (2019). Laser assisted direct energy deposition of Hastelloy-X[J]. Optics & Laser Technology 109: 14–19.
[20] Jinoop, A. N., Kumar, V. A., Paul, C. P., Ranjan, R. and Bindra, K. S. (2020). Hot deformation behavior of Hastelloy-X preforms built using directed energy deposition based laser additive manufacturing[J]. Materials Letters 270: 127737.
[21] Zhai, Y., Lados, D. A., Brown, E. J. and Vigilante, G. N. (2019). Understanding the microstructure and mechanical properties of Ti-6Al-4V and Inconel 718 alloys manufactured by Laser Engineered Net Shaping[J]. Additive Manufacturing 27: 334–344.
[22] Firoz, R., Basantia, S. K., Khutia, N., Bar, H. N., Sivaprasad, S. and Murthy, G. V. S. (2020). Effect of microstructural constituents on mechanical properties and fracture toughness of Inconel 718 with anomalous deformation behavior at 650°C[J]. Journal of Alloys and Compounds 845: 156276.
[23] Balan, A., Perez, M., Chaise, T., Cazottes, S., Bardel, D., Corpace, F. et al. (2021). Precipitation of γ''in Inconel 718 alloy from microstructure to mechanical properties[J]. Materialia 20: 101187.
[24] Ferreri, N. C., Vogel, S. C. and Knezevic, M. (2020). Determining volume fractions of γ, γ', γ'', δ, and MC-carbide phases in Inconel 718 as a function of its processing history using an advanced neutron diffraction procedure[J]. Materials Science and Engineering: A 781: 139228.
[25] Zhou, W., Tian, Y., Tan, Q., Qiao, S., Luo, H., Zhu, G. et al. (2022). Effect of carbon content on the microstructure, tensile properties and cracking susceptibility of IN738 superalloy processed by laser powder bed fusion[J]. Additive Manufacturing 58: 103016.
[26] Xu, J., Ding, Y., Gao, Y., Wang, H., Hu, Y. and Zhang, D. (2021). Grain refinement and crack inhibition of hard-to-weld Inconel 738 alloy by altering the scanning strategy during selective laser melting[J]. Materials & Design 209: 109940.
[27] Cloots, M., Uggowitzer, P. J. and Wegener, K. (2016). Investigations on the microstructure and crack formation of IN738LC samples processed by selective laser melting using Gaussian and doughnut profiles[J]. Materials & Design 89: 770–784.
[28] Henderson, M. B., Arrell, D., Larsson, R., Heobel, M. and Marchant, G. (2004). Nickel based superalloy welding practices for industrial gas turbine applications[J]. Science and Technology of Welding and Joining 9(1): 13–21.
[29] Tan, C., Weng, F., Sui, S., Chew, Y. and Bi, G. (2021). Progress and perspectives in laser additive manufacturing of key aeroengine materials[J]. International Journal of Machine Tools and Manufacture 170: 103804.
[30] Guo, C., Li, G., Li, S., Hu, X., Lu, H., Li, X. et al. (2023). Additive manufacturing of Ni-based superalloys: Residual stress, mechanisms of crack formation and strategies for crack inhibition[J]. Nano Materials Science 5(1): 53–77.
[31] Zhu, J., Kang, N., Li, D., Li, B., Feng, Z., Zhong, H. et al. (2023). Reducing directionally arranged substructure induced anisotropic mechanical properties of Hastelloy X superalloy fabricated by laser directed energy deposition[J]. Materials Science and Engineering: A 862: 144461.
[32] Parimi, L. L., Ravi, G. A., Clark, D. and Attallah, M. M. (2014). Microstructural and texture development in direct laser fabricated IN718[J]. Materials Characterization 89: 102–111.
[33] Zhang, S., Wang, L., Lin, X., Yang, H. and Huang, W. (2022). The formation and dissolution mechanisms of Laves phase in Inconel 718 fabricated by selective laser melting compared to directed energy deposition and cast[J]. Composites Part B: Engineering 239: 109994.
[34] Chen, Y., Guo, Y., Xu, M., Ma, C., Zhang, Q., Wang, L. et al. (2019). Study on the element segregation and Laves phase formation in the laser metal deposited IN718 superalloy by flat top laser and gaussian distribution laser[J]. Materials Science and Engineering: A 754: 339–347.
[35] Xiao, H., Li, S. M., Xiao, W. J., Li, Y. Q., Cha, L. M., Mazumder, J. et al. (2017). Effects of laser modes on Nb segregation and Laves phase formation during laser additive manufacturing of nickel-based superalloy[J]. Materials Letters 188: 260–262.
[36] Xiao, H., Xie, P., Cheng, M. and Song, L. (2020). Enhancing mechanical properties of quasi-continuous-wave laser additive manufactured Inconel 718 through controlling the niobium-rich precipitates[J]. Additive Manufacturing 34: 101278.

[37] Hosseini, E. and Popovich, V. A. (2019). A review of mechanical properties of additively manufactured Inconel 718[J]. Additive Manufacturing 30: 100877.

[38] Sun, S. H., Koizumi, Y., Saito, T., Yamanaka, K., Li, Y. P., Cui, Y. et al. (2018). Electron beam additive manufacturing of Inconel 718 alloy rods: Impact of build direction on microstructure and high-temperature tensile properties[J]. Additive Manufacturing 23: 457–470.

[39] Zhao, Y., Meng, F., Liu, C., Tan, S. and Xiong, W. (2021). Impact of homogenization on microstructure-property relationships of Inconel 718 alloy prepared by laser powder bed fusion[J]. Materials Science and Engineering: A 826: 141973.

[40] Alhuzaim, A., Imbrogno, S. and Attallah, M. M. (2021). Controlling microstructural and mechanical properties of direct laser deposited Inconel 718 via laser power[J]. Journal of Alloys and Compounds 872: 159588.

[41] Gu, D., Shi, Q., Lin, K. and Xi, L. (2018). Microstructure and performance evolution and underlying thermal mechanisms of Ni-based parts fabricated by selective laser melting[J]. Additive Manufacturing 22: 265–278.

[42] Gokcekaya, O., Ishimoto, T., Hibino, S., Yasutomi, J., Narushima, T. and Nakano, T. (2021). Unique crystallographic texture formation in Inconel 718 by laser powder bed fusion and its effect on mechanical anisotropy[J]. Acta Materialia 212: 116876.

[43] Ni, M., Chen, C., Wang, X., Wang, P., Li, R., Zhang, X. et al. (2017). Anisotropic tensile behavior of in situ precipitation strengthened Inconel 718 fabricated by additive manufacturing[J]. Materials Science and Engineering: A 701: 344–351.

[44] Bean, G. E., McLouth, T. D., Witkin, D. B., Sitzman, S. D., Adams, P. M. and Zaldivar, R. J. (2019). Build orientation effects on texture and mechanical properties of selective laser melting Inconel 718[J]. Journal of Materials Engineering and Performance 28: 1942–1949.

[45] Liu, F., Lin, X., Huang, C., Song, M., Yang, G., Chen, J. et al. (2011). The effect of laser scanning path on microstructures and mechanical properties of laser solid formed nickel-base superalloy Inconel 718[J]. Journal of Alloys and Compounds 509(13): 4505–4509.

[46] Buddaraju, K. M., Sastry, G. R. K. and Kosaraju, S. (2021). A review on turning of Inconel alloys[J]. Materials Today: Proceedings 44: 2645–2652.

[47] Madigana, C. S., Vaddula, A., Yerramsetti, S. D. and Buddaraju, K. M. (2023). Additive manufacturing of titanium and nickel-based superalloys: A review[J]. Materials Today: Proceedings.

[48] Akca, E. and Gürsel, A. (2015). A review on superalloys and IN718 nickel-based INCONEL superalloy[J]. Periodicals of Engineering and Natural Sciences (PEN) 3(1).

[49] Najmon, J. C., Raeisi, S. and Tovar, A. (2019). Review of additive manufacturing technologies and applications in the aerospace industry[J]. Additive Manufacturing for the Aerospace Industry 2019: 7–31.

[50] Blakey-Milner, B., Gradl, P., Snedden, G., Brooks, M., Pitot, J., Lopez, E. et al. (2021). Metal additive manufacturing in aerospace: A review[J]. Materials & Design 209: 110008.

[51] Gradl, P. R. and Protz, C. S. (2020). Technology advancements for channel wall nozzle manufacturing in liquid rocket engines[J]. Acta Astronautica 174: 148–158.

[52] Kanishka, K. and Acherjee, B. (2023). A systematic review of additive manufacturing-based remanufacturing techniques for component repair and restoration[J]. Journal of Manufacturing Processes 89: 220–283.

[53] Koehler, H., Partes, K., Seefeld, T. and Vollertsen, F. (2011). Influence of laser reconditioning on fatigue properties of crankshafts[J]. Physics Procedia 12: 512–518.

[54] Kelbassa, I., Gasser, A. and Wissenbach, K. (2004). Laser cladding as a repair technique for BLISKs out of titanium and nickel base alloys used in aero engines[C]. Pacific International Conference on Applications of Lasers and Optics. AIP Publishing.

[55] Koehler, H., Partes, K., Seefeld, T. and Vollertsen, F. (2010). Laser reconditioning of crankshafts: From lab to application[J]. Physics Procedia 5: 387–397.

[56] Guo, W. (2023). Laser powder fusion repair of Z-notches with inconel 713 powder: WO2005032755A2[P].

Chapter 3

Predicting 3D Temperature Field of Inconel 718 Multi-layer Deposition using Physics Informed Neural Network

3.1 Introduction

The temperature of the melt pool in the actual DED process is difficult to measure and is easily affected by factors such as powder beam and melt pool disturbance, resulting in significant noise in the measurement results. This makes it difficult to analyze the accumulation of heat and abstract theoretical mathematical formulas. To quantify the thermal accumulation effect in the multi-layer bidirectional scanning DED process and provide theoretical support for offline optimization of DED forming parameters, this section will analyze and abstract the thermal accumulation law in the multi-layer DED process through numerical simulation.

In addition, the model obtained by numerical simulation has a large amount of data and is only suitable for offline analysis, which cannot meet the needs of online prediction. Therefore, this paper adopts a new type of PINN (Physics-Informed Neural Networks) to model the temperature field, which combines the physical laws followed in the DED process with data-driven methods. With less data, this method can model the temperature field, making it easier to quickly obtain the temperature at a certain time and position.

3.2 The finite element model for multi-layer DED temperature field

3.2.1 Governing physics laws of heat transfer during the DED process

In the DED processing, transient heat conduction can be described by the following partial differential equation (PDE) according to Fourier's law of heat conduction [1]:

$$\frac{\partial(\rho c T)}{\partial t}-[\frac{\partial}{\partial x}(k\frac{\partial T}{\partial x})+\frac{\partial}{\partial y}(k\frac{\partial T}{\partial y})+\frac{\partial}{\partial z}(k\frac{\partial T}{\partial z})]=0 \tag{3.1}$$

Here, T is a function of temperature with respect to spatial coordinates (x, y, z) and time t, ρ is the material density (kg/m^3), c is the specific heat (J/kg·°C), and k is the thermal conductivity (W/m·°C) of the material. ρ, c, and k are all thermal properties parameters, and their values will be related to temperature. The thermal properties parameters used in the numerical simulation in this paper are obtained from the Engineering data database in ANSYS Workbench 19.0, as shown in Figure 3.1. In addition, the initial and boundary conditions are considered:

$$\begin{cases} T(x,y,z,t)\,|_{t=0}=T_0 \\ -k\dfrac{\partial T}{\partial n}\,|_{\partial V}=q-h_c(T-T_b)-\varepsilon\sigma(T^4-T_b^4) \end{cases} \tag{3.2}$$

DOI: 10.1201/9781003406327-3

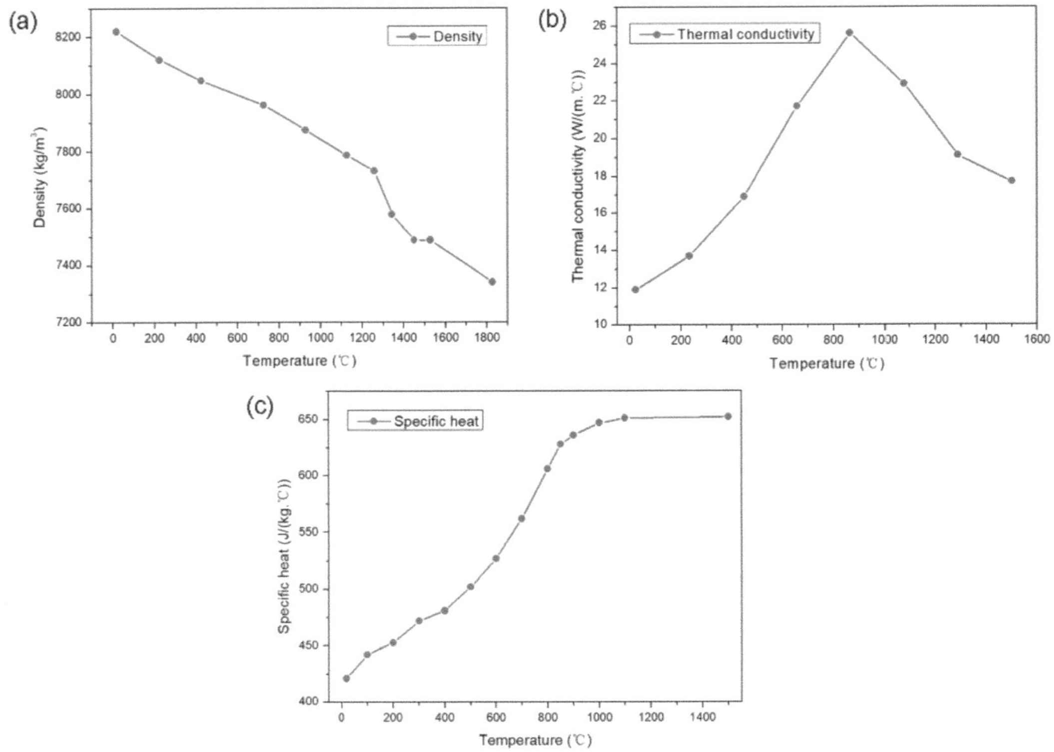

Figure 3.1. The thermos-physical properties of Inconel 718: (a) density; (b) heat conductivity; (c) specific heat.

Here, T_0 is the initial temperature, which is set to 22°C, and T_b is the temperature of the air, which is equal to T_0. h_c is the convective heat transfer coefficient, which is set to 25 W/(m²·°C), and ε is the radiative heat transfer coefficient, which is set to 0.5 [2], σ is the Boltzmann constant, and q is the laser heat input. This paper uses the following Gaussian function to describe the energy distribution of the laser spot:

$$q(x,y,t) = \frac{3\eta p}{\pi r^2}\exp(-3\times\frac{(x-x_0-v_s t)^2+(y)^2}{r^2}) \tag{3.3}$$

Here, p is the laser power, r is the effective radius of the laser spot on the workpiece plane, which is set to 0.8 mm. η is the laser energy utilization ate, which is set to 0.45 for Inconel 718 [3], and v_s refers to the scanning speed.

3.2.2 The establishment of a finite element model

The finite element model has a substrate of Inconel 718 alloy with dimensions of 60 mm × 15 mm × 10 mm, and the geometric size of the deposited layer is 40 mm × 2 mm × 0.4 mm. To analyze the temperature distribution during the DED process, while reducing the calculation time and hardware resource requirements, the finite element simulation in this study is conducted for six continuous layers of DED. After determining the geometric structure, the model is then meshed for subsequent solving. The granularity of the mesh partition will affect the accuracy of the solution. The finer the mesh, the more accurate the calculation results, but it will also increase the demand for solving time and memory. Under the condition of a fixed memory size, excessively fine mesh partitioning may even lead to insufficient memory for further calculation. Therefore, it is necessary to make a reasonable mesh partitioning. In this study, different mesh partitioning methods with different granularities are adopted for different calculation regions. As shown in Figure 3.2(a), the mesh

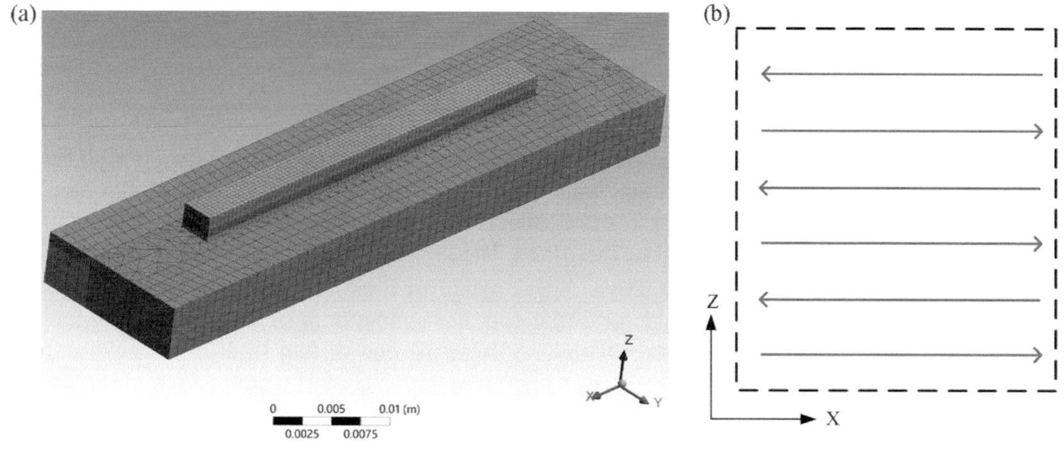

Figure 3.2. (a) Mesh generation and (b) the scanning strategy.

partitioning is the densest for the DED cladding area, which is the most concerned region. The mesh partitioning is also finer for the substrate near the cladding area because it needs to participate in heat conduction. However, for the substrate area far away from the cladding area, sparse mesh partitioning is used. The "birth-death element" method is used to load the DED laser heat source. The Gaussian heat source movement is simulated through APDL programming. When the heat source moves to a certain position, the corresponding element is activated, and the heat source acts on that element. The model calculation is completed on a 64-bit PC with a 6 GB GPU, a 6-core 2.9 GHz CPU, and 16 GB RAM. A typical bidirectional scanning method is used, as shown in Figure 3.2(b).

In this study, Gaussian heat sources with different laser powers and scanning speeds were implemented, and the corresponding temperature fields were obtained through heat source loading and finite element calculation. To verify the reliability of the finite element model calculation results, the center temperature of the molten pool was measured with a two-color infrared pyrometer as it changed over time, and it was compared with the molten pool temperature obtained from the finite element simulation. The results show that, as shown in Figure 3.3, the center temperature of the

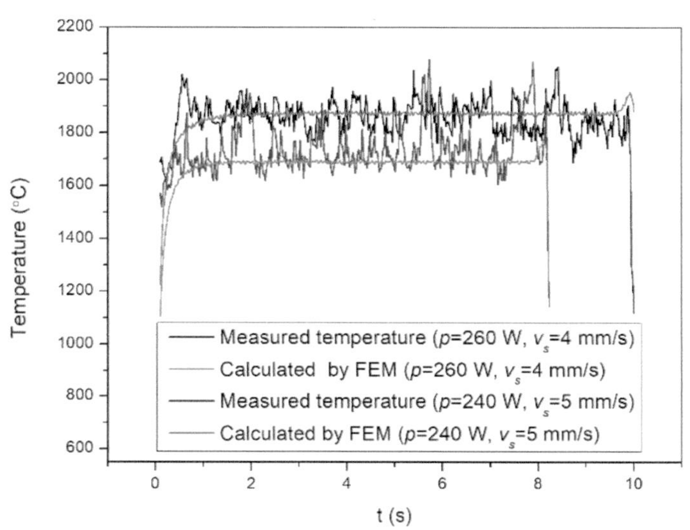

Figure 3.3. Validation for the result of FEM.

molten pool obtained through finite element simulation matches well, with that measured through DED experiments, which proves the accuracy of the finite element model.

3.3 Multi-layer DED thermal accumulation analysis

Figure 3.4 shows the overall temperature field distribution at three positions in the second, fourth, and sixth layers when using bidirectional continuous scanning with a laser power of 240 W and a scanning speed of 5 mm/s. From right to left in each layer, the positions are the starting point of the end, the middle position, and the end of that layer. The high-temperature area at the starting point of the end of the same layer has the largest volume and the most obvious thermal accumulation. As it moves away from the starting point of the end, the temperature of the molten pool and the overall high-temperature area decreases. However, near the end of that layer, the temperature of the molten pool and the volume of the high-temperature area increase, which is due to the change in the heat dissipation conditions relative to the middle position. At the end, heat conduction can only occur towards the middle and downward, so the temperature of the molten pool will increase. Comparing the temperature distribution at the same position in different layers, it can be found that, the volume of the high-temperature area and the temperature of the molten pool increases with the number of deposited layers. This is because of the continuous heating of the laser. When the heat generation rate is greater than the heat dissipation rate, heat accumulates, resulting in an increase in the temperature of the molten pool and the area of the high-temperature area, until the deposition height becomes higher and the heat dissipation conditions gradually improve, and the overall input energy and lost energy of the printed component reach a balance.

To more directly reflect the change in the temperature of the molten pool under thermal accumulation, Figure 3.5 shows the change in the temperature of the center of the molten pool over time. Due to heat accumulation, the temperature of the molten pool increases continuously with the number of deposited layers, and the temperature of the center of the molten pool gradually increases from 1980°C in the first layer to 2230°C in the sixth layer. However, the temperature of the center of the molten pool will not continue to increase rapidly layer by layer, and its rate of increase will

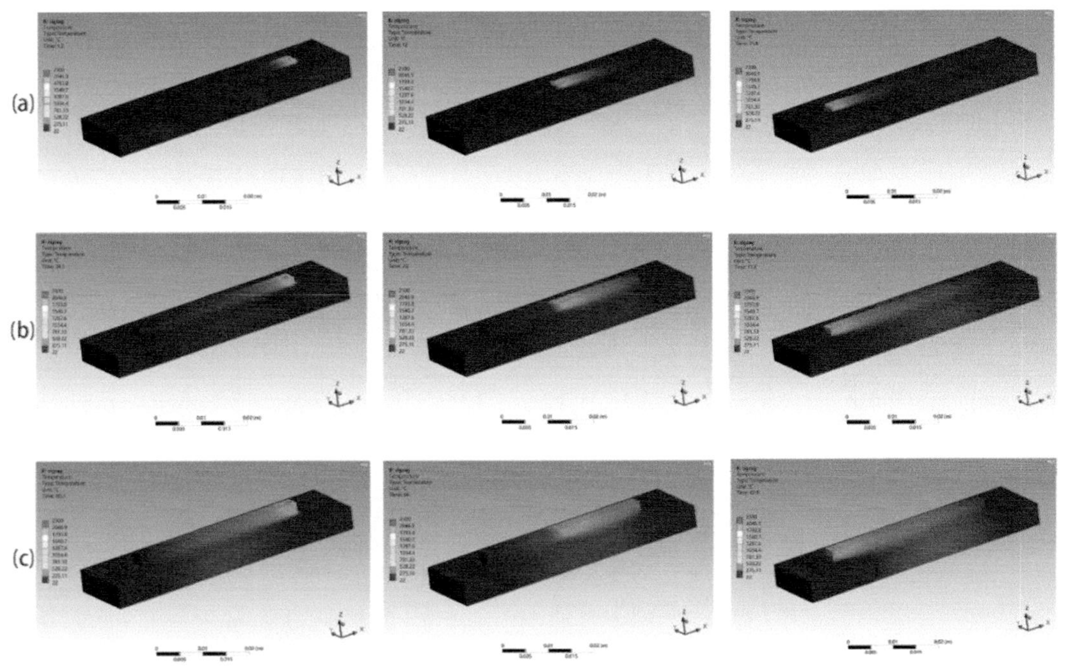

Figure 3.4. The temperature field in different DED layer: (a) layer 2; (b) layer 4; (c) layer 6.

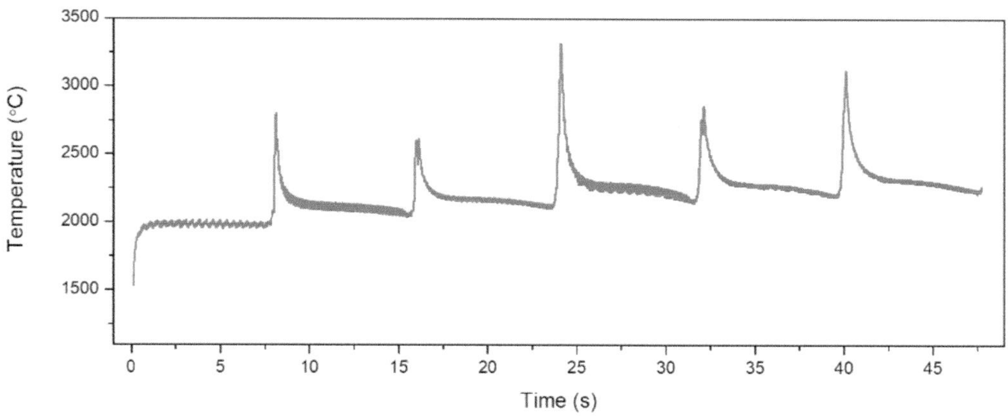

Figure 3.5. The molten pool temperature over time.

gradually slow down with the increase in the number of deposited layers. This is because the formed cladding layer will increase the overall heat dissipation area, thereby improving the heat dissipation conditions and ultimately making the heat dissipation and heat generation tend to balance.

Moreover, starting from the second layer, the temperature of the center of the molten pool near the starting point of the end of each layer is much higher than that in the middle position, and the thermal accumulation effect is obvious. This is because the bidirectional scanning method is used, and the area near the starting point of the end is acted on by the laser twice in a very short time, and the molten pool is heated again before it has cooled down, resulting in heat accumulation and a sharp increase in the temperature of the molten pool. As it moves away from the starting point of the end, the original middle position undergoes cooling for a certain period and has better heat dissipation conditions, so the temperature of the center of the molten pool is lower.

Next, further research was conducted on the temperature distribution of the molten pool in the starting point area. Figure 3.6 shows the surface temperature field of the molten pool at three

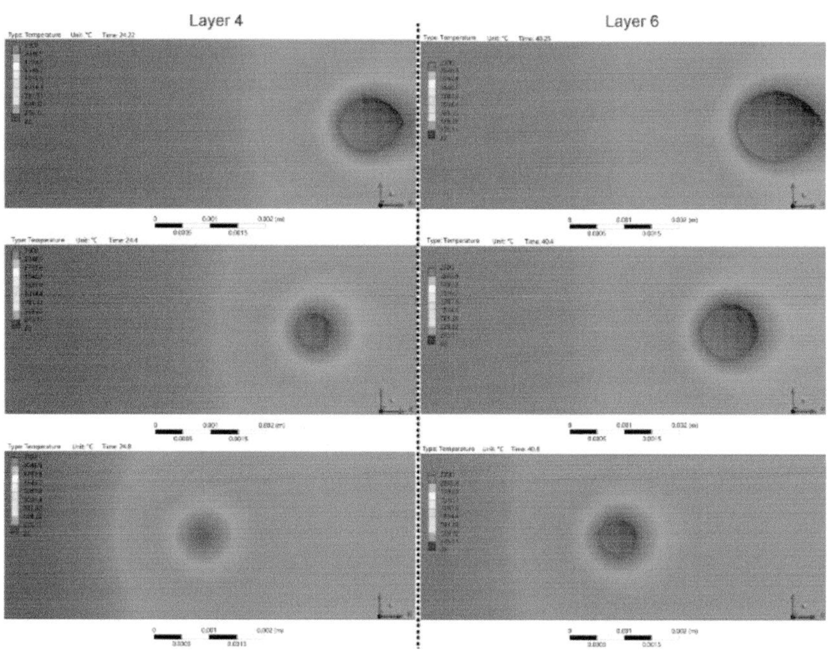

Figure 3.6. The temperature fields near endpoints of layer 4 and layer 6.

positions near the starting point of the end in the fourth and sixth layers. The laser scans from right to left. It can be clearly seen from the figure that the closer to the starting point of the end, the larger the area of the molten pool, and at the same horizontal position, the area of the molten pool in the sixth layer is larger than that in the fourth layer. This further illustrates the existence of layer-by-layer thermal accumulation in the vertical direction and thermal accumulation near the starting point of the end in the horizontal direction during bidirectional continuous scanning, and the phenomenon of thermal accumulation near the starting point of the end is obvious.

To analyze the rules of layer-by-layer thermal accumulation in the vertical direction and thermal accumulation near the starting point of the end in the horizontal direction, the temperature of the molten pool at the middle position of each deposited layer was extracted in turn. Because the temperature of the molten pool in the middle area of each layer changes less, it is conducive to analyzing the relationship between the temperature of the molten pool and the number of deposited layers. As shown in Figure 3.7, there is a certain quantitative relationship between the temperature of the center of the molten pool and the number of deposited layers, and this relationship can be quantitatively expressed by a mathematical relationship. This lays the foundation for this paper to formulate parameter offline optimization strategies based on the quantitative relationship between the temperature of the molten pool and the number of deposited layers in subsequent chapters.

In addition, this paper extracted the relationship between the temperature of the molten pool at a certain position near the starting point of the end in each deposited layer and its distance to the starting point of the end. As shown in Figure 3.8, there is a significant thermal accumulation effect at different positions near the starting point of the end in different deposited layers, and the closer

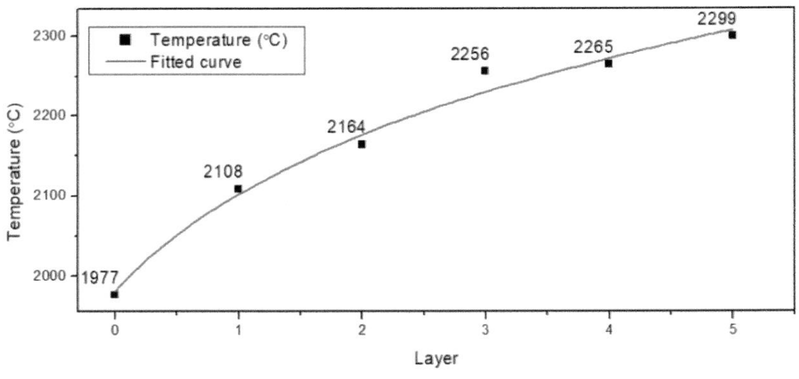

Figure 3.7. The changing rule of the molten pool temperature over the DED layer number.

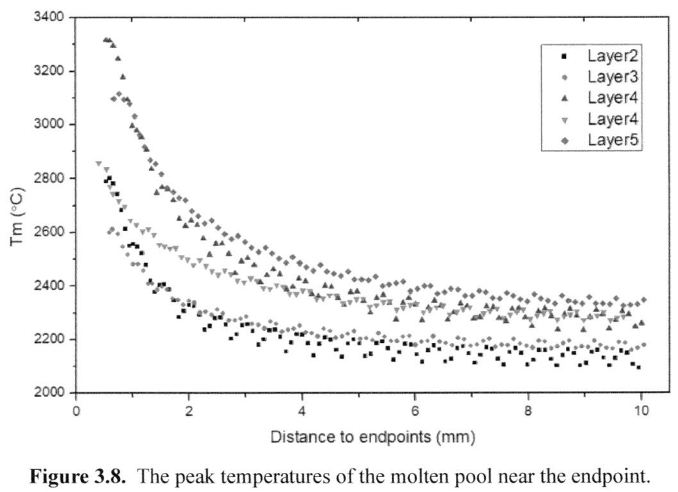

Figure 3.8. The peak temperatures of the molten pool near the endpoint.

to the starting point of the end, the higher the temperature of the molten pool is compared to the middle position. Moreover, the temperature of the molten pool in each layer shows a similar pattern of change with distance. As thermal accumulation can cause irreversible effects on DED forming, in subsequent chapters, this paper will further quantitatively analyze the change in the temperature of the molten pool near the starting point of the end and optimize the processing parameters offline based on the change in the temperature of the molten pool near the starting point of the end.

3.4 The temperature field prediction model based on PINN

The first two sections of this chapter used numerical simulation methods to calculate the temperature field in the multi-layer DED process and visualized the thermal accumulation under bi-directional continuous scanning. Although the finite element method can model the temperature field in the DED process, the results obtained by this method are generally suitable for offline analysis and cannot meet the requirements of real-time prediction. Firstly, the finite element model will obtain many elements after fine grid division, so the data volume at each time step is huge. For example, even if the model established in this paper is small and only contains 6 layers of DED, the temperature data obtained at each time step has reached nearly 200,000. Secondly, in numerical calculations, the time step is refined to about 0.02 s. The DED simulation in this paper lasted for 48 s, so there will be 2400 sets of data at different time points, totaling nearly 50 GB of temperature field data files. If you need to obtain the temperature at a certain position at a certain time, there are roughly two methods. (1) Use numerical simulation software to open the calculated model, export the temperature field data at a certain time, and then search for the temperature at a certain position. This method has complex operation steps and low automation. (2) Export the temperature field data of each time step to a file, then establish a temperature field database, and then read the data at a certain position at a certain time from the database. This method requires a lot of operations to export data in the early stage and occupies a lot of disk space to store tens or hundreds of GB of data. Moreover, searching for a certain data from billions of data is not efficient, and carrying a database file of hundreds of GB is not conducive to the development of offline software.

Therefore, it is necessary to model the temperature field and realize real-time prediction of the temperature field. By establishing a prediction model with process parameters, spatial coordinates, and time as inputs and temperature as output, the corresponding temperature can be quickly obtained according to the input, which is conducive to making offline adjustment strategies based on the temperature at different positions. Therefore, a new PINN algorithm will be used to model the temperature field below.

3.4.1 An overview of PINN

When solving the complex nonlinear relationship between the input [x, t] and the output u, data-driven methods are usually used, such as training a fully connected neural network to establish the mapping between the input and output (see Figure 3.9(a)). The parameter update of the neural network requires forward and backward propagation. In the backward propagation, the gradient descent method is used to optimize the parameters, so that the loss function converges, where the loss function represents the difference between the predicted value and the true value in supervised learning. The accuracy of the data-driven model depends on the quantity and representativeness of the training data. The larger the amount of data, the higher the accuracy of the model.

In fact, in real physical scenarios, [x, t] and u usually satisfy some underlying physical laws, and these physical laws can be described by relevant PDEs. However, purely data-driven methods often ignore this useful physical information and rely solely on a large amount of data to fit complex relationships. If we can use the underlying physical laws to constrain the machine learning process, then the training process of the model will not only be data-driven but also guided by physical relationships.

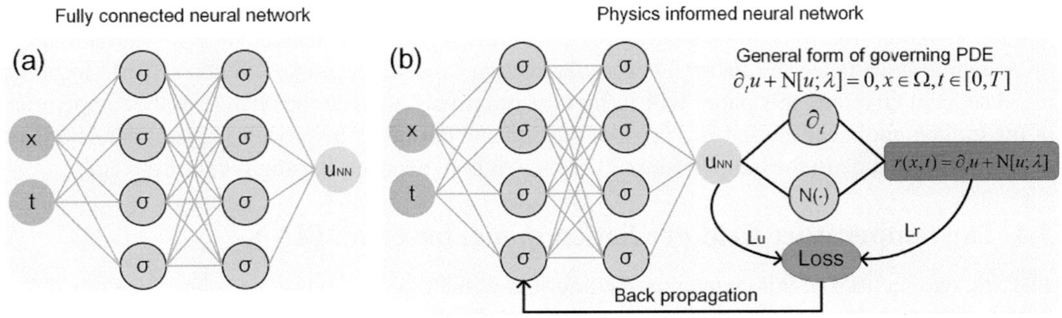

Figure 3.9. The schematic diagrams of (a) the fully connected neural network and (b) the general form of PINN.

PINN combines data-driven and physics-guided approaches to train neural networks. Specifically, assuming that the input [x, t] and output u satisfy the general form of the physical PDE [4]:

$$\frac{\partial u}{\partial t} + \mathrm{N}[u;\lambda] = 0, x \in \Omega, t \in [0,T] \tag{3.4}$$

where $u(x, t)$ represents the target function relationship between the input and output, and is also the solution that satisfies the PDE. $\mathrm{N}(u; \lambda)$ is a nonlinear partial differential operator, λ represents the parameters in the PDE, Ω is the computational domain of the system, and $[0, T]$ is the time range. To embed the physical laws represented by the PDE into the neural network, r(x, t) is used to represent the residual of the PDE:

$$r(x,t) = \frac{\partial u}{\partial t} + \mathrm{N}[u;\lambda] \tag{3.5}$$

If the output $u_{NN}(x, t)$ of the neural network is close to the feasible solution $u(x, t)$ of the PDE (Equation 3.4), then $r(x, t)$ will approach 0, indicating that the function relationship between the input and output satisfies the physical laws described by the PDE. As the predicted values of the model approach the true values, the input and output also gradually approach the feasible solution of the PDE. $r(x, t)$ is calculated by computing the partial derivatives of the model's output uNN(x, t) with respect to the input x and t. Therefore, the neural network that predicts $u(x, t)$ and the neural network that computes the residual $r(x, t)$ can share the same neural network parameters. They can be unified into one neural network by designing a loss function, and the neural network parameters are optimized by minimizing the newly designed loss function. The new loss function is constructed as follows:

$$L = L_u + L_r \tag{3.6}$$

$$L_u = \frac{1}{N_u}\sum_{i=1}^{N_u}\left|u(x_u^i, t_u^i) - u^i\right|^2 \tag{3.7}$$

$$L_r = \frac{1}{N_r}\sum_{i=1}^{N_r}\left|r(x_r^i, t_r^i)\right|^2 \tag{3.8}$$

Here, L_u represents the mean squared error between the model's output values and the actual values, which penalizes the difference between the predicted values and the actual values. L_r represents the mean squared error of the PDE residual, which penalizes the difference between the relationship between the current input and output and the feasible solution of the PDE. The schematic diagram of PINN is shown in Figure 3.9(b).

Compared with pure data-driven machine learning, models based on PINN have clearer physical meanings and interpretability. In traditional machine learning, the nonlinear relationship between the model's input and output is obtained by fitting a large amount of data, without considering the underlying physical laws in the actual scenario. However, PINN embeds the PDE that describes the physical laws into the loss function, guiding the relationship between the input and output to approach the feasible solution of the PDE during the learning iteration. Specifically, the purpose of each backward propagation of the neural network is to minimize the loss function and gradually approach 0. If the input value and predicted value deviate from the feasible solution of the PDE at this time, the absolute value of the PDE residual will be amplified, and the value of the loss function will also be amplified (see Eq. 3.6 and Eq. 3.8). Then, the optimizer based on gradient descent will update the parameters of the neural network, reduce the value of the loss function, and gradually approach the solution of the PDE in the next prediction. In summary, the learning process of the PINN model is guided and constrained by the PDE, so the PINN model is not a completely data-driven model, but a hybrid model driven by both data and physical laws.

3.4.2 DED temperature field modeling based on PINN

The basic theory of PINN has been introduced above. In the following, PINN will be used to model the temperature field during the DED deposition process. Section 3.2.1 of this chapter describes the heat transfer control equations in the DED process, which represent the physical laws of heat conduction. Therefore, it is possible to use the heat conduction PDEs satisfied in the DED process to establish a PINN temperature field prediction model.

Compared to pure data-driven machine learning, the core idea of PINN is to use relevant physical laws and given data to train neural networks [5]. To embed physical laws into the neural network, PINN uses the residual of the PDE as part of the loss function, so that the iterative learning process can be constrained by physical laws. In the actual DED process, the thermal properties parameters of the material, ρ, c, and k in Eq. 3.1, are not static but vary with temperature. If each of these parameters is modeled separately, the complexity will be very high. Instead, this paper uses the diffusion coefficient α to reflect the heat transfer ability of the material (see Eq. 3.9). The true value of α is difficult to determine through experimental measurement, and it is also difficult to define with a clear function. Even if α can be approximated by an approximate function, the function will be very complex, and an inaccurate function cannot reflect the true heat diffusion state of the DED process. Once inaccurate values or functions are embedded in the neural network, the problem of gradient explosion may occur during the learning process, causing the loss function to fail to converge and the model to fail to train successfully. Therefore, like the weights and biases in the neural network, α will be treated as a parameter and updated through data-driven methods to obtain the expected prediction accuracy.

$$\frac{\partial T}{\partial t} - \alpha \cdot (\frac{\partial^2 T}{\partial x^2} + \frac{\partial^2 T}{\partial y^2} + \frac{\partial^2 T}{\partial z^2}) = 0 \tag{3.9}$$

To embed the above heat conduction PDE into the neural network, let R_T represent the residual of the heat conduction PDE:

$$R_T = \frac{\partial T}{\partial t} - \alpha \cdot (\frac{\partial^2 T}{\partial x^2} + \frac{\partial^2 T}{\partial y^2} + \frac{\partial^2 T}{\partial z^2}) \tag{3.10}$$

Where T is the output of the neural network at each iteration. When T satisfies the heat conduction PDE, the value of R_T will approach 0, indicating that the predicted temperature field satisfies the heat conduction PDE. Therefore, the convergence process of RT is consistent with the process of the model's predicted temperature approaching the true temperature. Thus, R_T will be embedded into the loss function as an additional constraint in the learning process. Considering the initial and boundary conditions, the newly designed loss function consists of four parts: (1) data constraint L_{data}, (2) physical law constraint $L_{physics}$, (3) initial condition constraint L_{init}, and (4) boundary condition constraint $L_{boundary}$. Therefore, the loss function is defined as:

$$L = L_{data} + L_{phyics} + L_{init} + L_{boundary} \tag{3.11}$$

$$L_{data} = \frac{1}{m}\sum_{i=1}^{m}(\hat{T}^{(i)} - T^{(i)})^2 \tag{3.12}$$

$$L_{physics} = \frac{1}{m}\sum_{i=1}^{m} R_T^2 \tag{3.13}$$

$$L_{init} = \frac{1}{m_0}\sum_{i=1}^{m_0}(\hat{T}_0^{(i)} - T_0^{(i)})^2 \tag{3.14}$$

$$L_{boundary} = \frac{1}{m_b}\sum_{i=1}^{m_b}(\hat{T}_b^{(i)} - T_b^{(i)})^2 \tag{3.15}$$

Where $\hat{T}^{(i)}$ is the predicted temperature of the model, $T^{(i)}$ is the true temperature, $\hat{T}_0^{(i)}$ is the predicted temperature of the model at $t = 0$ s, $T_0^{(i)}$ is the true temperature at t = 0 s, $\hat{T}_b^{(i)}$ is the predicted temperature in the boundary region, and $T_b^{(i)}$ is the true temperature in the boundary region.

The DED temperature field prediction model based on PINN is shown in Figure 3.10. The model's inputs include process parameters, time, and spatial coordinates *x*, *y*, *z*. After forward propagation, the model outputs the predicted temperature. To obtain the numerical value of the loss function during backpropagation, automatic differentiation (AD) [6] is used to calculate the partial derivatives of *T* with respect to *t*, *x*, *y*, and *z*, respectively. The function relationship between the input and output has been established, and AD can calculate the partial derivatives based on the chain rule of differentiation. Finally, an optimizer is used to iteratively optimize the neural network parameters (weights, biases, and α) to minimize the loss function.

To determine the appropriate number of hidden layers, the number of neurons in each hidden layer, and the learning rate, this paper conducted a detailed study using 5000 training samples and 2000 testing samples. As shown in Table 3.1, this paper used grid search to determine the optimal

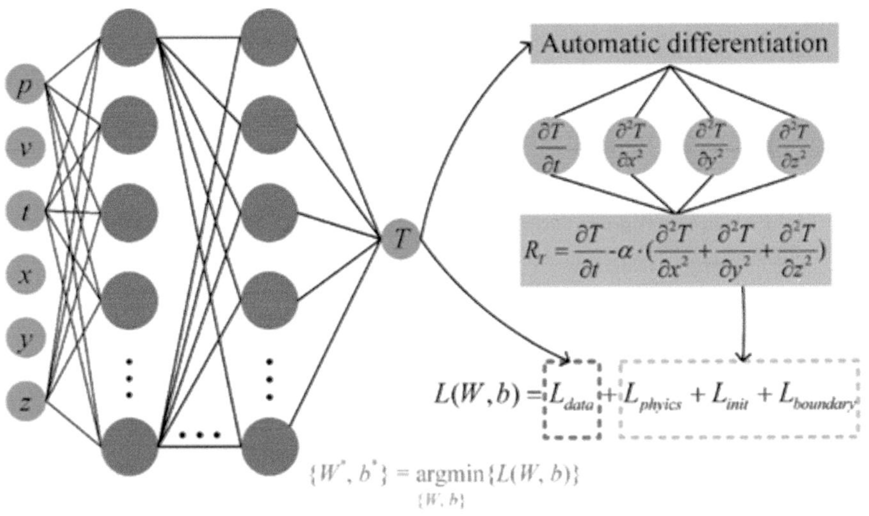

Figure 3.10. The temperature field prediction model using PINN.

combination of the number of hidden layers and neurons in each layer. The prediction error was evaluated using the relative L_2 norm error, $e = \|\hat{y} - y\|_2 / \|y\|_2$, Where $\hat{y}$ is the predicted value and y is the true value. In addition, the learning rate a lso affects the speed of model convergence and the accuracy of model prediction. As shown in Table 3.2, this paper tried different learning rates to find the appropriate value. Finally, the PINN contains 5 hidden layers, each with 50 neurons, and the learning rate of 0.0005 is the most suitable.

In PINN, the relationship between the front and back layers of each hidden layer can be expressed as follows:

$$a_i = \tanh(W_i * a_{i-1} + b_i) \tag{3.16}$$

Where a_{i-1} is the output of the i–1 layer, a_i is the output of the i-th layer, W_i represents the weight vector of the i-th layer, and b_i is the bias vector of the i-th layer. tanh is the activation function in the model. Compared with the rectified linear unit (ReLU) activation function, the tanh function is a smooth function centered at 0, which can accelerate the backpropagation process [7]. The Adam optimizer combines the advantages of the AdaGrad and RMSProp optimizers, is easy to implement, and has a small m 10^{-8} [8]. In the model, the loss function is first optimized by the Adam optimizer for 30,000 steps, and then further optimized by the L-BFGS optimizer for 10,000 steps to ensure that the model converges sufficiently.

Considering that the loss function can be composed of different combinations of Eq. 3.12~15, this study examined the prediction error of the following forms of loss function: L_{data}, $L_{physics}$, L_{data}+$L_{physics}$, L_{data}+L_{init}+$L_{boundary}$, $L_{physics}$+L_{init}+$L_{boundar}$. Grid search was used to find the optimal number of hidden layers, number of neurons, and learning rate for each combination. The results are presented in Table 3.3, and for comparison, the errors obtained using the loss function in the PINN model are also included in the table. The results indicate that using only the physical term $L_{physics}$ in the loss function cannot accurately predict the temperature because we cannot directly solve the PDE without boundary conditions. When the boundary condition $L_{boundary}$ and initial condition L_{init} are added, the prediction error is reduced. The model performs best when the data term L_{data} is further added to the loss function, indicating that the physics-informed and data-driven PINN model is more suitable for temperature field modeling.

Table 3.1. The prediction error (10-2) using the grid search method (the learning rate is 0.0005).

Layers \ Neurons	30	40	50	60
3	4.72	4.84	4.82	5.01
4	5.39	5.11	5.16	5.46
5	7.73	5.97	2.61	6.73
6	2.62	2.56	5.38	2.72

Table 3.2. The prediction error for different learning rates.

Learning rate	0.0001	0.0005	0.001	0.005	0.01
Error (10^{-2})	4.93	2.61	3.90	14.9	47.8

Table 3.3. Prediction errors using different forms of the loss function.

The loss function form	L_{data}	$L_{physics}$	$L_{data} + L_{physics}$	$L_{physics} + L_{init} + L_{boundary}$	$L_{data} + L_{init} + L_{boundary}$	$L_{data} + L_{physics} + L_{init} + L_{boundary}$
Error (%)	7.52	82.6	5.72	6.68	5.88	2.61
Learning rate	0.0005	0.0005	0.0005	0.0005	0.0005	0.0005
Hidden layer	5	2	5	5	4	5
Neurons	50	50	50	40	30	50

3.4.3 Model evaluation and experimental validation

This paper applies PINN to temperature field modeling in multi-layer DED to verify the reliability of the proposed method in the DED process. Using data generated by the finite element method, the training set consists of 6000 random samples from each sedimentary layer, and the test set consists of 2000 random samples from the remaining data. We use a 6 GB GPU, a 6-core 2.9 GHz CPU, and 16 GB RAM on a 64-bit PC to run the PINN model. To fully utilize the advantages of GPU in parallel computing, PINN uses the Tensorflow-GPU package for encoding.

This paper evaluates the performance of the temperature field prediction model based on PINN using the error evaluation indicators shown in Table 3.4. The prediction accuracy of the model is evaluated from three different perspectives: RMSE, MAE, and MRE. The smaller the values of these indicators, the closer the predicted results are to the actual values, and the smaller the error of the model. R^2 reflects the correlation between the predicted results and the actual values. When R^2 is close to 1, it indicates that the output of the prediction model is strongly correlated with the actual values.

This paper establishes a temperature field prediction model for multi-layer DED based on PINN and verifies the applicability of this physics-data hybrid method in complex DED situations. In multi-layer sedimentation processes, laser scanning strategies and substrate preheating are two important means of controlling workpiece quality. The temperature field under different sedimentation methods can lead to different distributions of thermal stress inside the workpiece, which is closely related to cracks and other defects. Here, the PINN method is used to model the DED temperature field under three different scanning strategies and preheating conditions (Table 3.5). The actual DED process is simulated for 48 seconds (6 layers for bidirectional scanning and 3 layers for unidirectional scanning), with a laser power of 240 W and a scanning speed of 5 mm/s.

Figure 3.11 shows the predicted temperature fields on the cross-section (see Figure 3.12) during the DED process under the three conditions. The thermal accumulation becomes more obvious

Table 3.4. Evaluation indices of the model.

Indices	Description
Root mean squared error	$RMSE = \sqrt{\frac{1}{m}\sum_{i=1}^{m}(\hat{y}^{(i)} - y^{(i)})^2}$
Mean absolute error	$MAE = \frac{1}{m}\sum_{i=1}^{m}\left\lvert\hat{y}^{(i)} - y^{(i)}\right\rvert$
Mean relative error	$MRE = \frac{1}{m}\sum_{i=1}^{m}\left\lvert\frac{\hat{y}^{(i)} - y^{(i)}}{y^{(i)}}\right\rvert$
R^2	$R^2 = 1 - \frac{\sum_{i=1}^{m}(\hat{y}^{(i)} - y^{(i)})^2}{\sum_{i=1}^{m}(\bar{y} - y^{(i)})^2}$

Table 3.5. Processing conditions of the three cases (240 W, 5 mm/s).

Cases	Scanning strategy	Preheating condition	Layer number
1	Bidirectional	No Preheating	6
2	Bidirectional	Preheated to 400°C	6
3	Unidirectional	No preheating	3

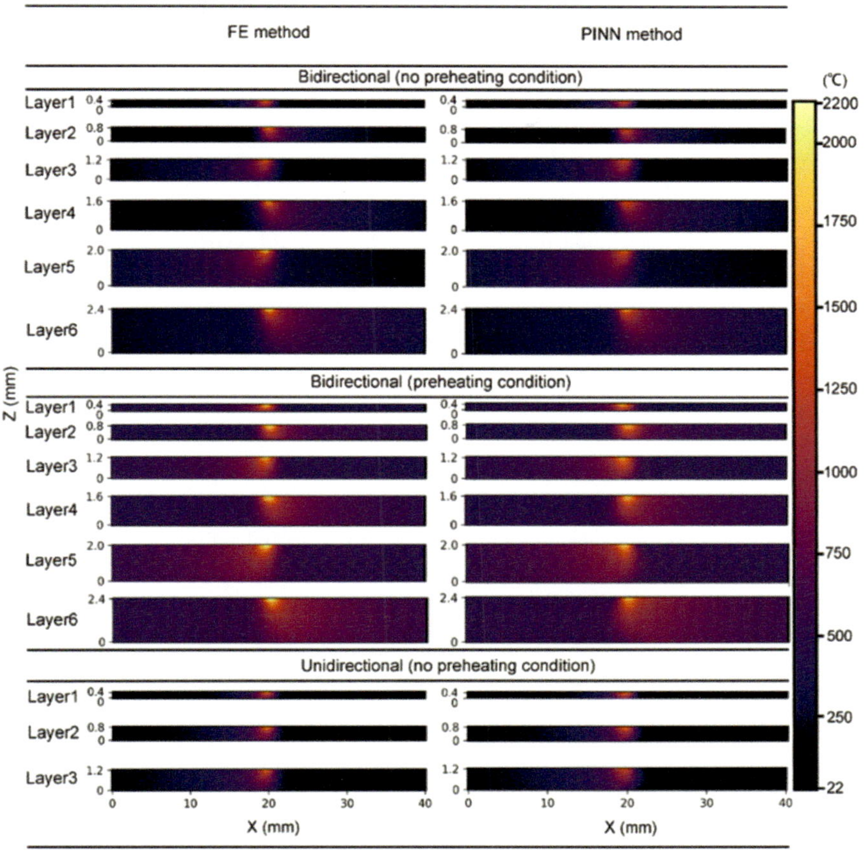

Figure 3.11. Comparison of FEM and PINN results.

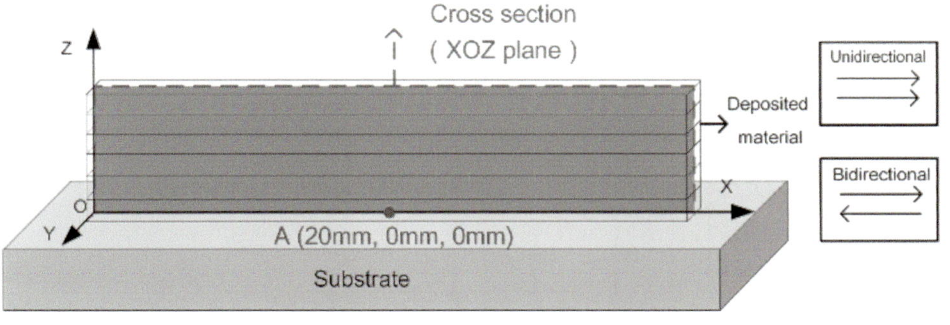

Figure 3.12. Schematic diagram of multi-layer DED.

with the increase of DED sedimentation layers, and the temperature of the molten pool gradually increases, while the heat-affected zone continues to expand. Due to the sufficient cooling of the formed sedimentary layer during the return stage of the laser processing head, the temperature of the unidirectional scanning is lower than that of the bidirectional scanning. The temperature field predicted by the PINN method matches well with the finite element calculation results. As shown in Table 3.6, the MRE of the model's predicted temperature under the three conditions are 0.8%, 1.85%, and 3.5%, respectively, and the R^2 values are 0.99, 0.97, and 0.99, respectively. Therefore, PINN can accurately and reliably model the temperature field in multi-layer DED.

Table 3.6. Prediction performance in different multi-layer DED cases.

Cases	RMSE (°C)	MAE (°C)	MRE	R^2
1	10.87	4.42	0.8%	0.99
2	28.30	12.85	1.85%	0.97
3	18.05	9.46	3.5%	0.99

Figure 3.13 shows the predicted thermal history at the fixed position (point A in Figure 3.12) on the sedimentary layer/substrate interface with and without preheating conditions. The MAE of the predicted temperature by PINN under the preheating and non-preheating conditions are 13.15°C and 8.59°C, respectively. Due to the preheating of the substrate, its peak temperature is higher than that without preheating, which is consistent with previous research results [9]. In addition, the temperature difference between the highest and lowest temperatures of each thermal cycle under the preheating condition is smaller than that without preheating. Therefore, substrate preheating can produce smaller thermal stress and deformation at the sedimentary/substrate interface [10, 11].

Figure 3.14 shows a comparison of the thermal history at the fixed position (point A in Figure 3.12) on the sedimentary layer/substrate interface between unidirectional and bidirectional scanning. The results show that the predicted thermal history by PINN is consistent with the finite element calculation in both trend and value. The MAE of the predicted temperature under unidirectional and bidirectional scanning are 13.15°C and 7.38°C, respectively. Compared with continuous heating under bidirectional scanning, each sedimentary layer under unidirectional scanning will be sufficiently cooled during the return stage of the laser head due to the heating stop. As shown in Figure 3.14, during each thermal cycle, the sedimentary layer/substrate interface under unidirectional scanning experiences greater temperature fluctuations compared to bidirectional scanning, which will result in greater residual stress [12].

To verify the reliability of the temperature predicted by the PINN model, this paper conducted three different DED experiments on Inconel 718. Multi-layer DED is widely used in the manufacturing of Inconel 718, and different laser scanning strategies are usually adopted to control the temperature distribution. This paper used bidirectional and unidirectional scanning to verify the predicted results of PINN. In addition, another DED experiment was conducted under the condition

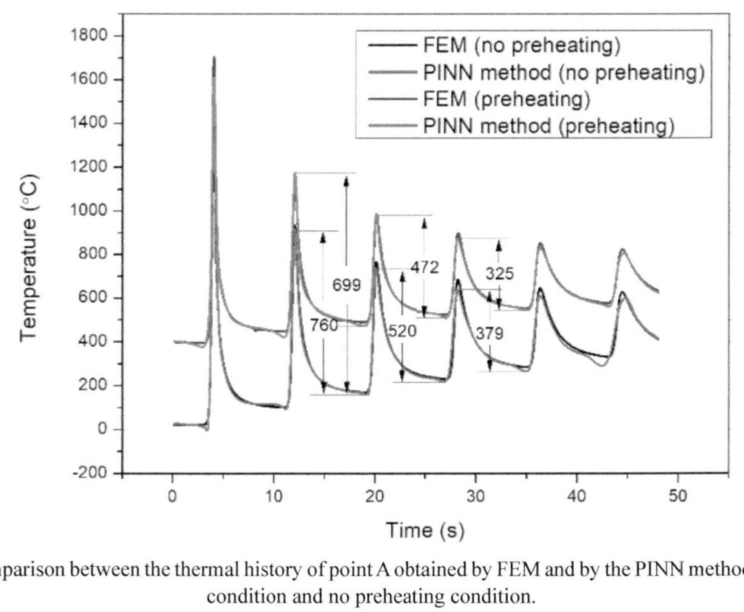

Figure 3.13. Comparison between the thermal history of point A obtained by FEM and by the PINN method under preheating condition and no preheating condition.

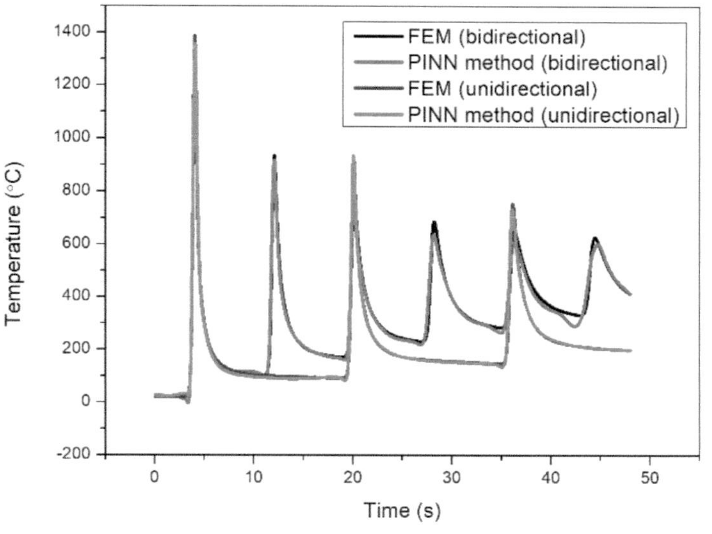

Figure 3.14. Comparison between the thermal history of point A obtained by FEM and by PINN.

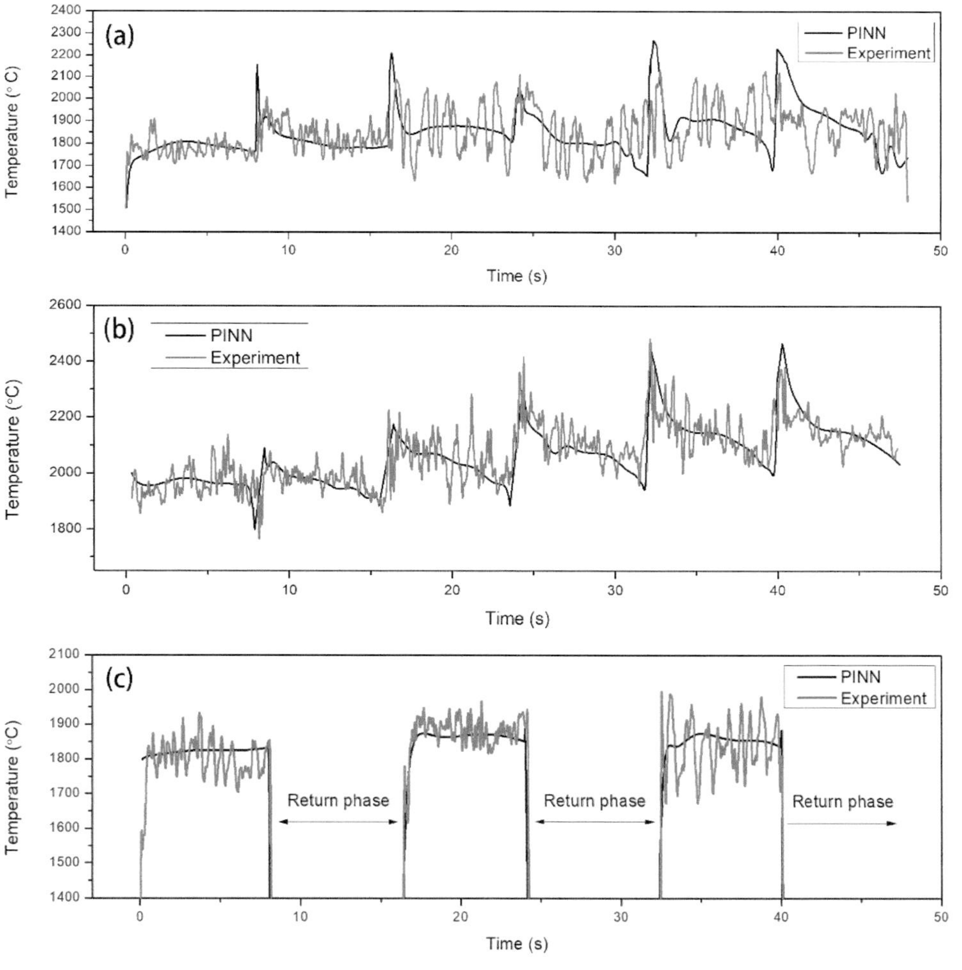

Figure 3.15. Predicted and measured molten pool temperature: (a) bidirectional scanning, (b) under preheating condition and (c) unidirectional scanning.

of bidirectional scanning, that is, preheating the substrate (the substrate was heated to 400°C with a laser before sedimentation). All the above experiments were conducted at a laser power of 240 W, a laser scanning speed of 5 mm/s, and a powder feeding rate of 7.6 g/min, which are suitable parameters for obtaining good forming quality. At the same time, a dual-color infrared thermometer was used to collect the real-time temperature of the molten pool, and the verification results are shown in Figure 3.15.

Figure 3.15(a) shows a comparison between the predicted molten pool temperature by PINN and the measured temperature in the DED experiment under bidirectional scanning. The predicted temperature by PINN is consistent with the measured molten pool temperature. Considering the fluctuation of the measured values, the predicted MRE is 4.83%. Figure 3.15(b) shows a comparison between the predicted and measured temperatures under the condition of preheating the substrate to 400°C before DED. The molten pool temperature gradually increases with the increase of sedimentation layers, indicating the existence of thermal accumulation in multi-layer DED. It can be seen from the figure that the predicted temperature is consistent with the actual temperature in both trend and value, and its MRE is 2.85%. Figure 3.15(c) shows a comparison between the predicted and measured temperatures under unidirectional scanning. In this case, due to the sufficient cooling time of the deposited layer during the return stage without laser heating, the thermal accumulation effect is not obvious. Under this scenario, the MRE of the predicted temperature by PINN is 3.21%. The experimental verification results show that modeling the temperature field in the DED process using PINN is effective and reliable.

3.4.4 The physical part in the loss function of PINN

To verify the applicability of the proposed physics-data hybrid modeling method to other materials besides Inconel 718, this paper also used PINN to model the temperature field of 316L stainless steel to check the residual term of the heat conduction PDE in the loss function of the trained PINN model. If the trained PINN model is constrained by the heat conduction law, then the residual of the heat conduction PDE in its loss function should tend to zero. To verify this, this paper randomly selected 1200 sample points from each test set of the two materials as the input of the model, and then calculated the predicted temperature T through forward propagation, and then calculated the partial derivatives of T with respect to x, y, z, and t through automatic differentiation (AD), and finally obtained the residual of the heat conduction PDE in the loss function. Figure 3.16 shows the PDE residual values obtained using randomly selected samples. The results show that the PDE residual values in the PINN models of Inconel 718 and 316L stainless steel are close to zero, and the mean absolute values are 0.0016 and 0.0017, respectively. The good convergence of the PDE residual indicates that the PINN-based model has been constrained by physical laws after training.

The true diffusion coefficient of a material is often difficult to determine experimentally. In addition, the measured diffusion coefficient cannot be directly substituted for α (see Eq. 3.9) because the physical units of the input data have been eliminated by normalization. This paper treats α as an unknown variable that can be inferred based on the temperature distribution data during the learning process. As shown in Figure 3.17, the value of α is continuously updated with the learning iterations of PINN until it reaches the default convergence criterion of the L-BFGS optimizer. Figure 3.18 shows the relationship between $\partial_t T$ and $\nabla^2 T$ (see Eq. 3.10) in the loss function after forward propagation of randomly selected samples in the test set through the trained PINN model. The results show that they follow a linear distribution, and the fitted slope represents α. The α obtained by fitting on the test set is almost identical to the α learned in Figure 3.17, and the linear relationship between $\partial_t T$ and $\nabla^2 T$ in the obtained model is consistent with the design in the initial loss function, which further demonstrates that the trained model is constrained by the heat conduction law. In addition, the evaluation of the PINN model in the previous section also shows that the strategy of dynamically updating α is effective and can improve the prediction accuracy of the model. Therefore, the main structure of the physical term in the loss function of the proposed

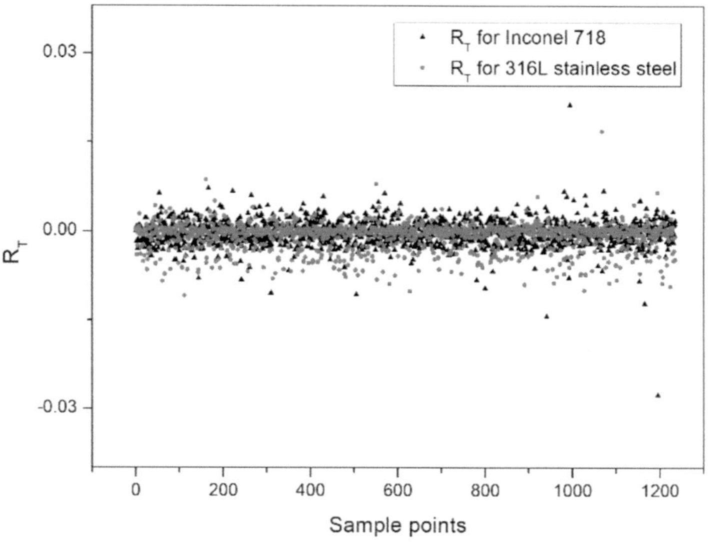

Figure 3.16. The residuals of PDE estimated with randomly chosen input data for the trained models of Inconel 718 and 316L stainless steel.

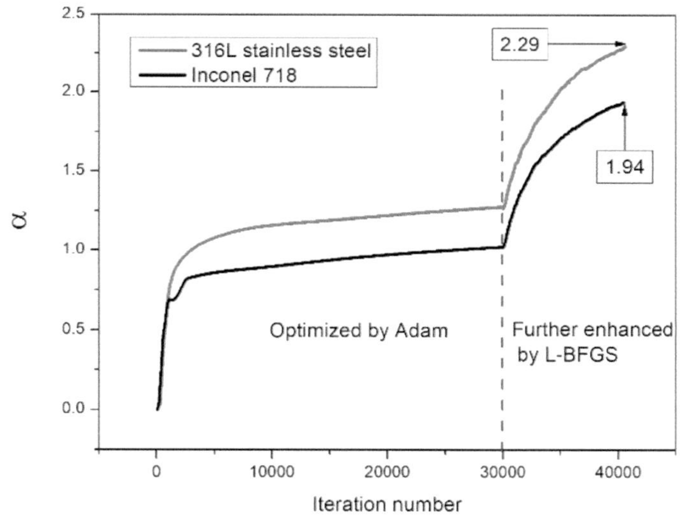

Figure 3.17. The updating process of α during learning.

PINN model is determined by the heat conduction PDE, but the involved parameter α is updated as a variable during the model training process through iteration. Therefore, the PINN temperature field prediction model mixes physical laws with data in two aspects. First, the loss function consists of the mean square error of the difference between the predicted temperature driven by data and the true temperature, and the mean square error of the residual of the heat conduction PDE representing physical laws. On the other hand, the definition of the physical term in the loss function is based on the heat conduction law, while the corresponding parameter α in the physical term is determined by data-driven methods.

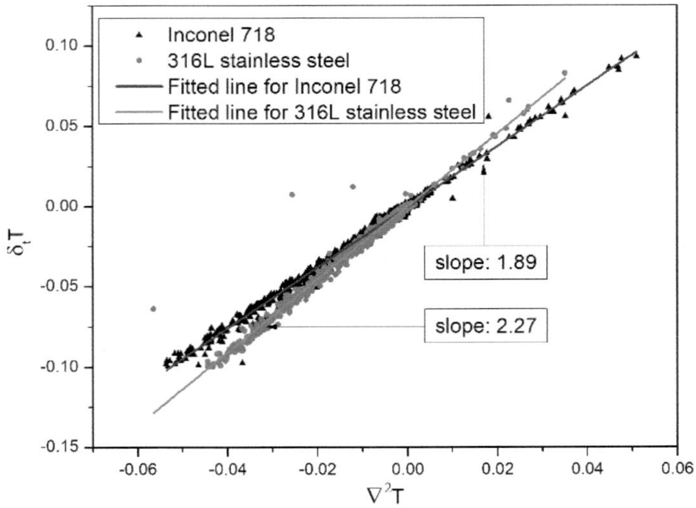

Figure 3.18. Predicted $\hat{\partial}_t T$ versus Ñ²T after implementing the forward propagation on the trained model.

3.4.5 Comparison with data-driven methods

To further investigate the performance of the PINN model, we compared it with other data-driven models, including DNN, LSTM, and XGBoost. DNN is a typical data-driven machine learning model that gradually approaches the actual value through repeated iterations. LSTM solves the problem of gradient disappearance by regulating gates (see Figure 3.19). XGBoost improves gradient boosting trees and has been widely used due to its high computational efficiency and excellent performance [13]. In XGBoost, each tree serves as a weak base learner and fits the residual of the previous learner to improve the performance of the model.

This paper investigated the dependence of the above models on the sample size of the training data. Multiple sets of input data were used to train PINN, DNN, LSTM, and XGBoost, and then the same test data was used for validation. The prediction accuracy of the test data set was evaluated using an error formula based on the relative L_2 norm [14]:

$$\text{Accuracy} = 1 - \frac{\sqrt{\sum_{i=1}^{m}(\hat{y}^{(i)} - y^{(i)})^2}}{\sqrt{\sum_{i=1}^{m}(y^{(i)})^2}} \tag{3.17}$$

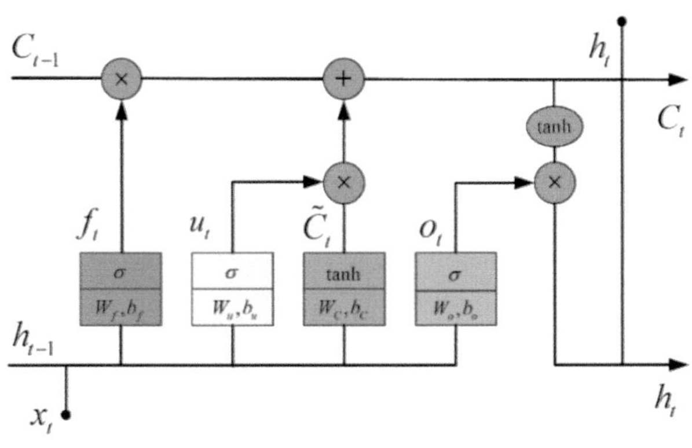

Figure 3.19. Schematic diagram of LSTM.

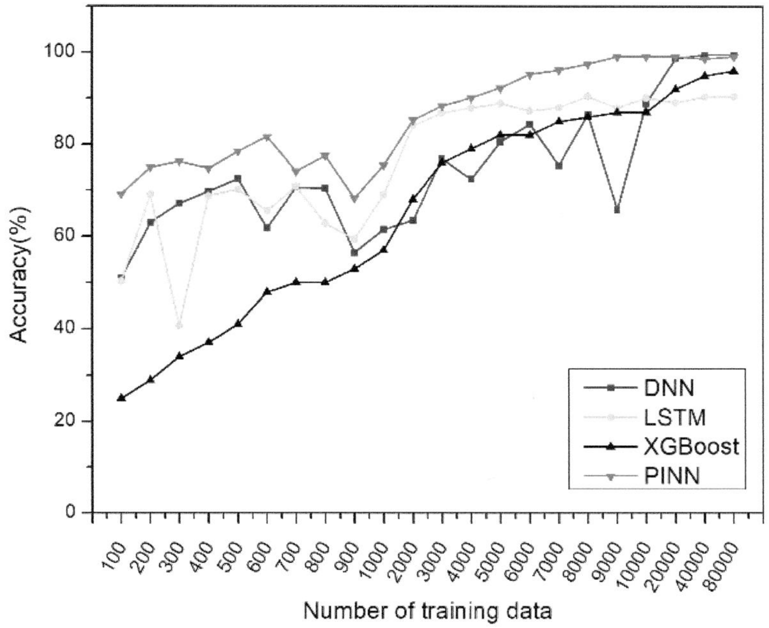

Figure 3.20. The accuracy of the temperature field prediction models with different number of training data.

Here, $\hat{y}$ is the predicted value, y is the true value. To obtain a reliable model, the training set not only needs to have enough samples but also needs to cover most cases. Figure 3.20 describes the changes in the prediction accuracy of the four models with the variation of the training sample size. A common feature of the PINN, LSTM, DNN, and XGBoost models is that using a larger training set can achieve higher prediction accuracy. In temperature field prediction, XGBoost is most dependent on the amount of training data. When the number of training data is less than 2000, XGBoost has the lowest prediction accuracy among the four algorithms. When the number of training data reaches 80,000, the prediction accuracy of XGBoost can reach 96%, which is higher than LSTM but still lower than PINN. Overall, the PINN model has higher prediction accuracy and less fluctuation than LSTM, XGBoost, and DNN, which is particularly significant when the training data is less than 10,000. In summary, the PINN model can achieve higher accuracy without a large amount of training data.

In addition, this paper evaluated the minimum training data required for the four modeling methods to achieve the same accuracy. As shown in Figure 3.21, the training data sample size required for PINN to achieve 80% accuracy is only 12% of that required for DNN. Specifically, DNN requires a data size of 5000, LSTM requires a data size of 2000, while PINN only requires a data size of 600. To achieve 90% accuracy, training DNN and XGBoost requires a data size of 20,000, training LSTM requires a data size of 8000, while training the PINN model only requires a data size of 4000. This means that PINN can achieve the same prediction accuracy as DNN and XGBoost with only 20% of the training data required. Therefore, the training data required for PINN is much less than that required for LSTM, XGBoost, and DNN. Since the loss function of PINN consists of physical and data terms, the parameters updated at each iteration not only make the predicted values close to the true values but also approach the feasible solution of the partial differential equation. Therefore, using a physics-data hybrid model based on PINN is very useful when the cost of collecting training data through experiments or numerical simulations is high, and the corresponding physical process can be defined by some physical PDEs.

To comprehensively evaluate the performance of the PINN temperature field prediction model, this paper compared the computational efficiency of the above four models when achieving the

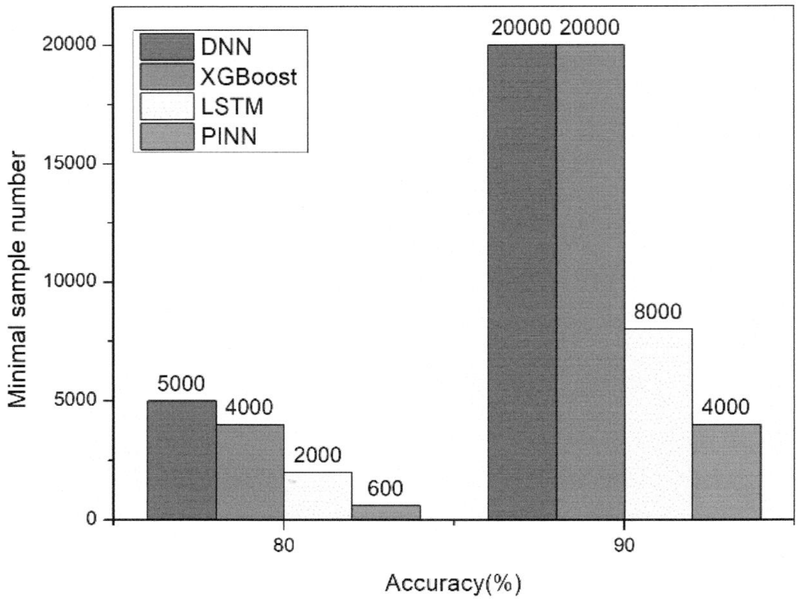

Figure 3.21. The required minimal number of training samples for achieving the given accuracy.

Table 3.7. Computational efficiencies for achieving the accuracy of 90%.

Methods	**DNN**	**LSTM**	**XGBoost**	**PINN**
Number of training data needed	20000	8000	20000	4000
Training time cost (s)	182.06	235.19	10.18	622.41
Mean prediction time cost (ms)	0.18	1.03	0.01	0.20

same 90% prediction accuracy, as shown in Table 3.7. XGBoost, as a decision tree algorithm, has the highest computational efficiency in model training and prediction, but it requires more training data than PINN to achieve the target accuracy. Compared with DNN, PINN requires a large amount of additional automatic differentiation operations to calculate the partial derivatives of the loss function, resulting in a significant increase in computational cost and training time. However, the difference in average prediction time cost between PINN and DNN is small. Among the three models, LSTM has the highest average prediction time cost. Overall, the PINN-based model can achieve higher prediction accuracy with less data and a few minutes of computation time. Considering the long time and high cost of training data collection in DED manufacturing, it is more advantageous than XGBoost, DNN, and LSTM.

3.5 Summary

This chapter studied the temperature distribution characteristics and temperature modeling methods in the multi-layer DED process of Inconel 718. The thermal accumulation in multi-layer DED was analyzed by finite element simulation, and the temperature field was modeled based on PINN. The following conclusions were obtained:

1) There is a layer-by-layer increase in thermal accumulation in the vertical direction in multi-layer DED, and the pool temperature increases layer by layer. Due to the improvement of heat dissipation conditions, the temperature rise rate becomes slower and slower. In addition, in the horizontal direction, the starting position of each layer's end point is repeatedly heated in a short time during the bidirectional scanning, and the thermal accumulation phenomenon

is obvious. The pool temperature is higher at the end, and then gradually decreases as it moves away from the end.

2) Analysis shows that there is a certain regularity in the layer-by-layer change of the pool temperature and the change of the pool temperature near the end of the same layer. Quantitative relationships can be proposed based on this change pattern in the future.
3) A DED temperature field prediction model based on PINN was proposed, which can accurately model the temperature field. The relative error of the experiment is within 4.81%.
4) The PINN model is more interpretable than traditional data-driven machine learning models and is supported by physical laws. The main structure of the physical term in the loss function is determined by the heat transfer law, and the unmeasurable parameters in the physical term are iteratively updated by data-driven methods. The additional embedded partial differential equation guides the predicted value to approach the feasible solution during the learning process.
5) Compared with data-driven machine learning methods such as DNN, LSTM, and XGBoost, PINN can achieve higher accuracy with fewer training samples due to the additional guidance of physical laws.

References

[1] Kundakcioglu, E., Lazoglu, I. and Rawal, S. (2016). Transient thermal modeling of laser-based additive manufacturing for 3D freeform structures[J]. The International Journal of Advanced Manufacturing Technology 85(1): 493–501.

[2] Ahn, D. G., Byun, K. W. and Kang, M. C. (2010). Thermal characteristics in the cutting of Inconel 718 superalloy using CW Nd: YAG laser[J]. Journal of Materials Science & Technology 26(4): 362–366.

[3] Denlinger, E. R., Jagdale, V., Srinivasan, G. V., El-Wardany, T. and Michaleris, P. (2016). Thermal modeling of Inconel 718 processed with powder bed fusion and experimental validation using *in situ* measurements[J]. Additive Manufacturing 11: 7–15.

[4] Raissi, M., Perdikaris, P. and Karniadakis, G. E. (2019). Physics-informed neural networks: A deep learning framework for solving forward and inverse problems involving nonlinear partial differential equations[J]. Journal of Computational Physics 378: 686–707.

[5] Mao, Z., Jagtap, A. D. and Karniadakis, G. E. (2020). Physics-informed neural networks for high-speed flows[J]. Computer Methods in Applied Mechanics and Engineering 360: 112789.

[6] Baydin, A. G., Pearlmutter, B. A., Radul, A. A. and Siskind, J. M. (2018). Automatic differentiation in machine learning: a survey[J]. Journal of Machine Learning Research 18(153): 1–43.

[7] Nwankpa, C., Ijomah, W., Gachagan, A. and Marshall, S. (2018). Activation functions: comparison of trends in practice and research for deep learning[J].

[8] Byrd, R. H., Lu, P., Nocedal, J. and Zhu, C. (1995). A limited memory algorithm for bound constrained optimization[J]. SIAM Journal on Scientific Computing 16(5): 1190–1208.

[9] Alimardani, M., Toyserkani, E. and Huissoon, J. P. (2008). Effect of preheating on the delamination and crack formation in laser solid freeform fabrication process[C] 2008: 29–37.

[10] Buchbinder, D., Meiners, W., Pirch, N., Wissenbach, K. and Schrage, J. (2014). Investigation on reducing distortion by preheating during manufacture of aluminum components using selective laser melting[J]. Journal of Laser Applications 26(1): 012004.

[11] Kim, H., Lee, K.-K., Ahn, D.-G. and Lee, H. (2021). Effects of deposition strategy and preheating temperature on thermo-mechanical characteristics of Inconel 718 super-alloy deposited on AISI 1045 substrate using a DED process[J]. 14(7): 1794.

[12] Ren, K., Chew, Y., Fuh, J. Y. H., Zhang, Y. F. and Bi, G. J. (2019). Thermo-mechanical analyses for optimized path planning in laser aided additive manufacturing processes[J]. Materials & Design 162: 80–93.

[13] Chen, T. and Guestrin, C. (2016). XGBoost: A scalable tree boosting system[J]. Proceedings of the 22nd ACM SIGKDD International Conference on Knowledge Discovery and Data Mining 2016: 785–794.

[14] Rao, C., Sun, H. and Liu, Y. (2020). Physics-informed deep learning for incompressible laminar flows[J].

Chapter 4

Off-line Processing Parameter Optimization for Excellent Inconel 718 Deposition Shaping

4.1 Introduction

The previous chapter analyzed the temperature field in multi-layer DED, and the simulation results and experiments both showed that there is a significant thermal accumulation effect in the multi-layer DED process. This effect is manifested in the fact that the pool temperature increases with the number of deposited layers, and the temperature at the end position is much higher than that in the middle position due to repeated heating. The increase in pool temperature will result in a larger amount of metal powder melted per unit time, resulting in thicker deposition layers, ultimately leading to uneven deposition layer thicknesses at different deposition layers or different positions in the same layer. To compensate for the uneven deposition layer thickness caused by the thermal accumulation effect, fixed parameters should not be used for processing throughout the entire process. Instead, the process parameters should be dynamically adjusted based on the changes in the pool temperature.

However, there is currently a lack of quantitative models to describe how to adjust process parameters to reduce the forming errors caused by the thermal accumulation effect, and only rough adjustment plans are available. To address this issue, this chapter will establish a mathematical model between the pool temperature and the DED forming thickness based on the geometric relationship, mass conservation relationship, energy conservation relationship, and other factors in multi-layer DED forming. Then, the change pattern of the pool temperature in the multi-layer DED process will be analyzed. Based on the established basic relationship between forming size, pool temperature, and process parameters, corresponding quantitative optimization strategies will be proposed from three different angles: Z-axis lifting amount, laser power, and scanning speed to reduce the deviation caused by thermal accumulation and ensure the uniform thickness of the deposition layer.

4.2 The analysis of the multi-layer DED formation characteristics

In the multi-layer DED process, the forming quality is affected by the processing parameters. Among the process parameters of multi-layer DED, laser power, scanning speed, powder feeding rate, and Z-axis lifting amount have the greatest impact on the forming effect, and there is a strong coupling relationship between these parameters. As shown in Figure 4.1, laser power and scanning speed determine the line energy. When the line energy input is too low, the powder entering the melt pool cannot be completely melted and solidified into a reliable deposition layer. Even if a reasonable Z-axis lifting amount and powder feeding rate are determined, the formed component will still have forming defects such as material loss and poor fusion. The powder feeding rate directly determines

DOI: 10.1201/9781003406327-4

Figure 4.1. The forming characteristics of the DED fabricated part with too low energy input.

the maximum amount of powder that can enter the melt pool per unit time. When the powder feeding rate is too low, even if the line energy is high, it is impossible to obtain the expected height and width of the deposition layer. On the contrary, when the powder feeding rate is too high, due to the attenuation effect of powder flow on the laser beam, such as reflection and refraction, it will cause defects such as poor fusion.

In addition, the rationality of the Z-axis lifting amount determines whether the final forming can be successful. As shown in Figure 4.2(a), the focusing characteristics of the powder beam and laser beam determine that the distance between the processed plane and the nozzle should be within a reasonable range. When the processing plane coincides with the focusing plane of the powder beam, the powder beam has the best focusing performance, and more powder can be fed into the melt pool to form a deposition layer. The value of the Z-axis lifting amount is related to the forming height of the previous deposition layer. As shown in Figure 4.2(b), if the Z-axis lifting amount is too high, higher than the thickness of the previous deposition layer, the processing plane may be located below the lower limit plane of the powder flow, and the powder flow becomes divergent. The powder entering the melt pool is very small, making it difficult to form the deposition layer. If the Z-axis lifting amount is too small, the powder will not be well focused, and the powder entering the melt pool will also be small, and the melt pool radius will be large, resulting in a small deposition height. Usually, the Z-axis lifting amount is determined by conducting process pre-experiments based on experimental parameters before the formal DED deposition, and then the value is determined based on experience. Later in the paper, we will derive the mathematical relationship between the Z-axis lifting amount and the thickness of the cladding layer from the physical and geometric perspectives of DED forming.

In addition to the fixed processing parameters that affect the DED forming effect, the existence of thermal accumulation in the multi-layer DED deposition process will also affect the final

Figure 4.2. (a) The powder flow of DED and (b) the fabricated parts when the Z-axis increment is too large.

forming. Below, we will qualitatively analyze the influence of thermal accumulation in multi-layer DED on the morphology of the cladding layer. Chapter 3 has visualized the existence of thermal accumulation in multi-layer DED through numerical calculations, that is, the temperature of the melt pool will gradually increase with the increase of the number of deposition layers, and during the multi-layer reciprocating printing, the temperature of the melt pool near the starting point will be significantly higher than the middle position due to being heated twice in a short time. Theoretically, the increase in the temperature of the melt pool will enable more metal powder to be captured and melted by the melt pool, thereby increasing the thickness and width of the deposition layer. The forming plane will gradually deviate from the focus plane of the powder beam, getting closer to the nozzle plane, causing the powder focusing performance to deteriorate, resulting in an increase in the width of the melt pool and a decrease in the layer thickness. However, multi-layer DED usually has a self-regulating mechanism. After the layer thickness of the current layer becomes smaller, when printing the next layer, the distance between the nozzle plane and the workpiece plane will be greater than the current distance, and the powder focusing performance will be better. Therefore, the width of the melt pool will decrease, and the layer thickness will increase, ultimately making the width and thickness of each layer of the thin-walled part uneven. Even if this potential layer thickness self-regulation mechanism can make the deposition process proceed normally, when it exceeds the self-regulation range, the thermal accumulation problem will directly and significantly reflect on the forming effect.

As shown in Figure 4.3, with the increase of the number of deposition layers, the temperature of the melt pool will gradually increase, and more metal will be melted per unit time. Therefore, the deposition width becomes larger in the upper layers, and there is a phenomenon of metal flowing to both sides at the end. In addition, when using bidirectional scanning, the area near the end is repeatedly heated in a short time without timely heat dissipation, making the thermal accumulation phenomenon in this area more obvious. This causes the thickness near the end to be significantly higher than the middle area, and this deviation will accumulate with the number of layers, eventually leading to the problem of excessive deposition at the end.

Therefore, to achieve multi-layer DED forming with uniform thickness, in addition to determining appropriate initial processing parameters, it is also necessary to consider how to compensate for the forming deviation caused by thermal accumulation effects. Since the thermal accumulation phenomenon occurs during the processing, it is necessary to adjust and optimize the parameters at different stages of DED, rather than using fixed parameters throughout the process. From the previous analysis, the most direct manifestation of thermal accumulation effect is the change in the temperature of the melt pool, which will affect the amount of melted metal and thus affect the height and width of the forming. Therefore, the parameters can be adjusted based on the change in the temperature of the melt pool to keep the forming thickness uniform and compensate for the effect of thermal accumulation on the forming. Chapter 3 has studied the temperature field

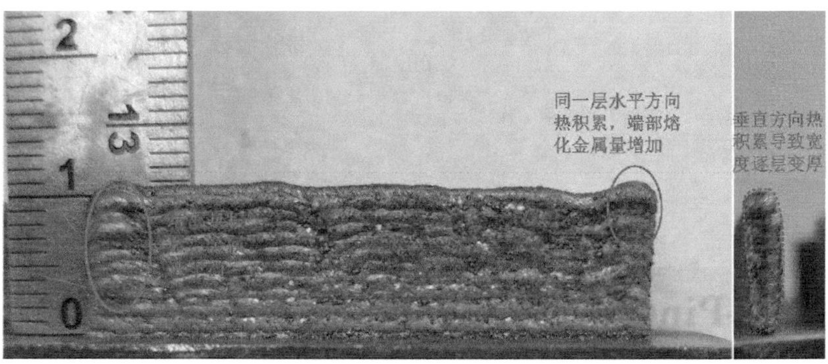

Figure 4.3. The multi-layer DED formation without dealing with the heat accumulation.

distribution in the multi-layer DED deposition process, so the following will establish a quantitative relationship between the forming height and the temperature of the melt pool through physical modeling. Then, we will study how to compensate for the change in forming height caused by temperature variation through optimization of processing parameters, thereby weakening the effect of thermal accumulation on forming.

4.3 Relationship between the geometric dimensions of DED forming and the temperature of the molten pool

The next task is to translate an academic paper from Chinese to English. The paper explores the relationship between the geometric dimensions of the formed object and the temperature of the molten pool from the perspective of the physical mechanism of DED. The study begins by examining the geometric relationship of the formed object. In the DED process, metal powder is melted by laser heating to form a molten pool, which is then shaped into the final contour under the combined effects of gravity, surface tension, protective gas flow, and other factors. According to literature [1] the cross-sectional geometry of the deposited layer can be represented by a crown-shaped model, as shown in Figure 4.4, taking into account factors such as surface tension. The center of the crown can be located above or below the substrate plane depending on the scanning speed. This basic model can be used to model DED.

During the deposition process, only a portion of the powder stream can be melted by the laser and enter the molten pool to form the deposited layer. According to the principle of mass conservation, the mass of the deposited layer that increases per unit length is equal to the mass of the powder that enters the molten pool per unit length.

$$\rho S = \beta \frac{f}{v} \tag{4.1}$$

Where ρ represents the density of the deposited metal, β represents the powder utilization rate, f represents the powder feeding rate, v represents the scanning speed, and S represents the area of the crown. As shown in Figure 4.4, the area of the crown can be expressed as:

$$S = (\frac{w^2}{4} + (R-h)^2)\theta - \frac{w}{2}(R-h) \tag{4.2}$$

In the above equation, w represents the width of the deposited layer, h represents the height, R represents the radius of the crown, and θ represents half of the central angle. The expression for the cross-sectional area S of the crown contains both the angle and the radius of the arc, which can pose significant mathematical difficulties in subsequent calculations. To facilitate mathematical calculations, it is best to convert S into an expression that is only a function of the height h and width w. According to the approximate formula for the crown area proposed by Harris [2] in 1998,

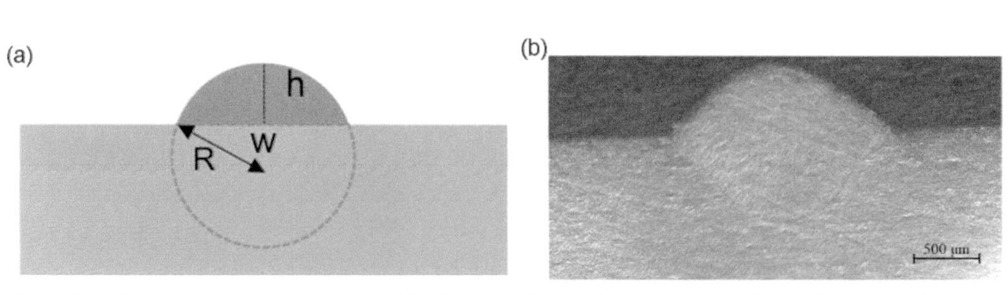

Figure 4.4. The cross-section morphology of single track: (a) the schematic diagram; (b) by the optical microscope.

which has an error of less than 0.8% in all calculation scenarios, the cross-sectional area S can be approximated as:

$$S = \frac{2wh}{3} + \frac{h^3}{2w} \tag{4.3}$$

After mathematical manipulation, the area of the crown can be expressed in terms of the height h and width w of the deposited layer. However, S is still a third-order equation in terms of h, and even though third-order equations can be solved using root-finding formulas, the results are still very complex and not conducive to clarifying the relationships between various physical quantities. It is already very difficult to simplify Eq. 4.3 mathematically. To further simplify the expression for S, this paper analyzed Eq. 4.3 based on the actual width w and height h of the deposited layer obtained. The paper substituted the width w and height h of 37 deposited layers obtained under different parameters into the terms of Eq. 4.3 and obtained the following results.

As shown in the results in Figure 4.5, it can be observed that the value of S is almost equal to the value of the simplified expression, with an average relative deviation of only 1.9%. Therefore, in the process of modeling the size of the deposited layer, to simplify the calculation results, Equation 4.3 can be approximated as:

$$S = \frac{2wh}{3} \tag{4.4}$$

Combining Eq. 4.1 and Eq. 4.4, we can obtain the relationship between the formed height h and width w, and the powder feeding rate f and scanning speed v:

$$h = \frac{3\beta f}{2v}\frac{1}{w} \tag{4.5}$$

Earlier, we established relationships between some parameters based on geometric and mass conservation principles. However, to study the relationship between the formed dimensions and the melt pool temperature, we need to analyze the thermal process of laser-directed energy deposition from an energy perspective. In a steady state, according to the law of energy conservation, the energy absorbed by the melt pool per unit time Q_a is balanced by the energy transferred to the

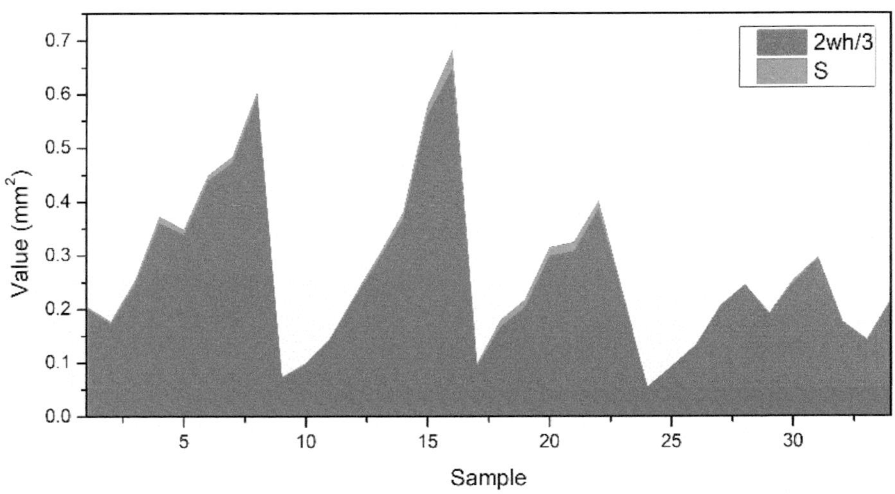

Figure 4.5. The comparison between $2wh/3$ and S.

substrate by thermal conduction Q_c, the energy exchanged with the air by convection Q_h, and the energy dissipated by radiation Q_e, i.e.,

$$Q_a = Q_c + Q_h + Q_e \tag{4.6}$$

Considering that in the actual process, the energy exchanged by convection and radiation is negligible compared to the energy transferred by thermal conduction, the above equation can be simplified as:

$$Q_a = Q_c \tag{4.7}$$

In addition, during the laser-directed energy deposition process, only a portion of the laser's actual power output can be absorbed by the melt pool after being reflected by the substrate, obstructed by the powder flow, and scattered. Therefore, the power absorbed by the melt pool is related to the laser's output power as follows:

$$Q_a = \eta p \tag{4.8}$$

Where η is the laser absorption rate. Under transient heat conduction conditions, according to the heat conduction formula proposed by Holman [3], the energy Q_c transferred to the substrate per unit time by the melt pool is:

$$Q = \frac{kA(T - T_0)}{\sqrt{\pi \alpha t}} \tag{4.9}$$

Among them, k is the thermal conductivity, α is the thermal diffusion coefficient of the material, t is the time that the molten pool exists, and A is the surface area of the molten pool. Due to the effect of surface tension, the entire molten pool can be assumed to be a sphere [4], and its surface area A can be expressed as

$$A = \gamma \pi w^2 \tag{4.10}$$

Where γ is the shape factor. If the molten pool solidifies immediately after the laser beam leaves the pool, the time that the molten pool exists can be expressed as:

$$t = \frac{w}{v} \tag{4.11}$$

Therefore, by combining Eq. 4.6 and Eq. 4.11, the expression for the width of the cladding layer w can be solved as follows:

$$w = \left(\frac{\eta p \sqrt{\pi \alpha}}{\pi k \gamma \sqrt{v}} \right)^{\frac{2}{3}} \left(T_m - T_0 \right)^{-\frac{2}{3}} \tag{4.12}$$

By combining equations 4.5 and 4.12, a quantitative relationship between the deposition height h and the molten pool temperature can be obtained as follows:

$$h = \frac{3f}{2v} \left(\frac{\eta \sqrt{\pi \alpha}}{\pi k \gamma \sqrt{v}} \right)^{-\frac{2}{3}} \beta \left(\frac{\left(T_m - T_0 \right)}{p} \right)^{\frac{2}{3}} \tag{4.13}$$

Finally, Eq. 4.13 is used to establish a quantitative relationship between the height of the DED deposition layer and the molten pool temperature and process parameters. Under the condition of constant processing parameters, i.e., when the laser power p, scanning speed v, and powder feeding rate f remain unchanged, there is a positive correlation between the height of the cladding layer h and the molten pool temperature T_m, that is, the higher the molten pool temperature, the thicker the cladding layer h, which quantitatively reflects the influence of thermal accumulation on the geometric dimensions of DED. Eq. 4.13 provides a theoretical basis for compensating for the thickness variation of the deposition layer caused by thermal accumulation through offline adjustment of process parameters, that is, it can quantitatively optimize the forming quality from the perspective of laser power, scanning speed, etc.

On the surface, Eq. 4.13 reflects a negative correlation between the height of the deposition layer h and the laser power p, that is, the higher the laser power, the smaller the deposition height, which is opposite to the relationship between laser power and forming height in actual DED processes. However, further analysis reveals that the powder utilization rate β increases with the increase of molten pool temperature, and the increase of laser power will improve the powder utilization rate while raising the molten pool temperature, thereby increasing the deposition thickness. Therefore, the laser power will ultimately have a positive effect on the DED forming height. In addition, reference [5] studied the influence of laser power p on the powder utilization rate β through numerical simulation, as shown in Figure 4.6, there exists a power function relationship of 0.8819 between the powder utilization rate and the laser power mathematically. After substituting it into Eq. 4.13, the height of the deposition layer and the laser power will become positively correlated even without considering the molten pool temperature. Therefore, increasing the laser power p will increase the height of the deposition layer h.

During the DED process, there is a thermal accumulation effect, and the molten pool temperature will increase due to thermal accumulation, which will increase the DED deposition height. If fixed parameters are still used for DED, poor forming results will occur. This section establishes a model for the relationship between deposition height and molten pool temperature from a mathematical and physical perspective, quantitatively analyzes the influence of thermal accumulation on the deposition layer height, and obtains a quantitative expression, providing a theoretical basis for offline parameter optimization to compensate for thermal accumulation effects. In addition, numerical analysis of the three-dimensional temperature field of multi-layer DED has been carried out in Chapter 3. Therefore, the following will combine the changes of molten pool temperature with the number of deposition layers and the position of each layer to optimize the process parameters of DED offline, adjust the process parameters according to the progress of DED, and reduce the dimensional deviation of forming. In the process parameters of robot-assisted DED, the Z-axis lifting amount, laser power, and scanning speed are variables that can be accurately adjusted during the process, while the powder feeding rate also has a significant impact on forming, but the control

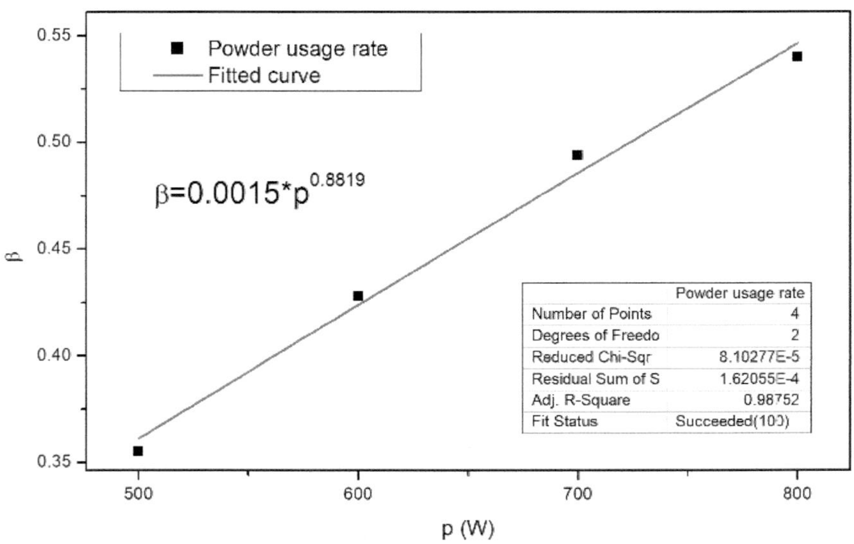

Figure 4.6. The theoretical relationship between β and p [5].

system cannot currently program it. Therefore, the following will focus on optimizing forming from the perspectives of adjusting the Z-axis lifting amount, laser power, and scanning speed.

4.4 Offline optimization strategy of Z-axis lifting capacity based on molten pool temperature change

4.4.1 Establishment of Z-axis lift model

The Z-axis lift amount is a crucial processing parameter in multi-layer DED, and determining the appropriate Z-axis lift amount is the foundation for obtaining good forming results by allowing the powder beam to focus well. The ideal value of the Z-axis lift amount is to ensure that the distance between the processing plane and the nozzle can always remain constant, and its value can precisely compensate for the reduction in distance between the surface of the deposited layer and the nozzle after the completion of the layer deposition.

If the Z-axis lift amount is too large, the distance between the processing plane and the nozzle in the printing process of the next layer will be greater than the initial set value, resulting in a larger defocus amount, a larger laser radius, a lower energy density, and a larger powder beam radius. This will reduce the amount of metal powder covering the molten pool and ultimately reduce the height of the deposited layer. If the Z-axis lift amount is set too small, the distance between the nozzle and the processing plane in the next layer deposition will be smaller than the initial set value. If the processing plane is higher than the laser beams focal plane, problems such as an increase in the molten pool radius and poor powder beam aggregation may occur, resulting in an increase in the width of the final deposited layer and a lower height than expected. If this deviation accumulates continuously, it may lead to forming failure.

In theory, the Z-axis lift amount can be increased in the region of thermal accumulation, which can not only reduce the energy density but also reduce the amount of powder entering the molten pool, thereby compensating for the increase in the thickness of the deposited layer caused by the increase in molten pool temperature. However, how to quantitatively describe the Z-axis lift amount based on the thermal accumulation degree at a certain position? In this article, based on the modeling results in the previous section and the geometric characteristics of multi-layer DED, a theoretical model of the Z-axis lift amount is established.

According to the previous assumption, the contour of the solidified deposition layer is a circular crown, and in multi-layer DED, the previously formed layer will serve as the substrate for the next layer deposition, i.e., the next layer is deposited on the surface of the previous layer. During the deposition of the next layer, laser heating not only melts the powder entering the molten pool but also causes partial remelting of the metal in the upper part of the previous deposition layer. The molten liquid metal is affected by the combined action of gravity, surface tension, and airflow, and flows to both sides. Eventually, the arc-shaped contour of the previous layer becomes approximately flat due to remelting. As shown in Figure 4.7, this article approximates the cross-sectional contour of the previous layer after remelting as a rectangle. The metal in the BCF region of the previous layer is melted and flows to fill the ABE and CDG regions on both sides, resulting in the following geometric relationship:

$$S_{BCF} = S_{ABE} + S_{CDG} \tag{4.14}$$

To ensure that the distance between the processing plane and the nozzle remains constant during the deposition of the next layer, the nozzle needs to be lifted by a height of Δz in the Z-axis direction for deposition on the BC plane. According to the law of conservation of mass, the mass of the first deposition layer before remelting is equal to the mass after remelting. Therefore, the following geometric relationship exists:

$$S_{AEGD} = S_{AFD} \tag{4.15}$$

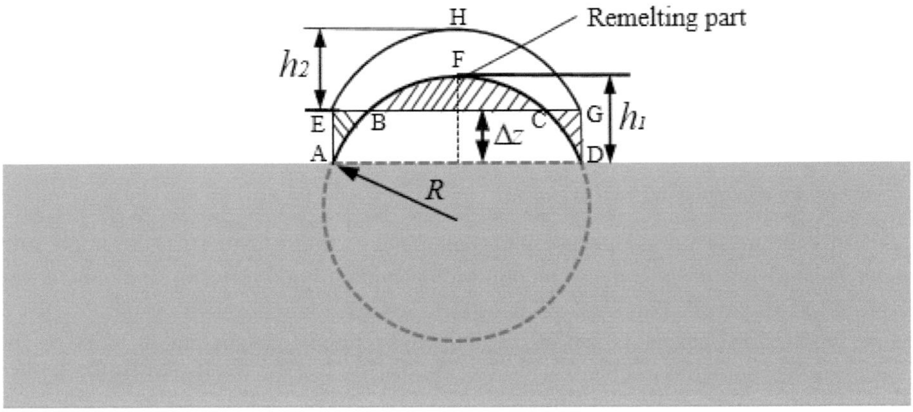

Figure 4.7. Theoretical cross section of the two-layer DED.

By substituting the expression for the circular crown area derived from formula 4.4 into formula 4.13, we can obtain:

$$w\Delta z = \frac{2wh_1}{3} \tag{4.16}$$

$$\Delta z = \frac{2}{3}h_1 = \frac{f}{v}\left(\frac{\eta\sqrt{\pi\alpha}}{\pi k\gamma p\sqrt{v}}\right)^{-\frac{2}{3}} \beta\left(T_m - T_0\right)^{\frac{2}{3}} \tag{4.17}$$

According to theoretical reasoning, the value of the Z-axis lift amount is not exactly equal to the thickness of the deposition layer but depends on 2/3 of the layer thickness. Equation 4.17 also indicates that the theoretical value of the Z-axis lift amount should be related to the temperature of the molten pool. To ensure that the distance between the processing plane and the nozzle remains constant at each position during multi-layer DED, it is necessary to adjust the Z-axis lift amount based on the thermal accumulation during the multi-layer deposition process. When the temperature of the molten pool increases, the Z-axis lift amount should be set larger. Based on Eq. 4.17, this article proposes a strategy for adjusting the Z-axis lift amount based on changes in the temperature of the molten pool. During the layer-by-layer printing process, the temperature of the higher layers is higher than that of the lower layers, so the Z-axis lift amount is gradually increased layer by layer. In the same layer, due to the bidirectional scanning of the laser, the temperature of the molten pool at the starting position of the end is higher, so the Z-axis lift amount needs to be increased.

4.4.2 Z-axis lift optimization strategy based on the temperature distribution characteristics of molten pool

In Chapter 3, numerical simulations were used to qualitatively analyze the relationship between the temperature of the molten pool and the number of deposition layers, as well as the temperature distribution at the ends of the deposition layers. In order to further quantitatively analyze the relationship between the Z-axis lift amount and the temperature of the molten pool from a mathematical perspective, the relationship between the temperature of the molten pool and the number of deposition layers will be analyzed. Firstly, in the vertical direction, the temperature of the molten pool of each layer will gradually increase layer by layer due to thermal accumulation. Therefore, the Z-axis lift amount of different deposition layers should be corrected in the vertical direction. When analyzing the temperature changes in the vertical direction, the middle position of each layer is taken for analysis. Under the conditions of fixed laser power p, scanning speed v, and powder feeding rate f, the Z-axis lift amount is mainly related to the temperature of the molten pool at the corresponding position of the next layer. Based on Eq. 4.17, the following relationship exists between the Z-axis lift amount after the end of the i-th layer and the Z-axis lift amount after the end of the initial first layer:

$$\frac{\Delta z^{(i)}}{\Delta z^{(0)}} = \frac{(T_m^{(i+1)} - T_0)^{\frac{2}{3}}}{(T_m^{(0+1)} - T_0)^{\frac{2}{3}}}, i = 0,1,2,3...,n \tag{4.18}$$

Here, $T_m^{(i+1)}$ represents the temperature of the molten pool of the i+1-th layer, $\Delta z^{(0)}$ represents the Z-axis lift amount after the end of the initial first layer, and according to the theoretical Eq. 4.17, $\Delta z^{(0)}$ is 2/3 of the deposition height of the first layer:

$$\Delta z^{(0)} = \frac{2}{3} h_0 \tag{4.19}$$

Here, h_0 represents the deposition height of the first layer. In Chapter 3, the temperature of the molten pool corresponding to the middle position of each deposition layer has been obtained. To propose the relationship between the Z-axis lift amount and the number of deposition layers for each layer, the temperature of the molten pool is processed according to Eq. 4.18 as follows:

$$\varepsilon_v = \frac{(T_m^{(i+1)} - T_0)^{\frac{2}{3}}}{(T_m^{(0+1)} - T_0)^{\frac{2}{3}}} \quad i = 0,1,2,...,n \tag{4.20}$$

Here, ε_v is the vertical gain coefficient, which represents the increase rate of the Z-axis lift amount of each layer relative to the Z-axis lift amount after the end of the initial first layer in the vertical direction. T_0 represents the initial temperature of 22°C. By substituting the temperature of the molten pool at the middle position of each layer into Eq. 4.20, the relationship shown in Figure 4.8 is obtained. Then, the expression for the vertical gain coefficient as a function of the number of deposition layers is obtained through data fitting:

$$\varepsilon_v = 0.998 \times (1+i)^{0.0376}, i = 0,1,2,...,n \tag{4.21}$$

Therefore, by combining Eq. 4.18 to 4.21, the Z-axis lift amount considering only the thermal accumulation effect in the vertical direction can be expressed as:

$$\Delta z = 0.665 h_0 (i+1)^{0.0376}, i = 0,1,2,...,n \tag{4.22}$$

The above analysis is based on the theoretical analysis of the basic lift amount of the Z-axis during the layer-by-layer deposition process, which is the lift amount at the middle position of each layer where the thermal accumulation is not significant. The Z-axis lift amount adjustment strategy in the vertical direction shown in Eq. 4.22 is summarized. Next, the Z-axis lift amount optimization strategy will be applied to the areas where the thermal accumulation is more significant at both ends of each layer. The results in Chapter 3 showed that in each deposition layer, the thermal accumulation phenomenon is more significant near the starting position of the end, and the temperature of the molten pool is higher. Therefore, the temperature of the molten pool at the end is processed as follows. Since the temperature of the molten pool near the end is higher, the Z-axis lift amount should be increased to a certain extent when approaching the end. Therefore, the horizontal gain coefficient ε_h is introduced based on Eq. 4.22 to obtain:

$$\Delta z = 0.655 \varepsilon_h h_0 (i+1)^{0.0376}, i = 0,1,2,...,n \tag{4.23}$$

The magnitude of the horizontal gain coefficient ε_h depends on the relative relationship between the temperature of the molten pool near the end of each layer and the temperature of the molten pool

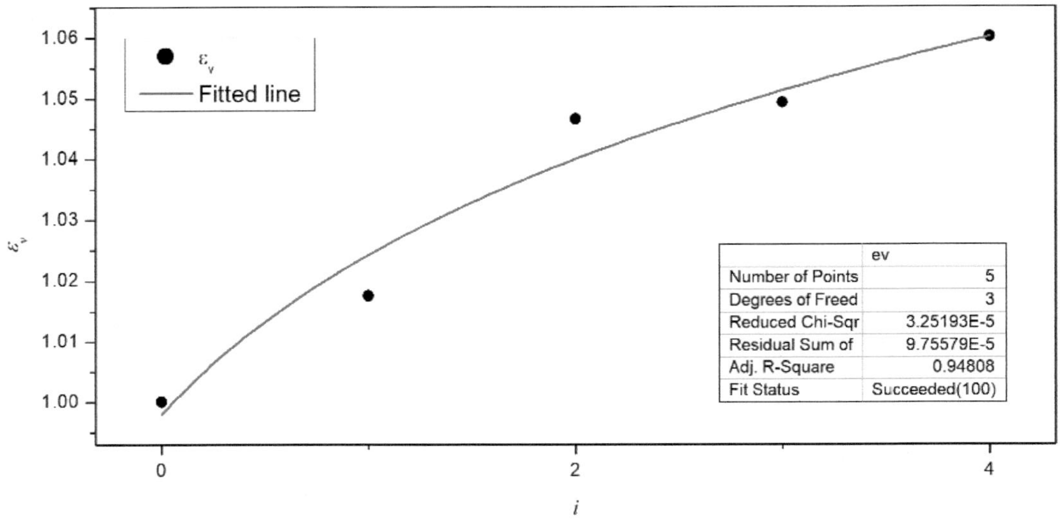

Figure 4.8. The fitted curve between ε_v and the layer number *i*.

at the middle position. Similar to the method used for the vertical gain coefficient, based on the basic relationship in Eq. 4.18, the temperature difference between the molten pool temperature at the end and the initial temperature (T_m–T_0) is divided by the temperature difference between the molten pool temperature at the middle position and the initial temperature (T_l–T_0), and then the result is raised to the power of 2/3 to obtain the relationship between the horizontal gain coefficient and the temperature of the molten pool:

$$\varepsilon_h = \frac{(T_m - T_0)^{\frac{2}{3}}}{(T_l - T_0)^{\frac{2}{3}}} \tag{4.24}$$

Here, T_l represents the theoretical temperature at the middle position of each layer. After the above processing, it is found that the gain coefficient ε_h at a certain position in each deposition layer

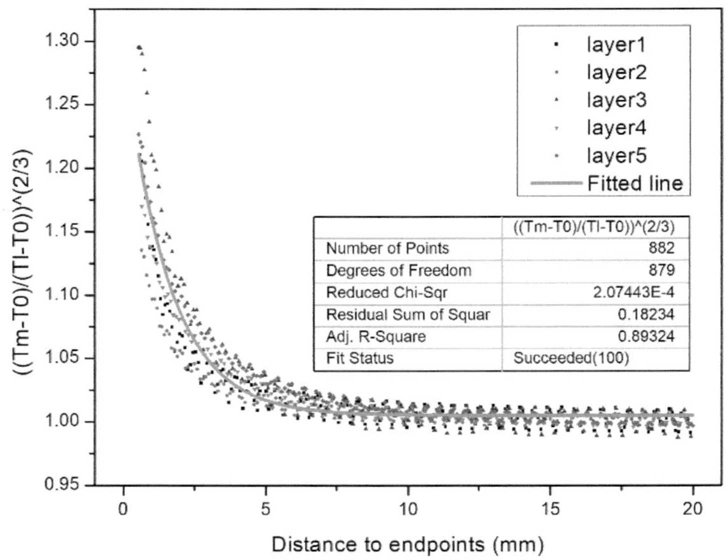

Figure 4.9. The horizontal gain coefficient ε_h over the distance to the end of DED layers.

has the same and similar characteristics as the Laplace distribution with respect to the distance d from that position to the end. Therefore, nonlinear curve fitting is performed on the data (as shown in Figure 4.9).

Therefore, the horizontal gain coefficient ε_h at a certain position can be expressed as a function of the distance d from that position to the end as follows:

$$\varepsilon_h = 1.005 + \frac{1}{2 \times 1.575} e^{\frac{-|d+0.147|}{1.575}} \tag{4.25}$$

By combining Eq. 4.22 and 4.25, the optimization algorithm for the Z-axis lift amount is obtained as follows:

$$\begin{aligned} \Delta z &= \varepsilon_h \varepsilon_v \Delta z_0 \\ &= 0.065 \times (1.005 + \frac{1}{2 \times 1.575} e^{\frac{-|d+0.147|}{1.575}}) \times (i+1)^{0.0376} h_0, (i = 0, 1, ..., n) \end{aligned} \tag{4.26}$$

4.4.3 Experimental validation

According to the proposed Z-axis lift amount optimization strategy, a DED single-track multi-layer deposition experiment was designed to verify the strategy. The first step is to determine the deposition height of the first layer: after determining the laser power p, scanning speed v, and powder feeding rate f for DED processing, a single-layer DED pre-experiment is performed, and then the deposition height h_0 of the first layer is measured using a point cloud device. The second step is to determine the Z-axis lift amount values at each position in each deposition layer: the mathematical model is obtained based on Eq. 4.26, and the Z-axis lift amount of each point is discretized. The third step is to generate the printing path offline. Figure 4.10 shows the Z-axis lift strategy at a laser power of 400 W, scanning speed of 6 mm/s, and powder feeding rate of 0.8 r/min. The initial Z-axis lift amount is 0.433 mm, which is measured by the point cloud device under this processing parameter, and the Z-axis lift amount at each position in the first 20 layers after the end of the first layer is obtained using Eq. 4.26. Finally, the KUKA robot trajectory is generated offline using Matlab. Firstly, the lift amount at the middle position of each layer gradually increases with the number of deposition layers, such as the initial Z-axis lift amount of 0.433 mm, and then the lift amount at the middle position of the second layer becomes 0.446 mm. In the same layer, the Z-axis lift amount at the end position is the largest, and then gradually decreases near the middle position.

The single-track multi-layer DED experiment was performed using the above parameter settings, and the number of deposition layers was 20. Figure 4.11(a) shows the forming effect using a fixed Z-axis lift amount. From the figure, firstly, the thickness between each deposition layer is not uniform enough, and the unevenness becomes more obvious with the increase of the number of deposition layers. Secondly, in the two end parts, because of thermal accumulation, the amount of deposited metal is significantly more than that in the middle area, and the width and thickness correspondingly increase. Especially in the left part, the molten metal tends to flow downward. Moreover, the thermal accumulation effect at the end accumulates layer by layer, making the deposition height at the last two ends significantly higher than that in the middle, showing a significant bulge at the top. Figure 4.11(b) shows the forming effect after Z-axis lift amount optimization. Firstly, in the vertical direction, the Z-axis lift amount is gradually increased layer by layer to compensate for the increase in deposition layer thickness caused by the thermal accumulation effect, ensuring that the distance between the laser jet plane and the workpiece plane can be maintained at the initial set value, so that the laser beam and powder jet can be well focused. Finally, the thickness of each layer is relatively uniform, and the side surface quality is better than before optimization. In the horizontal direction, the Z-axis lift amount is increased near the end to increase the distance between the laser nozzle and the processing plane in the end thermal accumulation area, so that the amount of powder entering

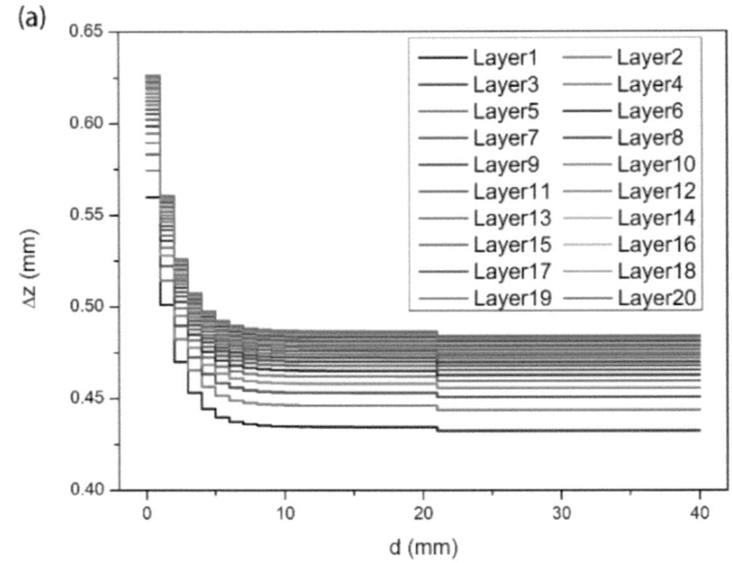

Figure 4.10. The planning of the Z-axis increment: the discretization result of the first 20-layer Z-axis increment.

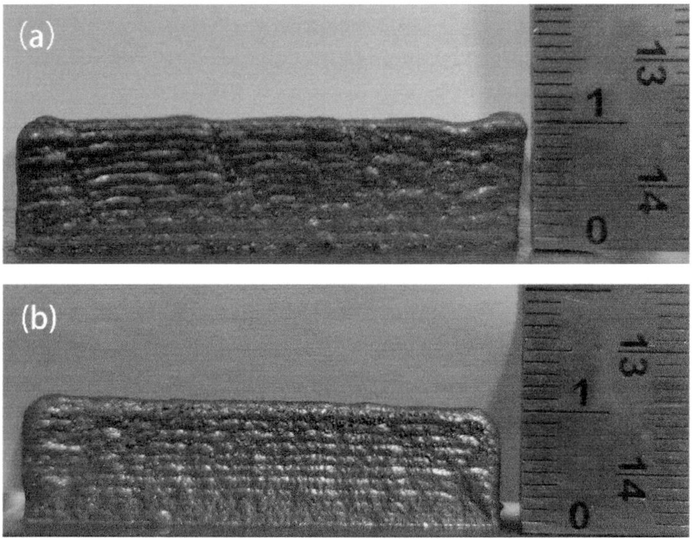

Figure 4.11. The formation comparison: (a) without; (b) without the Z-axis increment optimization.

the molten pool and the energy density both decrease, and the thickness of the deposition layer will not increase, thus making the thickness of the same deposition layer more uniform, and finally eliminating the bulge caused by the increase in deposition layer thickness at the end.

4.5 Laser power attenuation optimization strategy based on molten pool temperature change

4.5.1 Theoretical derivation of laser power attenuation strategy

In the previous section, a mathematical relationship between the Z-axis lift amount and the melt pool temperature was derived to propose an optimization strategy for the Z-axis lift amount to address the thermal accumulation issue in multi-layer DED deposition processes. In this section, we will start with the basic process parameters of DED to derive the corresponding offline optimization strategy for these parameters. Laser power is crucial in DED processing, as its magnitude directly affects the energy density applied to the melt pool. Increasing the laser power raises the melt pool temperature, allowing more metal powder to be melted per unit time, resulting in a thicker deposition layer. In multi-layer DED, the thermal accumulation caused by the increase in the number of deposition layers and the accumulation caused by the back-and-forth scanning of the same layer can be compensated for by appropriate laser power attenuation. However, there is currently a lack of theoretical support for how to quantitatively attenuate laser power. In the following, based on the melt pool temperature distribution characteristics of multi-layer DED, we will derive a theoretical explanation for power attenuation to address the issue of uneven deposition layer height caused by thermal accumulation, under the condition that the scanning speed v and powder feeding rate f remain constant. Equation 4.13 can be transformed as follows:

$$\frac{h}{h_0} = \frac{\beta}{\beta_0}\left(\frac{p}{p_0}\right)^{-\frac{2}{3}}\left(\frac{T_m - T_0}{T_m^0 - T_0}\right)^{\frac{2}{3}} \tag{4.27}$$

Where h_0 is the deposition layer height of the first layer, β_0 is the powder utilization rate of the first layer, p_0 is the set power, and $T_m{}^0$ is the melt pool temperature of the first layer. To achieve a uniform deposition height, it is desired that the deposition layer height at any position satisfies $h = h_0$, that is:

$$\frac{\beta}{\beta_0}\left(\frac{p}{p_0}\right)^{-\frac{2}{3}}\left(\frac{T_m - T_0}{T_m^0 - T_0}\right)^{\frac{2}{3}} = 1 \tag{4.28}$$

Laser power p can be derived from Eq. 4.28:

$$p = \left(\frac{\beta}{\beta_0}\right)^{\frac{3}{2}}\left(\frac{T_m - T_0}{T_m^0 - T_0}\right)p_0 \tag{4.29}$$

According to the previous discussion, there is a functional relationship between the powder utilization rate β and the laser power p. Therefore, further solving is required for Eq. 4.29. Based on the relationship between the powder utilization rate and the laser power shown in Figure 4.6, their mathematical relationship is:

$$\beta = 0.0015p^{0.8819} \tag{4.30}$$

The simultaneous Eq. 4.29 and 4.30 can solve the theoretical relationship between the laser power p and the molten pool temperature:

$$p = p_0\left(\frac{T_m - T_0}{T_m^0 - T_0}\right)^{-3.097} \tag{4.31}$$

From Eq. 4.31, the basic strategy for optimizing laser power attenuation to address thermal accumulation in multi-layer DED can be obtained. In multi-layer DED, the increase in the number of deposition layers and the back-and-forth scanning will cause the melt pool temperature of subsequent deposition layers, especially the end region, to be higher than that during the first layer deposition. To make the height of subsequent deposition layers the same as that of the first layer, the set power should decrease as the melt pool temperature increases. The relationship between the set laser power and the change in melt pool temperature and the initial set power should satisfy the above relationship.

With the basic theoretical relationship, this paper will model the quantitative attenuation of laser power based on the temperature change patterns of the melt pool at different deposition layers and different positions within the same deposition layer. According to Eq. 4.31, the adjustment of laser power depends on the relative change in melt pool temperature. Following the same research approach as in the previous section, this paper first analyzes the temperature change patterns of the melt pool at different layers and then analyzes the temperature change patterns of the melt pool within the same deposition layer. The vertical attenuation coefficient ω_v and horizontal attenuation coefficient ω_h are used to represent the degree of laser power attenuation due to the increase in the number of deposition layers and the reduction in the distance from the starting point to the end of the same layer, respectively:

$$p = \omega_v \omega_h p_0 \tag{4.32}$$

First, we need to obtain the relationship between the relative temperature change in different deposition layers in the vertical direction and the number of deposition layers. Let the vertical power attenuation coefficient be ω_v, and assume that the temperature change in the vertical direction is proportional to the number of deposition layers n, then we have:

$$\omega_v = \left(\frac{T_m^{(i)} - T_0}{T_m^{(0)} - T_0}\right)^{-3.097}, i = 0,1,2,\ldots,n \tag{4.33}$$

Here, i is the deposition layer number, and $T_m^{(i)}$ is the melt pool temperature at the middle position of the ith layer. Based on the numerical analysis results in Chapter 3, the melt pool temperature at the middle position of different deposition layers was extracted. By calculating using Eq. 4.33, the relationship between the vertical power attenuation coefficient and the number of deposition layers was obtained, as shown in Figure 4.12. The mathematical expression is:

$$\omega_v = 0.633 + \frac{1}{2\times 1.28} e^{\left(-\frac{|i-0.089|}{1.28}\right)}, i = 0,1,2,\ldots,n \tag{4.34}$$

Equation 4.34 shows that the laser power attenuation coefficient decreases layer by layer as the number of deposition layer increases. That is, the laser power should be set lower and lower to compensate for the impact of thermal accumulation on the forming process. As the number of deposition layers continues to increase, the attenuation coefficient gradually converges to around 0.633. This is because as the height of the formed part increases, the heat dissipation surface area of the part increases, and the heat dissipation conditions continue to improve. Therefore, the heat generated by laser heating and the heat dissipation of the formed part gradually reach a balance, and the laser power does not need to continue to attenuate.

Based on the layer-by-layer laser power attenuation strategy, the following analyzes the laser power attenuation strategy near the end region within the same deposition layer. Based on the temperature change pattern of the melt pool within the same layer summarized in Chapter 3, the closer to the end region, the higher the melt pool temperature. Therefore, it is necessary to attenuate the laser power in the end region to compensate for the increase in deposition layer thickness caused by the temperature rise. According to the basic relationship of Eq. 4.31, let the horizontal laser power attenuation coefficient be:

$$\omega_h = \left(\frac{T_m - T_0}{T_L - T_0}\right)^{-3.097} \tag{4.35}$$

Here, T_L is the melt pool temperature at the middle position of the deposition layer. The attenuation coefficient at a certain position within the same layer is related to its distance from the end region. Therefore, based on the relationship between the melt pool temperature and the distance from the end region within the same layer obtained in Chapter 3, the laser power horizontal attenuation coefficient ω_h at a certain position is calculated using Eq. 4.35. The relationship between the horizontal attenuation coefficient and the distance from the melt pool end region is obtained through mathematical fitting, as shown in Figure 4.13.

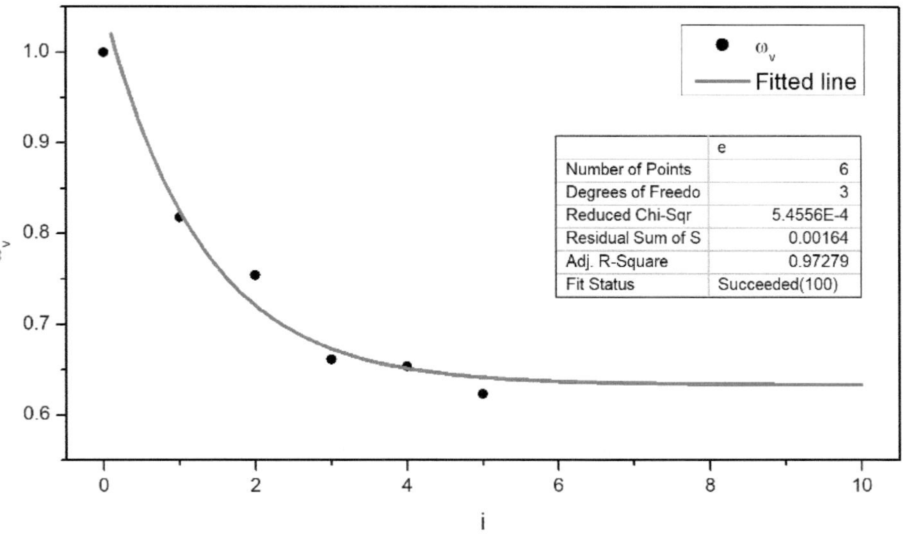

Figure 4.12. The relationship between the vertical laser power attenuation coefficient and the DED layer number.

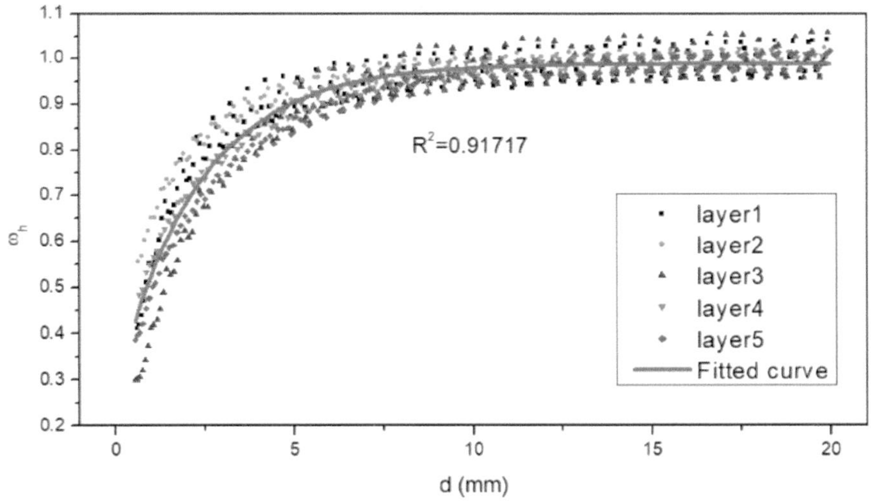

Figure 4.13. The relationship between the laser power horizontal attenuation coefficient and the distance to the end of the DED layers.

From Figure 4.13, even in different deposition layers, the relationship between the laser power attenuation coefficient at a certain position and its distance from the end region shows a similar pattern. That is, due to severe thermal accumulation near the end region, the laser power needs to be reduced to a greater extent, while away from the end region where thermal accumulation is not significant, the laser power should be basically the same as that at the middle position. Therefore, according to the fitting results in Figure 4.13, within the same deposition layer, the relationship between the laser power horizontal attenuation coefficient at a certain point and its distance from the end region can be expressed as:

$$\omega_h = 0.988 - 0.71e^{-0.428d} \tag{4.36}$$

Therefore, by combining Eq. 4.32, 4.34, and 4.36, the theoretical expression of the laser power attenuation strategy adopted to address the problem of uneven deposition layer height caused by thermal accumulation in multi-layer DED can be obtained:

$$p = (0.633 + \frac{1}{2\times 1.28} e^{(\frac{-|i-0.089|}{1.28})})(0.988 - 0.71e^{-0.428d})p_0, i = 0,1,2,...,n \tag{4.37}$$

4.5.2 Experimental validation

According to the multi-layer DED laser power attenuation strategy proposed above, this paper designed a corresponding DED single-pass multi-layer deposition experiment to verify the effectiveness of the strategy. In the verification experiment, the initial laser power was set to 400 W, the scanning speed was 6 mm/s, and the powder feeding rate was 0.8 r/min. A single-layer deposition experiment was conducted under these parameter settings, and then the layer thickness h_0 was measured using a Keyence point cloud device. According to the theoretical model of the Z-axis lifting amount, $2h_0/3$ was taken as the initial Z-axis lifting amount, and the final Z-axis lifting amount was 0.433 mm.

After determining the initial laser power p_0, the optimized power was discretized according to the laser power attenuation strategy in Eq. 4.37. The corresponding KUKA robot code was generated offline using Matlab programming, as shown in Figure 4.14. During the first layer printing, there was no interlayer thermal accumulation, and there was no problem of end region thermal accumulation caused by the reciprocating scanning. Therefore, the power of the first layer was set to 400 W. The laser power at the middle position of each layer would attenuate with the increase of the deposition layer number. The laser power near the end region within the same deposition layer would also gradually attenuate. However, in the actual robot DED process, when the robot code is used to modify the laser power, there is a response time because the robot needs to communicate with external devices. That is, the robot will stop and send instructions to the laser and then continue to move forward after the laser responds. During this short response time, the powder is still being injected into the melt pool, and the deposition height and thickness will increase accordingly. Therefore, the number of power adjustments should not be too many. This paper chooses to adjust the power only twice per layer. The first time, in the area where the end region thermal accumulation is severe, the laser power is attenuated, and then after moving forward a certain distance, the power is adjusted to the stable power of that layer.

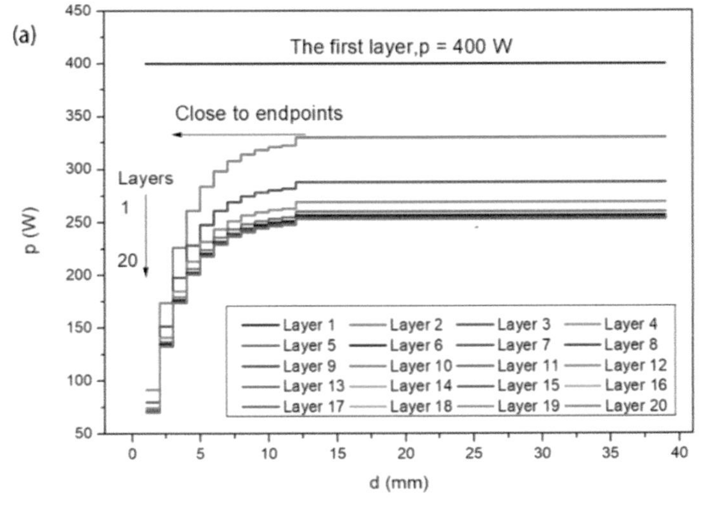

Figure 4.14. The strategy of the laser power optimization: the horizontal step.

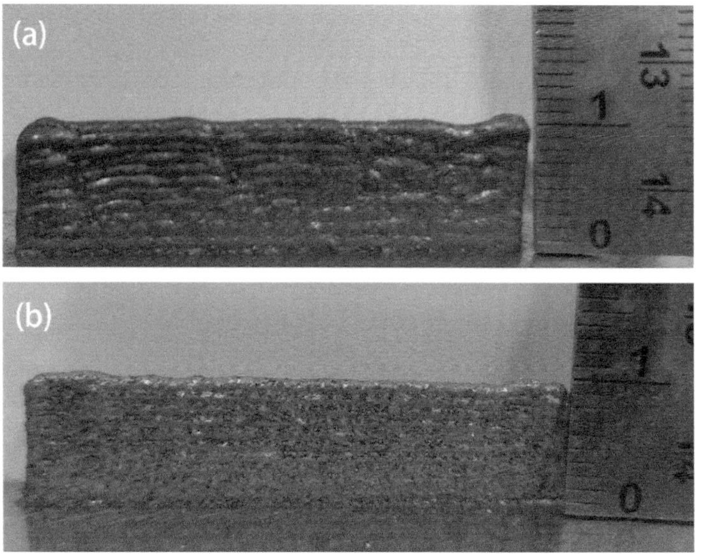

Figure 4.15. The formation comparison: (a) without optimization; (b) with the laser power optimization.

Then, the single-pass multi-layer DED deposition experiment was conducted according to the generated robot code, and the experimental results are shown in Figure 4.15. Through the layer-by-layer laser power attenuation and laser power attenuation in the areas where the end region thermal accumulation is severe, the multi-layer DED forming effect has been greatly improved compared to before optimization. As shown in Figure 4.15(b), in the vertical direction, the thickness and width between layers are more uniform, and the side surface is smoother; in the horizontal direction, by attenuating the power at the end region, the increase in deposition layer thickness caused by being heated twice in a short time at the starting point of the end region is basically eliminated. From the final forming result, the warping at both ends is eliminated, and the entire deposition layer is relatively flat. Therefore, the laser power attenuation strategy represented by Eq. 4.37 can effectively compensate for the uneven deposition layer thickness caused by thermal accumulation in the multi-layer DED process. However, in the actual robot automated DED process, controlling the external laser requires a certain response time, and the power attenuation operation should not be too frequent, otherwise, it will have a negative impact on the forming due to pauses.

4.6 Scanning speed optimization strategy based on molten pool temperature change

4.6.1 Theoretical derivation of scanning speed optimization

The previous section introduced the compensation of excessive deposition height caused by thermal accumulation in multi-layer DED by optimizing the Z-axis lift and laser power offline. During the automated DED process, the scanning speed is also a parameter that can be adjusted in real-time and precisely controlled. The theoretical derivation results of Eq. 4.13 and the experimental results of the influence of process parameters on the deposition height of a single layer and single pass both indicate that there is an inverse correlation between the deposition height and scanning speed. Therefore, to compensate for the uneven forming height caused by thermal accumulation in multi-layer DED, the scanning speed can be optimized offline. In other words, as the number of deposition layers increases or in the end region near the starting point of each layer, the thermal accumulation will increase the temperature of the melt pool and increase the thickness of the deposition layer. At this time, the laser scanning speed can be appropriately increased to reduce the line energy of the laser and compensate for the increased deposition layer thickness caused by thermal accumulation, making the final forming height more uniform. Next, a quantitative analysis will be conducted on how to optimize the laser scanning speed offline based on the thermal field distribution.

By appropriately transforming Eq. 4.13, the following relationship between the deposition height h and scanning speed v can be obtained:

$$h = \frac{3f}{2}\left(\frac{\eta\sqrt{\pi\alpha}}{\pi k\gamma}\right)^{-\frac{2}{3}} v^{-\frac{2}{3}}\beta\left(\frac{T_m - T_0}{p}\right)^{\frac{2}{3}} \tag{4.38}$$

As can be seen from the above formula, there is an inverse correlation between the deposition height and scanning speed. If the temperature of a certain area of the melt pool is high, it can be compensated by increasing the scanning speed to keep the deposition height constant. Therefore, it is necessary to obtain the relationship between the scanning speed and melt pool temperature when the deposition height is kept constant. Based on the results of Eq. 4.38, when the laser power p and powder feeding rate f are constant, the deposition height h and scanning speed v have the following mathematical relationship:

$$\frac{h}{h_0} = \left(\frac{v}{v_0}\right)^{-\frac{2}{3}}\left(\frac{T_m - T_0}{T_m^0 - T_0}\right)^{\frac{2}{3}} \tag{4.39}$$

In the formula, h_0 is the deposition height of the first layer, v_0 is the initial set scanning speed, and $T_m{}^0$ is the melt pool temperature of the first layer. In order to make the thickness of the subsequent deposition layer equal to that of the first layer, we hope that $h = h_0$, that is, Eq. 4.39 needs to satisfy the following relationship:

$$v = \left(\frac{T_m - T_0}{T_m^0 - T_0}\right)v_0 \tag{4.40}$$

That is, it is necessary to adjust the initial scanning speed v_0 based on the change in melt pool temperature to make the thickness of the deposition layer uniform. Equation 4.40 shows that the more severe the thermal accumulation phenomenon, that is, the higher the melt pool temperature, the larger the corresponding scanning speed should be. To obtain the optimal scanning speed values at different deposition layers and positions, the change in melt pool temperature with the number of deposition layers and the position within the same layer can be indirectly solved using Eq. 4.40. As in the previous two sections, Eq. 4.40 can be expressed as:

$$v = \mu_v \mu_h v_0 \tag{4.41}$$

Where v_0 is the initial set scanning speed; μ_v is the vertical gain coefficient of scanning speed, which represents the rate of change of scanning speed relative to the initial speed as the number of deposition layer increases; μ_h is the horizontal gain coefficient of scanning speed, which represents the rate of change of scanning speed at different positions within the same deposition layer relative to the scanning speed at the middle position. First, we study the change of the gain coefficient in the vertical direction with the number of deposition layers, and then study the relationship between the horizontal gain coefficient at different positions within the same deposition layer and the distance from a certain point to the starting point of the end. According to the basic relationship proposed by Eq. 4.40, the vertical gain coefficient is related to the ratio of the melt pool temperature of a certain layer to the melt pool temperature of the first layer, that is:

$$\mu_v = \frac{T_m^{(i)} - T_0}{T_m^{(0)} - T_0}, i = 0,1,2,\ldots,n \tag{4.42}$$

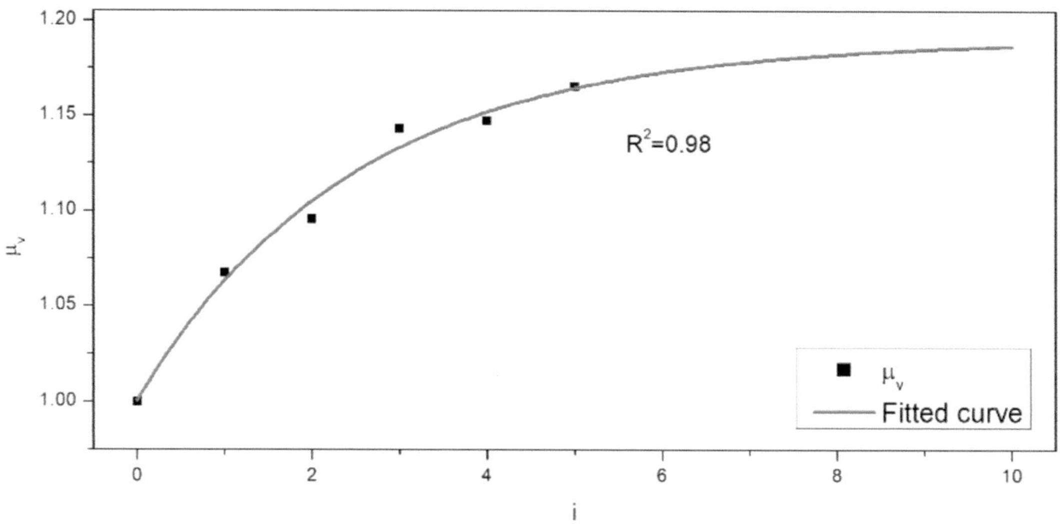

Figure 4.16. The relationship between μ_v and the DED layer number *i*.

Where $T_m^{(i)}$ represents the melt pool temperature at the middle position of the i-th layer. By substituting the melt pool temperature data of different deposition layers in Chapter 3 into Eq. 4.42, the relationship between the vertical attenuation coefficient of scanning speed and the number of deposition layers is obtained, as shown in Figure 4.16. As the number of deposition layer increases, the forming height increases, and the heat dissipation area gradually increases, resulting in a change in heat dissipation conditions and a gradual balance between heat dissipation and heat generation. Therefore, the speed gain coefficient eventually converges to 1.19.

Through data analysis and fitting, this paper derives the mathematical relationship between the velocity gain coefficients at the midpoints of different sedimentary layers and the sedimentary layers themselves.

$$\mu_v = 1.19\text{-}0.19\times e^{-0.399i}, i = 0,1,2,..,n \tag{4.43}$$

The following analysis will focus on how to compensate for the heat accumulation near the starting point at the ends caused by laser reciprocating scanning within the same sedimentary layer through the gain of the scanning speed. The horizontal attenuation coefficient is related to the change in the melt pool temperature, and according to the results of the numerical simulation in Chapter 3, the melt pool temperature at a given location within the same layer is related to its distance from the starting point. Therefore, based on the fundamental relationship of Equation 4.39, the horizontal gain coefficient can be expressed as:

$$\mu_h = \frac{T_m - T_0}{T_l - T_0} \tag{4.44}$$

By substituting the melt pool temperatures at various locations of each sedimentary layer into Equation 4.44, the results obtained are shown in Figure 4.17. The change in melt pool temperature at various locations of each sedimentary layer relative to the midpoint melt pool temperature exhibits the same variation relationship with the distance from the starting point at the end. Consequently, this paper establishes the relationship between the horizontal gain coefficient of the scanning speed and the distance. The closer to the end position, the more severe the heat accumulation effect, resulting in a higher melt pool temperature. Therefore, the scanning speed gain coefficient is larger, indicating that the scanning speed should be appropriately increased to compensate for the increase in sedimentary layer height caused by heat accumulation. Through data fitting, the relationship

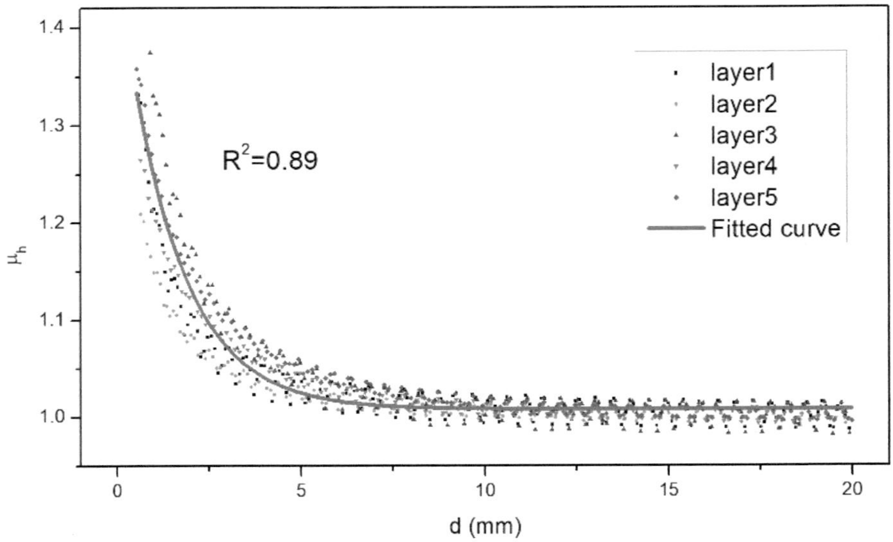

Figure 4.17. The relationship between μ_h and d.

between the horizontal gain coefficient of the scanning speed at a given location and its distance from the starting point at the end can be described by the following equation:

$$\mu_h = 1.007 + 0.466 \times e^{-0.663d} \tag{4.45}$$

Finally, through the theoretical derivation of scanning speed based on the changes in melt pool temperature mentioned above, the mathematical relationship between the optimized scanning speed at a given location and the initial speed during the multi-layer Directed Energy Deposition (DED) process is obtained. The expression is as follows:

$$v = (1.19 - 0.19 \times e^{-0.399i}) \times (1.007 + 0.466 \times e^{-0.663d}) v_0, i = 0,1,2,..,n \tag{4.46}$$

4.6.2 Experimental validation

To verify the actual effect of the scanning speed optimization strategy for compensating for the uneven deposition thickness caused by thermal accumulation, this paper designed a single-pass multi-layer DED experiment for comparison. First, the initial process parameters are determined, and then a pre-experiment is conducted based on the initial process parameters. The thickness of the first layer of the cladding layer is measured, and then the Z-axis lift amount is calculated. The experiment uses a laser power of 400 W, an initial scanning speed of 6 mm/s, a powder feeding rate of 0.8 r/min, and a Z-axis lift amount of 0.433 mm.

Then, the scanning speed optimization strategy proposed by Eq. 4.46 is used to offline optimize the scanning speed at each position of each layer. Finally, the scanning speed at each point is discretized using Matlab and robot code is generated, as shown in Figure 4.18. Since there is no thermal accumulation in the vertical direction for the first layer and no end thermal accumulation caused by reciprocating scanning, the scanning speed of the first layer is constant at 6 mm/s. The scanning speed in the vertical direction will gradually increase as the number of deposition layer increases, thereby compensating for the increase in cladding layer thickness caused by thermal accumulation. For example, the scanning speed at the middle position of the second layer increases to 6.4 mm/s. Near the end, the scanning speed will also be faster than the middle position to compensate for thermal accumulation. For example, in the second layer, the scanning speed gradually increases from 6.4 mm/s at the middle position to 9.4 mm/s at the end.

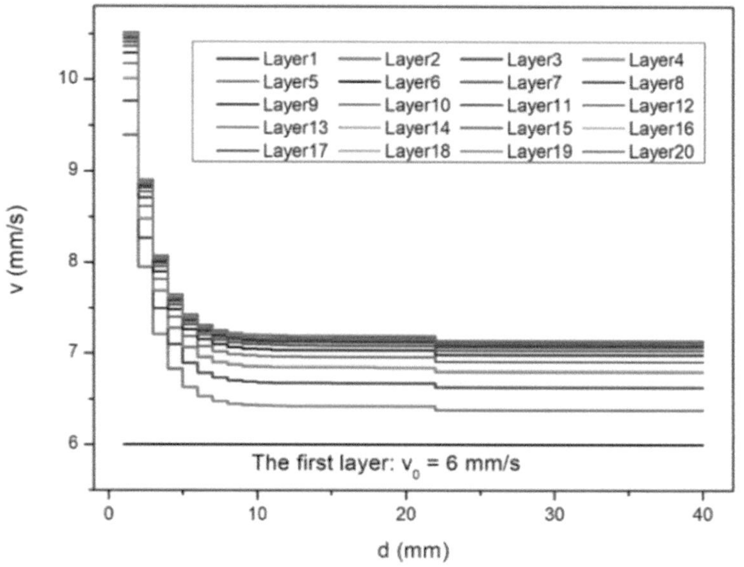

Figure 4.18. The strategy of the scanning speed optimization: the horizontal step of the first 20 layers.

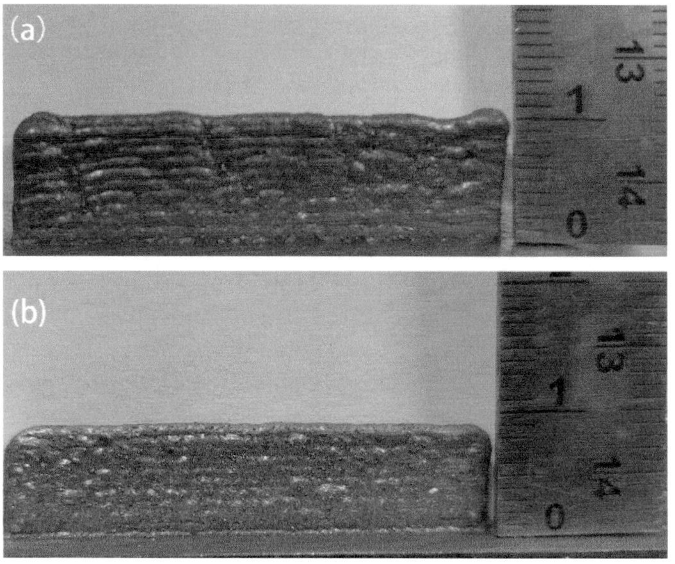

Figure 4.19. The formation comparison: (a) without; (b) with the scanning speed optimization.

Figure 4.19 shows a comparison of the multi-layer DED forming results before and after scanning speed optimization. As shown in Figure 4.19(b), overall, the forming quality after scanning speed optimization has been greatly improved compared to before optimization (Figure 4.19(a)). The side forming is very flat, and the fluctuation of the upper surface is very small. The thickness of each deposition layer is very uniform, and the protrusion caused by end thermal accumulation has been eliminated. The verification experiment results show that the scanning speed optimization strategy proposed by Eq. 4.46 is very effective in optimizing multi-layer DED forming, and the forming effect after optimization has been greatly improved. By gradually increasing the scanning speed in the vertical direction and increasing the scanning speed near the end in the horizontal direction, the energy input of the melt pool is reduced, and the amount of metal powder fed into the melt pool per unit time is also reduced, thereby compensating for the increase in cladding layer thickness caused by thermal accumulation and making the thickness of each deposition layer more uniform. Therefore, under the condition of a certain Z-axis lift amount, the distance between the processing plane and the laser nozzle can be maintained near the initial set value when printing the next layer, so that the focusing of the laser beam and powder jet flow is better. In addition, adjusting the scanning speed is directly done by modifying the $VEL.CP variable of the robot, which is much faster in response than adjusting the power of the external laser, and there is almost no pause, so the forming is very smooth and there is no trace of adjustment in the forming.

4.7 Summary

This chapter establishes a theoretical model of the quantitative relationship between the geometric characteristics of the deposition layer and the melt pool temperature and process parameters from the physical and mathematical aspects of DED forming, based on the temperature distribution characteristics of the melt pool in multi-layer DED. By analyzing the temperature distribution characteristics of the melt pool in the vertical and horizontal directions during multi-layer DED, three offline optimization strategies are quantitatively proposed, including Z-axis lift amount optimization, laser power attenuation, and scanning speed optimization. The following conclusions are obtained:

1) A quantitative theoretical model of the relationship between the geometric dimensions of forming and processing parameters and melt pool temperature is established by combining the geometric characteristics of DED forming with the laws of conservation of mass and energy. This lays the foundation for compensating for the uneven thickness of the deposition layer caused by thermal accumulation by adjusting the process parameters.
2) A quantitative relationship model is established from a geometric perspective, where the Z-axis lift amount is equal to 2/3 of the deposition layer height. Based on the temperature distribution characteristics of the melt pool, a Z-axis lift amount optimization model is established by combining theoretical analysis and mathematical fitting methods. Verification experiments show that by increasing the Z-axis lift amount layer by layer and increasing the Z-axis lift amount near the end of the same layer, the problem of uneven thickness of the deposition layer caused by thermal accumulation can be solved, and the forming quality can be optimized.
3) Based on the quantitative relationship between the deposition layer height and the melt pool temperature, the distribution characteristics of the melt pool temperature are analyzed, and a quantitative mathematical model of laser power attenuation is established through mathematical methods. Verification experiments show that by gradually attenuating the laser power layer by layer and attenuating the laser power at the end, the increase in forming thickness caused by thermal accumulation can be compensated, and the forming quality can be optimized. However, external adjustment of laser power is not suitable for frequent adjustments due to the required response time.
4) By analyzing the mathematical laws of the melt pool temperature at different positions in different layers and the same layer, an optimization model of scanning speed is established to address the problem of thermal accumulation. Verification experiments show that by increasing the scanning speed to compensate for the increase in deposition layer thickness caused by the rise in melt pool temperature, the forming effect can be significantly improved.

References

[1] Lin, J. (1999). A simple model of powder catchment in coaxial laser cladding[J]. Optics & Laser Technology 31(3): 233–238.
[2] Stocker, H. and Harris, J. W. (1998). Handbook of Mathematics and Computational Science[M]. Springer, New York, NY.
[3] Holman, J. P. (1976). Solutions Manual to Accompany Heat Transfer[M]. McGraw-Hill.

[4] El Cheikh, H., Courant, B., Hascoët, J. Y. and Guillén, R. (2012). Prediction and analytical description of the single laser track geometry in direct laser fabrication from process parameters and energy balance reasoning[J]. Journal of Materials Processing Technology 212(9): 1832–1839.
[5] KeYang, Liu. (2016). Simulation Study on Deposition Morphology of Coaxial Powder Feeding Laser Additive Manufacturing [D]. Hunan University.

Chapter 5

DED Process Parameter Decision-Making Model

5.1 Introduction

In the previous study, systematic research was conducted on the parameter optimization strategies for DED forming. However, there is still an issue that has not been resolved. According to the previous description, the appropriate Z-axis lift amount is crucial for improving the quality of multi-layer DED forming. When determining the initial Z-axis lift amount, DED pre-experiments should be conducted based on the corresponding process parameters, and then a three-dimensional profile scanner or manual preparation of metallographic samples should be used to measure the height of the first layer of the cladding, and finally, the appropriate initial Z-axis lift amount should be calculated based on this height. This method has low efficiency, low intelligence, and relies on additional measurement equipment, which is not suitable for practical DED processing. The ideal working state should be able to directly obtain the height of the cladding layer and calculate the Z-axis lift amount based on the laser process parameters. Therefore, it is necessary to establish a mapping between the process parameters and the height of the first layer, and to be able to reverse inference and obtain a set of corresponding process parameters based on the target deposition height. This is because when slicing and trajectory filling the model, the geometric dimensions of a single deposition layer need to be determined in advance, and then the corresponding process parameters need to be reverse inferred based on the dimensions of the deposition layer.

In addition, there is a strong coupling relationship between the basic process parameters and the horizontal overlap step size in multi-pass DED overlap. The basic process parameters determine the forming of a single DED pass, while the horizontal overlap step size determines the surface quality after overlap. Even though the theoretical optimal overlap step size can be obtained through the equivalent area method, the actual optimal overlap size is related to the true cross-sectional profile. Using an approximate function to represent the cross-sectional profile will result in errors, as the geometric shape of the cross-sectional profile will change with different process parameters, and there may also be overlap offset phenomena. Therefore, the reliability of the overlap model established based on simple geometric relationships is poor. In addition, in the absence of a reliable optimal step size decision-making model, the optimal overlap step size can only be determined through many DED pre-experiments by enumerating different overlap step sizes under the same process parameters. When switching to different parameters, the corresponding experimental operations need to be repeated. Therefore, it is necessary to establish a process decision-making model to solve the problem of how to match the optimal overlap step size under a certain process parameter. Finally, to implement the above achievements and improve the intelligence level of DED manufacturing, this paper developed a software system from prototype design, layered slicing, initial process parameter decision-making, offline optimization of process parameters, to robot code generation.

DOI: 10.1201/9781003406327-5

5.2 Experimental design

To establish process parameter decision-making models and optimal overlap step size matching models, this paper conducted many DED experiments under different process parameters and different overlap step sizes. As shown in Figure 5.1, to simultaneously obtain the geometric dimensions of single DED forming and multi-pass overlap forming as well as the overlap amount, this paper designed a DED horizontal overlap experiment with unequal length cladding paths. In Figure 5.1(a), Section A is used to measure the forming dimensions of a single pass, Section B is used to measure the overlap amount, and Section C is used to measure the average height of the deposited layer after overlap and evaluate the forming effect of the overlap. Figure 5.1(b) shows some DED experiments under different processing parameters and overlap step sizes.

To establish a reliable process parameter decision-making model, this paper designed DED experiments under different combinations of process parameters to obtain the forming dimensions. A total of 576 different parameter combinations of DED overlap experiments were designed and completed based on different laser power, scanning speed, powder feeding rate, and overlap step size. Table 5.1 shows the values of each process parameter.

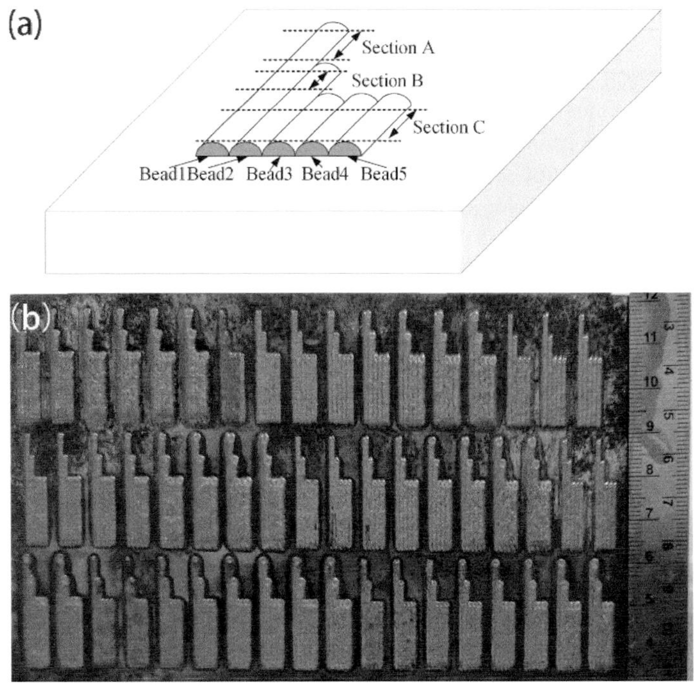

Figure 5.1. The experiment design for overlapping: (a) the schematic diagram; (b) the experimental result.

Table 5.1. The processing parameters setting.

Level No.	**1**	**2**	**3**	**4**	**5**	**6**
Laser power (W)	300	400	500	600	700	800
Scanning speed (mm/s)	3	4	5	6	-	-
Feeding speed (r/min)	0.3	0.4	0.5	0.6	0.7	0.8
Overlapping step (mm)	1.0	1.1	1.2	1.3	-	-

5.3 DED single-layer single-pass process parameter decision model

The purpose of establishing a process parameter decision-making model for single-layer single-pass DED mainly includes the following aspects. Firstly, it can predict the deposition layer height based on the process parameters, and further determine the corresponding Z-axis lift amount, which can avoid the need to measure the deposition height of the first layer through pre-experiments before the formal experiment. Secondly, by establishing a process parameter decision-making model, in the case of a known target deposition layer height, it can fix certain parameters and infer the reasonable values of the remaining process parameters and even output the ideal process parameter combination without fixing any parameters. In some scenarios, it is necessary to determine the corresponding process parameters based on the target deposition height. Conducting many pre-experiments to determine the process parameters through trial and error is time-consuming and laborious. The basis of the process parameter decision-making model is to establish a reliable deposition height prediction model. Therefore, this section will first establish a reliable forward height prediction model and then establish a reverse process parameter decision-making model.

5.3.1 Influence of process parameters on single-pass deposition height

The main factors affecting the height of a single DED pass include laser power, scanning speed, powder feeding rate, defocus amount, spot size, and shielding gas flow rate. According to the analysis of the powder jet in Chapter 4, to ensure the focusing of the laser beam and powder jet, there is an optimal distance between the work surface and the laser nozzle, which is generally determined to ensure the forming quality. In all experiments in this paper, this distance was set to 12 mm, so the defocus amount and spot size were fixed. Therefore, the variables affecting the thickness of the DED cladding layer are laser power, scanning speed, and powder feeding rate.

Figure 5.2(a) shows the variation of DED thickness with laser power. The height of the cladding layer increases nonlinearly with laser power. The heat input of the melt pool is determined by the laser power, and higher laser power can melt more metal powder, resulting in a thicker cladding layer. However, as the laser power increases, the melt pool temperature also increases, and the spreadability of the melt pool becomes stronger. Under the combined action of surface tension and gravity, the deposition height does not increase linearly with laser power.

Figure 5.2(b) shows the variation of DED cladding layer height with scanning speed. Scanning speed mainly affects the height of the deposition layer in two ways. Firstly, an increase in scanning speed reduces the line energy and decreases the heat input of the melt pool, resulting in less metal being melted. In addition, as the scanning speed increases, less metal powder is applied per unit length, resulting in less powder entering the melt pool. Therefore, an increase in scanning speed will decrease the height of the deposition layer. Moreover, as the scanning speed increases, the thickness of the cladding layer decreases more slowly, and there is also a nonlinear pattern.

Figures 5.2(c) and 5.2(d) show the effect of powder feeding rate on deposition height. The powder feeding rate directly controls the maximum amount of metal that can be captured by the melt pool. When the powder feeding rate is relatively low, the change in deposition height is slow because the amount of powder that can be utilized is limited, and therefore, the contribution to the height change is not high. However, when the powder feeding rate is fast, the amount of metal that can be melted by the laser is sufficient, and therefore, the effect on the deposition height is more significant.

Based on the above analysis, it is found that each process parameter has a significant impact on the deposition height of a single-layer single-pass DED, and the relationship between them is almost always nonlinear. Chapter 4 of this paper modeled the single-layer deposition height from a physical perspective, and the results also showed a nonlinear relationship between the deposition height and various process parameters. However, many physical parameters in the modeling results of Chapter 4 are difficult to determine their specific values through experiments, so they are mainly used for theoretical derivation in the subsequent chapters. To obtain a quantitative mathematical

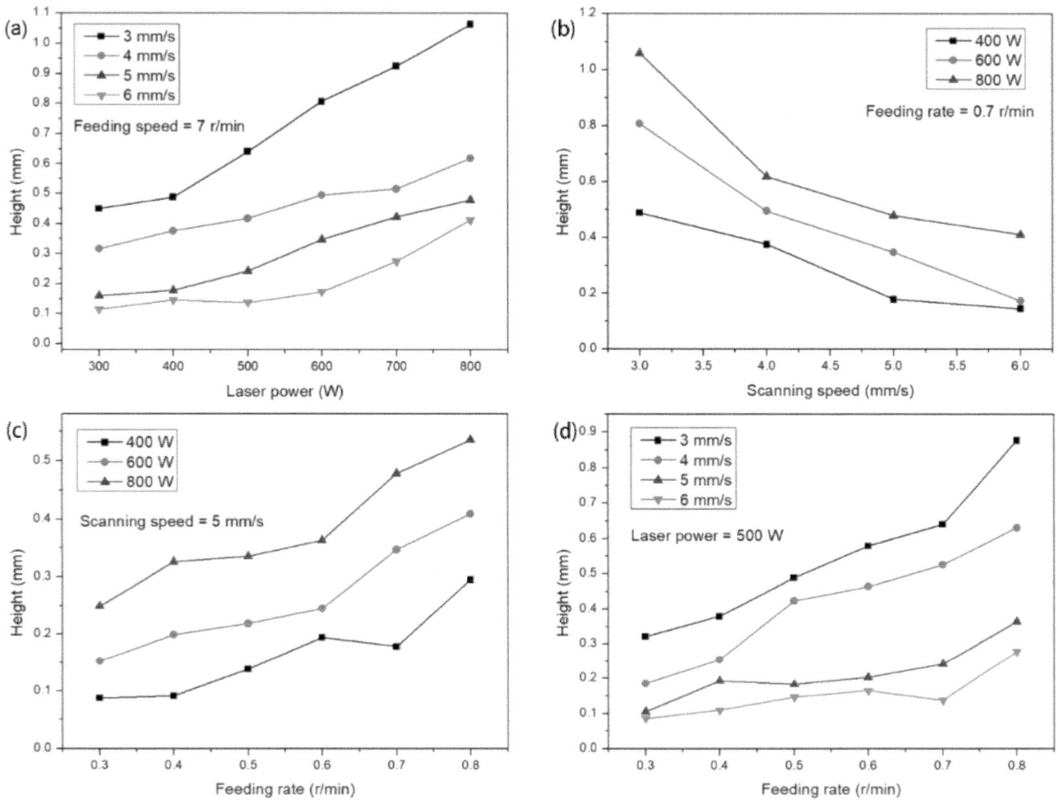

Figure 5.2. The effect of processing parameters on the track height.

model between the deposition layer and process parameters, other methods need to be selected for modeling. To describe the complex nonlinear relationship between the deposition height and process parameters, this paper chooses to use data-driven methods suitable for nonlinear modeling to model it.

5.3.2 Single-layer DED deposition height prediction model based on XGBoost

After analysis, it is found that there is a complex nonlinear relationship between the deposition height of a single-layer DED and process parameters. Machine learning algorithms, as powerful fitting tools, can fit complex mappings. XGBoost, as a tool that can perform parallel gradient boosting trees, has the advantages of high accuracy, fast training speed, and requiring fewer training samples in solving regression prediction problems. It is very suitable for manufacturing scenarios where large-scale training data is difficult to obtain. Therefore, this paper uses XGBoost to establish a forward prediction model for DED deposition layer height.

XGBoost is a specific engineering implementation of the gradient decision boosting tree (GBDT) algorithm. GBDT uses gradients to replace residuals in the boosting tree (BDT). BDT combines the forward stepwise algorithm and additive model and adds the results of multiple weak learners as the final strong learner. Each weak learner is a decision tree. Starting from the second decision tree, the input of each decision tree is the residual of the previous one (as shown in Figure 5.3).

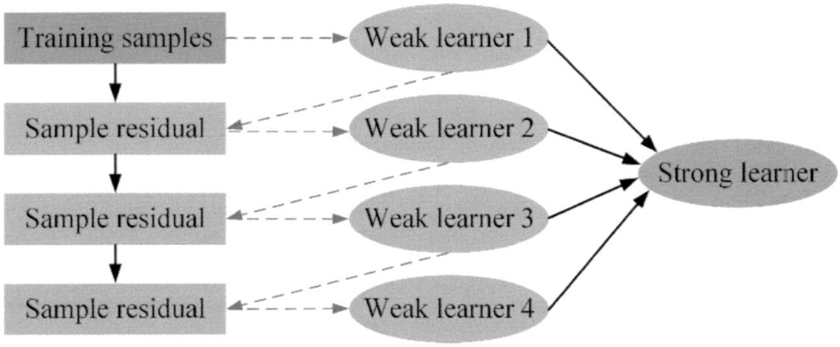

Figure 5.3. The schematic diagram of BDT.

The calculation process of BDT is as follows, enter the training dataset, and initialize the boosted tree firstly:

$$f_0(x) = 0 \tag{5.1}$$

The strong learner is composed of M ascending trees, and the forward stepwise algorithm is to iterate M steps to solve M weak learners. When iterating to the mth tree, a promoted tree containing the first m trees can be represented as:

$$f_m(x) = f_{m-1} + T(x, \theta_m) \tag{5.2}$$

In regression analysis, each step of the solution is minimizing the squared error loss function by updating the parameters:

$$\begin{aligned} L &= \sum_{i=1}^{N} (y_i - (f_{m-1}(x_i) + T(x_i, \theta_m)))^2 \\ &= \sum_{i=1}^{N} (y_i - f_{m-1}(x_i) - T(x_i, \theta_m)))^2 \\ &= \sum_{i=1}^{N} (r - T(x_i, \theta_m)))^2 \end{aligned} \tag{5.3}$$

Where $T(x, \theta_m)$ is the mth decision tree and r is the residual of the predicted result of the previous m–1 tree, so in fact, the mth tree is fitted to this residual term. Thus, a promotion tree containing M decision trees can be expressed as:

$$f_M = \sum_{m=1}^{M} T(x, \theta_m) \tag{5.4}$$

GBDT replaces residuals by employing a negative gradient. Because multiplying by 1/2 before Eq. 5.3 and then to $f_m(x_i)$ gets:

$$r = -\left[\frac{\partial L(y_i, f_m(x_i))}{\partial f_m(x_i)} \right] \tag{5.5}$$

In addition, each weak learner in GBDT has its own parameters and weights. Therefore, the calculation process of the entire GBDT is as follows: for the m-th decision tree, first calculate the negative gradient:

$$\tilde{y}_i = -\left[\frac{\partial L(y_i, f_m(x_i))}{\partial f_m(x_i)} \right], i = 1, 2, ..., N \tag{5.6}$$

Then optimize the value of the loss function:

$$w_m = \arg\min_{w_m} \sum_{i=1}^{N} (\tilde{y}_i - h_m(x_i; w_m))^2 \tag{5.7}$$

Unlike the BDT that directly adds each weak learner, the GBDT algorithm introduces the weights of each base learner and uses the same method to obtain the weights of the base learner

$$\alpha_m = \arg\min_{\alpha_m} \sum_{i=1}^{N} L(y_i, f_{m-1}(x_i) + \alpha_m h_m(x_i; w_m)) \tag{5.8}$$

It is then iterated to the mth decision tree, which can be updated to

$$f_m(x) = f_{m-1}(x) + \alpha_m h_m(x; w_m) \tag{5.9}$$

XGBoost implements GBDT, through forward step-by-step learning of an additive model containing M decision trees, because a tree's learning fitting ability is limited, so multiple trees are integrated to predict, and the final prediction result is the sum of each tree's prediction results:

$$\hat{y}_i = \sum_{m=1}^{M} f_m(x_i) \tag{5.10}$$

To learn the corresponding learner, XGBoost defines the objective function, which contains the loss function and the regular term, and the expression is as follows:

$$Obj(\Theta) = \sum_{i=1}^{N} L(y_i, \hat{y}_i) + \sum_{m=1}^{M} \Omega(f_m) \tag{5.11}$$

where f_m is a decision tree rather than a numerical vector, and the objective function cannot be optimized directly using gradient descent. However, combined with the previous forward stepwise algorithm, the objective function can be approximated using the Taylor expansion [1]:

$$\begin{aligned} Obj^{(m)} &= \sum_{i=1}^{N} L(y_i, \hat{y}_i^{(m-1)} + f_m(x_i)) + \Omega(f_m) \\ &= \sum_{i=1}^{N} \left(L(y_i, \hat{y}_i^{(m-1)}) + g_i f_m(x_i) + \frac{1}{2} h_i f_m^2(x_i) \right) + \Omega(f_m) \end{aligned} \tag{5.12}$$

where,

$$g_i = \frac{\partial L(y_i, \hat{y}_i^{(m-1)})}{\partial \hat{y}_i^{(m-1)}}, h_i = \frac{\partial^2 L(y_i, \hat{y}_i^{(m-1)})}{\partial^2 \hat{y}_i^{(m-1)}} \tag{5.13}$$

Since the inputs to the L function in step m are all numeric values and therefore constant, the constant term can be removed from the objective function because the purpose is to optimize the objective function. Therefore, the objective function can be simplified to

$$Obj^{(m)} = \sum_{i=1}^{N} \left(g_i f_m(x_i) + \frac{1}{2} h_i f_m^2(x_i) \right) + \Omega(f_m) \tag{5.14}$$

Now discussing the regular term in the objective function, because each tree is different in complexity, to prevent the learned decision tree from being too complex and overfitting, the regular term is used to measure the complexity of the decision tree, and its expression is:

$$\Omega(f_m) = \gamma T + \frac{1}{2} \lambda \sum_{j=1}^{T} s_j^2 \tag{5.15}$$

Where T is the number of leaf nodes, and s_j outputs the score for each leaf node. Finally, the objective function can be expressed as:

$$Obj^{(m)} = \sum_{i=1}^{N}\left(g_i f_m(x_i) + \frac{1}{2}h_i f_m^2(x_i)\right) + \gamma T + \frac{1}{2}\lambda\sum_{j=1}^{T} s_j^2 \tag{5.16}$$

The set $I_i = \{i \mid q(x_i) = j\}$ that finally falls on the leaf node j can further process the objective function:

$$\begin{aligned} Obj^{(m)} &= \sum_{i=1}^{N}\left(g_i f_m(x_i) + \frac{1}{2}h_i f_m^2(x_i)\right) + \gamma T + \frac{1}{2}\lambda\sum_{j=1}^{T} s_j^2 \\ &= \sum_{i=1}^{N}\left(g_i s_{q(x_i)} + \frac{1}{2}h_i s_{q(x_i)}^2\right) + \gamma T + \frac{1}{2}\lambda\sum_{j=1}^{T} s_j^2 \\ &= \sum_{j=1}^{T}\left(\sum_{i\in I_j} g_i s_j + \sum_{i\in I_j} h_i s_j^2\right) + \gamma T + \frac{1}{2}\lambda\sum_{j=1}^{T} s_j^2 \\ &= \sum_{j=1}^{T}\left(G_j s_j + \frac{1}{2}(H_j + \lambda)s_j^2\right) + \gamma T, G_j = \sum_{i\in Ij} g_i, H_j = \sum_{i\in Ij} h_i \end{aligned} \tag{5.17}$$

When optimizing the objective function, it is necessary to obtain the output score of each leaf node that minimizes the objective function, so the Eq. 5.17 derives the fraction S_j and takes the extreme value of the derivative, and the value of the output fraction of the leaf node can be found:

$$s_j = -\frac{G_j}{H_j + \lambda} \tag{5.18}$$

Combining the above formulas, the objective function can be derived:

$$Obj^{(m)} = -\frac{1}{2}\sum_{j=1}^{T}\left(\frac{G_j^2}{H_j + \lambda}\right) + \gamma T \tag{5.19}$$

XGBoost achieves the final tree structure by splitting the nodes, splitting one node each time based on the greedy algorithm that makes the most gain change. According to the expression of the objective function in Eq. 5.19, XGBoost uses the following equation to calculate the gain of the left subtree plus the right subtree minus the fraction before splitting after the node splits:

$$Gain = \frac{1}{2}\left[\frac{G_L^2}{H_L + \lambda} + \frac{G_R^2}{H_R + \lambda} - \frac{(G_L + G_R)^2}{H_L + G_L + \lambda}\right] - \gamma \tag{5.20}$$

The greater the gain produced by node splitting, the more the objective function decreases, and the better the learner works.

We will now model the height of a single-layer single-pass DED using XGBoost. The model takes basic process parameters as input and outputs the cladding layer height. Based on the experimental data collected earlier, a total of 144 sets of experimental data for single-layer single-pass DED were obtained, and the training set and test set were divided in a 9:1 ratio. XGBoost has many hyperparameters that affect the performance of the model. To obtain the optimal prediction model, this paper uses grid search to determine the optimal combination of hyperparameters. For each hyperparameter, we test its values within a certain range and select the value that can achieve the highest accuracy. The final hyperparameter settings are shown in Table 5.2.

Table 5.3 shows the comparison between the predicted results of the model and the measured heights. The average error of the model's predicted results is 0.056 mm, and the maximum error is –0.139 mm. Therefore, the deposition height prediction model based on the XGBoost algorithm

Table 5.2. The hyperparameter settings of XGBoost.

Hyper-parameters	Value	Description
learning_rate	0.3	The contraction step during weight updates
n_estimators	144	Number of weak learners
max_depth	3	The depth of XGBoost's lift tree
min_child_weight	1	The weight threshold for the smallest leaf node
subsample	0.7	Controls the proportion of random sampling, the smaller it is to avoid overfitting
colsample_bytree	1	The proportion of randomly sampled columns
gamma	0	Controls the loss function threshold for node splitting
reg_lamda	40	The weight coefficient of the L2 regularization term
seed	33	Random seeds

Table 5.3. The predicted track height by XGBoost.

Power (W)	Scanning speed (mm/s)	Powder feeding rate (r/min)	Actual height (mm)	Predicted height (mm)	Deviation (mm)
400	3	0.4	0.332	0.317	0.015
400	4	0.4	0.246	0.218	0.028
1000	5	0.7	0.476	0.552	–0.076
900	5	0.8	0.620	0.534	0.086
500	4	0.7	0.526	0.484	0.042
300	5	0.8	0.235	0.283	0.048
1000	5	0.8	0.473	0.560	0.087
700	4	0.8	0.509	0.600	0.091
600	5	0.7	0.347	0.323	0.024
900	5	0.5	0.392	0.379	0.013
300	3	0.7	0.458	0.597	–0.139
400	6	0.7	0.145	0.130	0.015
500	3	0.7	0.640	0.772	–0.132
800	4	0.6	0.455	0.494	–0.039
700	5	0.7	0.421	0.443	–0.022
500	6	0.6	0.164	0.121	0.043

established in this paper can accurately predict the height based on process parameters, which lays the foundation for the establishment of the reverse inference model of process parameters.

This paper analyzed the importance of the input features of the model. XGBoost selects the feature that maximizes the gain represented by Eq. 5.20 as the node to perform the split. The result of feature importance is the result of statistics on the features corresponding to the nodes of each tree, which reflects the importance of each feature in the process of constructing the decision tree. As shown in Figure 5.4, the laser power has the greatest impact on the deposition height of single-layer single-pass DED compared to the powder feeding rate and scanning speed.

XGBoost is essentially a decision tree algorithm. To comprehensively evaluate the performance of the model, this paper compared the XGBoost-based model with models based on other different types of algorithms. First, neural network algorithms were used. FCNN is widely used because of its simple structure and relatively small computational complexity. In addition, support vector regression (SVR) is also a machine learning algorithm suitable for a small number of samples. It

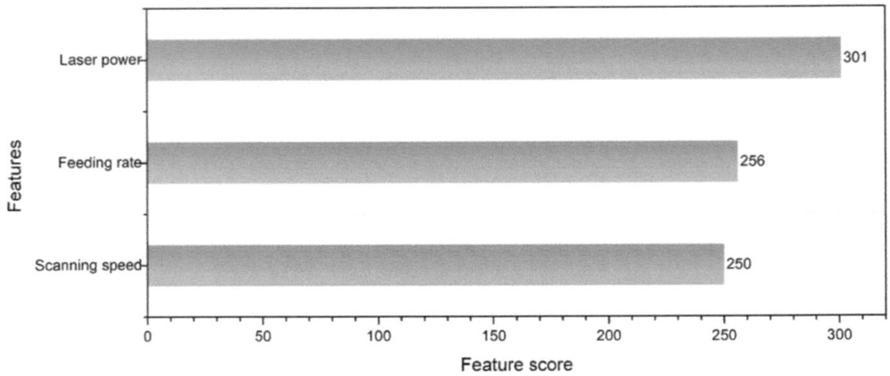

Figure 5.4. The importance of the features.

Table 5.4. The performance of the three models.

Models	MSE (mm×mm)	RMSE (mm)	MAE (mm)	MRE	R2
XGBoost	0.0047	0.0686	0.0561	0.1384	0.7722
FCNN	0.0167	0.1292	0.0772	0.1815	0.1925
SVR	0.0174	0.1317	0.1054	0.3068	0.6682

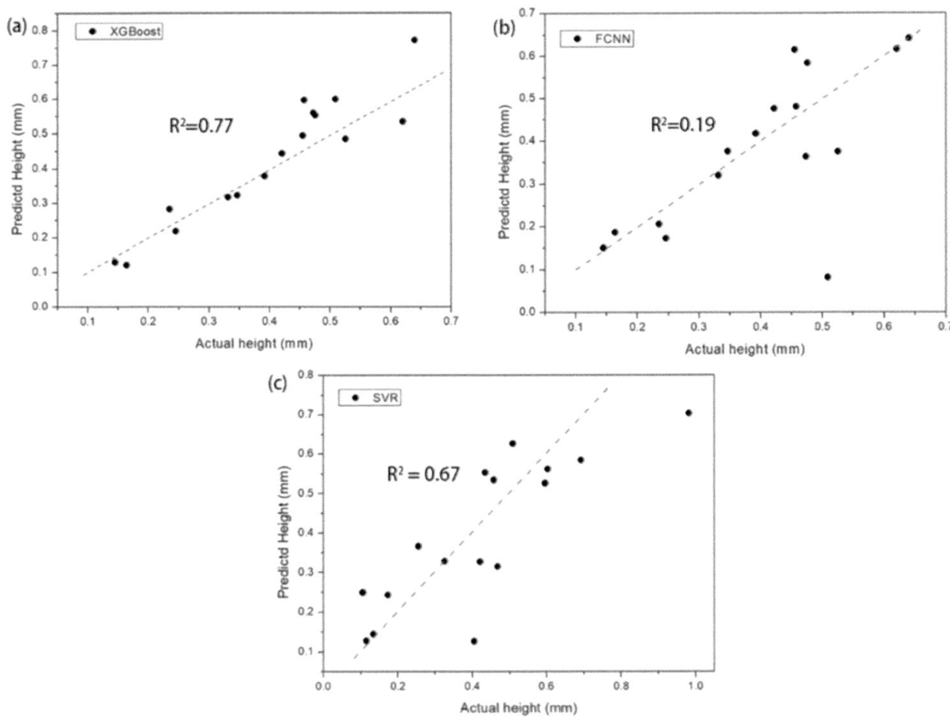

Figure 5.5. The track height comparison for the predicted and the measured.

has strong generalization ability, can suppress overfitting, and can achieve high accuracy with less training data. The three models were trained and evaluated based on the same training and test sets. Table 5.4 shows the comparison of the performance of the three models. When the number of training data samples is small, XGBoost has the highest accuracy, and the correlation between the predicted values and the true values is the strongest. Figure 5.5 shows the comparison between the true height and the predicted height. The predicted height of the XGBoost algorithm is very close to the true height, while the predicted results of FCNN are generally highly correlated with the true height, but some individual predicted results differ greatly, which lowers the overall correlation coefficient. The prediction deviation of SVR is also larger than that of XGBoost. Therefore, in terms of prediction accuracy and model regression correlation, XGBoost performs better and is more suitable for modeling the relationship between process parameters and DED deposition layer height.

5.3.3 The process parameters inverse inference model

With a reliable forward inference model for height, this paper established a process parameter reverse inference model based on a greedy iterative algorithm, which can infer the corresponding process parameters based on the target DED deposition layer height. The algorithm's process is shown in Figure 5.6. The core of this model is based on the XGBoost-based deposition height forward inference model. By continuously iterating and updating each process parameter, the difference between the predicted height and the target height is reduced, and a set of process parameters

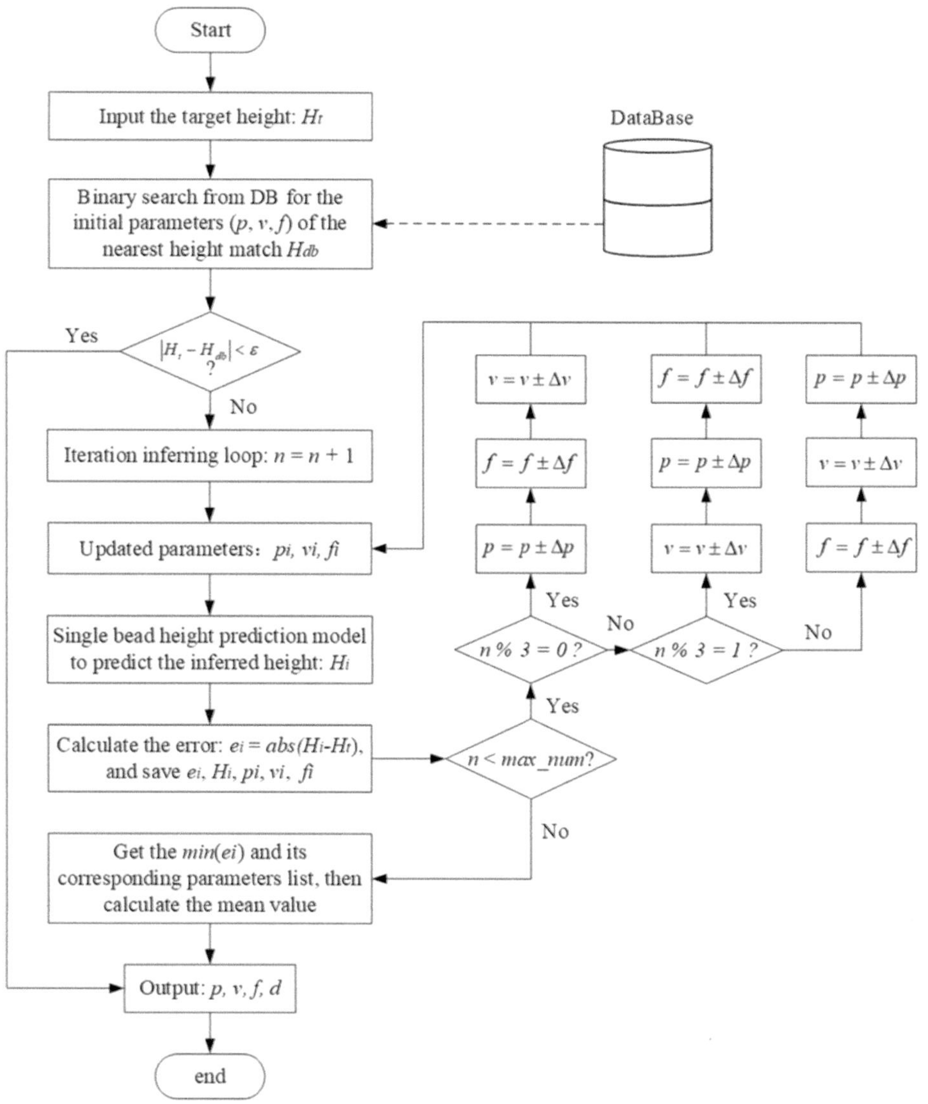

Figure 5.6. The algorithm of processing parameters inversely inferring for the target track height.

corresponding to the target deposition height is obtained, which assists in decision-making for DED process parameters. There are three main modes in practical application scenarios: (1) no fixed parameters, that is, laser power, scanning speed, and powder feeding rate are all unknown, and a set of process parameters needs to be inferred based on the target deposition height; (2) fix a certain process parameter, for example, some materials have certain requirements for laser power, so the laser power needs to be fixed, and the other two parameters are iteratively updated; (3) known two parameters, only one parameter is updated.

For the reverse inference model without fixed process parameters, there are three degrees of freedom for parameter updates. First, to quickly find the process parameters corresponding to the target deposition height, the most efficient binary search method is used to match the corresponding process parameters of the nearest target height from the database based on the target height, and then this set of parameters is used as the starting point for the reverse iterative search. After each parameter update, the forward single-layer height prediction model is used to calculate the inferred height and then compared with the target height. If the inferred height is greater than the target height, the laser power will be reduced, the scanning speed will be increased, or the powder feeding rate will be reduced to update the parameters. Conversely, if the inferred height is less than the target height, the laser power will be increased, the scanning speed will be slowed down, or the powder feeding rate will be increased to update the parameters, and then the next loop will be performed. When updating parameters, it is necessary to determine whether the current parameter has reached the set upper and lower thresholds. For example, when increasing the laser power p, if p has reached the threshold, the scanning speed v will be updated. If v has reached the threshold, the powder feeding rate f will be updated.

During reverse inference, the target height may correspond to multiple different combinations of process parameters. In order to prevent only increasing or decreasing a certain parameter and make the parameter adjustment more uniform, so as not to move towards extreme values and make the iteratively obtained parameters more valuable, this paper adopts a cyclic parameter update method. Specifically, the preferred updated parameter is constantly changed. If the laser power is updated first in the first iteration, then the speed will be updated first in the next iteration, and the powder feeding rate will be updated first in the next iteration, so that each parameter can be updated synchronously.

To verify the reliability of the reverse inference model without fixing any parameters, this paper obtained the process parameters of 10 target heights through the reverse inference model and then conducted DED experiments based on the inferred process parameters. The height of the deposition layer was measured, and the true height was compared with the target height, as shown in Table 5.5.

By analyzing the results in Table 5.5, the average absolute error of the process parameter reverse inference model is 0.039 mm, and the maximum error is 0.058 mm. The model can reliably infer the process parameter combination based on the target height, providing guidance for process decision-making.

Table 5.6 shows the inference results of the reverse iterative inference model with a single fixed process parameter (taking fixed laser power as an example). In order to obtain the cladding layer of the target height, the scanning speed and powder feeding rate are inferred based on the fixed laser power. After experimental verification, the maximum absolute error of inferring process parameters with fixed laser power is 0.029 mm, and the average absolute error is 0.017 mm.

Table 5.7 shows the results of inferring the powder feeding rate in reverse when setting the laser power and scanning speed. After fixing two parameters, the degree of freedom for reverse inference becomes 1, and only the powder feeding rate needs to be adjusted to achieve the target DED deposition height. The experimental results show that when the laser power and scanning speed are fixed, the average error between the actual deposition height and the target deposition height of the

Table 5.5. The result of processing parameters inversely inferring without any fixed parameters.

Cases	Target height (mm)	Inferred laser power (W)	Inferred scanning speed (mm/s)	Inferred powder feeding rate (r/min)	Inferred height (mm)	Actual height (mm)	Error (mm)
1	0.100	300	6.0	0.50	0.158	0.158	0.058
2	0.200	560	5.3	0.33	0.229	0.193	–0.007
3	0.300	539	5.7	0.69	0.293	0.337	0.037
4	0.400	874	5.2	0.38	0.403	0.371	–0.029
5	0.500	740	3.3	0.34	0.451	0.534	0.034
6	0.600	740	3.3	0.44	0.592	0.625	0.025
7	0.700	540	3.4	0.67	0.683	0.727	0.027
8	0.800	660	3.4	0.69	0.772	0.804	0.004
9	0.900	720	3.0	0.51	0.907	0.891	–0.009
10	1.000	800	3.0	0.80	0.983	1.058	0.058

Table 5.6. The result of processing parameter inversely inferring with the known laser power.

Cases	Target height (mm)	Set laser power (W)	Inferred scanning speed (mm/s)	Inferred powder feeding rate (r/min)	Inferred height (mm)	Actual height (mm)	Error (mm)
1	0.400	500	3.0	0.40	0.421	0.427	–0.027
2	0.500	400	4.7	0.70	0.477	0.503	0.003
3	0.500	500	5.4	0.74	0.486	0.482	–0.018
4	0.500	600	5.4	0.76	0.503	0.487	–0.013
5	0.600	700	4.0	0.54	0.587	0.629	0.029
6	0.700	600	4.0	0.74	0.688	0.709	0.009
7	0.700	700	3.0	0.44	0.692	0.678	–0.022

Table 5.7. The inversely inferring result with fixed laser power and scanning speed.

Cases	Expected height (mm)	Set laser power (W)	Set scanning speed (mm/s)	Inferred powder feeding rate (r/min)	Inferred height (mm)	Actual height (mm)	Error (mm)
1	0.100	300	6.0	0.50	0.128	0.128	0.028
2	0.200	400	5.0	0.33	0.195	0.199	–0.001
3	0.300	500	4.0	0.44	0.277	0.299	–0.001
4	0.400	600	4.0	0.54	0.388	0.430	0.030
5	0.500	700	4.0	0.60	0.509	0.513	0.013
6	0.600	500	3.0	0.66	0.572	0.594	–0.006
7	0.700	600	4.0	0.72	0.672	0.725	0.025
8	0.800	800	5.0	0.76	0.801	0.785	–0.015
9	0.900	700	3.0	0.70	0.926	0.884	–0.016
10	1.000	600	3.0	0.80	1.007	0.983	–0.007

inferred process parameters is 0.014 mm, and the maximum error is 0.028 mm. This indicates that the process decision-making model can achieve higher accuracy, and fixing two process parameters results in lower errors compared to no fixed parameters and fixed single parameters.

5.4 DED single-layer multi-channel optimal lap step matching model

The essence of DED three-dimensional stereoscopic forming is the overlap of a single cladding track in the horizontal direction and the stacking in the vertical direction. The previous section studied the influence of process parameters on the height of a single DED track and proposed a process parameter decision-making model for a single track. In addition, Chapter 4 conducted offline optimization of process parameters for vertical multi-layer DED forming. Therefore, the forming law in the horizontal direction still needs further study. This section will establish a horizontal overlap average height prediction model and establish an optimal overlap step size reverse iterative inference model based on the forward prediction model.

5.4.1 Effect of horizontal lap step on forming quality

In the horizontal overlap DED process, there is a strong coupling between process parameters and the overlapping offset distance. As shown in Figure 5.7(a) and (b), if the overlapping offset distance is too large, there will be problems such as poor overlap, uneven surface, and the average height of the formed layer being smaller than the height of a single track, which will affect the next layer of DED. If the overlapping offset distance is too small (Figure 5.7(e) and (f)), most of the metal in the overlapping track will fall on the previous track, and the deposition height will gradually accumulate, causing the distance between the work surface and the nozzle to deviate from the optimal working distance. This tilted forming is also not conducive to the next layer of DED processing. The ideal horizontal overlap forming should be that the material in the overlapping area between the overlapping tracks can just compensate for the grooves between the adjacent overlapping tracks, making the final overlapping forming surface approximately flat (Figure 5.7(c) and (d)). Therefore, the optimal horizontal overlap step size depends on the geometric dimensions of a single track. The smaller the geometric dimensions of a single track, the smaller the overlapping offset distance should be set, and vice versa. The geometric dimensions of a single track are determined by the

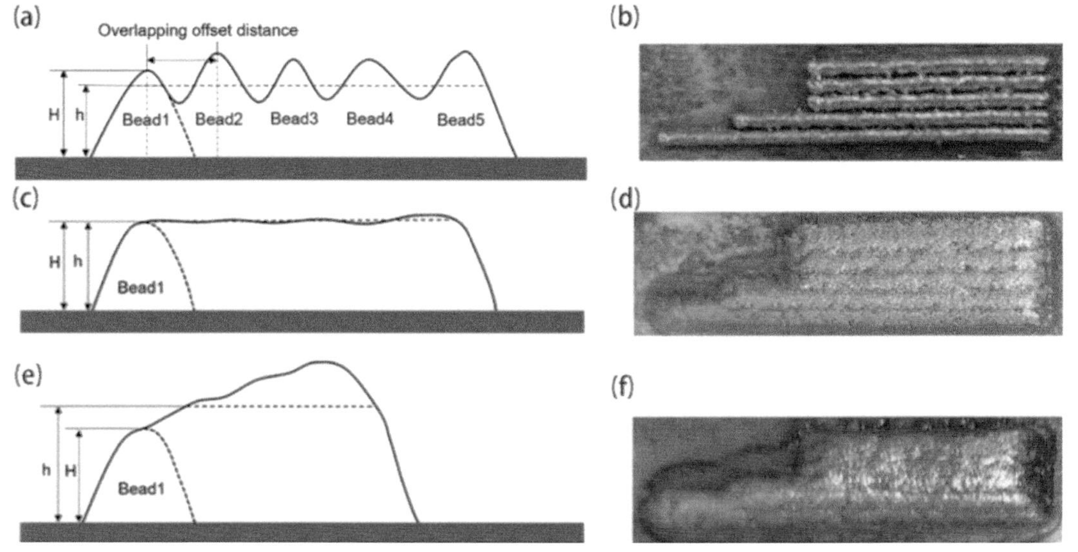

Figure 5.7. The effect of the horizontal overlapping offset on the DED formation.

process parameters, so there is a close relationship between the optimal overlapping offset distance and the process parameters.

For the theoretical overlapping offset distance, reference [2] derived an empirical calculation formula through geometric derivation, but this model requires knowledge of the geometric characteristics of a single track, which means that pre-experiments are still needed to measure the geometric characteristics of a single track before calculating the overlapping offset distance. This is not efficient. The ideal way is to obtain the optimal overlapping offset distance based on the process parameter decision-making after determining the process parameters.

5.4.2 Multi-channel horizontal lap average height prediction model

Before establishing the optimal horizontal overlap decision-making model for a single-layer multi-track, it is necessary to establish a forward prediction model for the average overlap height. The deviation between the average height and the height of the first track can be used as a reference for measuring the quality of the overlap forming. As shown in Figure 5.8, this paper samples the contour height at a certain interval and then takes the average value of the heights of each sampled point as the average height of the overlapping cladding layer.

Through the process experiments in Section 5.2 of this chapter, a total of 576 sets of horizontal overlap experimental data under different parameters were obtained, and a multi-track horizontal overlap average height prediction model based on XGBoost was constructed. The model takes laser power, scanning speed, powder feeding rate, and overlapping offset distance as inputs and average height as output. The ratio of the training set to the test set is 9:1. The test set is used to evaluate the prediction accuracy of the model. As shown in Table 5.8, the average absolute error of the XGBoost-based average height prediction model is 0.065 mm, and the accuracy is higher than the other two models. Figure 5.9 shows the comparison between the predicted values and the true values of the three models. The deviation of XGBoost is smaller and more suitable for the prediction modeling of the average height of single-layer multi-track horizontal overlap in this study.

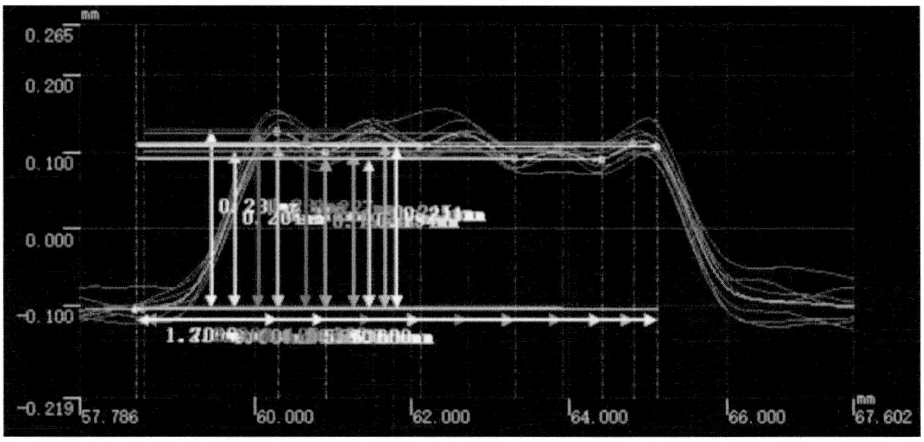

Figure 5.8. The profile of the overlapped tracks.

Table 5.8. The performance of the mean height of the overlapped tracks.

Models	MSE (mm×mm)	RMSE (mm)	MAE (mm)	MRE	R2
XGBoost	0.0228	0.1509	0.0451	0.2118	0.9291
FCNN	0.0288	0.1696	0.0872	0.2677	0.9104
SVR	0.0710	0.2664	0.1579	0.4757	0.7789

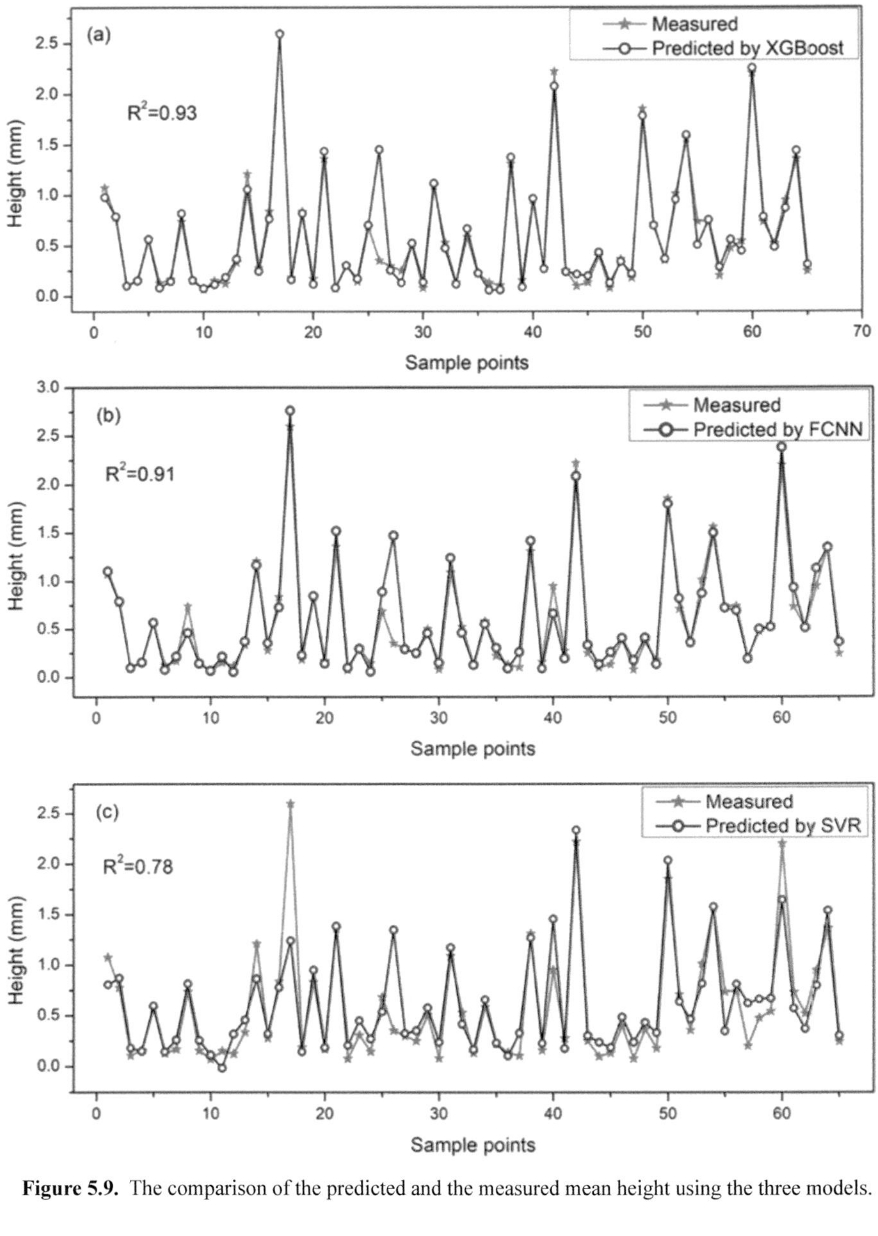

Figure 5.9. The comparison of the predicted and the measured mean height using the three models.

5.4.3 Horizontal lap step reverse iterative optimization

The purpose of the reverse inference model for the optimal overlapping offset distance is to be able to obtain the suggested horizontal overlapping offset distance through iterative inference based on the input process parameters (laser power, scanning speed). Figure 5.10 shows the influence of the overlapping offset distance on the average height under different process parameters. Using the same overlapping offset distance under different process parameters will change the relationship between the average height and the height of the first track. After determining the process parameters, the size of the first track can be determined, and the ideal overlapping forming should have an average height close to the height of the first track. If the overlapping offset distance is too large, the adjacent cladding tracks will not be in contact or in contact too little, and grooves will appear in the overlapping forming, so the average height will be smaller than the deposition height of the first

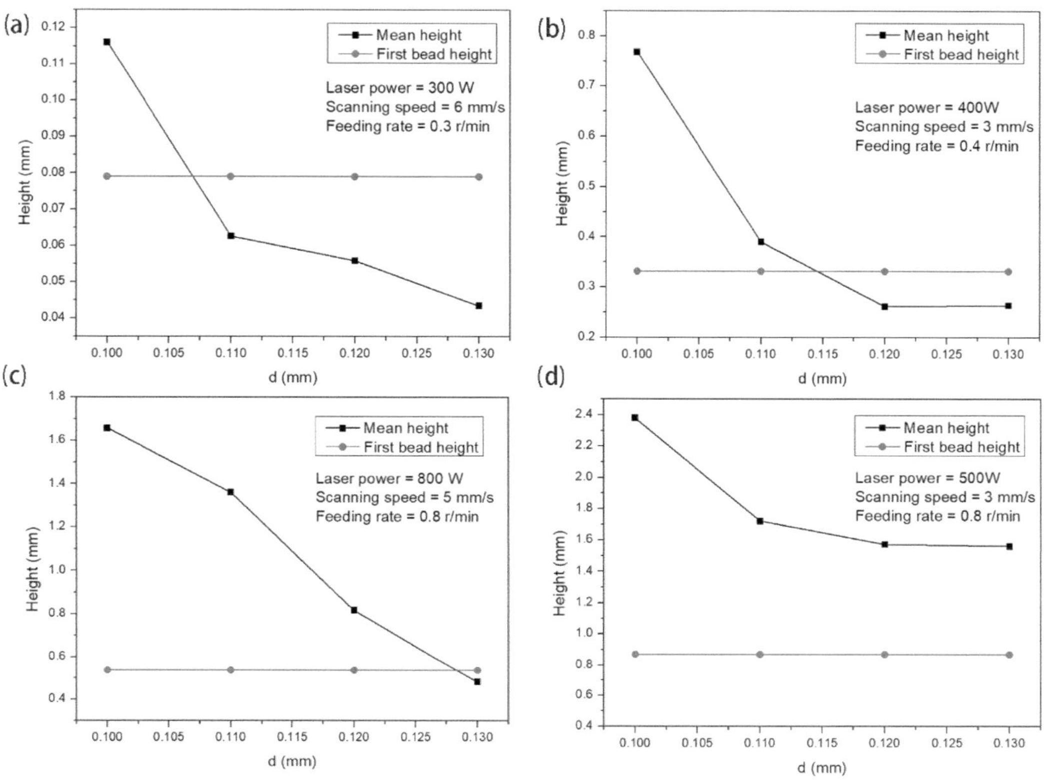

Figure 5.10. The mean height of the clad layer and the height of the first track under different processing parameters.

track; conversely, it will be greater than the deposition height of the first track. Therefore, this basic relationship can be used to establish the optimal overlapping offset distance decision-making model.

This paper establishes the optimal horizontal overlapping offset distance inference model based on the method of forward and reverse synchronous iteration, and the algorithm process is shown in Figure 5.11. First, input the basic process parameters (laser power p, scanning speed v, powder feeding rate f), call the single-layer single-track height prediction model established in Section 5.3 of this chapter to obtain the predicted height h_f of the first track; then call the multi-track horizontal overlap average height prediction model, use the process parameters and the initial overlapping offset distance d as inputs to obtain the predicted average height h_m of the overlap. Then, calculate the deviation e between the average height and the deposition height of the first track, and gradually update the overlapping offset distance d through iterative updates to reduce the deviation between hm and h_f. When hm is less than h_f, it means that the overlapping offset distance d is too large and should be reduced, and vice versa. Then, input the process parameters (p, v, f) and the updated overlapping offset distance d into the average height prediction model to obtain the predicted average height h_m, and then calculate the deviation between h_m and h_f until the absolute value of the deviation e is less than the set threshold, and then output the optimal horizontal overlapping offset distance d. According to the analysis of the iterative process, the inference model needs to call the single-layer single-track height prediction model once and call the overlapping average height prediction model multiple times, so the accuracy of the inference depends on the accuracy of the previous two forward prediction models.

In order to verify the reliability of the proposed reverse iterative inference model for the horizontal overlapping offset distance, this paper designed multiple sets of horizontal overlapping control experiments (Table 5.9). The design idea of the experiment is as follows: first, input the process parameters and the initial overlapping offset distance into the decision-making model to

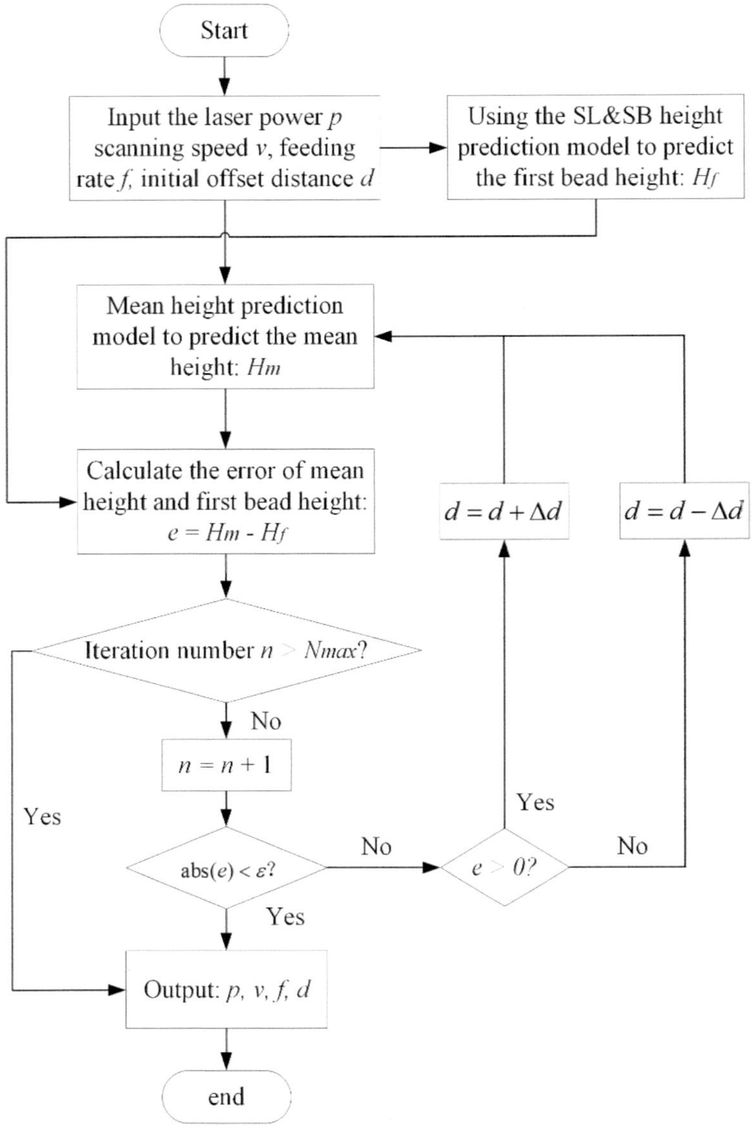

Figure 5.11. The inversely inferring model for the optimal overlapping offset.

Table 5.9. The validation experiment setting.

Cases		Laser power (W)	Scanning speed (mm/s)	Powder feeding rate (r/min)	Horizontal deviation distance (mm)
Case 1	Without optimization	800	6	0.3	1.3
	With optimization	800	6	0.3	1.14
Case 2	Without optimization	800	5	0.4	1.00
	With optimization	800	5	0.4	1.16

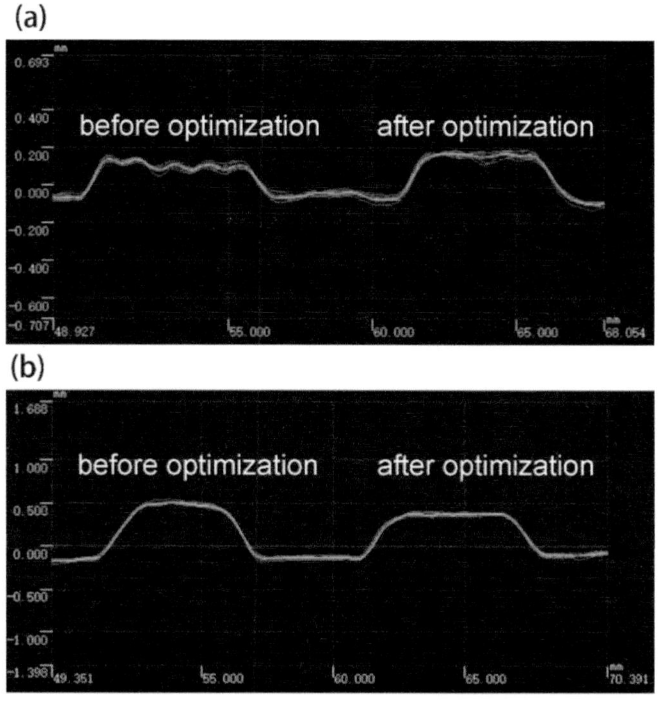

Figure 5.12. The comparison of the contour without/with the overlapping offset optimization: (a) case 1; (b) case 2.

obtain the iteratively optimized offset distance, and then perform DED experiments based on the set process parameters and optimized overlapping offset distance; finally, compare the DED forming before and after optimization.

Figure 5.12 shows the comparison of the cross-sectional profiles of the horizontal overlapping obtained through DED experiments under two different situations shown in Table 5.9. Figure 5.12(a) represents the case where the overlapping offset distance is too large. When the initial overlapping offset distance is 1.3 mm, the profile of the overlapping forming has high and low undulations due to too little overlapping between the cladding tracks. However, after optimization through the optimal overlapping offset distance matching model, a reference overlapping offset distance of 1.14 mm is obtained. Then, the optimized parameters are used for DED experiments, and the profile is extracted using a point cloud contour scanner. It can be seen from the figure that the optimized surface is smoother and eliminates the previous surface undulations. Figure 5.12(b) represents the case where the overlapping offset distance is too small. In this case, the height of the deposition layer of adjacent cladding tracks first increases and then gradually becomes constant due to the excessive overlap, and the average height of the deposition layer is greater than the height of the first track, which is not conducive to the subsequent deposition of cladding layers. After optimization through the proposed optimal overlapping offset distance matching model, the overlapping offset distance is optimized from 1.00 mm to 1.16 mm. Then, a DED comparison experiment is performed, and the cross-sectional profile is measured. The results show that the optimized forming is smoother and the average height is closer to the height of the first cladding track.

In summary, based on the single-layer single-track height prediction model and the multi-track horizontal overlapping average height prediction model, combined with the forward and reverse synchronous iterative algorithm, the optimal overlapping offset distance matching model is established, and its reliability is verified through DED experiments, thereby avoiding the use of a large number of trial-and-error experiments to determine the optimal overlapping offset distance.

5.5 Summary

This chapter focuses on improving the efficiency and intelligence level of practical DED manufacturing. Based on the research needs of the entire text, a single-layer single-path height prediction model, DED process parameter decision model, single-layer multi-path average height prediction model, and optimal overlay step matching model were established. The main conclusions are summarized as follows:

1) A horizontal overlay DED experimental scheme was designed that can simultaneously obtain the key geometric dimensions of single-path and multi-path overlays. A total of 576 sets of experimental data were obtained through DED experiments and 3D point cloud scanners. Based on these data, the impact of process parameters on single-path DED forming and the strong coupling relationship between process parameters and overlay step were analyzed.
2) A forward prediction model of single-layer single-path height based on XGBoost was established. After testing, its average absolute error was 0.056 mm. Compared with the other two types of algorithms, the prediction results of the XGBoost-based model are the most accurate and have the best correlation.
3) A process decision model was established to reverse-engineer process parameters based on single-layer single-path height. The reverse iteration algorithm is based on the XGBoost height prediction model and introduces optimization methods such as database matching based on binary search and cyclic updates of each parameter. It can reliably reverse-engineer the corresponding process parameters based on the height of the cladding layer. Experimental verification results show that without giving any process parameters, the maximum forming error of the inferred process parameters is 0.058 mm, and the average absolute error is 0.039 mm; given a single process parameter, the maximum forming error of the inference result is 0.029 mm, and the average absolute error is 0.017 mm; given two process parameters, the maximum forming error of the inference result is 0.028 mm, and the average absolute error is 0.014 mm.
4) The influence of horizontal overlay step on overlay forming was analyzed, and a horizontal overlay average height prediction model was established based on XGBoost. The average absolute error of the model is 0.065 mm.

References

[1] Friedman, J., Hastie, T. and Tibshirani, R. (2000). Additive logistic regression: a statistical view of boosting (with discussion and a rejoinder by the authors)[J]. The Annals of Statistics 28(2): 337–407.

[2] BAI Jiuyang, WANG Jihui, LIN Sanbao, YANG Chunli and FAN Chenglei. (2016). Model for multi-beads overlapping calculation in GTA-additive manufacturing[J]. Journal of Mechanical Engineering 52(10): 97–102.

Chapter 6

Tailoring Laves Phase and Mechanical Properties of Directed Energy Deposited Inconel 718 Thin-Wall via a Gradient Laser Power Method

6.1 Introduction

Inconel 718, a precipitation-strengthened nickel-based superalloy, has found extensive applications in the aerospace industry. Its exceptional combination of high tensile strength, resistance to creep, oxidation, and corrosion at high temperatures makes it a preferred material for manufacturing high-precision aviation components, particularly turbine blades [1]. However, turbine blades often experience severe damage, such as breakages and dimensional deformations, after prolonged exposure to harsh operating conditions involving high temperatures, pressures, and vibrations [2]. Instead of completely replacing the damaged blades, repairing them offers a more economical, time-saving, and resource-efficient solution, which is highly advocated in the maintenance, repair, and overhaul (MRO) industry. The advent of laser additive manufacturing has made it possible to reconstruct and repair damaged turbine blades through layer-by-layer material deposition. Among various laser additive manufacturing methods, directed energy deposition (DED) has emerged as a particularly promising technique for rebuilding damaged aviation components.

It is worth noting that turbine blades have a thin-wall structure with thicknesses of only a few millimeters. As the laser deposits material layer by layer, the thin walls are subjected to repetitive vertical thermal cycles in a small area (mm^2), leading to complex microstructural transformations involving primary solidification, remelting, and solid-state precipitation induced by multiple thermal cycles [3]. Thus, gaining a comprehensive understanding of the thermal cycles during DED is crucial for effectively tailoring the microstructure and mechanical properties of the deposited components. Infrared thermal cameras (IR) have become invaluable tools for non-contact temperature measurement, as they can convert the invisible infrared energy emitted by an object into visualized thermal images and collect thermal information during the layer-by-layer additive manufacturing process without disturbing the temperature field. In recent years, IR has been widely utilized in welding and additive manufacturing fields to investigate thermal cycles [4–6]. By analyzing parameters such as cooling rate, average temperature, and temperature gradient obtained from IR measurements, researchers can gain insights into the thermal behavior of as-deposited samples, enabling them to optimize process parameters and enhance the quality of the deposited components [7–9]. The integration of IR technology with laser additive manufacturing offers valuable opportunities for real-time monitoring and control of thermal cycles, facilitating the production of high-quality repaired turbine blades with tailored microstructures and enhanced mechanical performance.

DOI: 10.1201/9781003406327-6

Extensive research has been conducted on laser additive manufacturing of Inconel 718 alloy, as evidenced by numerous studies [10–12]. The process of depositing Inconel 718 superalloys through laser additive manufacturing involves intricate thermal cycles characterized by rapid heating, rapid cooling, directional heat extraction, and repeated melting. These thermal cycles give rise to a distinct microstructure that sets it apart from the microstructure obtained through traditional casting and wrought processes. The as-deposited structures exhibit several noteworthy microstructural characteristics, including micro-segregation, inter-dendritic Laves phases, restrained nano-strengthening precipitates, and highly oriented columnar structures. These characteristics collectively contribute to the unique properties of the as-deposited Inconel 718 superalloys. In order to meet the stringent service requirements of superalloy aerospace parts, mitigating the micro-segregation of alloying elements and tailoring the formation and distribution of Laves phase are key to improving the performance of repaired Inconel 718 parts.

In this chapter, we detail the microstructural features of laser direct energy deposited Inconel 718 thin-wall. To simulate the real restoration of the turbine blade tip, we examine the effects of laser power and thermal cycles on as-repaired Inconel 718 thin-wall. By gradually lowering the laser power within the processing window available, the gradient laser power deposition (GLP) method is firstly proposed to optimize the Laves phase. In-depth research has been done on the thermal data from the infrared thermal camera, the microstructure, texture, and mechanical characteristics of the as-repaired Inconel 718 samples after restoration.

6.2 Microstructural characteristics of as-deposited Inconel 718

6.2.1 Micro-segregation

During the laser deposition process of Inconel 718, the rapid heating and cooling conditions induce a phenomenon where chemical elements with a distribution coefficient (k) less than 1, such as niobium (Nb), molybdenum (Mo), and silicon (Si), have a pronounced propensity to segregate towards the remaining liquid phase during solidification. This segregation phenomenon leads to the accumulation of solute elements in the final solidified structure, thus influencing its composition and properties. The fast solidification rate in laser deposition restricts the diffusion of these elements, exacerbating their tendency to concentrate in certain regions. Consequently, the resulting microstructure of the Inconel 718 alloy exhibits variations in solute content, which can have a significant impact on its mechanical and thermal behavior.

The as-deposited Inconel 718 sample showcases prominent micro-segregation characteristics, particularly evident through the findings obtained from Energy-Dispersive X-ray Spectroscopy (EDS) mappings (refer to Figure 6.1). The regions where Laves phases precipitate between dendritic structures exhibit a notable enrichment of Nb and Mo elements, accompanied by a deficiency in chromium (Cr) and iron (Fe) elements. This severe micro-segregation of Nb during the laser powder deposition process gives rise to the formation of numerous elongated and brittle Laves phases.

In a study by Mantri et al. [13], the precipitation process in a compositionally inhomogeneous Inconel 718, deposited using laser additive manufacturing, was investigated. The utilization of 3D atom probe tomography (APT) confirmed an increase in niobium (Nb) content from the dendrite core to the periphery of dendrites and further to the inter-dendritic channels. This observation provided valuable insights into the spatial distribution of Nb within the deposited material.

In a different approach, Kumara et al. [14] employed a multi-component and multi-phase-field modeling approach to study the microstructure evolution during directed energy deposition (DED) of Inconel 718. The research highlighted that significant micro-segregation can alter the local thermodynamics of Inconel 718, consequently affecting the driving forces for the formation of different phases. This modeling approach provided valuable insights into the complex phase transformations occurring during the laser deposition process.

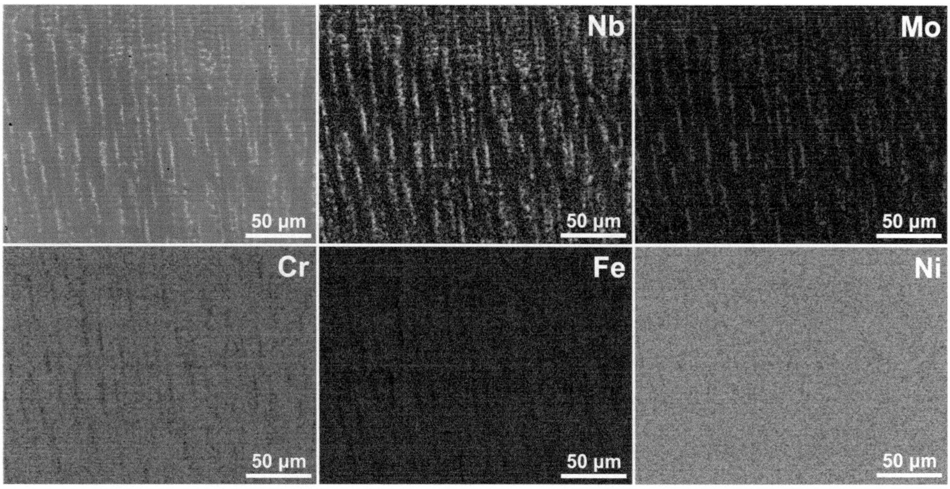

Figure 6.1. EDS mappings of severe micro-segregation in as-deposited Inconel 718 thin-wall.

Furthermore, Zhang et al. [15] conducted an analysis of the as-deposited Inconel 718 samples using differential thermal analysis (DTA) curves. Their findings indicated that the rapid cooling experienced during the laser deposition process caused compositional changes in the deposited material, which, in turn, affected the phase transition temperatures. This observation emphasized the influence of the deposition process on the thermal behavior and phase transformations of Inconel 718.

6.2.2 Inter-dendritic laves phases

The presence of severe micro-segregation in laser-deposited Inconel 718 alloys gives rise to the formation of eutectic products, such as γ + NbC and γ + Laves phases, in the inter-dendritic regions [16]. Laves phases ($(Ni,Fe,Cr)_2(Nb,Mo,Ti)$), known for their hardness and brittleness, can significantly diminish the mechanical properties of the as-deposited Inconel 718. The hexagonal Laves phase, particularly when it forms coarse and continuous long-chain structures, promotes crack initiation and propagation, leading to premature failure. These phases detrimentally affect the strength, ductility, stress rupture, fatigue, and creep rupture properties of the material. When the Laves phase content exceeds 2 ~ 3 wt%, the room-temperature strength and ductility of the material can be significantly compromised, resulting in approximately 20% and 60% losses, respectively [17].

Figure 6.2 shows the microstructure of as-deposited Inconel 718, as observed under scanning electron microscopy (SEM) in both the normal secondary electron mode and backscattered electron (BSE) mode. The as-deposited state of Inconel 718 primarily comprises the solid solution γ matrix, hexagonal Laves phases, and cubic MX ((Nb, Ti)(C, N)) particles, which are commonly referred to as MC particles.

Laves phases are brittle intermetallic compounds, belonging to the topologically close-packed (TCP) phase family, that typically form during the rapid solidification process of Inconel 718. Towards the final stages of solidification, elements with distribution coefficients less than 1 accumulate in the remaining liquid phase, leading to the precipitation of Laves phases within the inter-dendritic regions of columnar dendrites through eutectic reactions. These Laves phases, indicated by yellow arrows (Figure 6.2), exhibit elongated chain-like morphologies aligned parallel to the direction of columnar dendrites. Furthermore, the microstructure of as-deposited Inconel 718 contains MX particles, which predominantly consist of carbides and nitrides. These MX particles,

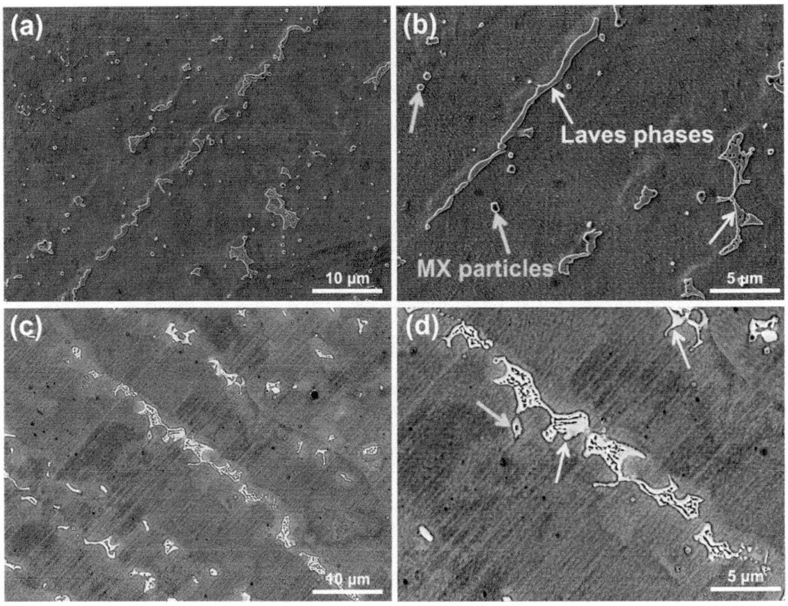

Figure 6.2. Distribution of Laves phases in the inter-dendritic region and random MX particles: (a, b) SEM images and (c, d) BSE images.

marked by green arrows, exhibit a mixture of block-shaped and spherical morphologies. They are non-uniformly distributed within the grains as well as along the grain boundaries.

Note that the formation of these long-chain Laves phases not only alters the microstructure of the Inconel 718 alloy but also negatively impacts its mechanical behavior and overall performance. The resultant weakened material properties highlight the significance of mitigating Nb micro-segregation during the laser powder deposition process in order to enhance the alloy's structural integrity and improve its resistance to crack formation and propagation.

6.2.3 Restrained strengthening precipitates

The strengthening mechanisms of Inconel 718 alloy primarily involve solid-solution strengthening with refractory metal elements, primarily Nb and Mo, as well as precipitation strengthening with nano-scale γ'' and γ' strengthening phases within the face-centered cubic (fcc) austenite γ matrix. In Inconel 718, the primary strengthening phases are the γ'' phases (Ni_3Nb) with a disc morphology. These phases contribute to the overall strengthening of the alloy due to their comparatively large crystal misfit. Additionally, the γ' phases ($Ni_3(Al,Ti)$) with a circular morphology act as auxiliary strengthening phases. The formation of massive nano-scale γ'' and γ' strengthening phases occurs within the temperature range of 600–900°C, with the corresponding peak-aged volumes reaching up to 15% and 4%, respectively [18].

These strengthening phases play a crucial role in enhancing the mechanical properties of Inconel 718. They contribute to increased strength, improved resistance to deformation, and enhanced high-temperature performance. The control and optimization of the precipitation of these strengthening phases are essential for tailoring the material's properties and ensuring its suitability for specific applications.

Figure 6.3 depicts the distribution of dislocations within the as-deposited γ matrix of Inconel 718. The microstructure reveals the presence of Laves phases and MX particles dispersed within the γ matrix. Transmission electron microscopy (TEM) investigations indicate that no nano-scale γ'' and γ' strengthening phases are observed. The rapid cooling rates during the laser powder deposition (LPD) process effectively suppress the precipitation of these nano-strengthening phases.

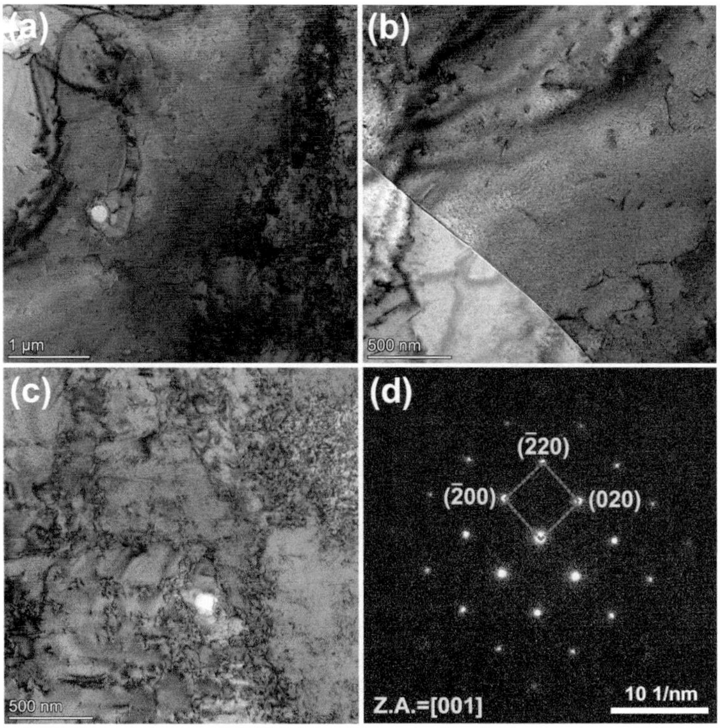

Figure 6.3. TEM analysis showing a lot of dislocations in the as-deposited γ matrix: (a, c) the dislocation distribution close to the intragranular region, (b) the dislocation distribution close to the intergranular region, and (c) corresponding SAED pattern of the γ matrix.

Furthermore, the severe micro-segregation and formation of Laves phases contribute to the depletion of crucial alloying elements like Nb, Mo, and Ti, which are primarily responsible for strengthening the Inconel 718 alloy. Consequently, the γ matrix becomes devoid of these strengthening elements, leading to weakened solid-solution strengthening and precipitation strengthening [19]. Additionally, the precipitation of γ'' and γ' strengthening phase particles during subsequent aging heat treatment is adversely affected [20].

Significantly, a considerable number of dense dislocations are observed in the vicinity of the secondary phases and throughout the γ matrix, as demonstrated in Figure 6.3(a–c). The high mechanical constraints and rapid cooling rates experienced during the thermal cycles of the LPD process generate substantial residual stresses, thereby inducing the formation of numerous dislocations. Tucho et al. [21] noted that rapid solidification shrinkage, which occurs during additive manufacturing processes, leads to increased residual stress. The presence of clustered dislocations indicates the extent of plastic deformation resulting from the extreme thermal cycles and enables accommodation of the large residual thermal strain generated by the fast cooling process of LPD. Furthermore, Figure 6.3(d) presents the corresponding selective area electron diffraction (SAED) pattern of the γ matrix. The solid solution γ matrix possesses a fcc crystal structure with lattice parameters of a = 0.372 nm, conforming to the Fm-3m space group.

6.2.4 Strongly-orientated columnar structures

The morphology and crystallographic texture of grains are known to have a significant impact on the mechanical properties of Inconel 718 samples. Understanding the crystallographic orientations and grain growth patterns is essential for evaluating the anisotropic properties and mechanical performance of laser-deposited Inconel 718 alloys. The growth direction of columnar grains is determined by a combination of factors including heat flow, epitaxial/preferred growth, nucleation, and competitive growth [22]. Studies conducted by Yang et al. [23] and Wang et al. [24] have shed light on the growth behavior of columnar grains in laser additive manufacturing. It has been observed that the growth direction of columnar grains is primarily influenced by the competitive growth characteristics of epitaxial growth, which can be characterized by the angle between the local heat flow and the preferred growth direction (<001> direction). A smaller angle between the local heat flow and the preferred growth direction accelerates grain growth and promotes epitaxial growth, leading to a more pronounced preferred texture. On the other hand, a larger angle hinders epitaxial growth, resulting in a more random texture. The competition between epitaxial growth and random growth determines the final crystallographic texture of the Inconel 718 sample.

Figure 6.4 depicts the electron backscatter diffraction (EBSD) crystallographic orientation results obtained from two different directions of the as-deposited Inconel 718 sample. In the longitudinal cross-section (Figure 6.4(a)), which is parallel to the scanning direction, the growth of the coarse columnar grains primarily aligns along the deposition direction (Z direction) and extends across multiple deposited layers. The dominant crystallographic orientation in this direction is <001>. As the deposition process proceeds, heat accumulation becomes more severe. The heat dissipation within the plane diminishes gradually, resulting in the highest thermal gradient aligning along the deposition direction towards the substrate. Remarkably, this thermal gradient direction closely aligns with the preferred growth direction of fcc structure materials. As a result, epitaxial growth is facilitated, promoting the nearly vertical growth orientation of the coarse columnar grains exhibiting the <001> texture. It demonstrates that this alignment between the thermal gradient and the favored growth direction plays a pivotal role in directing the growth behavior of the grains during the laser additive manufacturing process of Inconel 718.

In the transverse cross-section (Figure 6.4(b)), which is perpendicular to the scanning direction, the grain growth still exhibits epitaxial growth characteristics. The grains near the center of the Inconel 718 thin-wall parts appear coarse and grow predominantly along the Z direction. Conversely, the grains near the surface of the Inconel 718 thin-wall parts are finer and exhibit growth closer to the center. In terms of crystallographic orientation, the Inconel 718 deposited layers primarily exhibit the <101> and <111> texture features in this transverse direction.

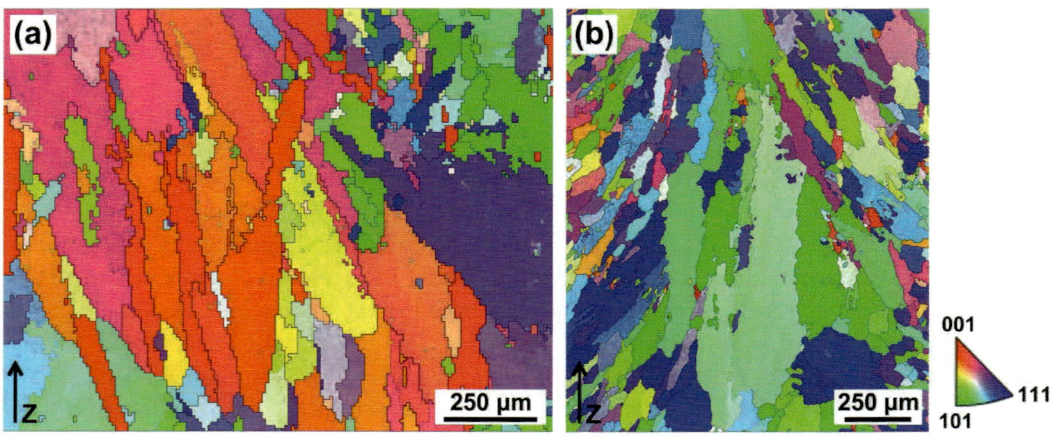

Figure 6.4. EBSD inverse pole figures of as-deposited thin-wall Inconel 718 samples along (a) longitudinal cross-section and (b) transverse cross-section.

6.3 A novel gradient laser power deposition method proposed and thermal cycle regulation

6.3.1 Gradient laser power deposition method

Based on the analysis of the microstructural characteristics of as-deposited Inconel 718, it has been observed that modifying the energy input during the deposition process can effectively influence the microstructure and mechanical properties of the Inconel 718 thin-wall samples. Specifically, by reducing the heat accumulation and adjusting the local solidification conditions, it is possible to achieve desired modifications. Among the various process parameters, regulating the laser power during deposition proves to be a favorable choice due to its convenience and efficiency compared to adjusting the scanning speed.

In this chapter, we propose a novel deposition method called the gradient laser power (GLP) method, which aims to precisely control the local solidification conditions by gradually decreasing the laser powers within the available processing window. The GLP method involves a uniform reduction of laser power layer by layer, starting from 410 W for the first layer and gradually decreasing to 220 W for the final layer (71st layer) as the deposition height increases. As a point of reference, we also employ the conventional constant laser power (CLP) deposition method, where the laser power remains constant throughout the deposition process. All other process parameters, except for the laser power, are kept identical for both the GLP and CLP methods.

By implementing the GLP and CLP methods and comparing their outcomes, we aim to evaluate the influence of laser power modulation on the microstructure and mechanical properties of the Inconel 718 thin-wall samples. Through systematic analysis and characterization, we can gain insights into the effectiveness of the GLP method in tailoring the solidification behavior and subsequently optimizing the material properties of the deposited components.

6.3.2 Thermal cycle regulation

The typical temperature distribution obtained by the IR camera is described in Figure 6.5(a), which reveals clear distinctions between the as-deposited part and the substrate. Cooling rates are measured by tracking the temperature decrease within the first second after reaching the highest temperature at each measurement point. Cooling rates from the 12th, 36th, and 60th layers, located in the middle of the deposited length, are shown in Figure 6.5(b). Notably, the cooling rate exhibits a negative correlation with both laser powers and deposition heights for samples fabricated using the conventional CLP method, consistent with findings reported by Bennett et al. [8]. In contrast, the gradient laser power (GLP) sample demonstrates an increasing cooling rate as the deposition height increases, indicating a rapid cooling process.

The estimation of average temperatures is based on the average surface temperature of each deposited layer and the as-deposited parts (excluding the substrate) at the completion of every two deposited layers. These two types of average temperature coverage areas are depicted by yellow strips in Figure 6.6. For all samples, the average temperatures of each layer exhibit a sharp increase in the initial 10 layers. A higher laser power results in a steeper and higher average temperature. In the case of samples prepared using the CLP method, the average temperature of each layer demonstrates an increasing trend during the middle and late stages of deposition. Conversely, for the GLP sample, the average temperature of each layer shows a relatively slow rise and rapid decline in the early and late stages of deposition, remaining relatively constant between the 12th and 44th layers. The average temperature of the as-deposited parts follows similar trends to the average temperature of each layer in the early stage of deposition. As the volume of as-deposited parts gradually increases, the average temperature of these parts decreases after reaching its maximum value as shown in Figure 6.6(b). Compared to the CLP method, the GLP method mitigates heat accumulation, elevates the layer reaching the maximum average temperature of as-deposited parts, and accelerates the overall cooling process.

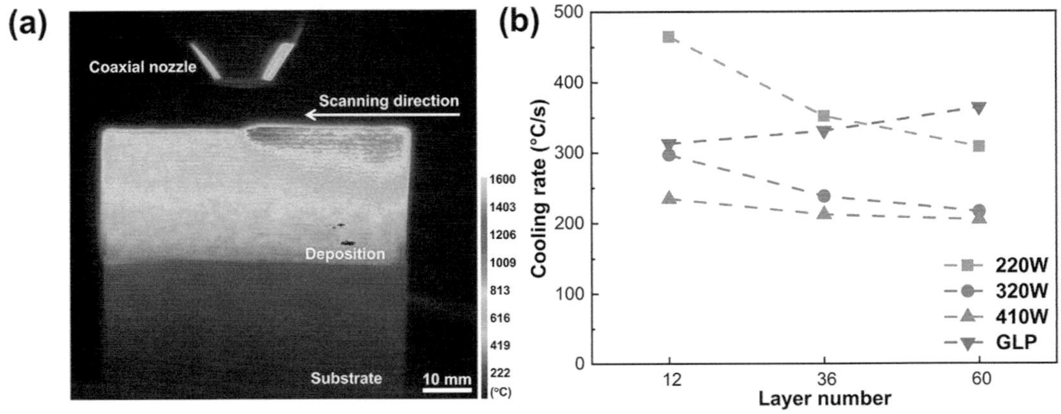

Figure 6.5. Thermal information of as-deposited samples: (a) the typical temperature distribution of the as-repaired thin-wall sample by IR camera, and (b) cooling rates under various laser powers and deposition heights.

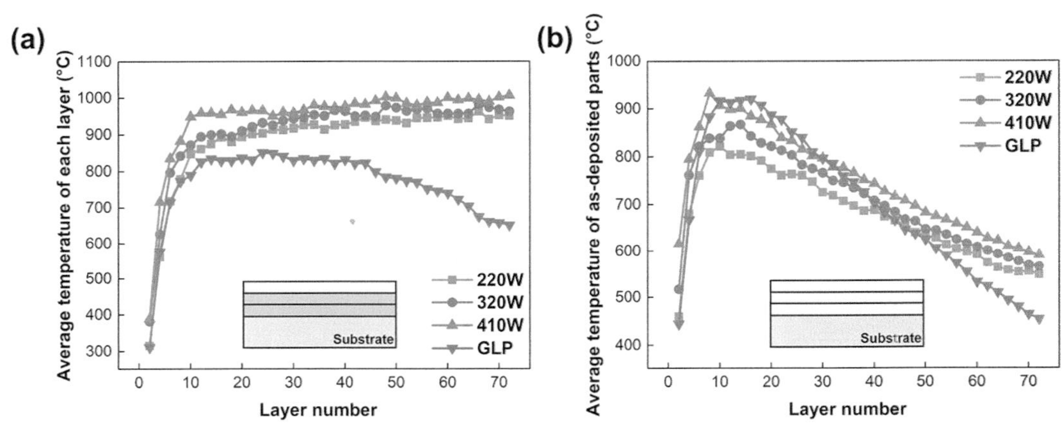

Figure 6.6. The average temperatures of (a) each layer, and (b) as-deposited parts.

Figure 6.7(a) and (b) respectively depict the typical isothermal diagrams of the middle length of the 36th deposited layer under different laser powers and different deposition heights under the 220 W laser power. The temperature gradient along the horizontal direction (laser scanning direction) is relatively small, while it is larger along the vertical direction (building direction). Increasing the laser power results in a decrease in the temperature gradient at the same as-deposited position. Similarly, under the same laser power, the temperature gradient gradually decreases as the distance from the substrate increases. The presence of small irregularly shaped isotherms in the isothermal maps indicates an uneven surface caused by powder adherence to the high-temperature as-deposited layers.

The average temperature gradients between 1200 and 800°C were determined based on the measured data, as depicted in Figure 6.7(c) and (d). This temperature range encompasses the solvus temperatures of Laves phase, γ'', and γ' in Inconel 718 [25]. It is observed that increasing the laser power results in a decrease in both the horizontal and vertical temperature gradients. As the height of the deposited part increases, both the vertical and horizontal temperature gradients initially decrease and then moderately increase at the top of the deposited region. This can be attributed to the substantial cooling effect of the substrate, acting as a heat sink, which causes a significant temperature gradient at the beginning of the deposition [26]. However, as heat accumulates layer

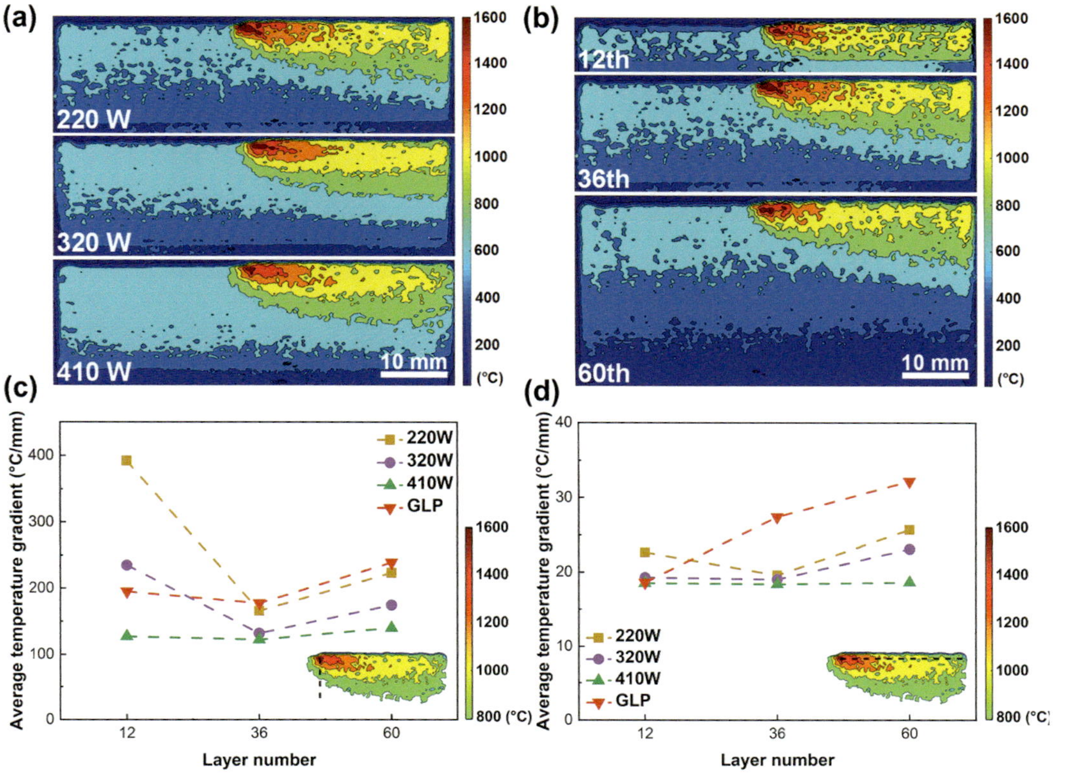

Figure 6.7. Temperature gradient distribution of as-deposited samples: isothermal diagrams of (a) the 36th deposited layer with various laser powers and (b) various deposition heights at 220 W, the average temperature gradients in (c) vertical direction and (d) horizontal direction with various laser powers and deposition heights.

by layer, the temperature gradient gradually decreases during the deposition process. In the later stages, heat dissipation at the top is enhanced due to a balanced effect between increased heat rejection by the already thick deposited layers and heat accumulation by the newly added layer. Furthermore, it has been observed that temperature gradients for high laser powers remain stable and are less influenced by the height of the deposited parts [27]. In the case of GLP samples, the temperature gradient tends to increase with the deposition height, particularly in the horizontal direction. This suggests that gradually reducing the laser powers weakens heat accumulation in the vertical direction while strengthening lateral heat dissipation, resulting in an amplified horizontal temperature gradient.

6.4 Microstructural regulation by a gradient laser power deposition method

6.4.1 Typical features of the microstructure

Figure 6.8 illustrates the typical microstructure of the as-deposited Inconel 718 thin-wall component in the Y-Z plane at different laser powers. Based on the microstructural characteristics, the specimen can be clearly divided into three distinct zones: the substrate zone (SZ), bonding zone (BZ), and deposited zone (DZ), as depicted in Figure 6.8(a). Since the polycrystalline base materials and the deposited materials have the same composition, the deposited parts nucleate directly on the un-melted substrate. Initially, columnar dendrites form at the edge of the BZ and subsequently undergo epitaxial growth during the deposition process.

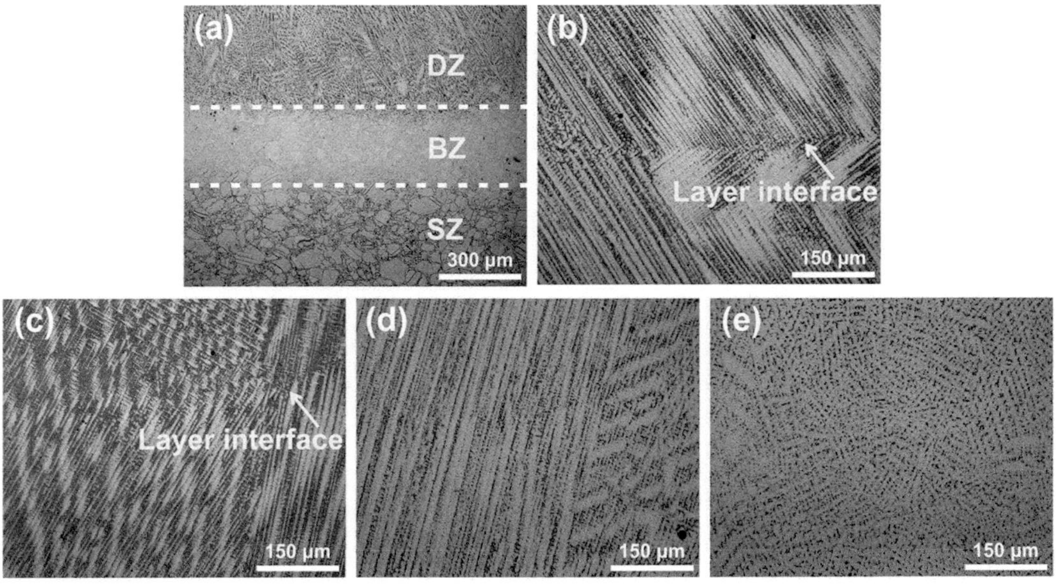

Figure 6.8. Optical metallography of as-deposited Inconel 718 thin-wall: (a) the typical microstructure of 220 W and DZ under (b) 220 W, (c) 320 W, (d) 410 W, and (e) GLP.

The CLP method produces samples with distinct columnar dendritic structures. When the laser power is low (Figure 6.8(b)), the columnar dendrites in adjacent layers form an approximate 90° angle due to bidirectional scanning. The layer interfaces, indicated by yellow arrows, are clearly visible. The growth direction of the columnar dendrites is always perpendicular to the trailing edge of the molten pool and parallel to the opposite direction of local heat flow. The secondary dendrites from the previously deposited layer, which are mostly oriented perpendicularly to the primary

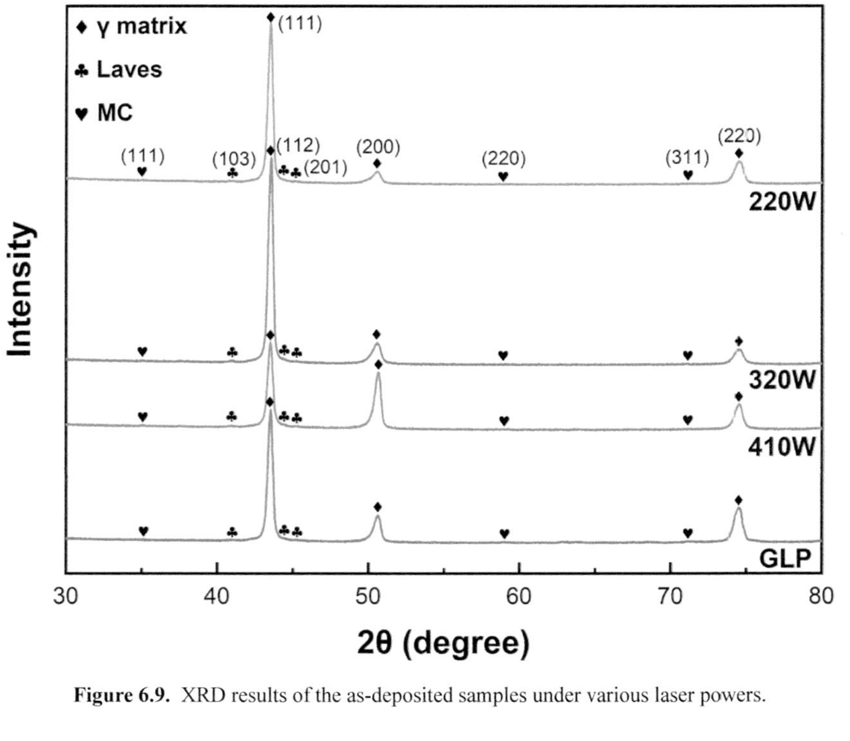

Figure 6.9. XRD results of the as-deposited samples under various laser powers.

dendrites in fcc crystals, serve as growth sites for the primary dendrites of the next deposited layer under low laser powers [28]. As the laser power increases, grain growth predominantly occurs in an epitaxial manner from the previous layer, rather than through re-nucleation across layers. The columnar dendrites grow epitaxially along the deposition direction, extending even across several deposited layers (refer to Figure 6.8(c) and (d)). In Figure 6.8(e), the microstructure of GLP samples primarily consists of columnar dendrites exhibiting random growth directions, with the layer interface being nearly imperceptible. This unique microstructure is a result of the continuous variation in thermal cycles employed in the GLP method.

Figure 6.9 displays the X-ray diffraction (XRD) outcomes for the Inconel 718 samples in their as-deposited state, subjected to varying laser powers. The diffraction patterns exhibit distinct peaks corresponding to the Ni-Cr-Fe γ matrix. Additionally, peaks associated with the Laves phase and MC carbide are observed. Notably, the intensity of the MC carbides peak is significantly lower compared to other phases. This is related to their lower content. Remarkably, as the laser power increases in the case of the CLP method, the (200) peak becomes more pronounced. This phenomenon aligns with the observation that columnar grains tend to grow along the <001> crystallographic orientation under conditions of high heat input.

6.4.2 The morphology and distribution of the Laves phase

Figure 6.10 illustrates two typical morphologies observed in Laves phases, namely the long-chain and discrete Laves phases corresponding to the CLP and GLP methods, respectively. EDS analysis was conducted to determine the chemical compositions of different phases in the as-deposited samples. Table 6.1 presents the composition results for the as-deposited Inconel 718, revealing that it primarily consists of the γ matrix (A and D), Laves phases (B and E), and MC particles such as Ti-rich nitride (C) and Nb-rich carbide (F). In CLP samples, long-chain interconnected Laves phases are typically formed in the inter-dendritic regions. Conversely, GLP samples exhibit fine and discrete Laves phases as the dominant morphology. Figure 6.10(b) and (d) provide a magnified view of the two types of Laves phases. The darker-colored regions surrounding the Laves phases indicate areas rich in elements. Figure 6.10(e) shows the presence of nano-scale strengthening phase particles (γ'' or γ') around the discrete Laves phases, whereas the region around the long-chain

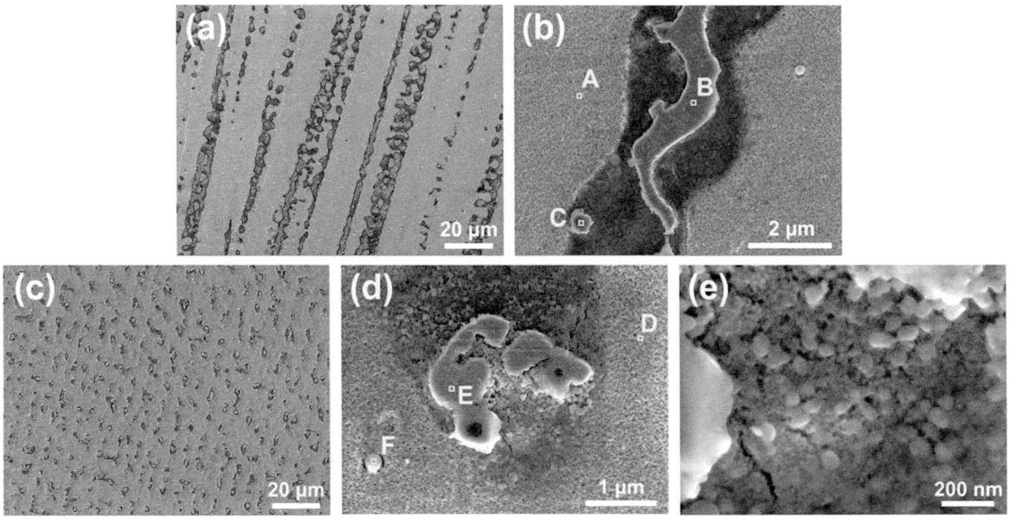

Figure 6.10. The different morphologies of the Laves phases in the Inconel 718 thin-wall: (a, b) long-chain Laves phases in the CLP sample, (c, d) fine and discrete Laves phases in the GLP sample, and (e) distribution of γ'' or γ' particles around the Laves phase shown in (d).

Table 6.1. EDS analysis results of the chemical compositions of different phases in the as-deposited samples (in wt%).

Position	Ni	Nb	Ti	Al	Cr	Fe	Co	Mo	Ta	C
A(Matrix)	50.01	2.13	0.58	0.58	19.09	19.48	0.12	2.45	0.63	4.93
B(Laves)	32.52	25.60	0.88	0.23	12.74	12.09	0.01	6.48	0.61	8.83
C(MC)	45.97	8.94	6.06	3.76	16.67	14.07	0.00	3.70	0.84	0.00
D(Matrix)	48.98	4.60	1.01	0.57	18.63	17.50	0.17	2.84	0.80	4.89
E(Laves)	35.95	20.30	1.32	0.35	12.30	11.24	0.10	4.06	0.70	13.68
F(MC)	29.30	22.29	2.65	0.51	10.94	9.82	0.12	1.93	0.38	22.06

Laves phases contains almost no strengthening phase particles. The high cooling rates during the DED process are believed to inhibit the *in-situ* precipitation of strengthening phases, resulting in very limited amounts of γ'' and γ'.

Varying thermal conditions during the DED process can result in significant disparities in micro-segregation and in-process precipitation [29]. In order to examine micro-segregation and its correlation with the morphology of Laves phases, the distribution of elements in γ matrices (A1 and A2), Nb- and Mo-rich regions (B1 and B2), and Laves phases (C1 and C2) was analyzed. Figure 6.11 illustrates the measurement locations and results for long-chain Laves phases and discrete Laves phases. EDS mapping conducted in these regions (Figure 6.11(c) and (d)) reveals distinct micro-segregation of Nb and Mo alloying elements within the Laves phases formed during the

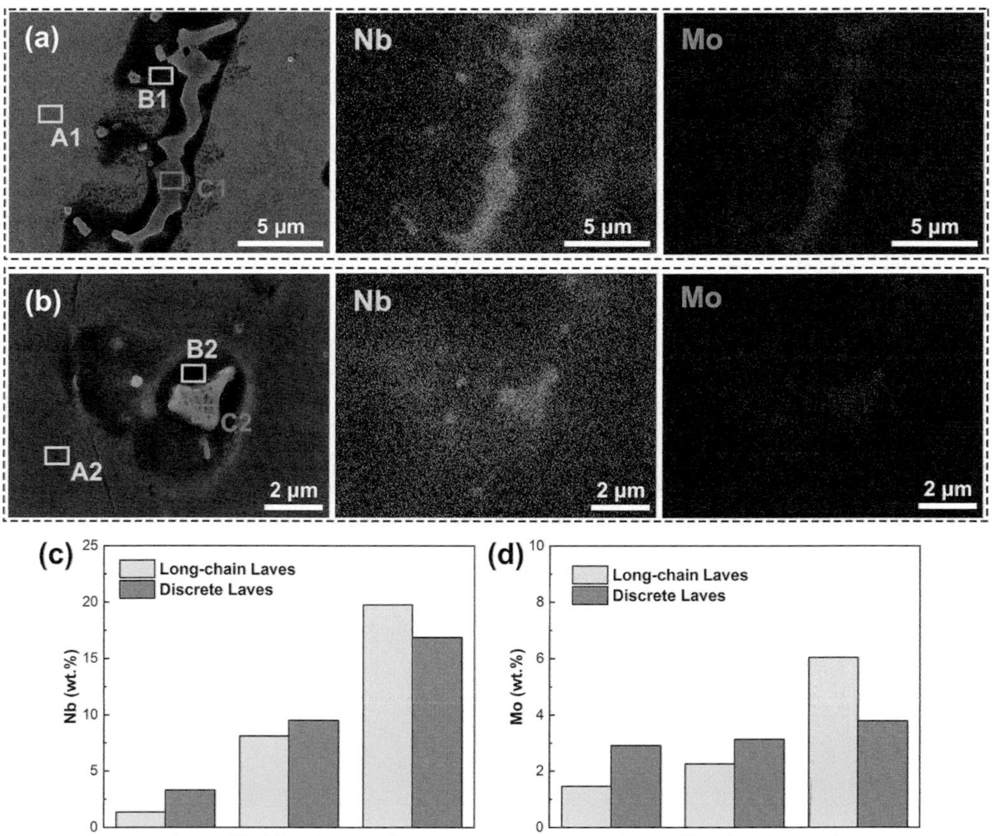

Figure 6.11. Element distributions of the two kinds of Laves phases and their surrounding regions: (a, b) EDS mapping results, and (c) Nb content and (d) Mo content in the γ matrix, Nb- and Mo-rich regions and Laves phases.

solidification process. While the concentration of Nb in the γ matrix is only 2–5 wt%, the concentration of Nb in Laves phases reaches 16–25 wt%, surpassing the nominal chemical composition of Inconel 718 (~ 5 wt%). In comparison, the micro-segregation level of Mo is less pronounced, which can be attributed to its low diffusivity and high density [30]. Contrasting with the CLP sample characterized by long-chain Laves phases, the GLP sample exhibits a higher dissolution of Nb and Mo in the γ matrix and the element-rich regions surrounding the discrete Laves phases. Consequently, the Nb and Mo contents in discrete Laves phases are significantly lower than those in the long-chain Laves phases.

The size and shape of dendrites are greatly influenced by the cooling rates. The primary dendrite arm spacing (PDAS) was determined by analyzing more than five SEM images for each deposition condition. The average PDAS can be used to estimate the cooling rate using the equation [31]:

$$\lambda = 80\varepsilon^{-0.33} \tag{6.1}$$

where, λ represents the PDAS in μm, and ε denotes the cooling rate in °C/s. The measured average PDAS and calculated cooling rates for different laser powers in the bottom, middle, and top regions are presented in Figure 6.12. In the case of the CLP sample, the PDAS increases with the deposition height and laser power. Conversely, for the GLP sample, the PDAS decreases with the deposition height, reaching only 7.55 μm in the top region. The calculated cooling rate exhibits a negative correlation with the PDAS and aligns with the trend of cooling rates measured by the IR camera.

Figure 6.13 illustrates the distribution of Laves phases in the DZ of Inconel 718 thin-wall structures fabricated using various laser powers and deposition heights. Noticeable variations in Laves phases are observed from the bottom to the top regions in samples deposited at 220 W, 320 W, and 410 W. Coarse and interconnected Laves phases tend to form in the middle and top regions of the fabricated parts. The bottom region of the DZ undergoes repeated thermal cycles due to subsequent layer deposition, resulting in partial dissolution of Laves phases and diffusion of Nb atoms into the γ matrix [32]. In the GLP sample, the distribution of Laves phases is more uniform and discrete, with the Laves particles becoming finer at higher deposition heights. Additionally, the area fraction of Laves phases was quantified using Image Pro Plus software. The average volume fractions of Laves phases are summarized in Table 6.2. For samples deposited at 220 W, 320 W, and 410 W, the volume fraction of Laves phases gradually increases with the laser powers, with a noticeable increase towards the top region. The volume fraction of Laves phases in all samples exceeds 3%, and the Inconel 718 sample deposited at 410 W exhibits the highest volume fraction of Laves phases near the top region at 4.95%. However, in the GLP sample, the volume fraction decreases with the deposition height, reaching a minimum of 3.14% in the top region.

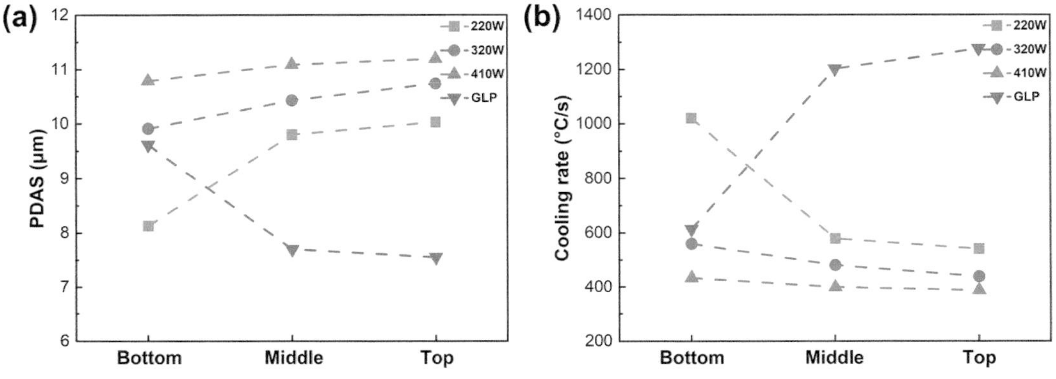

Figure 6.12. The relationship between the dendrite size and cooling rate: (a) measured average PDAS, and (b) calculated cooling rates for various laser powers in various regions.

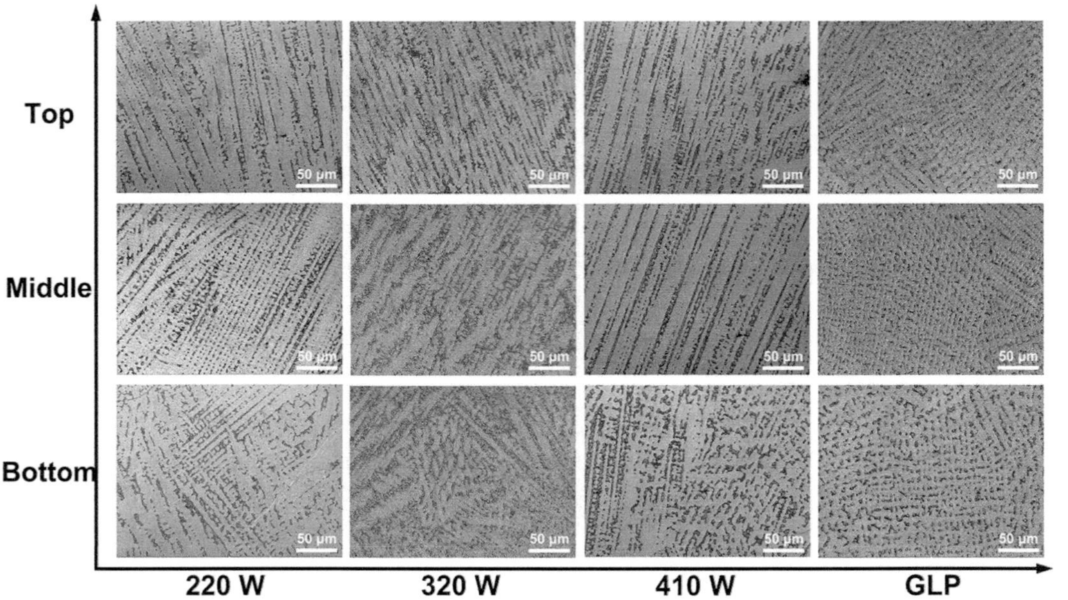

Figure 6.13. The SEM images of DZ obtained at various laser powers and deposition heights.

Table 6.2. The average volume fractions of Laves phases with various laser powers and deposition heights.

Laser power (W)	Bottom region (%)	Middle region (%)	Top region (%)
220	3.22	3.51	3.87
320	3.55	4.17	4.28
410	4.18	4.77	4.95
GLP	3.59	3.36	3.14

Both micro-segregation and the formation of harmful Laves phases in deposited Inconel 718 are influenced by the cooling rate, which is closely tied to thermal cycles. The morphology and size of Laves phase particles are rationalized based on the solidification rate (R) and the temperature gradient (G), considering the local thermal conditions. The cooling rate (dT/dt), represented by G×R, determines the size of the solidified structure, while the G/R ratio determines the morphology of the solidified structures. These parameters can be described as follows [33, 34]:

$$\frac{dT}{dt} = -2k\pi\left(\frac{V}{P}\right)\Delta T^2 \tag{6.2}$$

$$\frac{G}{R} = \frac{2k(T-T_0)^2}{\varepsilon PV\cos\theta} \tag{6.3}$$

In these equations, k represents thermal conductivity, V is the laser scanning speed, P is the laser power, ΔT is the temperature change during cooling, T is the liquid temperature of the alloy, T_0 is the initial temperature of the substrate, ε is the absorption coefficient of the laser, and θ is the angle between V and R.

It has been demonstrated that the fraction of Laves phases is inversely related to the cooling rate. Under low cooling rates and relatively high G/R ratios, continuous long-chain Laves phase particles are formed, while under high cooling rates, high G×R, and low G/R ratios, discrete and fine Laves phase particles are formed [20]. Low cooling rates, associated with high laser powers and increased deposition heights, not only lead to dendrite coarsening but also worsen the micro-segregation of Nb [35]. The cooling rates during DED have a significant impact on the inter-miscibility of Nb and Mo. Low cooling rates reduce the ability of the γ matrix to trap solute atoms and provide sufficient time for Nb, Mo, and Ti atoms to diffuse into the residual liquid phase between dendrites. The enrichment of these atomic groups increases nucleation sites for Laves phases. Increasing laser powers in the CLP method decreases G×R, resulting in the formation of coarse dendrites and numerous long-chain Laves phases.

In the GLP method, reducing the laser power layer by layer promotes uniform heat dissipation, alleviates heat accumulation, and accelerates the cooling process. The incremental G×R and reduced G/R ratio in the GLP method promote the formation of fine dendrites and mitigate the micro-segregation of Nb. Wang et al. [36] confirmed that long-chain Laves phases tend to form in inter-columnar dendrites, while discrete or block Laves phases are more likely to precipitate in equiaxed dendrites. Furthermore, the GLP method enhances lateral heat dissipation in the scanning direction, strengthens the horizontal temperature gradient, and suppresses the epitaxial growth of dendrites. This promotes interconnection among dendrites and the formation of misoriented grains. In the final stage of solidification, the remaining Nb-rich liquid is interrupted and divided by randomly grown dendrites, resulting in the formation of evenly dispersed and fine Laves phase particles.

6.4.3 Crystallographic orientation

The inverse pole figures of the γ matrix at the same height in the central region of as-deposited Inconel 718 samples with different laser powers are displayed in Figure 6.14. It can be observed that the sample deposited with 220 W laser power exhibits fine grains with distinct layer interfaces. On the other hand, the samples deposited with 320 W and 410 W laser powers predominantly consist of coarse and elongated columnar grains, many of which grow epitaxially across the deposited layer without being interrupted by the re-melting process, resulting in blurred layer interfaces. In the case of the GLP sample, fine equiaxed grains are present at the layer interface, along with irregular misoriented grains within the layer. The size and shape of the grains are evaluated using the mean fitted ellipse major diameter and grain aspect ratio (AR), as shown in Figure 6.14(e). An increase in the mean fitted ellipse major diameter from 49.66 μm to 77.76 μm is observed with the increment of laser powers for CLP deposition. The GLP sample exhibits a mixture of fine equiaxed grains and irregular misoriented grains, with a mean fitted ellipse major diameter of 57.10 μm. The AR, which represents the ratio of the long to the short diameters of a fitted ellipse, can indicate the extent of equiaxed grains. The presence of coarse columnar grains across the deposited layers in the 410 W sample results in a maximum AR of 2.51, while the GLP sample has the smallest AR of 2.01, similar to the substrate's AR (2.08).

The crystallographic texture is clearly revealed in the EBSD pole figures (Figure 6.15). As the laser power increases from 220 W to 410 W, the microstructure transitions from fine columnar grains with weak texture to coarse and elongated columnar grains with a strong <001> texture along the build direction, consistent with previous research [37–39]. The <001> direction becomes more pronounced with higher laser power, and the maximum polar density reaches 8.27 in the 410 W sample. This enhancement of the <001> direction is also confirmed by XRD analysis. Nevertheless, the GLP sample exhibits a weak texture, with a maximum polar density of 3.98. For GLP sample, the increased cooling rate due to the gradual laser power reduction enhances the stirring effect in the molten pool, and the remelted broken dendrites provide additional nucleation sites. The combination of high constitutional supercooling and heterogeneous nucleation promotes the transition from columnar to equiaxed growth. The microstructure tends to form dendrites with multiple growth directions, increasing the presence of equiaxed dendritic structures. For GLP process, besides the heat flow that occurs parallel to the deposition direction, there is an intensified heat flow in other

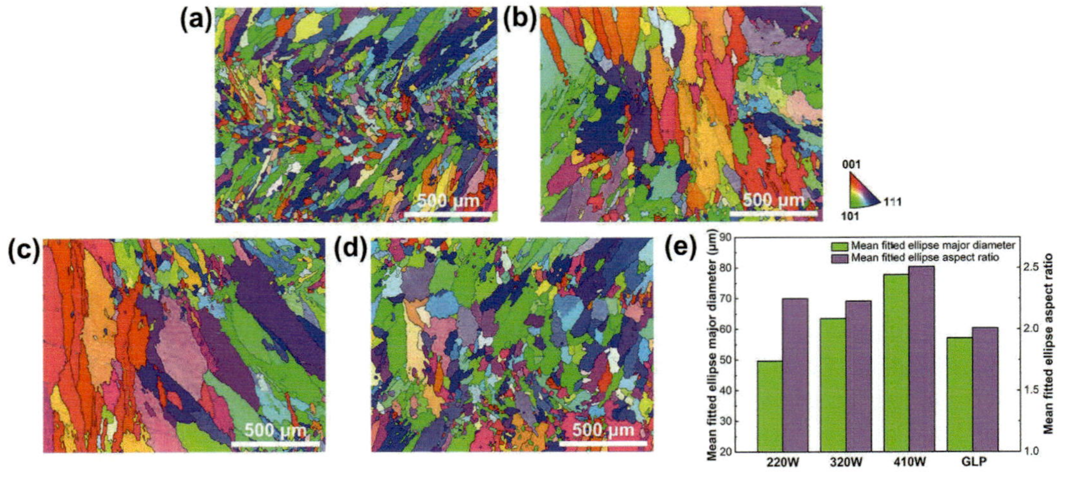

Figure 6.14. EBSD inverse pole figures for Inconel 718 samples deposited under (a) 220 W, (b) 320 W, (c) 410 W, (d) GLP, and (e) the results of mean fitted ellipse major diameter and mean fitted ellipse aspect ratio.

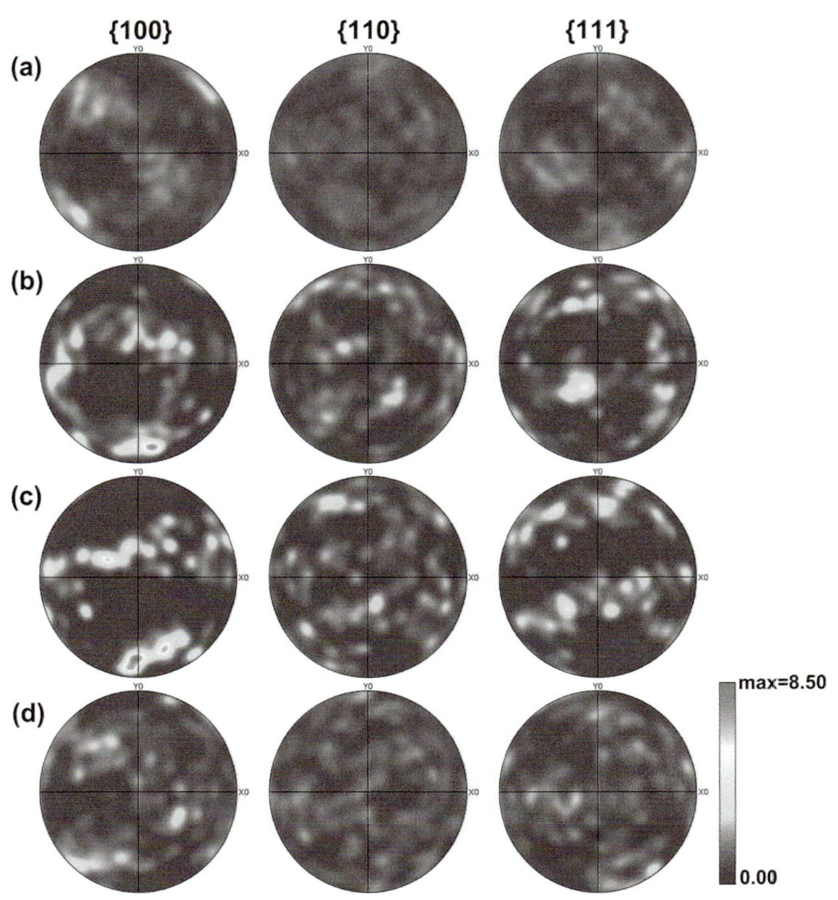

Figure 6.15. EBSD pole figures for Inconel 718 samples deposited under (a) 220 W, (b) 320 W, (c) 410 W, and (d) GLP.

directions. This is primarily caused by variations in laser power between layers. Consequently, the direction of grain growth deviates from the typical <001> orientation. GLP also exhibits reduced dilution rate and re-melting of each layer. The unique characteristics of GLP lead to the formation of columnar dendrites with multiple growth directions. This is attributed to the process of heterogeneous nucleation and subsequent growth, which is primarily based on the un-melted portions of the previous layer. As a result, the relationship between the grain growth direction and the deposition direction for GLP is not clearly defined. Therefore, the GLP sample contains more fine equiaxed grains and exhibits no obvious texture.

6.5 Mechanical property improvement

6.5.1 Residual stress

During laser additive manufacturing, thermal strain and residual stress often occur in the as-deposited samples due to high thermal gradients and repeated local heat transfer. The presence of severe residual tensile stress negatively affects the performance of the repaired samples. Residual stresses in crystalline metals or alloys can be categorized into three types based on different analyzed scales: Type I stress on the macro-scale (macro-stress), Type II stress on the micro-scale (micro-stress) at the grain level, and Type III stress on the nano-scale [40]. To evaluate the level of residual stress, we analyzed and discussed the results from both macro and micro scales.

The Grain Orientation Spread (GOS) obtained from EBSD analysis reflects the magnitude of plastic strains and residual stress. GOS indicates the distribution of micro-strain (micro-stress) at the grain-size level, which corresponds to Type II stress distribution. The mean GOS values are calculated based on the degree of orientation changes between each pixel within the grain and the mean grain orientation. The mean GOS value of a grain i is defined as follows [41]:

$$GOS(i) = \frac{1}{J(i)} \sum\nolimits_{j} \omega_{ij} \tag{6.4}$$

where $J(i)$ represents the numbers of pixel in grain i, and ω_{ij} is the misorientation angle between the orientation of pixel j and the mean orientation of grain i.

Based on the GOS distribution shown in Figure 6.16, the sample deposited with high laser power exhibits a higher mean GOS value, while the sample from the GLP method has the lowest mean GOS value. The GOS map highlights grain-scale differences in geometrically necessary dislocations, residual plastic strain, and the degree of grain distortion. Due to significant thermal expansion and contraction caused by localized heating and cooling, the sample deposited with 410 W laser power shows a high mean GOS value, indicating an uneven distribution of plastic strains and residual stresses. The increase in texture contributes to higher mean GOS values in samples deposited with high laser power [42]. The GLP method, with its continuously reducing laser powers, alleviates micro-scale plastic strain and residual stress in the deposited parts by ensuring uniform heat dissipation, resulting in the lowest GOS value of 1.24°.

Type I residual stress, which is on the macro-level, can be measured using XRD. Figure 6.17 displays the residual stress results of Inconel 718 samples deposited with various laser powers in the longitudinal (Y) and normal (Z) directions. For samples deposited using the CLP method, as the laser power increases, the longitudinal stress (σ_y) undergoes a transition from tensile to compressive, while the normal stress (σ_z) follows the opposite trend. Interestingly, the GLP sample exhibits compressive residual stresses in both directions, particularly with σ_y reaching up to –250 MPa. It is well-known that rapid heating and cooling in DED processes induce residual stresses. These stresses primarily arise due to the non-uniform deformation of materials during thermal cycles. The tensile residual stresses observed in most CLP deposited samples align with the general situation of residual stresses in the DED process. During DED, the previously deposited layer constrains the free expansion of the newly deposited layer, leading to compressive plastic deformation of the

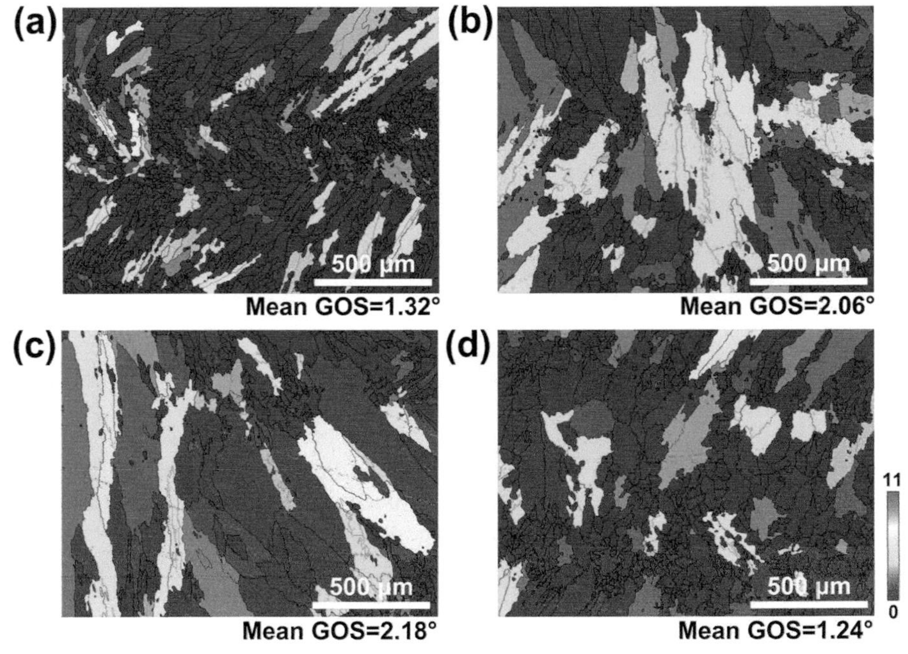

Figure 6.16. GOS maps of the as-deposited Inconel 718 samples with different laser powers: (a) 220 W, (b) 320 W, (c) 410 W, and (d) GLP.

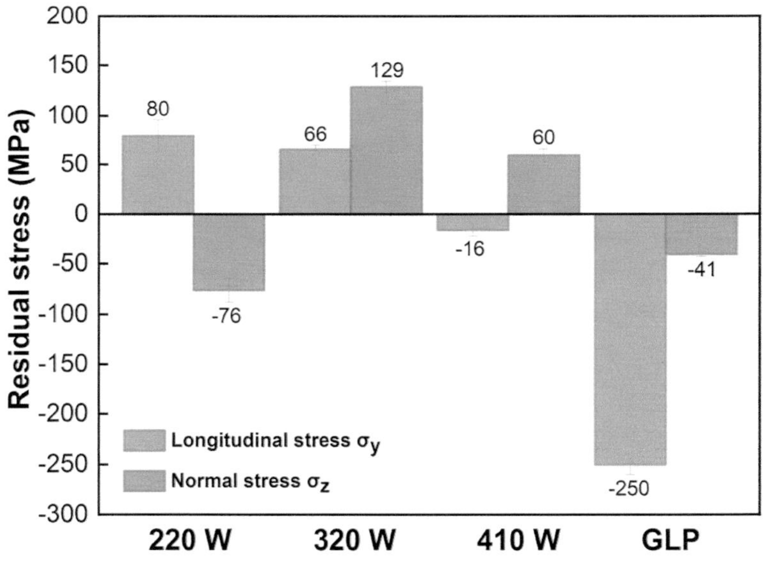

Figure 6.17. Normal and longitudinal residual stress results of the as-deposited Inconel 718 samples with various laser powers.

newly deposited layer at high temperatures. Upon subsequent solidification and cooling, the newly deposited layer undergoes contraction, which is restricted by the pre-deposited layer and results in tensile residual stress within the newly deposited layer [43].

The sign of residual stresses is primarily determined by the thermal expansion coefficients and the effects of thermal contraction. Meanwhile, the magnitude of residual stresses is mainly influenced by plastification effects, thermal gradients, and cooling rates [44, 45]. Additionally, the distribution of temperature fields plays a crucial role in the cyclic deformation behavior and relaxation of residual stresses. In the case of GLP samples, the gradual reduction of laser power layer by layer leads to a decrease in the size of the molten pool. This reduction weakens the shrinkage effect and promotes the formation of compressive stresses. Furthermore, the GLP method enhances cooling rates, creates a more uniform heat distribution field, and facilitates heat dissipation in the longitudinal direction. As a result, the GLP process contributes to a significant increase in compressive stress (σ_y). Studies have shown that the presence of compressive residual stress can improve the fatigue resistance of materials and extend their service life [46]. Therefore, it is reasonable to expect that the presence of compressive stresses in GLP-deposited samples would be advantageous for enhancing their mechanical properties.

6.5.2 Microhardness

Figure 6.18 illustrates the distribution of average hardness along the longitudinal cross-section, specifically focusing on the bottom, middle, and top regions. As the laser power is increased, the hardness initially rises and then declines. Moreover, as the distance from the substrate increases, there is a noticeable decrease in hardness across all samples. The presence of coarse microstructures with lower hardness in the top region can be attributed to the lower cooling rates associated with greater deposition height, in accordance with the Hall-Petch relation [47]. The higher hardness observed at the bottom of the deposited zone can be attributed to the enhanced precipitation hardening resulting from repeated thermal cycles. In the bottom region, the partial dissolution of Laves phases releases free Nb atoms that diffuse into the matrix, accelerating γ" precipitation in the inter-dendritic region. Conversely, the GLP sample demonstrates a different behavior compared to the previously mentioned CLP samples. The hardness of the GLP sample remains high and exhibits uniformity, although there is a slight reduction with increasing deposition height. The high cooling rate associated with the GLP process suppresses Nb micro-segregation and increases the amount of Nb in the γ matrix. The incorporation of Nb within the γ matrix induces a distorted elastic stress field due to the atomic size difference and the short-range internal stress resulting from the difference in elasticity modulus between the γ matrix and solute atoms [15]. This enhanced solid solution strengthening contributes to the overall increase in microhardness. Additionally, the uniformly dispersed discrete Laves phases act as dispersion strengthening agents, further improving the hardness to some extent.

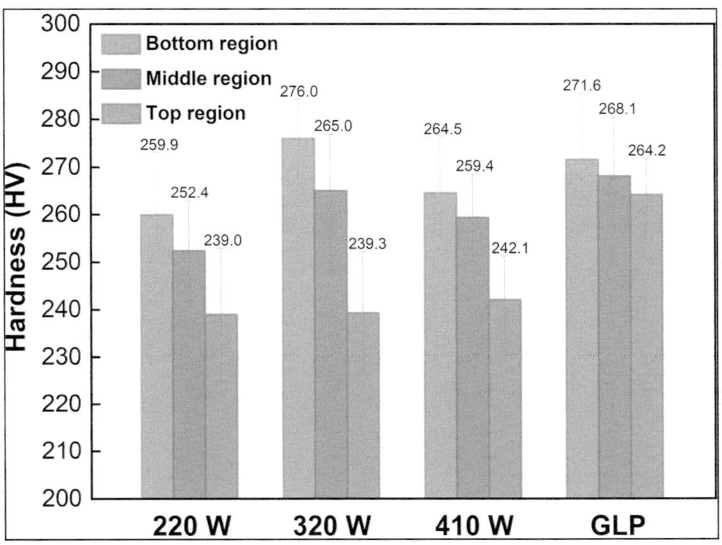

Figure 6.18. Average hardness of DZs from the bottom to the top region.

6.5.3 Tensile properties

Tensile specimens comprising a 50% deposition and 50% substrate were meticulously fabricated to assess both the repair effectiveness and the strength of the repaired interfaces. Notably, the fracture occurrence in all specimens took place within the deposited zone, indicating the structural integrity of the deposition-substrate combination and the robust strength of the bonding zone. The representative tensile curves of different samples are illustrated in Figure 6.19(a), while a comprehensive summary of the yield strength (YS), ultimate strength (UTS), and elongation is presented in Figure 6.19(b). Observing the 220 W, 320 W, and 410 W deposited samples, it can be found that the YS initially increases and subsequently decreases, with no significant discrepancy in UTS. Notably, the YS and UTS of the 320 W deposited sample exhibit a slightly higher value. Comparatively, the GLP samples demonstrate similar YS and UTS values to those of the conventionally CLP samples. However, a notable disparity arises in terms of elongation when comparing samples deposited under different parameters. As the laser power increases, the elongation displays a gradual increase after a marginal decline. In the case of the CLP method, the 320 W deposited samples yield the lowest elongation, averaging at 18.50%. Conversely, the GLP sample showcases remarkably high elongation, with an average value of 30.09% and a maximum elongation of 35.79%.

The variation in ductility can be primarily attributed to two factors: Laves phases and texture. It is reported that the ductility of the material is closely associated with the relative strength of the inter-dendritic phases and the matrix phases [48]. The Laves phase is characterized as brittle and hard, exhibiting poor plastic deformation capability. The interface between Laves phase and the γ matrix is weak due to the incoherent relationship. Within the inter-dendritic zone, numerous Laves phases precipitate, acting as obstacles that hinder the movement and multiplication of dislocations, thereby causing stress concentration. In comparison to the CLP sample, which contains a large amount of long-chain Laves phase, the GLP sample with a smaller amount of discrete Laves phase demonstrates better ductility.

Apart from the Laves phases, the formation of MC particles along the grain boundaries can also serve as strengthening phases by impeding the migration of grain boundaries. This phenomenon reduces ductility but may improve creep resistance [36]. However, considering the small number of MC particles and their similar distribution in both the CLP and GLP deposited samples, MC particles are not the main factor contributing to significant differences in the ductility of these samples.

Moreover, the tensile properties of the material are influenced by the angle between the tensile axis and the texture direction. When samples are deposited using a higher laser power (410 W), they exhibit improved ductility but lower strength. This can be attributed to the presence of a strong <001> texture and a coarse columnar grain morphology. The columnar grain boundaries provide a pathway for preferential accumulation of damage, leading to failure when a tensile load

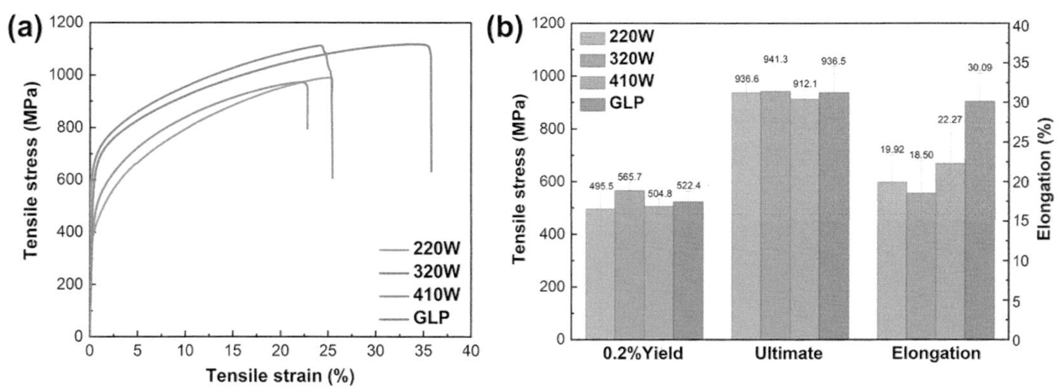

Figure 6.19. Results of tensile tests at room temperatures: (a) tensile stress-strain curves, and (b) the statistic results.

is applied perpendicular to these boundaries. Conversely, when a tensile load is applied parallel to the columnar grain with the strong <001> texture, failure is delayed due to fewer short axes of the grains experiencing dislocation pile-up [49]. For GLP sample, a favorable balance between strength and ductility is observed. This is attributed to a combination of relatively fine columnar grains and equiaxed grains. The morphology of the grains prepared using the GLP method tends to be equiaxed, with a low aspect ratio. This characteristic facilitates the adjustment to applied loads through appropriate movement and deformation, resulting in enhanced mechanical properties with high ductility [50].

6.5.4 Fracture morphology

According to the microstructural analysis above, the CLP method primarily produces Laves phases in the form of long-chain configurations. In contrast, the GLP sample exhibits discrete, fine, and uniformly dispersed Laves phases. When a sample is exposed to tensile load, the presence of Laves phases makes it susceptible to the initiation of micro-voids and microcracks. These initial defects gradually propagate and interconnect with each other, ultimately leading to the final failure of the material.

Figure 6.20 depicts the fracture morphologies of these two types of Laves phases near the fracture zone, specifically at the Y-Z plane. The yellow arrows indicate the fragmentation of Laves phases, while the red arrows represent the debonding of Laves phases from the γ matrix. The long-chain Laves phases are observed to break into smaller parts along a specific direction, leaving behind micro-voids as they separate from the γ matrix. On the other hand, the small granular Laves phases show minimal fractures and can be directly separated from the matrix without fragmentation (Figure 6.20(d)). When magnified (Figure 6.20(e) and (f)), slip bands can be observed surrounding the micro-voids caused by the debonding of small Laves phases and MC carbides. This confirms the high plastic deformation ability of granular Laves phases.

Based on the fractural characteristics of two types of Laves phases observed in Figure 6.20, the long chain Laves phase and the granular Laves phase exhibit distinct failure mechanisms. The long-chain Laves phase typically experiences internal fractures, whereas the granular Laves phase tends to undergo interfacial decohesion. These findings align with the conclusions drawn by Sui

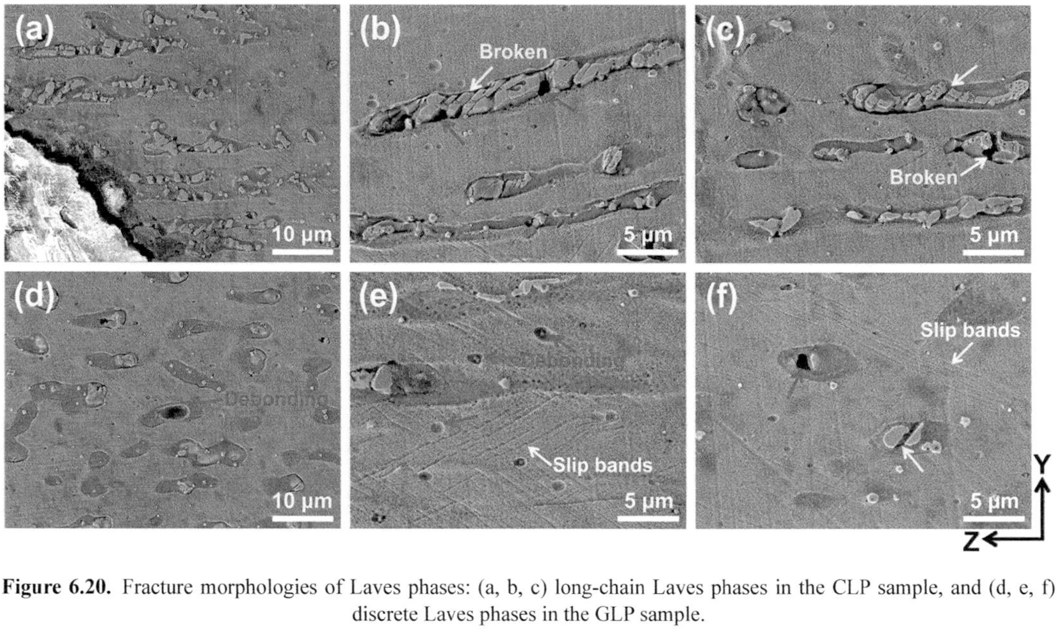

Figure 6.20. Fracture morphologies of Laves phases: (a, b, c) long-chain Laves phases in the CLP sample, and (d, e, f) discrete Laves phases in the GLP sample.

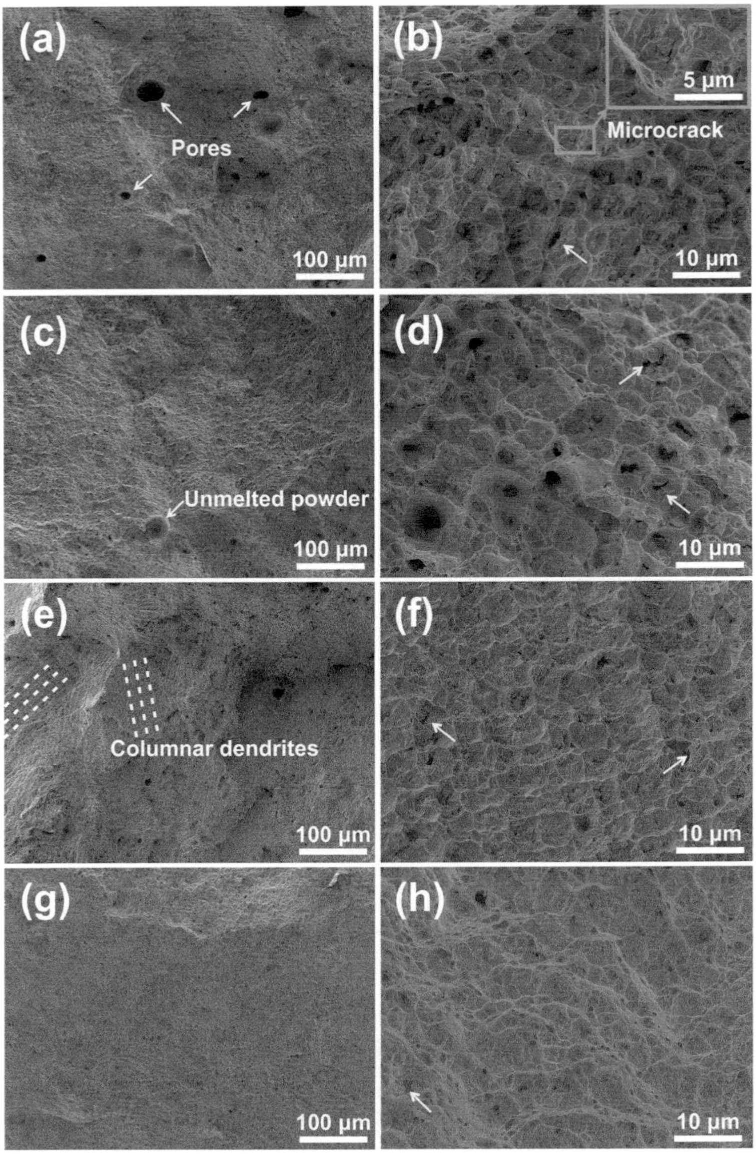

Figure 6.21. Fracture morphologies of Inconel 718 samples deposited under various laser powers: (a, b) 220 W, (c, d) 320 W, (e, f) 410 W, and (g, h) GLP.

et al. [20, 51]. The large size of the long-chain Laves phases hinders their movement within the γ matrix during plastic deformation. Consequently, these phases are prone to slipping and breaking into smaller fragments, which facilitates deformation with the γ matrix. However, this process carries a high risk of crack propagation [52]. The fragmentation of long-chain Laves phases and their separation from the γ matrix contribute to the presence of numerous fractured Laves phases and cavities within the dimples. In contrast, fine, discrete, and granular Laves phases have a positive impact on enhancing ductility. They possess a strong affinity for the γ matrix and can maintain their original morphologies or debond from the γ matrix without fracturing. The critical stress required for fracture in the discrete Laves phase surpasses that of its long-chain counterpart.

Figure 6.21 illustrates the typical fracture surface morphologies of the Inconel 718 samples. In all samples, a distinct dendritic morphology is evident, accompanied by characteristic dimpled surfaces indicative of ductile fracture. In the sample deposited using low laser power (Figure 6.21(a)), numerous pores are observed due to insufficient fusion. Additionally, Figure 6.21(c) reveals the presence of un-melted powder particles, which are well-known inherent defects in the DED process. These defects can act as stress concentrators, contributing to the occurrence of final failures.

The presence of residual fractured Laves phases can be identified by yellow arrows in Figure 6.21(b), (d), (f), and (h), where they appear at dimple nucleation points. Inside the dimples (Figure 6.21(b)), microcracks are observed as a result of the fragmentation of Laves phases. In the 320 W sample, deep cavities caused by the separation of large-sized Laves phases from the γ matrix during the tensile process are depicted by the red arrows. These cavities are particularly prominent in the 320 W sample. As the laser power increases, the size of the dimples gradually increases. In the 410 W sample, the dimples are arranged in a regular pattern along columnar dendrites, indicating excellent ductility at high laser powers.

Conversely, for the GLP sample, the fracture surface exhibits distinct columnar dendrites with random directions, and no obvious defects such as holes are observed. The fracture surface of the GLP sample reveals an abundance of fine dimples, indicating micro-void coalescence during the ductile fracture process. Among the dimples, residual Laves phases within the γ matrix and cavities resulting from the debonding of granular Laves phases and MC carbides from the γ matrix can be found. It is worth noting that these cavities are smaller in size and more dispersed compared to those caused by long-chain Laves phases.

6.6 Summary

The as-deposited microstructures of Inconel 718 superalloys possess several remarkable microstructural features, encompassing micro-segregation, inter-dendritic Laves phases, restrained nano-strengthening precipitates, and highly oriented columnar structures. These factors collectively underscore the critical role of mitigating micro-segregation and controlling the formation of Laves phases in order to enhance the overall mechanical performance and reliability of laser-deposited Inconel 718 alloys. Based on the deep investigation of microstructural characteristics, a novel GLP method is proposed to optimize the Laves phases and enhance the mechanical properties of as-repaired Inconel 718 thin-wall parts during the DED process.

The traditional CLP method produces interconnected long-chain Laves phases among coarse dendrites. As the laser power increases, the size and aspect ratio of the columnar grains increase, and the <001> texture is strengthened. However, the GLP method employs decreasing laser power layer by layer, which alleviates heat accumulation, increases cooling rates and promotes horizontal heat dissipation. This results in the formation of fine columnar dendrites with disordered growth directions. The GLP samples exhibit more equiaxed grains and no obvious texture. The micro-segregation is reduced, and the fine Laves phase particles are discrete and uniformly distributed in the γ matrix.

The comprehensive mechanical properties of the GLP samples are superior. In comparison to the CLP samples, the GLP-deposited samples exhibit high and uniform hardness. The GLP method also enables the attainment of excellent ductility while maintaining good strength. The presence of compressive residual stresses in the longitudinal and normal directions of GLP samples is expected to improve their overall service performance. The failure modes of the long-chain Laves phases primarily involve fragmentation and debonding, whereas the granular Laves phases demonstrate strong coordination and can be retained or separated from the γ matrix during the tensile process. The granular Laves phases exhibit a higher ability for plastic deformation compared to the long-chain counterparts.

The GLP method demonstrates the feasibility of optimizing the Laves phases and enhancing the repair performance of Inconel 718 components by reducing the laser powers. This makes it possible to adjust the processing parameters layer by layer or even track by track during the deposition process. In the future, there is a need for further exploration and development of in-depth theories and principles regarding the flexible *in-situ* regulation of processing parameters. Ultimately, the goal is to establish a comprehensive framework that elucidates the intricate relationship between processing parameters, thermal cycles, microstructure, and mechanical properties. This framework will provide a solid foundation for the development of advanced laser deposition techniques that offer unprecedented flexibility in adjusting microstructure and material properties of additively manufactured components.

References

[1] Liu, R., Wang, Z., Sparks, T., Liou, F. and Newkirk, J. (2017). Aerospace applications of laser additive manufacturing. pp. 351–371. *In*: Laser Additive Manufacturing. Elsevier.

[2] Pinkerton, A. J., Wang, W. and Li, L. (2008). Component repair using laser direct metal deposition. Proc. Inst. Mech. Eng. B J. Eng. Manuf. 222: 827–836.

[3] Stevens, E. L., Toman, J., To, A. C. and Chmielus, M. (2017). Variation of hardness, microstructure, and Laves phase distribution in direct laser deposited alloy 718 cuboids. Mater Des. 119: 188–198. https://doi.org/10.1016/j.matdes.2017.01.031.

[4] Frappier, R., Benoit, A., Paillard, P., Baudin, T., Le Gall, R. and Dupuy, T. (2014). Quantitative infrared analysis of welding processes: temperature measurement during RSW and CMT-MIG welding. Science & Technology of Welding & Joining 19: 38–43.

[5] Farshidianfar, M. H., Khajepour, A. and Gerlich, A. P. (2016). Effect of real-time cooling rate on microstructure in laser additive manufacturing. J. Mater Process Technol. 231: 468–478. https://doi.org/10.1016/j.jmatprotec.2016.01.017.

[6] Muvvala, G., Patra Karmakar, D. and Nath, A. K. (2017). Online monitoring of thermo-cycles and its correlation with microstructure in laser cladding of nickel based super alloy. Opt Lasers Eng. 88: 139–152. https://doi.org/10.1016/j.optlaseng.2016.08.005.

[7] Yang, D., Wang, G. and Zhang, G. (2017). Thermal analysis for single-pass multi-layer GMAW based additive manufacturing using infrared thermography. J. Mater Process Technol. 244: 215–224. https://doi.org/10.1016/j.jmatprotec.2017.01.024.

[8] Bennett, J. L., Kafka, O. L., Liao, H., Wolff, S. J., Yu, C., Cheng, P. et al. (2018). Cooling rate effect on tensile strength of laser deposited Inconel 718. Procedia Manuf. 26: 912–919. https://doi.org/10.1016/j.promfg.2018.07.118.

[9] Rodrigues, T. A., Duarte, V., Avila, J. A., Santos, T. G., Miranda, R. M. and Oliveira, J. P. (2019). Wire and arc additive manufacturing of HSLA steel: Effect of thermal cycles on microstructure and mechanical properties. Addit. Manuf. 27: 440–450. https://doi.org/10.1016/j.addma.2019.03.029.

[10] Alhuzaim, A., Imbrogno, S. and Attallah, M. M. (2021). Controlling microstructural and mechanical properties of direct laser deposited Inconel 718 via laser power. J. Alloys Compd. 872: 159588. https://doi.org/10.1016/j.jallcom.2021.159588.

[11] Jones, J., Whittaker, M., Buckingham, R., Johnston, R., Bache, M. and Clark, D. (2017). Microstructural characterisation of a nickel alloy processed via blown powder direct laser deposition (DLD). Mater Des. 117: 47–57. https://doi.org/10.1016/j.matdes.2016.12.062.

[12] Tian, Y., McAllister, D., Colijn, H., Mills, M., Farson, D., Nordin, M. et al. (2014). Rationalization of microstructure heterogeneity in INCONEL 718 builds made by the direct laser additive manufacturing process. Metall Mater Trans. A Phys. Metall. Mater. Sci. 45: 4470–4483. https://doi.org/10.1007/s11661-014-2370-6.

[13] Mantri, S. A., Dasari, S., Sharma, A., Alam, T., Pantawane, M. V., Pole, M. et al. (2021). Effect of micro-segregation of alloying elements on the precipitation behaviour in laser surface engineered Alloy 718. Acta Mater 210. https://doi.org/10.1016/j.actamat.2021.116844.

[14] Kumara, C., Segerstark, A., Hanning, F., Dixit, N., Joshi, S., Moverare, J. et al. (2019). Microstructure modelling of laser metal powder directed energy deposition of alloy 718. Addit. Manuf. 25: 357–364. https://doi.org/10.1016/j.addma.2018.11.024.

[15] Zhang, Y., Li, Z., Nie, P. and Wu, Y. (2013). Effect of cooling rate on the microstructure of laser-remelted INCONEL 718 coating. Metall. Mater Trans. A Phys. Metall. Mater. Sci. 44: 5513–5521. https://doi.org/10.1007/s11661-013-1903-8.

[16] DuPont, J. N., Robino, C. V. and Marder, A. R. (1998). Solidification of Nb-bearing superalloys: part II. Pseudoternary solidification surfaces. Metallurgical & Materials Transactions Part A 29A: 2797–2806.

[17] Popovich, V. A., Borisov, E. V., Popovich, A. A., Sufiiarov, V. S., Masaylo, D. V. and Alzina, L. (2017). Functionally graded Inconel 718 processed by additive manufacturing: Crystallographic texture, anisotropy of microstructure and mechanical properties. Mater Des. 114: 441–449. https://doi.org/10.1016/j.matdes.2016.10.075.

[18] Sun, S.-H., Koizumi, Y., Saito, T., Yamanaka, K., Li, Y.-P., Cui, Y. et al. (2018). Electron beam additive manufacturing of Inconel 718 alloy rods: Impact of build direction on microstructure and high-temperature tensile properties. Addit. Manuf. 23: 457–470.

[19] Chen, Y., Guo, Y., Xu, M., Ma, C., Zhang, Q., Wang, L. et al. (2019). Study on the element segregation and Laves phase formation in the laser metal deposited IN718 superalloy by flat top laser and gaussian distribution laser. Materials Science and Engineering A 754: 339–347. https://doi.org/10.1016/j.msea.2019.03.096.
[20] Sui, S., Chen, J., Ma, L., Fan, W., Tan, H., Liu, F. et al. (2019). Microstructures and stress rupture properties of pulse laser repaired Inconel 718 superalloy after different heat treatments. J. Alloys Compd. 770: 125–135. https://doi.org/10.1016/j.jallcom.2018.08.063.
[21] Tucho, W. M., Cuvillier, P., Sjolyst-Kverneland, A. and Hansen, V. (2017). Microstructure and hardness studies of Inconel 718 manufactured by selective laser melting before and after solution heat treatment. Materials Science and Engineering: A 689: 220–232.
[22] Holland, S., Wang, X., Chen, J., Cai, W., Yan, F. and Li, L. (2019). Multiscale characterization of microstructures and mechanical properties of Inconel 718 fabricated by selective laser melting. J. Alloys Compd. 784: 182–194. https://doi.org/10.1016/j.jallcom.2018.12.380.
[23] Yang, H., Jing, G., Gao, P., Wang, Z. and Li, X. (2020). Effects of circular beam oscillation technique on formability and solidification behaviour of selective laser melted Inconel 718: From single tracks to cuboid samples. J. Mater Sci. Technol. 51: 137–150. https://doi.org/10.1016/j.jmst.2019.09.044.
[24] Wang, Y., Shi, J. and Liu, Y. (2019). Competitive grain growth and dendrite morphology evolution in selective laser melting of Inconel 718 superalloy. J. Cryst Growth 521: 15–29. https://doi.org/10.1016/j.jcrysgro.2019.05.027.
[25] Ferreri, N. C., Vogel, S. C. and Knezevic, M. (2020). Determining volume fractions of γ, γ′, γ″, δ, and MC-carbide phases in Inconel 718 as a function of its processing history using an advanced neutron diffraction procedure. Materials Science and Engineering A 781: 139228. https://doi.org/10.1016/j.msea.2020.139228.
[26] Yang, H., Meng, L., Luo, S. and Wang, Z. (2020). Microstructural evolution and mechanical performances of selective laser melting Inconel 718 from low to high laser power. J. Alloys Compd. 828: 154473. https://doi.org/10.1016/j.jallcom.2020.154473.
[27] Yao, X. X., Ge, P., Li, J. Y., Wang, Y. F., Li, T., Liu, W. W. et al. (2020). Controlling the solidification process parameters of direct energy deposition additive manufacturing considering laser and powder properties. Comput Mater Sci. 182. https://doi.org/10.1016/j.commatsci.2020.109788.
[28] Ramakrishnan, A. and Dinda, G. P. (2019). Direct laser metal deposition of Inconel 738. Materials Science and Engineering A 740–741: 1–13. https://doi.org/10.1016/j.msea.2018.10.020.
[29] Kürnsteiner, P., Wilms, M. B., Weisheit, A., Barriobero-Vila, P., Gault, B., Jägle, E. A. et al. (2017). In-process precipitation during laser additive manufacturing investigated by atom probe tomography. Microscopy and Microanalysis 23: 694–695. https://doi.org/10.1017/S1431927617004135.
[30] Manikandan, S. G. K., Sivakumar, D., Rao, K. P. and Kamaraj, M. (2014). Effect of weld cooling rate on Laves phase formation in Inconel 718 fusion zone. J. Mater Process Technol. 214: 358–364. https://doi.org/10.1016/j.jmatprotec.2013.09.006.
[31] Segerstark, A., Andersson, J., Svensson, L. E. and Ojo, O. (2018). Microstructural characterization of laser metal powder deposited Alloy 718. Mater Charact. 142: 550–559. https://doi.org/10.1016/j.matchar.2018.06.020.
[32] Wang, X. and Chou, K. (2017). Effects of thermal cycles on the microstructure evolution of Inconel 718 during selective laser melting process. Addit. Manuf. 18: 1–14. https://doi.org/10.1016/j.addma.2017.08.016.
[33] Song, H., Lei, J., Xie, J., Wu, S., Wang, L. and Shou, W. (2019). Laser melting deposition of K403 superalloy: The influence of processing parameters on the microstructure and wear performance. J. Alloys Compd. 805: 551–564. https://doi.org/10.1016/j.jallcom.2019.07.102.
[34] Shao, J., Yu, G., He, X., Li, S., Chen, R. and Zhao, Y. (2019). Grain size evolution under different cooling rate in laser additive manufacturing of superalloy. Opt. Laser Technol. 119: 105662. https://doi.org/10.1016/j.optlastec.2019.105662.
[35] Yuan, K., Guo, W., Li, P., Wang, J., Su, Y., Lin, X. et al. (2018). Influence of process parameters and heat treatments on the microstructures and dynamic mechanical behaviors of Inconel 718 superalloy manufactured by laser metal deposition. Materials Science and Engineering A 721: 215–225. https://doi.org/10.1016/j.msea.2018.02.014.
[36] Wang, K., Liu, Y., Sun, Z., Lin, J., Lv, Y. and Xu, B. (2020). Microstructural evolution and mechanical properties of Inconel 718 superalloy thin wall fabricated by pulsed plasma arc additive manufacturing. J. Alloys Compd. 819: 152936. https://doi.org/10.1016/j.jallcom.2019.152936.
[37] Parimi, L. L., Ravi, G., Clark, D. and Attallah, M. M. (2014). Microstructural and texture development in direct laser fabricated IN718. Mater Charact. 89: 102–111. https://doi.org/10.1016/j.matchar.2013.12.012.
[38] Liu, Z., Kim, H., Liu, W., Cong, W., Jiang, Q. and Zhang, H. (2019). Influence of energy density on macro/micro structures and mechanical properties of as-deposited Inconel 718 parts fabricated by laser engineered net shaping. J. Manuf. Process 42: 96–105. https://doi.org/10.1016/j.jmapro.2019.04.020.
[39] Liu, S. Y., Li, H. Q., Qin, C. X., Zong, R. and Fang, X. Y. (2020). The effect of energy density on texture and mechanical anisotropy in selective laser melted Inconel 718. Mater Des. 191. https://doi.org/10.1016/j.matdes.2020.108642.
[40] Chen, H., He, Z. and Lu, L. (2020). Correlation of surface features with corrosion behaviors of interstitial free steel processed by temper rolling. J. Mater Sci. Technol. 36: 37–44. https://doi.org/10.1016/j.jmst.2019.06.011.

[41] Zhao, Y., Li, K., Gargani, M. and Xiong, W. (2020). A comparative analysis of Inconel 718 made by additive manufacturing and suction casting: Microstructure evolution in homogenization. Addit. Manuf. 36: 101404. https://doi.org/10.1016/j.addma.2020.101404.
[42] Nadammal, N., Cabeza, S., Mishurova, T., Thiede, T., Kromm, A., Seyfert, C. et al. (2017). Effect of hatch length on the development of microstructure, texture and residual stresses in selective laser melted superalloy Inconel 718. Mater Des. 134: 139–150. https://doi.org/10.1016/j.matdes.2017.08.049.
[43] Liu, X., Huang, L., Wang, Y. and Li, J. (2020). Effect of forged substrate geometry on temperature and stress field in additive manufacturing. J. Manuf. Process 52: 79–95. https://doi.org/10.1016/j.jmapro.2020.01.054.
[44] Sun, G., Zhou, R., Lu, J. and Mazumder, J. (2015). Evaluation of defect density, microstructure, residual stress, elastic modulus, hardness and strength of laser-deposited AISI 4340 steel. Acta Mater 84: 172–189. https://doi.org/10.1016/j.actamat.2014.09.028.
[45] Balbaa, M., Mekhiel, S., Elbestawi, M. and McIsaac, J. (2020). On selective laser melting of Inconel 718: Densification, surface roughness, and residual stresses. Mater Des. 193: 108818. https://doi.org/10.1016/j.matdes.2020.108818.
[46] Lesyk, D. A., Martinez, S., Mordyuk, B. N., Dzhemelinskyi, V. V., Lamikiz, A. and Prokopenko, G. I. (2020). Post-processing of the Inconel 718 alloy parts fabricated by selective laser melting: Effects of mechanical surface treatments on surface topography, porosity, hardness and residual stress. Surf Coat Technol. 381: 125136. https://doi.org/10.1016/j.surfcoat.2019.125136.
[47] Yang, H., Yang, J., Huang, W., Wang, Z. and Zeng, X. (2018). The printability, microstructure, crystallographic features and microhardness of selective laser melted Inconel 718 thin wall. Mater Des. 156: 407–418. https://doi.org/10.1016/j.matdes.2018.07.007.
[48] Qi, H., Azer, M. and Ritter, A. (2009). Studies of standard heat treatment effects on microstructure and mechanical properties of laser net shape manufactured INCONEL 718. Metall Mater Trans A Phys. Metall. Mater Sci. 40: 2410–2422. https://doi.org/10.1007/s11661-009-9949-3.
[49] Ni, M., Chen, C., Wang, X., Wang, P., Li, R., Zhang, X. et al. (2017). Anisotropic tensile behavior of in situ precipitation strengthened Inconel 718 fabricated by additive manufacturing. Materials Science and Engineering A 701: 344–351. https://doi.org/10.1016/j.msea.2017.06.098.
[50] Gu, D., Shi, Q., Lin, K. and Xi, L. (2018). Microstructure and performance evolution and underlying thermal mechanisms of Ni-based parts fabricated by selective laser melting. Addit. Manuf. 22: 265–278. https://doi.org/10.1016/j.addma.2018.05.019.
[51] Sui, S., Tan, H., Chen, J., Zhong, C., Li, Z., Fan, W. et al. (2019). The influence of Laves phases on the room temperature tensile properties of Inconel 718 fabricated by powder feeding laser additive manufacturing. Acta Mater 164: 413–427. https://doi.org/10.1016/j.actamat.2018.10.032.
[52] Sui, S., Chen, J., Zhang, R., Ming, X., Liu, F. and Lin, X. (2017). The tensile deformation behavior of laser repaired Inconel 718 with a non-uniform microstructure. Materials Science and Engineering A 688: 480–487. https://doi.org/10.1016/j.msea.2017.01.110.

Chapter 7

Achieving Superior Mechanical Property of Inconel 718 Thin-Wall using Gradient-Laser-Power Deposition

7.1 Introduction

The maintenance, repair, and overhaul (MRO) industry in the aviation sector plays a critical role in resource conservation and enhancing the sustainability of manufacturing processes. Laser powder deposition (LPD), a form of laser-based additive manufacturing, has emerged as a highly promising and practical method for rebuilding damaged aviation components like turbine blades. LPD enables the rapid fabrication of near-net-shape products and offers flexible repair capabilities through coaxial powder feeding and layer-by-layer material deposition.

Inconel 718, a precipitation-hardening superalloy based on nickel, has extensive application in the production of aerospace turbine blades, nuclear reactors, rocket motors, and spacecraft. This alloy is chosen for its excellent weldability, high-temperature strength, and exceptional resistance to creep, corrosion, and oxidation at elevated temperatures. However, during LPD, a characteristic microscopic feature of as-deposited Inconel 718 samples is the formation of brittle and hard phases called the Laves phases. This phenomenon occurs due to pronounced micro-segregation of niobium (Nb) in the inter-dendritic regions. The presence of the Laves phase is undesirable as it depletes the strengthening elements and promotes the initiation and propagation of cracks [1]. It is widely acknowledged that the formation of Laves phase, particularly in the form of coarse and continuous long-chain structures, leads to premature failure and has a detrimental effect on mechanical properties such as tensile ductility, tensile strength, stress rupture life, and fatigue life [2].

In as-deposited Inconel 718, the occurrence of micro-segregation and the formation of Laves phase are inevitable due to non-equilibrium solidification. However, it is possible to reduce the amount and size, as well as modify the morphology of the Laves phase in order to mitigate its negative impact and potentially enhance the mechanical properties. Previous studies have shown that as-deposited Inconel 718, which contains discrete and fine Laves phase particles, exhibits superior mechanical properties compared to samples with coarse and elongated Laves phases [3]. Hence, it is crucial to precisely control the content, size, and morphology of Laves phases to improve the performance of laser additive manufactured Inconel 718.

Currently, there are two primary methods to regulate the formation of Laves phases during Directed Energy Deposition (DED) processes. One approach involves post-heat treatment, while the other focuses on controlling the conditions of solidification. Regarding the restoration process, the components requiring repair through material deposition are joined to the substrate. Performing high-temperature heat treatment, which eliminate Laves phases, can lead to significant recrystallization and grain coarsening of the substrate. As a result, the performance of the repaired components may be compromised. Currently, the selection of heat treatment for repairing samples

DOI: 10.1201/9781003406327-7

primarily revolves around aging heat treatment, which does not effectively dissolve Laves phases. Given the limitations of heat treatment in all repair scenarios, effective control of solidification conditions during the material deposition process represents a viable approach for modifying the microstructure and mechanical properties.

In the previous chapter, the feasibility of enhancing the performance of Inconel 718 repair specimens was demonstrated through the gradient laser power (GLP) deposition method. This method involves reducing the laser power layer by layer. Unlike traditional laser additive manufacturing, the GLP method achieves the formation of fine and discrete Laves particles instead of coarse long-chain Laves phase. The findings indicate that by optimizing the microstructures in a synergistic manner, the desired mechanical properties can be attained. However, these results also highlight the need for further comprehensive investigations into the influence of the laser power change rate on dendrite and Laves phase regulation, as well as the impact of solidification conditions on the resulting microstructure and mechanical properties.

In this chapter, we summarize methods for manipulating the Laves phases by controlling the solidification conditions during deposition. We further investigated the GLP deposition method proposed in the previous chapter. We utilized Inconel 718 thin-walls as the substrates to accurately simulate real repair scenarios. Our focus was on investigating the impact of varying laser power change rates on the modification of Laves phase, with the ultimate goal of optimizing the microstructure to enhance mechanical properties. We conducted a systematic analysis by comparing Inconel 718 samples deposited using the GLP method, which involved different laser power change rates, with samples deposited using the conventional constant laser power (CLP) deposition method. Our study encompassed the examination of macroscopic features, geometrical characteristics (such as layer widths and layer heights), and porosity of the repaired thin-wall samples. We also conducted a comprehensive investigation into the formation of the distinct Laves phase and its relationship with microstructure and performance. Based on our findings, we can confidently conclude that the GLP method enables the customization of the unfavorable Laves phase and facilitates the improvement of mechanical properties in repaired components by strategically adjusting the laser power throughout the LPD process.

7.2 Tuning Laves phases in laser additive manufactured Inconel 718 alloy

Figure 7.1 presents a summary of the morphological distribution of Laves phases under different solidification conditions, based on the research conducted by Sui et al. [2]. In the context of local solidification conditions, the temperature gradient (G) and solidification rate (R) emerge as the most crucial parameters that influence the morphology and size of Laves phases. The cooling rate, represented by G×R, and the shape of the solidified microstructure, determined by G/R, play significant roles in determining the size and shape of the Laves phases.

As illustrated in Figure 7.1, the green and blue areas correspond to the primary laser additive manufacturing methods, namely Selective Laser Melting (SLM) and Laser Solidification Processing (LSP), respectively. The orange area represents the specific solidification parameter used to obtain discrete and fine Laves phases. These discrete and fine Laves phases can be achieved during the blown powder laser power deposition process when the G×R exceeds 10^4 K/s, while keeping the G/R below 10^7 K s/m^2. This finding confirms the existence of a theoretical basis for controlling the formation of Laves phases by modifying the local solidification conditions and increasing the cooling rates.

Numerous researchers have conducted investigations into controlling the formation of the Laves phase by modifying the solidification conditions in various ways. Among these approaches, some scholars have focused on adjusting the local solidification conditions by altering the laser mode to regulate the Laves phase. In the study conducted by Xiao et al. [4], it was observed that the quasi-continuous wave (QCW) mode exhibited a lower average energy input, better heat dissipation conditions, and consequently higher cooling rates compared to the continuous wave (CW) mode.

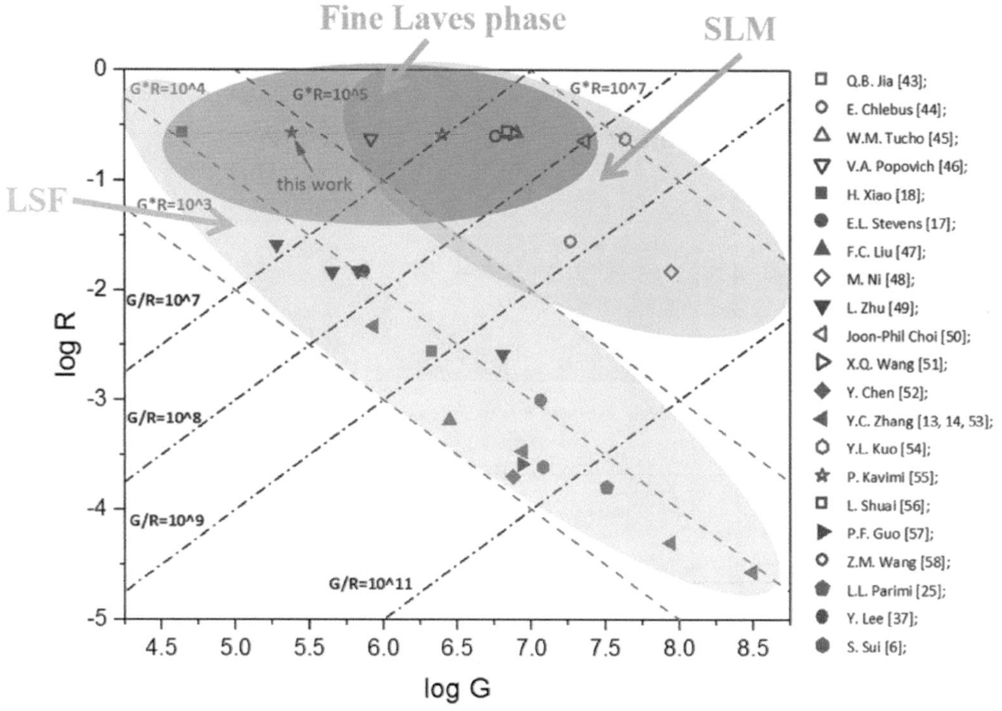

Figure 7.1. Morphology distribution map of Laves phases with different solidification conditions [2].

Figure 7.2 illustrates the stark contrast between the CW and QCW samples. The CW sample displayed coarse and interconnected Laves phase particles, whereas the QCW sample showcased fine and discrete Laves phase particles. The periodic energy input associated with the QCW mode resulted in periodic temperature fluctuations, thereby intensifying the disturbance in heat flow within the molten pool. As a result, the deposited samples exhibited a microstructure characterized by fine equiaxed dendrites and discrete Laves phase particles.

Imbrogno et al. [5] conducted a study on the impact of laser pulsing frequency on the Inconel 718 thin-walls fabricated by LPD. The results, depicted in Figure 7.3, illustrate the varying micro-segregation morphology corresponding to different pulsed frequencies. The interrupted heat source led to the cyclic temperature fluctuations within the molten pool, increased cooling rates and fluctuating thermal gradients. Consequently, this generated fine dendrite nucleation and isolated Laves phases, as opposed to the formation of interconnected long ones. In another investigation by Chen et al. [6], it was demonstrated that the utilization of a flat top laser beam (FTLB) resulted in weaker micro-segregation and reduced precipitation of Laves phases compared to a Gaussian distribution laser beam (GDLB). The FTLB was distributed evenly, effectively lowering the molten pool temperature and temperature gradient along the deposition height. This distribution aided in enhancing lateral heat dissipation and alleviating accumulated heat buildup. Furthermore, Yang et al. [7] employed a circular beam oscillation technique to mitigate the presence of Laves phases. By oscillating the laser beam along a predetermined path at a high frequency, the molten pool was enlarged, convection within the pool was enhanced, and the stirring effect during the solidification process was reinforced. The introduction of controlled turbulence within the molten pool facilitated the uniform diffusion of solute atoms, thus alleviating micro-segregation.

Furthermore, various physical external fields, such as ultrasonic fields and electromagnetic fields, have been incorporated into the deposition process to regulate the solidification conditions and enhance the microstructure. Wang et al. [8] employed a novel technique called self-induced ultrasonic arc welding to control the formation of Laves phases. By applying ultrasonic pulse

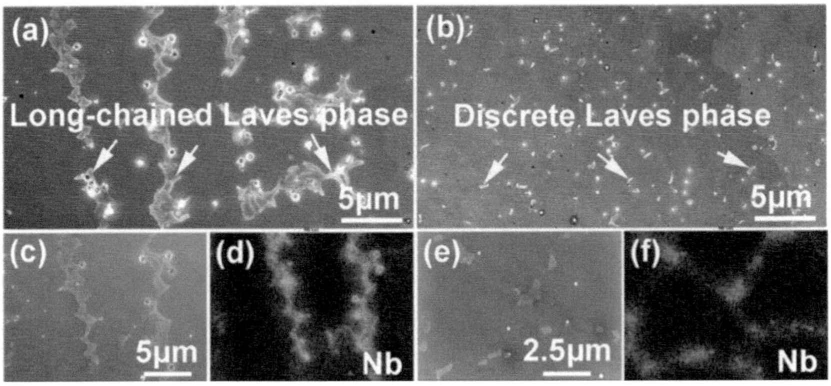

Figure 7.2. Laves phase morphology and Nb distribution in the interdendritic region (a, c, d) CW samples; (b, e, f) QCW samples [4].

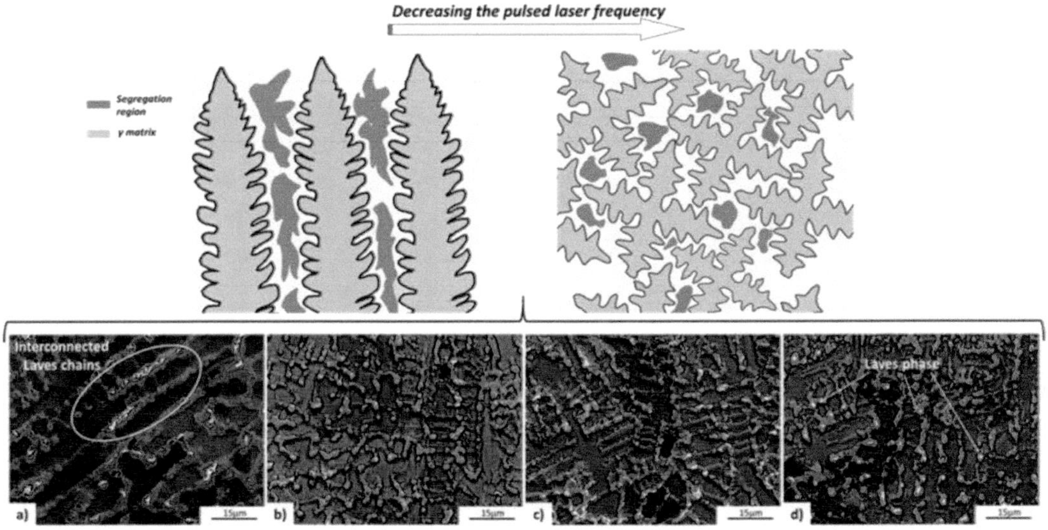

Figure 7.3. Micro-segregation morphology at different pulsed frequencies: (a) CW, (b) 1000 Hz, (c) 100 Hz and (d) 10 Hz [5].

current, the presence of Laves phase particles was reduced while the grain refinement with a high orientation was achieved. These effects played a crucial role in improving both the tensile strength and ductility of the material. Another study conducted by Liu et al. [9] combined laser solid forming with electromagnetic stirring to manipulate the microstructure and mechanical properties of the as-deposited Inconel 718 superalloy. The vigorous convection of liquid metals induced by electromagnetic stirring significantly influenced the growth mode of the solid-liquid interface. As a result, it mitigated the micro-segregation of alloying elements at the interface and reduced the degree of constitutional supercooling. By increasing the intensity of the magnetic field, the size and quantity of Laves eutectic phases were diminished, and alloying elements were more uniformly distributed.

In addition to the utilization of external fields, modifying the cooling conditions of the surrounding environment has also been employed to enhance the cooling rate during the deposition process. Zhang et al. [10] investigated the impact of ultra-high cooling rates achieved by employing liquid nitrogen in the laser cladding of Inconel 718 coatings. The study revealed that the ultra-high cooling rates resulted in the reduction and refinement of Laves phases with a dispersed distribution. This rapid cooling approach helped mitigate constitutional supercooling and suppressed the segregation of Nb into Laves phases. It is important to note that although ultra-high cooling rates achieved through ambient temperature adjustments can yield these benefits, they may also increase the risk of cracking during the actual processing, which adversely affects the mechanical properties. Therefore, it is crucial to select appropriate methods for controlling the solidification conditions based on the specific circumstances at hand.

The methods described above can be quite complex due to the demanding requirements on the laser and the need for additional auxiliary facilities. Moreover, it is widely recognized that the laser process parameters also play a crucial role in the solidification of the deposition process and allow for the optimization of Laves phase formation. However, most current research focuses on analyzing the impact of processing parameters on the formation of Laves phases, while there are limited studies on reducing the occurrence of Laves phase by adjusting these parameters [11–14].

In contrast to the aforementioned methods that have high demands on the laser generator and intricate auxiliary facilities, adjusting the processing parameters provides a convenient and flexible approach to modify the solidification conditions in laser additive manufacturing. Xie et al. [15] investigated the effects of laser power and scanning speed on solidification conditions and Laves phase formation using a novel multi-scale numerical model validated through experiments. They found that high cooling rates achieved by employing high scanning speeds or low laser powers could suppress composition undercooling, decrease the spacing between secondary dendrite arms, and consequently yield fine and isolated Laves phase precipitates. Popovich et al. [16] produced functionally graded Inconel 718 components by alternately employing laser powers of 250 W and 950 W. The resulting graded microstructure exhibited regions with fine grains and highly oriented coarse grains. The occurrence of Laves phase was primarily observed in the regions deposited with the high-energy laser source. Liu et al. [17] employed a hybrid processing strategy to deposit Inconel 718 thin-walls by alternating two sets of laser scanning speeds layer by layer. By adjusting and perturbing the growth behaviors of primary and secondary dendritic arms, they were able to control dendritic morphology, grain texture, and precipitate formation. These methods demonstrate the potential to customize the microstructure and mechanical properties by intentionally manipulating and combining different sets of processing parameters.

However, limited research has been conducted on the regulation of the Laves phase through variations in laser power during LPD. Additionally, in the studies investigating the LPD repair process, a high laser power is commonly employed to enhance repair efficiency. When repairing thin-walled structures like turbine blades, the repair area is typically on the millimeter scale. The significant heat input from the high laser power can lead to the collapse, tilting, and deformation of the deposited thin-wall parts. There are scarce studies available in the public domain that have utilized low heat input for the repair of thin-wall components.

7.3 Build quality

7.3.1 GLP method with different laser power change rates

The GLP method is characterized by a uniform decrease in laser power layer by layer as the deposition height increases. In the initial GLP experiment, the laser power employed was 420 W for the first layer and subsequently reduced by 1 W for each subsequent layer until reaching 370 W for the final layer. This particular set of experiments is referred to as d-1. Four additional sets of GLP experiments, namely d-2, d-3, d-4, and d-5, were conducted. In these experiments, the laser power used for the first layer remained fixed at 420 W, while the reduction in laser power for each subsequent layer was 2 W for d-2, 3 W for d-3, 4 W for d-4, and 5 W for d-5. Accordingly, the laser power employed for the final layer was calculated as 320 W for d-2, 270 W for d-3, 220 W for d-4, and 170 W for d-5. The energy input of the GLP deposition method is quantified by the volumetric energy density E (J/mm^3). This is calculated as the laser energy per unit volume of material deposited using the equation [18]:

$$E = \frac{P}{v_s \times D \times t} \tag{7.1}$$

where P represents the laser power (W), v_s is the scanning speed (5 mm/s), D is the laser spot diameter (1.6 mm), and t is the layer thickness (0.42 mm). The laser power and corresponding volumetric energy density variations across subsequent layers for the GLP experiments are illustrated in Figure 7.4. The region between the two gray dotted lines denotes the position for tensile sampling. Given that the other parameters remain constant, the volumetric energy density is directly proportional to the laser power. For comparison, samples deposited using the traditional CLP LPD method, utilizing fixed laser powers of 170 W, 220 W, 320 W, and 420 W, were employed. Additionally, the powder mass flow rate was set at 7.2 g/min.

For a moving point heat source, we can estimate the cooling rates (dT/dt) using the Rosenthal solution [19]:

$$\frac{dT}{dt} = -2k\pi\left(\frac{v_s}{P}\right)\Delta T^2 \tag{7.2}$$

where k represents thermal conductivity (11.4 W/m·°C) and ΔT denotes the temperature change during cooling. In this calculation, we assume that the liquidus temperature (1336°C) is the maximum temperature of the molten pool, while the substrate temperature is 25°C. The estimated cooling rates for various laser powers are 1465.58°C/s, 1663.64°C/s, 1923.58°C/s, 2279.80°C/s, 2797.93°C/s, and 3620.86°C/s, corresponding to the laser powers 420 W, 370 W, 320 W, 270 W, 220 W, and 170 W, respectively. It is worth noting that there is an inversely proportional relationship between the laser power (volumetric energy density) and cooling rates. However, it is important to mention that this cooling rate estimation does not consider the heat accumulation effect, and the reported values are approximate rather than exact.

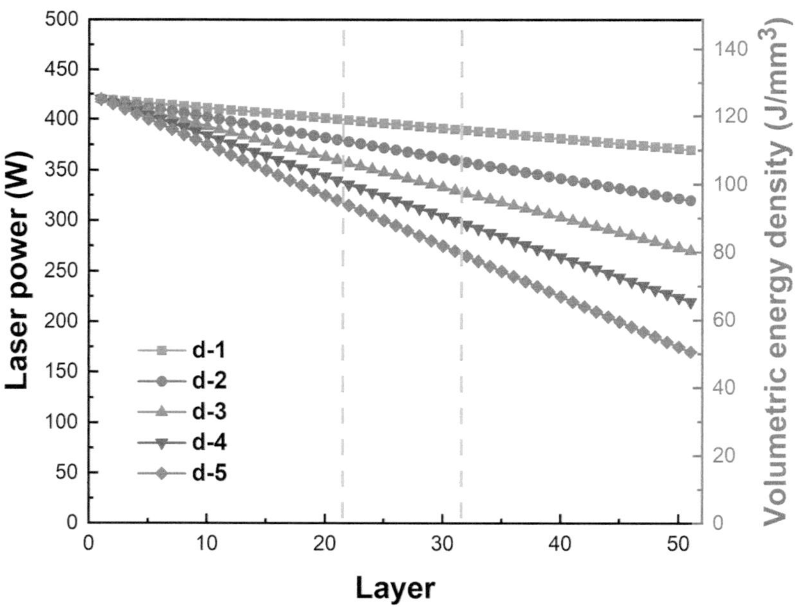

Figure 7.4. The laser power and corresponding volumetric energy density variations as functions of the deposited layer during the GLP deposition.

7.3.2 Macroscopic features

When considering the geometrical formation of repaired parts, it is crucial to examine the macroscopic characteristics of thin-walls formed. Figure 7.5 illustrates the typical macroscopic morphologies of samples deposited using the CLP and GLP methods. As depicted in Figure 7.5(a), the CLP sample deposited at a low laser power (170 W) exhibits noticeable agglomeration of unmelted powders and defects caused by a lack of fusion. Due to interfacial tension, the molten metal tends to shrink into spherical shapes, resulting in the formation of clustered particles known as the balling effect [20, 21]. Insufficient and discontinuous fusion occurs due to the poor flowability of the molten pool at low laser powers. Conversely, when the laser power is set too high (420 W), significant heat accumulation takes place during the deposition process. This leads to uneven and undulating surfaces due to the splattering of droplets within the molten pool. In contrast, the GLP method yields smooth and uniform macroscopic surface features for samples d-1 and d-2 (not shown), as depicted in Figure 7.5(b). These GLP samples exhibit satisfactory macroscopic morphologies similar to the CLP sample morphology achieved using optimal processing parameters, namely laser powers of 220 W and 320 W.

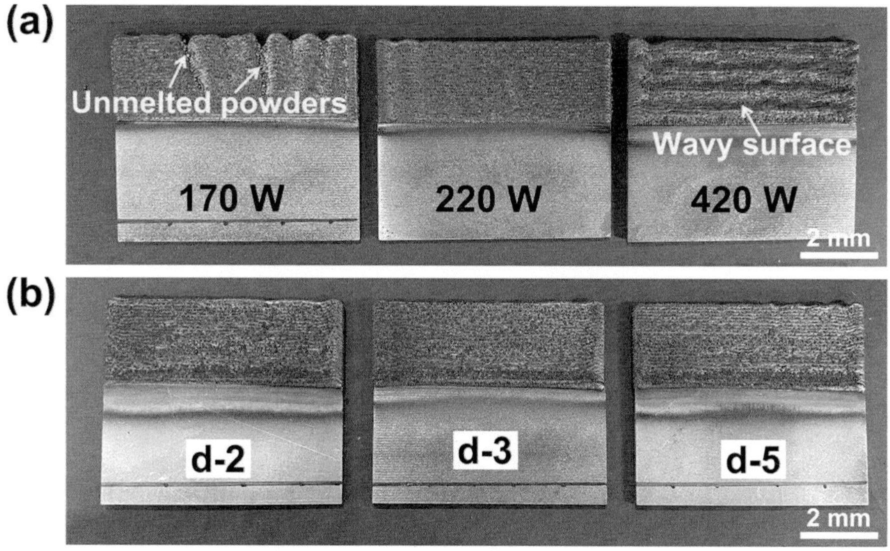

Figure 7.5. The macroscopic morphologies of as-deposited Inconel 718 samples: (a) 170 W, 220 W and 420 W for CLP deposition, and (b) d-2, d-3 and d-5 for GLP deposition.

7.3.3 Layer width and layer height

To investigate the impact of varying laser power on the widths and heights of thin-wall structures, we captured optical micrographs of the Y-Z plane of etched thin-wall samples. Figure 7.6(a) depicts a representative transverse section of the d-4 sample, while Figure 7.6(b) illustrates the measurement method based on the reference [22]. Each sample underwent three measurements. The total heights of the deposited parts range between 21.4 and 21.9 mm, as shown in Figure 7.6(c). Notably, the total height of the CLP sample slightly increases with higher laser powers.

Table 7.1 presents the average values of layer width and layer height. As the laser power change rates increase for the GLP samples and laser powers decrease for the CLP samples, the average layer widths gradually decrease. Conversely, the average layer heights remain relatively constant for all samples. Figure 7.6(d, e) and (f, g) display the variations in layer width and height with the number of deposited layers for GLP and CLP samples, respectively. In Figure 7.6(d), the layer widths of

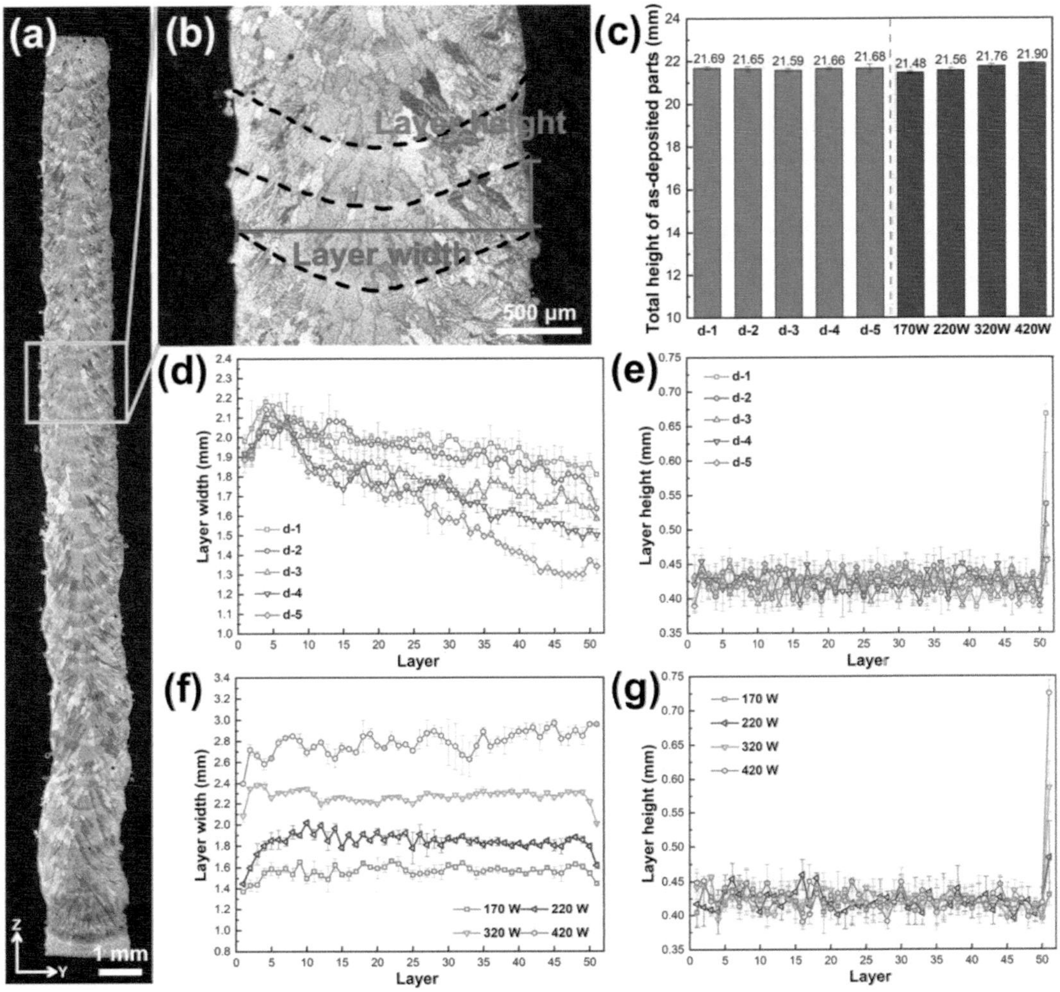

Figure 7.6. Measurement results of layer widths and heights for as-deposited Inconel 718 thin-wall samples: (a) optical micrograph of the d-4 sample in the transverse (Y-Z) plane, (b) schematic diagram of measuring layer widths and heights, (c) total heights of as-deposited parts, (d) layer widths of GLP samples, (e) layer heights of GLP samples, (f) layer widths of CLP samples, and (g) layer heights of CLP samples.

the GLP samples exhibit a downward trend as the number of deposited layers increases, and this downward trend becomes more pronounced with higher laser power change rates. For CLP samples (Figure 7.6(f)), the layer widths increase with increasing constant laser powers. While the layer widths remain mostly stable as the deposition height increases for low laser powers, they exhibit an increasing trend along the deposition height due to significant heat accumulation at high laser power (420 W). Figure 7.6(e) and (g) demonstrate that the layer heights of both GLP and CLP samples fluctuate with the number of deposited layers but remain around 0.42 mm (equal to the predetermined increment for each layer in the Z-axis). The height of the last layer increases due to the lack of remelting of the subsequent layer. Higher laser powers correspond to greater heights for the last layer.

Numerous researchers have reported that increasing laser power has a more significant effect on layer width than on layer height during multilayer thin wall deposition [23, 24]. As the laser powers decrease layer by layer for the GLP samples, the energy densities decrease, resulting in reduced

Table 7.1. Measured average layer widths (mm) and heights (mm) for as-deposited Inconel 718 thin-wall samples.

	d-1	d-2	d-3	d-4	d-5	170 W	220 W	320 W	420 W
Layer width	1.97 ± 0.10	1.93 ± 0.12	1.82 ± 0.14	1.74 ± 0.17	1.65 ± 0.25	1.56 ± 0.07	1.84 ± 0.11	2.27 ± 0.07	2.79 ± 0.14
Layer height	0.425 ± 0.04	0.425 ± 0.03	0.423 ± 0.03	0.425 ± 0.03	0.425 ± 0.02	0.421 ± 0.02	0.423 ± 0.03	0.427 ± 0.03	0.429 ± 0.05

widths and heights of the deposited layers. This reduction can be attributed to a smaller molten pool size, reduced energy input per unit area, and decreased powder catchment rate [25]. Despite the fixed Z increment per layer, the reduction in laser power leads to a decrease in clad height, molten pool depth, and dilution rate. Consequently, the overlap between layers is reduced. From the perspective of the fusion line, the layer spacing remains similar, and the overall layer height remains unchanged. Therefore, GLP deposition is considered a stable process, despite minor fluctuations in layer heights.

7.3.4 Porosity

Pores, which are significant material defects in Laser Powder Deposition (LPD), have a detrimental impact on the mechanical properties of the material. In this study, over 20 optical micrographs were utilized for each sample to accurately quantify the porosity using Image-Pro-Plus software. The software employed a suitable threshold for binarization processing, enabling the effective differentiation of pores within the deposited samples.

Figure 7.7(a, c) and (b, d) display the typical optical micrographs and statistical findings regarding porosity for the deposition processes of GLP and CLP deposition, respectively. In the case of GLP

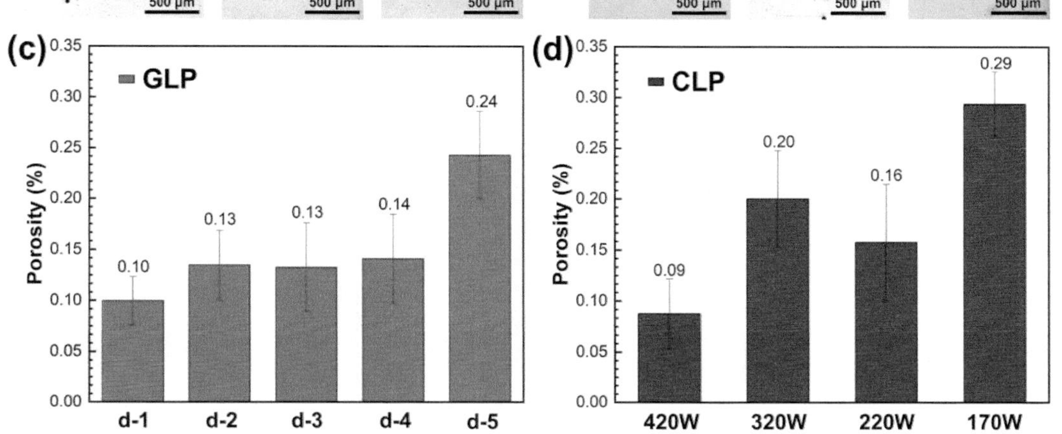

Figure 7.7. Porosity of as-deposited Inconel 718: (a) metallographic images of d-1, d-3 and d-5 for GLP deposition, (b) metallographic images of 420 W, 220 W and 170 W for CLP deposition, statistical results of porosity in (c) GLP samples and (d) CLP samples.

samples, the porosity gradually increases as the laser power change rates rise. Conversely, for CLP samples, an inverse trend is observed where the porosity decreases with an increase in laser powers. Notably, as the deposition height increases, the GLP sample (d-5) subjected to excessive reduction in laser power exhibits larger and more numerous pores. A similar phenomenon is witnessed in CLP samples deposited under low laser powers. It has been reported that the presence of pores is primarily attributed to factors such as lack of fusion, hollow powder particles, and entrapped gas bubbles within the molten pool. The utilization of higher energy density proves beneficial in reducing porosity, as it leads to the creation of a larger molten pool characterized by low viscosity and enhanced wettability. Furthermore, higher energy density allows for an extended residence time, enabling the escape of gas bubbles [26]. However, it should be noted that high energy density can result in unfavorable macroscopic morphology and coarse microstructure due to excessive heat accumulation. To mitigate the adverse effects of severe heat accumulation, it is crucial to establish more suitable deposition parameters.

7.4. Microstructure

7.4.1 Dendritic microstructure

Figure 7.8 illustrates the optical dendritic microstructure of the as-deposited Inconel 718 samples using the GLP method at various deposition heights. In the bottom region of the d-1 samples (Figure 7.8(a)), characterized by a low laser power change rate, the epitaxial growth of columnar dendrites is observed, with dendrites crossing multiple layers due to thermal accumulation. The presence of coarse columnar dendrites growing epitaxially along the building direction aligns with the morphology of dendrites produced under a high constant laser power. As the deposition height increases, randomly oriented columnar dendrites become dominant in the d-1 sample. Furthermore, as the laser power change rates increase, the epitaxially grown coarse columnar dendrites gradually diminish in the bottom regions. Instead, fine columnar dendrites with random orientations prevail in the middle and top regions, as depicted in Figure 7.8(b–d). Notably, equiaxed dendrites can be observed at the interfaces between layers, as indicated by the red arrows. Conversely, in the top region of the d-5 sample, excessively reduced laser powers lead to the formation of dendrites that

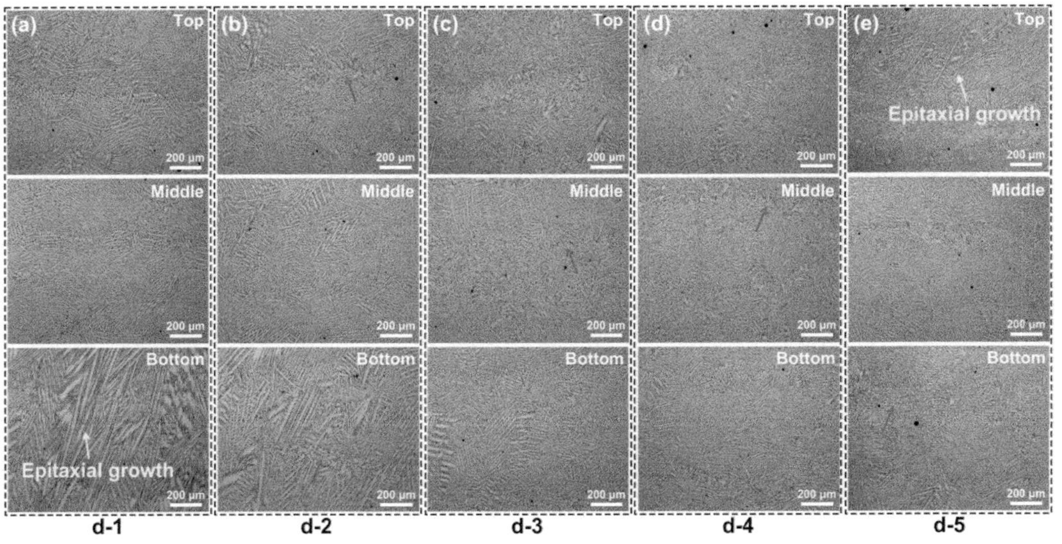

Figure 7.8. Optical dendritic morphologies in as-deposited Inconel 718 thin-walls with various deposition heights: (a) d-1 sample, (b) d-2 sample, (c) d-3 sample, (d) d-4 sample, and (e) d-5 sample.

exhibit epitaxial growth in a specific direction within the layer, disrupting the growth pattern of disordered dendrites (see Figure 7.8(e)). This particular direction of dendrite growth is influenced by the bidirectional scanning process, which mainly operates at low constant laser powers [27].

The primary dendrite arm spacing (PDAS) is introduced to characterize the size of the dendrites [28]. The cooling rates can be estimated based on the PDAS using this equation [12]:

$$\lambda = 80\varepsilon^{-0.33} \tag{7.3}$$

where λ represents the PDAS (μm) and ε represents the cooling rate (°C/s). In order to measure the PDAS, different regions at various deposition heights are selected, and the PDAS values are determined using the Image-Pro-Plus software. The obtained PDAS results and calculated cooling rates are presented in Table 7.2. It is evident that as the deposition height increases, the PDAS demonstrates a decreasing trend. A similar trend is observed with an increase in the laser power change rates. The size of the dendrites is closely related to the cooling rates. Decreasing the laser power leads to higher cooling rates, which in turn reduces the time available for the development and growth of dendritic structures, resulting in smaller PDAS values.

Furthermore, the trends in cooling rates estimated using PDAS and the Rosenthal solution (Equation (7.2) in Section 7.3.1) show good agreement. However, the cooling rates determined based on PDAS are lower than those estimated by Equation (7.2). This discrepancy arises from the fact that the resulting PDAS is influenced by the heat accumulation during the actual laser powder deposition (LPD) process, which is not taken into account in the Rosenthal solution.

It is widely acknowledged that various thermal cycles occurring during the deposition repair process can give rise to the development of diverse microstructures. Consequently, the resulting differences in dendrite appearance exert a significant influence on the morphology and size of the Laves phase. In the case of traditional CLP deposition, the conductive thermal resistance to the substrate progressively increases as the thin-wall is built up, leading to a gradual accumulation of heat. This gradual heat accumulation, coupled with a reduced cooling rate, results in the formation of coarse dendrites. Parimi et al. [27] reported that high molten pool temperatures and continuous heat fluxes from high laser power deposition can cause complete re-melting of successive layers and facilitate the epitaxial growth of coarse columnar dendrites. In contrast, for GLP samples, the reduction of laser powers on a layer-by-layer basis can induce significant changes in the local thermal conditions during the solidification process. This layer-by-layer reduction in laser powers, combined with an increasing laser power change rates, creates a condition of rapid heat dissipation and higher cooling rates for the GLP sample. These conditions are favorable for the formation of fine dendrites.

Furthermore, the variation in the depth of the molten pool plays a crucial role in the development of disordered columnar dendrites during the process of GLP deposition. Within a single deposited layer, columnar grains emerge from the outer region of the molten pool and grow towards its center, while equiaxed grains form in the upper part of the molten pool upon complete solidification [29]. In a previous investigation conducted by Luo et al. [30], it was demonstrated that the size and

Table 7.2. Measured average PDAS (μm) and calculated cooling rates (°C/s) for various laser power change rates in the bottom, middle and top regions.

Deposited region		d-1	d-2	d-3	d-4	d-5
Bottom region	PDAS	9.6 ± 0.7	9.0 ± 0.8	8.6 ± 0.8	8.1 ± 0.7	7.6 ± 0.8
	Cooling rate	617.11	750.40	861.24	1032.65	1252.59
Middle region	PDAS	9.2 ± 0.6	8.3 ± 0.7	8.0 ± 0.7	7.5 ± 0.5	6.9 ± 0.8
	Cooling rate	702.05	959.08	1072.27	1303.88	1678.70
Top region	PDAS	9.1 ± 0.8	8.1 ± 0.8	7.6 ± 0.7	7.3 ± 0.6	6.6 ± 0.6
	Cooling rate	725.69	1032.65	1252.59	1415.17	1920.76

temperature of the molten pool increase as the building height rises due to heat accumulation in the multilayer laser cladding, with laser parameters held constant. At the initial stages of deposition, steering dendrites can be observed on the uppermost portion of the deposited layers. However, they are mostly absent in subsequent layers due to remelting caused by the larger size of the remolten pool. Consequently, the epitaxial growth of columnar dendrites gradually prevails, with a preferred <001> direction emerging over time. Yi et al. [31] confirmed that an increase in laser power leads to a deeper molten pool. As evidenced by the analysis of layer height in Section 7.3.3, the depth of the molten pool and the dilution rate of the GLP sample decrease as the laser power is reduced layer by layer. Consequently, the amount of overlap between layers is reduced. This implies that the equiaxed dendrites present on the top surface of the previously deposited layer are not sufficiently remelted by the subsequent layers as the deposition height increases. During solidification, the dendrites of the newly deposited layer grow based on the non-melted equiaxed dendrites from the previous layer. Moreover, the layer-by-layer adjustment of laser powers during GLP deposition intensifies the disturbance within the molten pool, thereby enhancing the Marangoni heat flow to a certain extent. This promotes multi-directional heat dissipation and disrupts the dendritic growth. The increased presence of residual equiaxed dendrites and remelted fragmented dendrites contribute to the generation of additional heterogeneous nucleation sites, consequently impeding epitaxial growth. Consequently, the GLP method enables the production of fine columnar dendrites with disordered growth orientations.

7.4.2 Laves phase

Figure 7.9 illustrates the typical morphologies of Inconel 718 in as-deposited condition under the backscattered electron (BSE) mode, along with the corresponding energy-dispersive X-ray spectroscopy (EDS) line scanning maps. In traditional CLP samples, the presence of long-chain

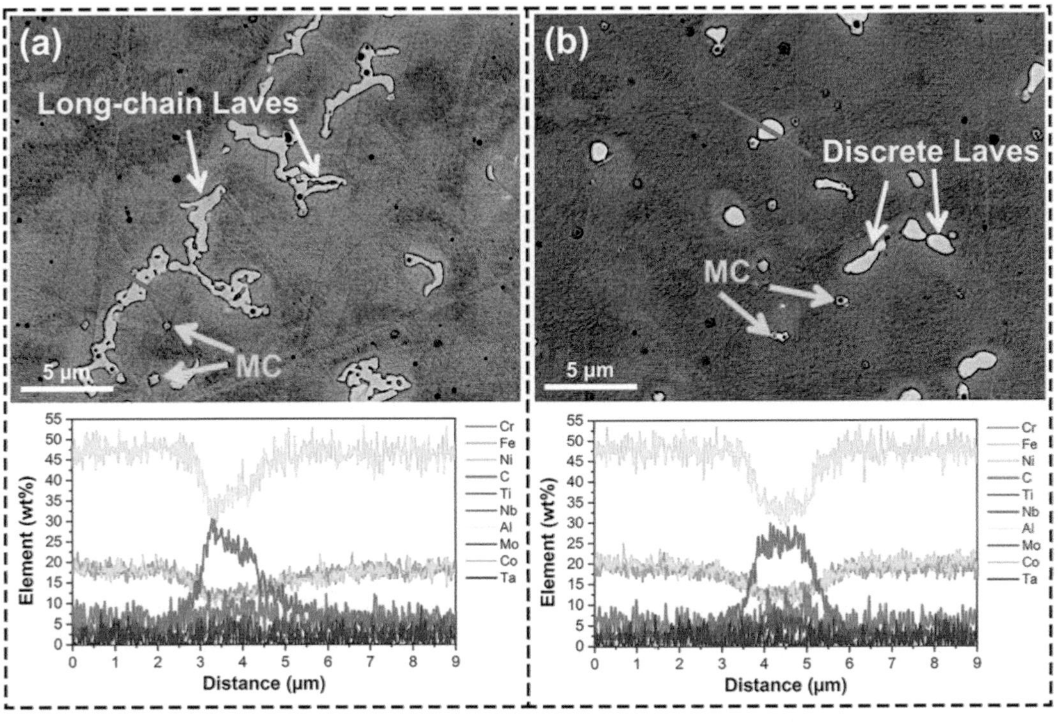

Figure 7.9. The typical SEM micrographs of as-deposited Inconel 718 in BSE mode and corresponding EDS line scanning maps: (a) long-chain Laves phases in CLP sample, and (b) fine and discrete Laves phases in GLP sample.

Laves phases is widespread (Figure 7.9(a)). However, by utilizing the GLP deposition method, we can obtain fine discrete Laves phases (Figure 7.9(b)). The occurrence of severe micro-segregation can be attributed to the concentration of constituents with low melting points in the inter-dendritic regions. The Laves phases, represented by the white regions located between dendrites (indicated by yellow arrows), are accompanied by offwhite regions surrounding them, which indicate the presence of element-rich areas. Additionally, a few MC particles (highlighted by green arrows) are dispersed within the γ matrix. Notably, the line scanning results reveal clear evidence of micro-segregation of Nb, Mo, and Ti, elements with high atomic numbers. Conversely, Ni, Fe, and Cr exhibit deficiencies in the inter-dendritic regions.

The morphology and distribution of Laves phase particles in the bottom, middle, and top regions of the d-2 and d-5 samples are depicted in Figure 7.10(a–f). Figure 7.10(a) and (d) exhibit the presence of elongated long-chain Laves phases in the bottom region of both the d-2 and d-5 samples. During the initial stages of the deposition process, heat gradually accumulates, and the cooling effect of the substrate diminishes, resulting in lower cooling rates over deposition time. Consequently, in the final stages of solidification, the concentrated Nb and Mo liquids form elongated Laves phases within the dendrite arms, making it difficult to avoid their presence in the bottom region of the as-deposited samples.

As the laser power is reduced layer by layer, discrete, fine, and granular Laves phases are dispersed among the dendrites in random directions, as shown in Figure 7.10(b), (c), and (e). The morphology of the Laves phases is closely linked to the morphology of the dendrites. In the top region of the d-5 sample (Figure 7.10(f)), due to a significant reduction in laser power, a fine long-chain Laves phase appears in the columnar inter-dendritic region along a specific direction. This excessive reduction in laser power weakens the discrete and uniform distribution of Laves phase particles.

To quantitatively analyze the impact of changes in laser power rates on Laves phases, we employed the Image-Pro-Plus software to conduct a quantitative analysis of the volume fraction and geometric characteristics of the Laves phases. The data presented here represent the mean value obtained from multiple measurements. More than twenty BSE micrographs with a magnification of 3000 were obtained across entire as-deposited parts of each sample. These micrographs were subjected to image processing techniques to enhance contrast, enabling clear differentiation between

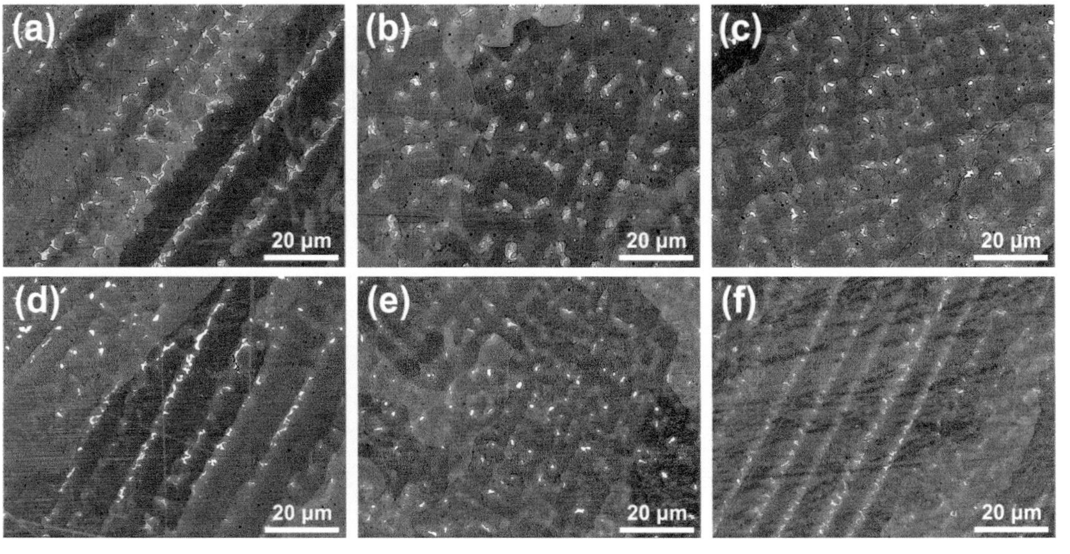

Figure 7.10. Laves phase morphology and distribution in the (a) bottom, (b) middle and (c) top regions of d-2 sample and (d) bottom, (e) middle and (f) top regions of d-5 sample.

the boundaries of the Laves phases and the γ matrix. This preprocessing step was necessary to ensure accurate delineation for subsequent image analysis procedures [32]. Manual adjustments were made to the thresholds in order to encompass the Laves phases, after which the micrographs were binarized. Subsequently, we calculated the area, length, width, and aspect ratio of each individual Laves phase. To mitigate the potential impact of manually selected thresholds on image binarization, we averaged at least five measurements for each BSE image and eliminated any measured outliers. The area data were accumulated and divided by the total area of the micrograph to calculate the area fraction. For the determination of volume fraction, we employed the areal method based on the stereological Cavalieri-Hacquert relation [33].

Figure 7.11 illustrate the statistical outcomes. It is evident that the volume fraction, mean area per Laves phase, length, and width of Laves phases exhibit a general downward trend as the laser power change rates increase. In the case of the d-5 sample, the volume fraction and size of the Laves phase can decrease significantly, reaching as low as 1.66% and 0.67 μm^2, respectively. However, an unusual pattern is observed in the d-2 parameter. Specifically, the aspect ratio of the Laves phase initially decreases and then increases with the rise in laser power change rates, with a minimum value of 2.32 observed at the d-2 parameter. The statistical analysis outcomes indicate that the escalation of laser power change rates not only diminishes the size and quantity of Laves phases but also influences their aspect ratio. Consequently, there is a tendency for Laves phases to transform from granular to long-chain configurations in samples with high laser power change rates.

The morphology and distribution of the Laves phase exhibit a strong dependence on the previously discussed dendrite appearance. Figure 7.12 provides a detailed schematic representation of the relationship between dendrite morphology and the Laves phase in both GLP samples and CLP samples. In the case of CLP samples, coarse and long-chain Laves phases tend to form among the coarse columnar dendrites that grow in a specific direction. On the other hand, disorderly growing dendrites and equiaxed dendrites play a crucial role in the formation of discrete and fine Laves phases in GLP samples. In a study by Nie et al. [34], it was noted that the morphology of Laves phases is greatly influenced by the temperature gradient (G) and the solidification rate (R). Fine and discrete Laves phases are more likely to form under high cooling rates (G·R) and a low ratio of the temperature gradient to solidification rate (G/R).

Considering the estimated cooling rate and measurements of PDAS, it is observed that reducing the laser power leads to an increase in the cooling rate. For the GLP samples, the PDAS decreases as the laser power decreases layer by layer, demonstrating the enhanced cooling effect. Additionally, higher laser power change rates for GLP samples lead to even higher cooling rates. These elevated cooling rates can alleviate the segregation of Nb in the deposited zone due to insufficient time for solute redistribution. Consequently, fine and disordered dendrites impede solute diffusion and

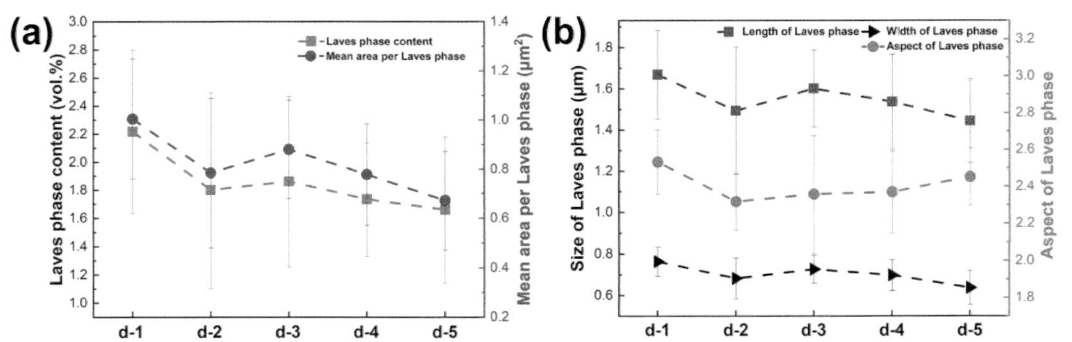

Figure 7.11. Statistical results of the Laves phases under various laser powers: (a) the volume fraction and mean area per Laves phase, and (b) length, width, and aspect ratio of the Laves phases.

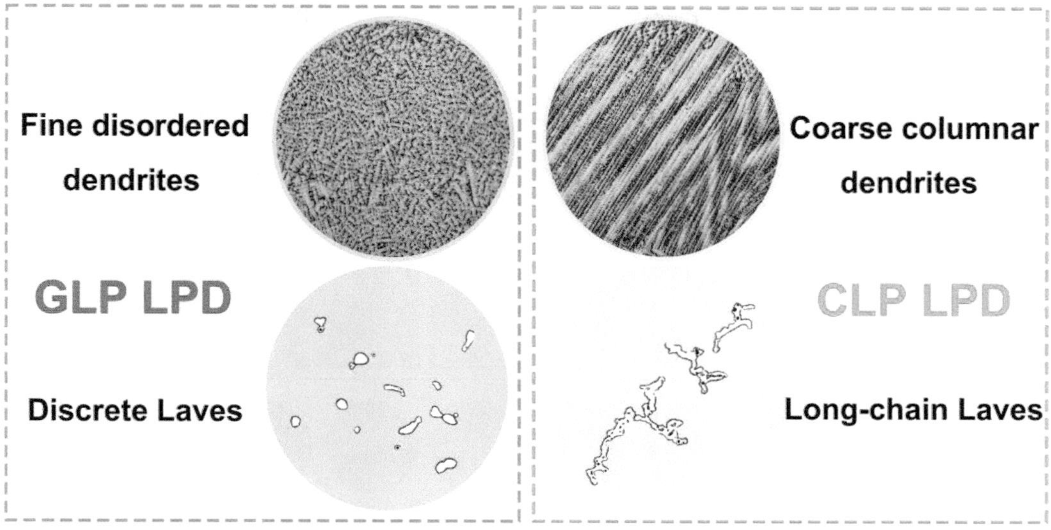

Figure 7.12. Schematic diagram of the relationship between the Laves phase and dendrite morphology for GLP and CLP deposition methods.

promote the separation of the remaining liquid zone into isolated Nb-rich regions. Both of these factors limit the formation and growth space for Laves phases. As a result, eutectic reactions give rise to the formation of fine and discrete Laves phase particles with lower content and smaller size.

The size of Laves phases is primarily influenced by the cooling rate, and the shape of these phases is sensitive to the growth direction of dendrites [35]. The growth direction of dendrites or grains is determined mainly by the combined effects of the preferred growth direction of face-centered-cubic (FCC) crystal structure (<001>) and the direction of maximum heat flux [36]. When there is multidirectional heat flux, the growth of columnar dendrites deviates significantly from the intended direction, resulting in dendrites with no specific preferred orientation. The GLP samples exhibit fine, discrete, and granular Laves phase particles due to superior lateral heat dissipation and a high proportion of dendrites with various orientations. However, the GLP method does have certain limitations on the regulation of dendrites and Laves phases. On one hand, near the bottom of the deposited part, where severe heat accumulation occurs early on, epitaxial growth of columnar dendrites and long-chain Laves phases can be observed. On the other hand, when a GLP sample is deposited with a high laser power change rate (d-5), columnar dendrites within the layer shift towards the scanning direction, resulting in a zigzag structure in the grain morphology. This shift causes the granular Laves phases to transform into long-chain ones.

7.4.3 Grain morphology and crystallographic texture

The EBSD inverse pole figures and pole figures presented in Figure 7.13(a1–e1) and (a2–e2), respectively, show the grain morphologies and crystallographic textures in the middle regions of GLP samples along longitudinal sections (X-Z plane). It is important to note that the GLP method has the capability to generate a mixed structure consisting of equiaxed grains and fine elongated columnar grains. To statistically evaluate the grain morphology, the mean equivalent circle diameter and mean fitted ellipse aspect ratio (AR) are employed, as depicted in Figure 7.13(f1). As the laser power change rates increase, the grain sizes initially decrease from 50.53 μm (d-1) to 36.65 μm (d-3), and then increase to 39.58 μm (d-5). Meanwhile, the AR of the grains exhibits a similar trend. The d-1 sample possesses the largest AR, reaching up to 2.45, whereas the d-2 sample exhibits the smallest AR of 1.97. In the case of a low laser power change rate (d-1), where the laser power is reduced by only 50 W from the first layer to the last layer, the heat accumulation effect becomes

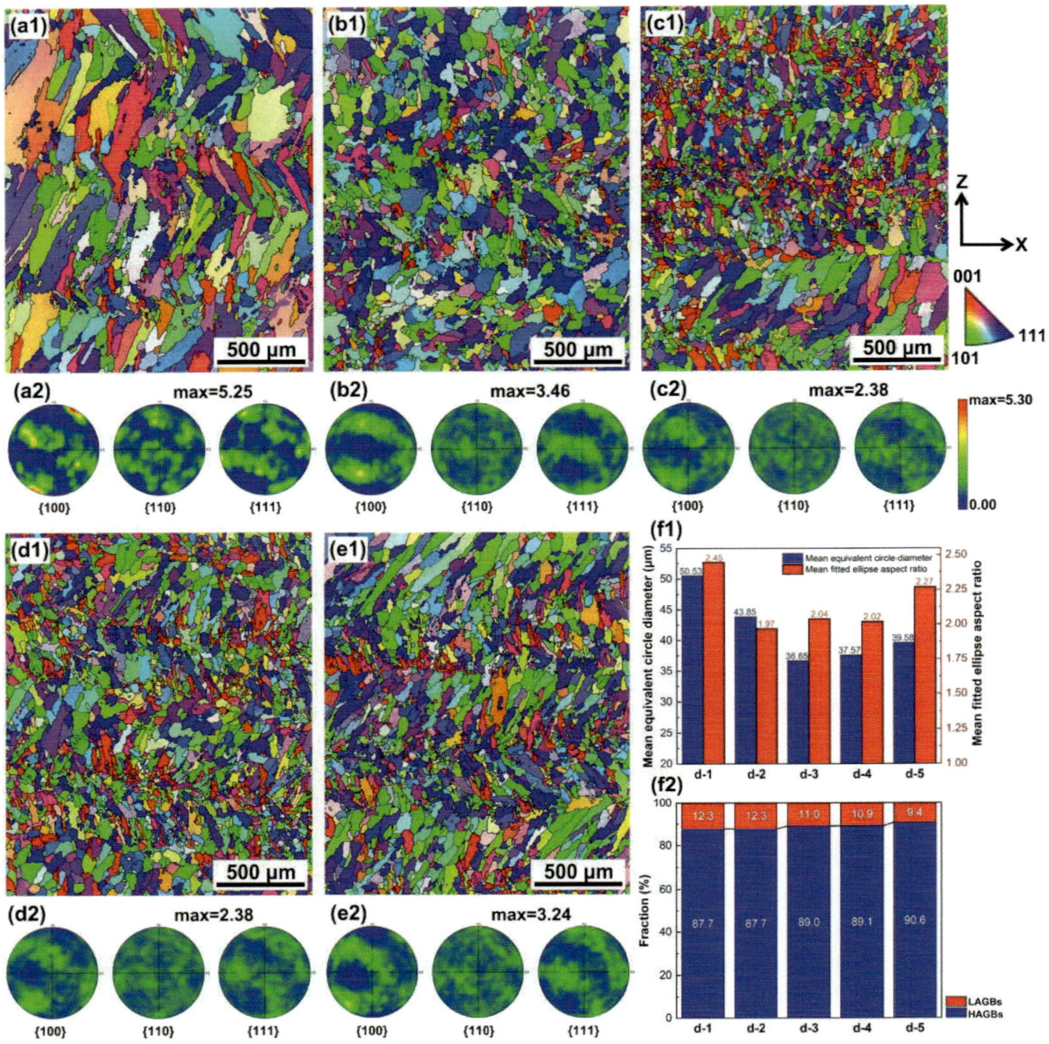

Figure 7.13. EBSD texture analysis and grain characteristics of GLP samples: inverse pole figures and pole figures with (a1, a2) d-1, (b1, b2) d-2, (c1, c2) d-3, (d1, d2) d-4, (e1, e2) d-5, (f1) results of mean equivalent circle diameter and mean fitted ellipse aspect ratio, and (f2) fractions of LAGBs and HAGBs.

prominent. Consequently, coarse and elongated columnar grains growing epitaxially along the building direction and extending across the layer interface can be clearly observed. The largest grain in this sample reaches approximately 690 µm in length. Furthermore, the d-1 sample demonstrates a relatively strong <001> texture with a maximum polar density of 5.25.

As the laser power decreases layer by layer to a great extent, resulting in an increased cooling rate, a greater number of fine equiaxed grains form, as observed in Figure 7.13(b1–d1). These fine grains exhibit disorderly growth without any noticeable preferred orientation. The smallest AR observed in the d-2 sample indicates a high fraction of equiaxed grains within it. In the d-5 sample, the grain morphology is characterized by a zigzag arrangement of columnar grains between the layers, and distinct layer boundaries can be observed due to reciprocating deposition. This finding is in good agreement with the morphologies obtained using the CLP method with low laser powers. The increased fraction of columnar grains in the d-5 sample contributes to an enlargement of the AR (2.27), and the maximum polar density reaches 3.24.

The grain boundary misorientation angle plays a crucial role in reflecting the microstructural characteristics of materials. Grain boundaries are classified into two types based on their misorientation angles: high-angle grain boundaries (HAGBs) with angles greater than 15°, and low-angle grain boundaries (LAGBs) with angles less than 15°. The distribution of grain boundary misorientation types is illustrated in Figure 7.13(f2). It can be observed that the fraction of HAGBs increases gradually as the laser power change rates increase, while the fraction of LAGBs decreases. The abundance of LAGBs suggests that the grains are arranged uniformly along a specific direction, forming a local strong texture [37]. These LAGBs consist of densely packed dislocations, which arise from high internal stress and significant residual thermal strain caused by the rapid solidification process of LPD [38]. On the other hand, most of the GLP samples exhibit a high proportion of HAGBs and a weak crystallographic texture. The low fraction of LAGBs in GLP samples can be attributed to the high cooling rates and uniform heat dissipation resulting from a gradual reduction of heat input layer by layer. This process facilitates a lower level of residual internal stress and residual thermal strain in the as-deposited samples.

Unlike the epitaxially grown coarse columnar grains obtained through the CLP deposition method, the GLP samples deposited using high laser power change rates exhibit a significant presence of fine grains. Figure 7.14(a–e) presents the kernel average misorientation (KAM) maps of grains with mean equivalent circle diameters smaller than 50 μm, while Figure 7.14(f) illustrates the grain size distribution of the GLP samples. With increasing laser power change rates, the proportion of these fine grains initially increases and then decreases. Despite some fluctuations in the layer

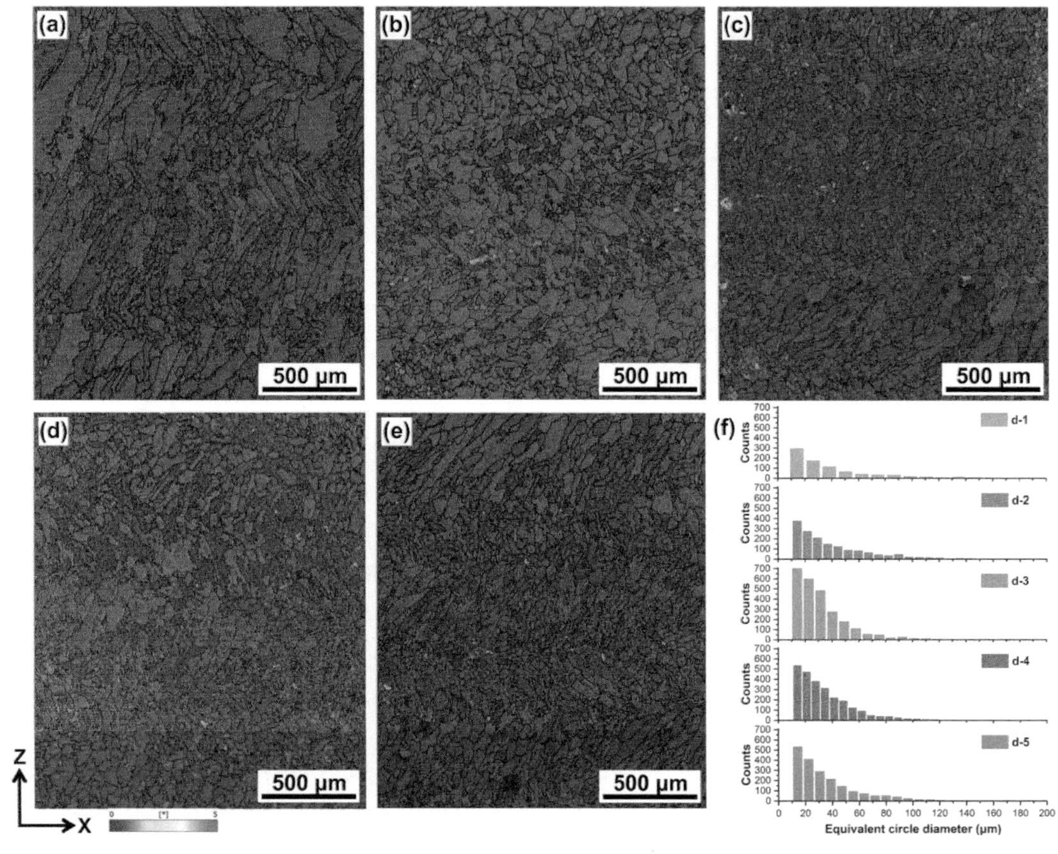

Figure 7.14. KAM maps of fine grains (the mean equivalent circle diameters ≤ 50 μm) for (a) d-1, (b) d-2, (c) d-3, (d) d-4, (e) d-5, and (f) grain size distribution map of GLP samples.

height, the fine grains tend to predominantly distribute at the interface of the layers. The KAM value is determined by calculating the average misorientation angle between a given non-boundary point (< 5°) and its neighboring points. These fine grains in the GLP samples exhibit low KAM values, which indicates a low dislocation density and low stored excess energy within the grains. The KAM value is associated with local crystal misorientation, lattice distortion, stored energy from distortion, and dislocation density within the grains [39].

7.5. Mechanical properties

7.5.1 Microhardness

In order to investigate the changes in mechanical properties of a single GLP sample, we conducted microhardness measurements on transverse sections. Each sample underwent three tests, and the distance between adjacent measurement points was set at 500 μm. Figure 7.15 illustrates the variations in micro-hardness for GLP samples with different laser power change rates as the deposition height increases. Overall, we observed a slight decrease in hardness as the deposition height increased for the GLP samples. However, the difference in hardness distribution among the different GLP samples was found to be negligible. The average hardness values for these GLP samples were measured to be 263.4 ± 8.7 HV (d-1), 264.2 ± 7.4 HV (d-2), 262.4 ± 4.3 HV (d-3), 262.7 ± 4.8 HV (d-4), and 264.1 ± 6.6 HV (d-5) respectively. It is worth noting that the bottom region of the as-deposited parts undergoes multiple thermal cycles. These cycles facilitate the diffusion of refractory elements, the dissolution of Laves phase, and the precipitation of γ'' and γ' phases. As a result, the relative hardness in this region is relatively high [40].

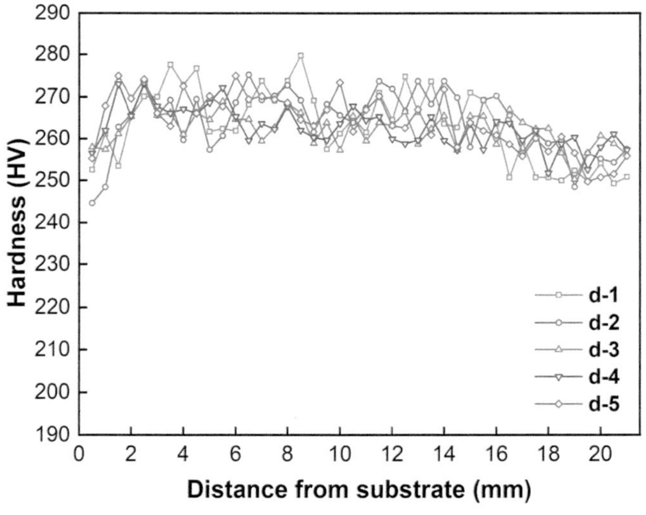

Figure 7.15. Micro-hardness of GLP samples along the deposition height.

7.5.2 Room temperature tensile properties

To evaluate the impact of deposition mode and laser power change rate on the mechanical properties, room temperature tensile tests were carried out. Figure 7.16 displays the typical tensile curves of the samples, along with the statistical data for yield strength (YS), ultimate strength (UTS), and elongation. It is evident that the samples deposited using the GLP method exhibit noteworthy YS, high UTS, and superior ductility when compared to their counterparts deposited through the CLP method. In the case of CLP samples with low laser powers, the presence of lack-of-fusion defects and high internal porosity contributes to reduced strength. On the other hand, CLP samples

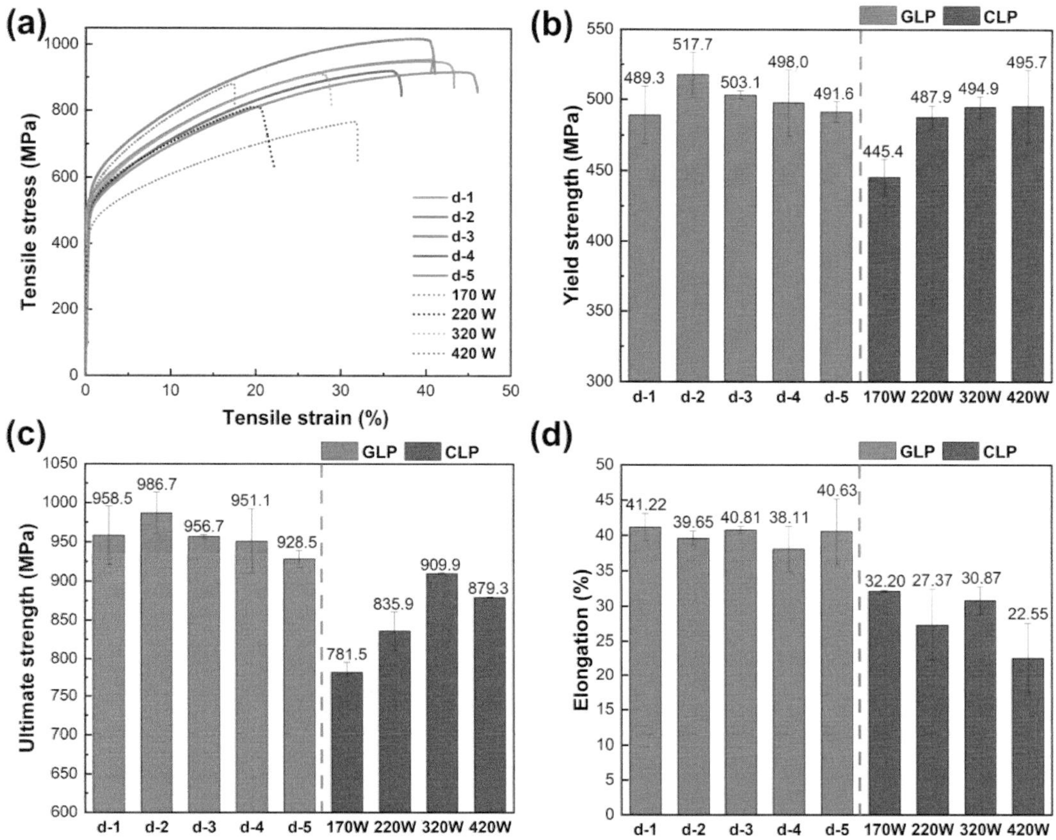

Figure 7.16. Results of room temperature tensile tests: (a) tensile stress-strain curves, (b) YS, (c) UTS, and (d) elongation.

deposited with high laser powers exhibit coarse dendrites and long-chain Laves phases, which lead to diminished strength and poor ductility. In the GLP samples, as the laser power change rate increases, both YS and UTS initially increase and then decrease. The d-2 samples demonstrate the highest YS and UTS values, reaching up to 537.4 MPa and 1018.3 MPa, respectively. Notably, the elongation of GLP samples remains consistently around 40%, with the highest value reaching 46.09%. Compared to the CLP samples, the elongation of GLP samples shows an increase ranging from 18% to 83%.

As is widely recognized, strain hardening is a phenomenon in which stress increases as deformation increases during the plastic deformation phase. To further analyze plastic behavior and quantify the response of strain hardening, the strain hardening rate ($\theta = dS/de$, where S represents true stress and e represents true strain) is plotted against true strain for GLP and CLP samples in Figure 7.17(a) and (b). The curve of strain hardening rate can be divided into three distinct stages. During stage I, known as the strong strain hardening stage, the strain hardening rate is relatively high and decreases sharply as true strain increases due to the proliferation of dislocations [41]. As true strain continues to increase, the strain hardening rate decreases slightly and stabilizes in stage II. This steady strain-hardening stage represents the main process of uniform plastic deformation and corresponds to the linear section observed in the true stress-true strain curves. Following necking, the weak strain hardening stage (region III) occurs, characterized by a rapid decline in the strain hardening rate with increasing true strain.

The length of the steady strain hardening stage determines the capacity for uniform plastic deformation. The longer the uniform deformation stage lasts, the stronger the ability for uniform plastic deformation and the higher the level of plasticity [42]. It is evident that the GLP samples

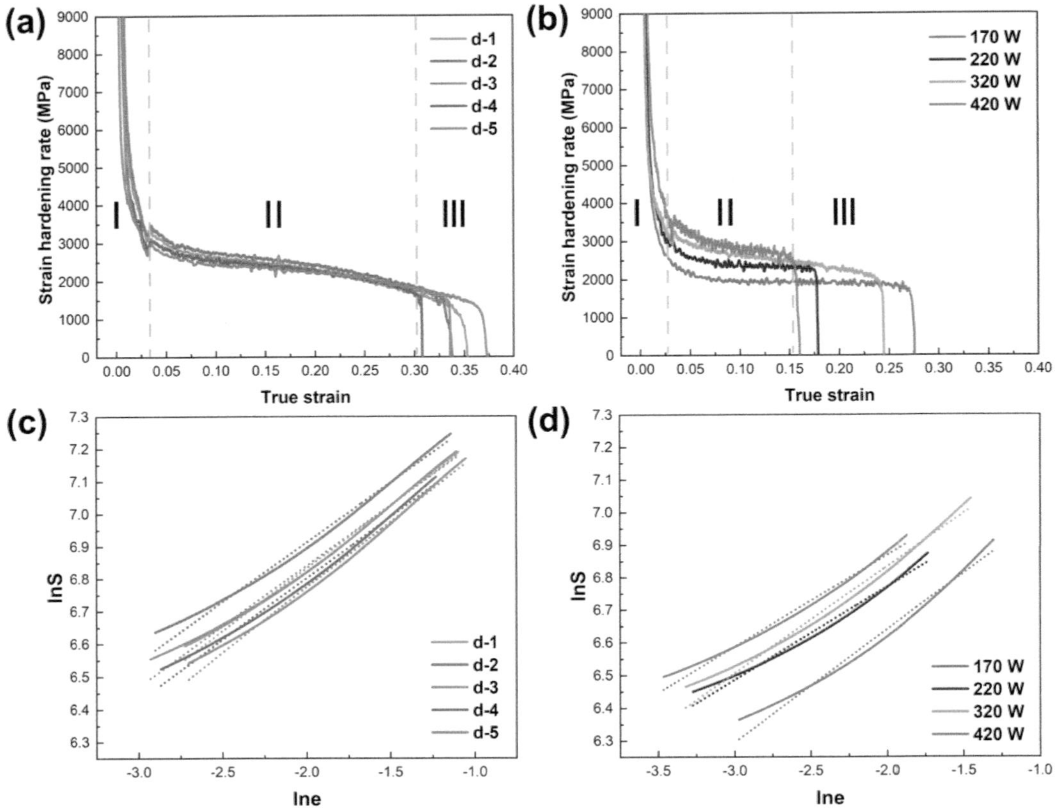

Figure 7.17. Strain hardening curves of different Inconel 718 samples: the strain hardening rate vs. true strain curves of (a) GLP samples, (b) CLP samples, and the ln*S*-ln*e* curves of (c) GLP samples, (d) CLP samples.

exhibit a longer stage II compared to the CLP samples, indicating a superior ability to resist further plastic deformation.

The Hollomon equation, which is commonly used in the analysis of tensile tests, provides a relationship between true stress and true strain during the stage of uniform plastic deformation. It is expressed as [43]:

$$S = K \cdot e^n \tag{7.4}$$

$$lnS = lnK + nlne \tag{7.5}$$

where S and e represent the true stress and the true strain, respectively. K represents the hardening coefficient. n is the strain hardening exponent.

Figure 7.17(c) and (d) depict the *lnS-lne* curves for GLP and CLP samples during the uniform plastic deformation stage. The slopes of these curves correspond to the strain hardening exponents. In order to provide a clearer understanding, several characteristic parameters, such as the strain hardening exponent (n), hardening coefficient (K), and goodness of fit (R^2), are estimated and presented in Table 7.3. It is evident from the results that all GLP samples exhibit larger values of n compared to CLP samples. For instance, the d-5 sample has an n value of 0.4010, whereas the 420 W-sample has a value of only 0.2814. The strain hardening exponent, n, reflects the material's ability to resist continuous plastic deformation. A higher n indicates a more ductile behavior, allowing for greater elongation before the onset of necking [44]. The results of strain hardening rates and strain hardening exponents clearly demonstrate the superior ability of GLP samples to undergo uniform plastic deformation.

Table 7.3. Strain hardening exponents and corresponding fitting results under various process parameters.

	d-1	d-2	d-3	d-4	d-5	170 W	220 W	320 W	420 W
n	0.3859	0.3665	0.3718	0.3769	0.4010	0.3445	0.2848	0.3272	0.2814
K	2007.35	2091.30	1967.15	1912.01	1951.40	1523.78	1541.64	1786.37	1687.98
R^2	0.9960	0.9924	0.9918	0.9922	0.9939	0.9839	0.9852	0.9863	0.9881

The enhancement of mechanical properties in laser-deposited Inconel 718 samples is attributed to a combination of factors, encompassing porosity, dendritic structure, Laves phase, grains, texture, and more. Porosity, a prominent material defect in LPD, has an adverse impact on the mechanical properties of the material. During the tensile test, these pores within the samples act as crack initiators. Stress concentration typically leads to the initiation and propagation of micro-cracks along these vulnerable positions, ultimately accelerating the final failure and compromising the mechanical properties of the as-fabricated samples. However, controlling the porosity can be achieved through two approaches: maintaining a small laser power change rate in the GLP deposition and utilizing a high constant laser power in the CLP deposition. Both methods contribute to reducing porosity, leading to an improvement in mechanical properties.

From the perspective of enhancing mechanical properties, the GLP deposition method plays a vital role in obtaining a unique microstructure. With increasing laser power change rates in GLP samples, finer columnar dendrites with random orientations are attained, and there is generally a decrease in the content and size of Laves phase. The grain size and aspect ratio initially decrease and then increase, exhibiting an opposite trend compared to the strength.

The grain size and width of dendritic structures play a crucial role in the strengthening mechanism. In the as-deposited samples, where there is minimal precipitation of strengthening phases (γ'' and γ'), the fine microstructure contributes to high YS through the Hall-Petch relationship and the Kurz and Fischer relationship [45]. For GLP samples, an increase in laser power change rates leads to higher cooling rates, resulting in the formation of finer, disordered columnar dendrites with smaller PDAS. Due to the size effect associated with grain size and PDAS, the presence of fine dendrites and small grain boundary distances impedes dislocation movement, thereby enhancing the mechanical properties. However, contrary to expectations, the tensile test results show that the strength does not continuously increase with decreasing PDAS. Instead, a minor drop in mechanical properties occurs. Consequently, the variation in mechanical properties cannot be solely attributed to the grain size of γ-dendrites.

Among the GLP samples, the d-2 sample exhibits the smallest aspect ratio of Laves phases and the smallest grain aspect ratio (AR), while displaying the highest YS and UTS. This suggests that the morphology of Laves phases (granular or long-chain shapes) and the morphology of grains significantly influence the strength. Fine precipitates dispersed in the matrix contribute to strengthening through two main mechanisms: the shearing mechanism, where dislocations cut over precipitates, and the Orowan mechanism, where dislocations bypass precipitates [46]. Considering the incoherent relationship between Laves phases and the γ matrix, the Orowan mechanism is considered the primary mechanism. The strength increment resulting from precipitates ($\Delta\sigma_P$) can be expressed using the following equation [47]:

$$\Delta\sigma_P = \frac{0.7MGb\sqrt{f}}{d} \tag{7.6}$$

where M represents the Taylor factor, G denotes the shear modulus, b represents the Burgers vector, f indicates the volume fraction of the precipitates, and d represents the mean size of the precipitates. Based on Eq. (7.6), when the fraction of the strengthening phases increases and their size decreases, the effect of precipitation strengthening becomes more pronounced. Taking into account the morphology of the strengthening phases, the Laves phase is simplified to a rod shape.

The strengthening effect of the Laves phase ($\Delta\sigma_{Laves(rods)}$) can be determined using the following calculation [48]:

$$\Delta\sigma_{Laves(rods)} = \frac{0.85MGb}{2\pi(1-v)^{1/2}} \frac{1}{2R\left(\frac{2}{3A}\right)^{1/3}\left[\sqrt{\left(\frac{\pi}{6f}\right)}-\sqrt{\frac{2}{3}}\right]} \ln\left(\frac{2\sqrt{2}R\left(\frac{2}{3A}\right)^{1/3}}{r_0}\right) \tag{7.7}$$

where v is the Poisson ratio, r_0 is the radius of dislocation core ($r_0 = b$), A is the aspect ratio, and R is the rod radius of Laves phases. The shape enhancement factor $2R\left(\frac{2}{3A}\right)^{1/3}$ is introduced in this equation. It should be noted that the shape enhancement factor increases when the rod diameter increases and the aspect ratio decreases. The d-2 sample demonstrates a favorable outcome in terms of shape enhancement factor due to the presence of a moderate amount of fine, discrete Laves phase particles with the smallest aspect ratio. This favorable condition contributes to the improvement of YS. Moreover, the grain distribution in the d-2 samples tends to be more uniform and regular, characterized by a high proportion of equiaxed grains and the lowest AR. This distribution further aids in achieving greater YS and minimizing tensile anisotropy.

Furthermore, it is important to note that the UTS of a material is greatly influenced by the level of strain it can withstand before fracture, also known as ductility. The UTS is dependent on the YS and the strain hardening exponent (n). When the YS remains constant, a higher value of n corresponds to a higher UTS. In comparison to CLP samples with lower YS and n, GLP samples exhibit higher UTS. On the other hand, GLP samples with different laser power change rates exhibit similar values of n. Consequently, the trend of UTS in GLP samples aligns with the trend observed in YS.

The ductility of a material is primarily influenced by factors such as brittle phases, grain morphology, and texture. It is widely acknowledged that hard and brittle Laves phases can significantly degrade the mechanical properties of nickel-based superalloys by inducing stress concentration and providing preferential sites for crack initiation and propagation. Hu et al. [43] stated that the ductility of as-deposited superalloys is closely associated with the distribution characteristics of Laves phases, including their spacing and radius, as well as the strain hardening exponent. Moreover, Sui et al. [49] emphasized that the presence of coarse and elongated Laves phases with a continuous distribution has a detrimental effect on the tensile ductility and the value of n in LPD Inconel 718 alloys. In contrast, the existence of fine and spherical Laves phases with a dispersed distribution, achieved through controlled solidification conditions, can mitigate embrittlement and enhance ductility. Therefore, the GLP method, which produces fine, granular, and evenly distributed Laves phases, positively influences the ductility and plastic deformation behavior of the material. In terms of strain hardening, the dislocation proliferation during plastic deformation plays a crucial role. A lower initial dislocation density promotes greater dislocation proliferation, leading to a higher value of n. The presence of fine grains with low KAM values and a low fraction of LAGBs in GLP samples indicates a low initial dislocation density and the existence of low-strain grains with low lattice distortion. Consequently, these characteristics improve the material's ability to undergo plastic deformation during subsequent tensile tests, resulting in a larger value of n [50].

Furthermore, GLP samples exhibit a fine-grained microstructure, with an increasing proportion of equiaxed grains and no discernible texture. In contrast to the coarse columnar grains that epitaxially grow along the <001> direction in the conventional CLP LPD process, the fine equiaxed grains generated through GLP methods possess smaller grain boundary curvature. This characteristic proves advantageous in reducing deformation resistance and extending the stage of uniform plastic deformation, thereby enhancing coordination deformation ability among grains and leading to improved ductility.

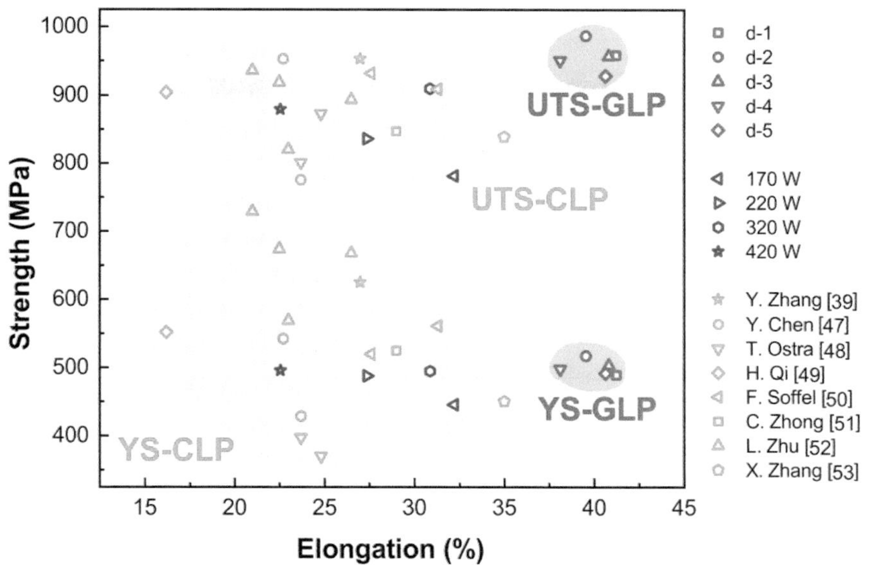

Figure 7.18. Tensile properties distribution of as-deposited Inconel 718 by various LPD methods: YS vs. elongation, and UTS vs. elongation.

To illustrate the level of the tensile test values in GLP LPD Inconel 718 in this chapter, Figure 7.18 presents the YS versus elongation and the UTS versus elongation for as-deposited Inconel 718 fabricated using both GLP LPD and CLP LPD [6, 35, 51–56] methods. Different colored data points represent YS, UTS, and elongation values corresponding to different LPD methods for comparison. It is evident that the ductility of GLP samples exhibits a significant enhancement, surpassing that reported by other researchers utilizing CLP LPD processing. The YS of GLP samples is comparable to that of traditional as-deposited samples produced by CLP LPD and falls within the intermediate range compared to other studies. Thus far, based on the available literature, it appears that the UTS of GLP samples has nearly exceeded that of conventionally manufactured as-built samples. The substantial difference between YS and UTS in the GLP process further underscores the superior strain hardening ability exhibited by GLP samples.

7.6 Summary

Laser powder deposition (LPD) of Inconel 718 superalloy is a promising technique for repairing aerospace components. A novel method called gradient laser power (GLP) deposition is introduced to construct Inconel 718 thin-wall structures by gradually reducing laser powers. This method controls the solidification conditions, alleviates micro-segregation, and modifies the content, size, and morphology of Laves phases. The chapter provides an overview of manipulating Laves phases during deposition by controlling solidification conditions and investigates the GLP deposition method, focusing on different laser power change rates. The study examines macroscopic features, geometrical characteristics, porosity, microstructure, microhardness, and tensile properties, and discusses the relationships among processing parameters, microstructure, and mechanical properties.

The GLP samples exhibit low porosity and smooth macroscopic surfaces. The layer-by-layer reduction of laser powers in GLP samples primarily affects layer width rather than layer height during LPD deposition. Overall, the GLP deposition method is stable despite fluctuations in layer heights.

The morphology and distribution of Laves phases depend on the dendritic growth pattern. Coarse and long-chain Laves phases form with coarse columnar dendrites in conventional CLP samples, while GLP samples with disorderly growing columnar dendrites and equiaxed dendrites result in discrete and fine Laves phases. GLP samples exhibit discrete fine Laves phases, a mixture of equiaxed and fine columnar grains, and weak texture. The size and content of Laves phases decrease with higher laser power change rates, while grain size, aspect ratio (AR), and maximum pole density first decrease and then increase. Excessive reduction in laser power strengthens the influence of bidirectional scanning, leading to a transformation of Laves phases from granular shapes to long-chain counterparts.

Compared to samples deposited using the conventional CLP method, GLP samples demonstrate an excellent combination of strength and ductility. They possess considerable YS, high UTS, prominent strain hardening exponent, and superior ductility of 40%. The GLP method allows tailoring of microstructures and improvement of mechanical properties through intentional manipulation of processing parameters. The influence of Laves phases on mechanical properties is not only determined by their content and size but also by their morphology. Among the various GLP samples, the one prepared by reducing laser power from 420 W to 320 W at a rate of 2 W per layer exhibits the highest strength and excellent ductility. This is attributed to the presence of fine and granular Laves phases with the lowest aspect ratio and a high fraction of equiaxed grains.

The feasibility of the GLP deposition method is demonstrated in this investigation. It enables flexible manipulation of local solidification conditions and tailoring of microstructures, thereby enhancing the mechanical properties of Inconel 718 thin-wall structures. This deposition method, involving layer-by-layer modification of laser power, may also find applications in additive manufacturing of other metallic materials. Purposeful design and adjustment of processing parameters provide the possibility for customized laser additive manufacturing. Overcoming heat accumulation and achieving dynamic thermal equilibrium using the GLP method with an appropriate laser power change rate pose significant challenges. Nonlinear and dynamic regulation of laser powers based on thermal conditions is an interesting topic for further research.

References

[1] Mantri, S. A., Dasari, S., Sharma, A., Alam, T., Pantawane, M. V., Pole, M. et al. (2021). Effect of micro-segregation of alloying elements on the precipitation behaviour in laser surface engineered Alloy 718. Acta Mater 210. https://doi.org/10.1016/j.actamat.2021.116844.

[2] Sui, S., Chen, J., Ma, L., Fan, W., Tan, H., Liu, F. et al. (2019). Microstructures and stress rupture properties of pulse laser repaired Inconel 718 superalloy after different heat treatments. J. Alloys Compd. 770: 125–135. https://doi.org/10.1016/j.jallcom.2018.08.063.

[3] Zhang, D., Niu, W., Cao, X. and Liu, Z. (2015). Effect of standard heat treatment on the microstructure and mechanical properties of selective laser melting manufactured Inconel 718 superalloy. Materials Science and Engineering A 644: 32–40. https://doi.org/10.1016/j.msea.2015.06.021.

[4] Xiao, H., Li, S. M., Xiao, W. J., Li, Y. Q., Cha, L. M., Mazumder, J. et al. (2017). Effects of laser modes on Nb segregation and Laves phase formation during laser additive manufacturing of nickel-based superalloy. Mater Lett. 188: 260–262. https://doi.org/10.1016/j.matlet.2016.10.118.

[5] Imbrogno, S., Alhuzaim, A. and Attallah, M. M. (2021). Influence of the laser source pulsing frequency on the direct laser deposited Inconel 718 thin walls. J. Alloys Compd. 856. https://doi.org/10.1016/j.jallcom.2020.158095.

[6] Chen, Y., Guo, Y., Xu, M., Ma, C., Zhang, Q., Wang, L. et al. (2019). Study on the element segregation and Laves phase formation in the laser metal deposited IN718 superalloy by flat top laser and gaussian distribution laser. Materials Science and Engineering A 754: 339–347. https://doi.org/10.1016/j.msea.2019.03.096.

[7] Yang, H., Jing, G., Gao, P., Wang, Z. and Li, X. (2020). Effects of circular beam oscillation technique on formability and solidification behaviour of selective laser melted Inconel 718: From single tracks to cuboid samples. J. Mater Sci. Technol. 51: 137–150. https://doi.org/10.1016/j.jmst.2019.09.044.

[8] Wang, Y., Liu, W., Wang, D., Yu, C., Xu, J., Lu, H. et al. (2021). Simultaneously enhanced strength and ductility of TIG welds in Inconel 718 super-alloy via ultrasonic pulse current. Materials Science and Engineering A 807: 140894. https://doi.org/10.1016/j.msea.2021.140894.

[9] Liu, F., Cheng, H., Yu, X., Yang, G., Huang, C., Lin, X. et al. (2018). Control of microstructure and mechanical properties of laser solid formed Inconel 718 superalloy by electromagnetic stirring. Opt Laser Technol. 99: 342–350. https://doi.org/10.1016/j.optlastec.2017.09.022.

[10] Zhang, Y. C., Li, Z. G., Nie, P. L. and Wu, Y. X. (2013). Effect of ultrarapid cooling on microstructure of laser cladding IN718 coating. Surface Engineering 29: 414–418. https://doi.org/10.1179/1743294413Y.0000000142.

[11] Manikandan, S. G. K., Sivakumar, D., Rao, K. P. and Kamaraj, M. (2014). Effect of weld cooling rate on Laves phase formation in Inconel 718 fusion zone. J. Mater Process Technol. 214: 358–364. https://doi.org/10.1016/j.jmatprotec.2013.09.006.

[12] Segerstark, A., Andersson, J., Svensson, L. E. and Ojo, O. (2018). Microstructural characterization of laser metal powder deposited Alloy 718. Mater Charact. 142: 550–559. https://doi.org/10.1016/j.matchar.2018.06.020.

[13] Wang, X. and Chou, K. (2017). Effects of thermal cycles on the microstructure evolution of Inconel 718 during selective laser melting process. Addit. Manuf. 18: 1–14. https://doi.org/10.1016/j.addma.2017.08.016.

[14] Yuan, K., Guo, W., Li, P., Wang, J., Su, Y., Lin, X. et al. (2018). Influence of process parameters and heat treatments on the microstructures and dynamic mechanical behaviors of Inconel 718 superalloy manufactured by laser metal deposition. Materials Science and Engineering A 721: 215–225. https://doi.org/10.1016/j.msea.2018.02.014.

[15] Xie, H., Yang, K., Li, F., Sun, C. and Yu, Z. (2020). Investigation on the Laves phase formation during laser cladding of IN718 alloy by CA-FE. J. Manuf. Process 52: 132–144. https://doi.org/10.1016/j.jmapro.2020.01.050.

[16] Popovich, V. A., Borisov, E. V., Popovich, A. A., Sufiiarov, V. S., Masaylo, D. V. and Alzina, L. (2017). Functionally graded Inconel 718 processed by additive manufacturing: Crystallographic texture, anisotropy of microstructure and mechanical properties. Mater Des. 114: 441–449. https://doi.org/10.1016/j.matdes.2016.10.075.

[17] Liu, Z., Zhu, Q. and Zhang, X. (2021). Tailoring microstructure and mechanical property of laser powder deposited Inconel 718 superalloy via hybrid processing strategies. J. Manuf. Process 68: 1837–1848. https://doi.org/10.1016/j.jmapro.2021.07.005.

[18] Liu, S. Y., Li, H. Q., Qin, C. X., Zong, R. and Fang, X. Y. (2020). The effect of energy density on texture and mechanical anisotropy in selective laser melted Inconel 718. Mater Des. 191. https://doi.org/10.1016/j.matdes.2020.108642.

[19] Song, H., Lei, J., Xie, J., Wu, S., Wang, L. and Shou, W. (2019). Laser melting deposition of K403 superalloy: The influence of processing parameters on the microstructure and wear performance. J. Alloys Compd. 805: 551–564. https://doi.org/10.1016/j.jallcom.2019.07.102.

[20] Jelvani, S., Razavi, R. S., Barekat, M., Dehnavi, M. R. and Erfanmanesh, M. (2019). Evaluation of solidification and microstructure in laser cladding Inconel 718 superalloy. Opt Laser Technol. 120: 105761. https://doi.org/10.1016/j.optlastec.2019.105761.

[21] Liu, Z., Kim, H., Liu, W., Cong, W., Jiang, Q. and Zhang, H. (2019). Influence of energy density on macro/micro structures and mechanical properties of as-deposited Inconel 718 parts fabricated by laser engineered net shaping. J. Manuf. Process 42: 96–105. https://doi.org/10.1016/j.jmapro.2019.04.020.

[22] Pinkerton, A. J. and Li, L. (2004). Modelling the geometry of a moving laser melt pool and deposition track via energy and mass balances. J. Phys. D Appl. Phys. 37: 1885–1895. https://doi.org/10.1088/0022-3727/37/14/003.

[23] Tan, H., Shang, W., Zhang, F., Clare, A. T., Lin, X., Chen, J. et al. (2018). Process mechanisms based on powder flow spatial distribution in direct metal deposition. J. Mater Process Technol. 254: 361–372. https://doi.org/10.1016/j.jmatprotec.2017.11.026.

[24] Gharbi, M., Peyre, P., Gorny, C., Carin, M., Morville, S., Le Masson, P. et al. (2013). Influence of various process conditions on surface finishes induced by the direct metal deposition laser technique on a Ti-6Al-4V alloy. J. Mater Process Technol. 213: 791–800. https://doi.org/10.1016/j.jmatprotec.2012.11.015.

[25] Huang, Y., Khamesee, M. B. and Toyserkani, E. (2019). A new physics-based model for laser directed energy deposition (powder-fed additive manufacturing): From single-track to multi-track and multi-layer. Opt Laser Technol. 109: 584–599. https://doi.org/10.1016/j.optlastec.2018.08.015.

[26] Ni, M., Chen, C., Wang, X., Wang, P., Li, R., Zhang, X. et al. (2017). Anisotropic tensile behavior of in situ precipitation strengthened Inconel 718 fabricated by additive manufacturing. Materials Science and Engineering A 701: 344–351. https://doi.org/10.1016/j.msea.2017.06.098.

[27] Parimi, L. L., Ravi, G., Clark, D. and Attallah, M. M. (2014). Microstructural and texture development in direct laser fabricated IN718. Mater Charact. 89: 102–111. https://doi.org/10.1016/j.matchar.2013.12.012.

[28] Jones, J., Whittaker, M., Buckingham, R., Johnston, R., Bache, M. and Clark, D. (2017). Microstructural characterisation of a nickel alloy processed via blown powder direct laser deposition (DLD). Mater Des. 117: 47–57. https://doi.org/10.1016/j.matdes.2016.12.062.

[29] Stevens, E. L., Toman, J., To, A. C. and Chmielus, M. (2017). Variation of hardness, microstructure, and Laves phase distribution in direct laser deposited alloy 718 cuboids. Mater Des. 119: 188–198. https://doi.org/10.1016/j.matdes.2017.01.031.

[30] Luo, K. Y., Xu, X., Zhao, Z., Zhao, S. S., Cheng, Z. G. and Lu, J. Z. (2019). Microstructural evolution and characteristics of bonding zone in multilayer laser cladding of Fe-based coating. J. Mater Process Technol. 263: 50–58. https://doi.org/10.1016/j.jmatprotec.2018.08.005.

[31] Yi, J. H., Kang, J. W., Wang, T. J., Wang, X., Hu, Y. Y., Feng, T. et al. (2019). Effect of laser energy density on the microstructure, mechanical properties, and deformation of Inconel 718 samples fabricated by selective laser melting. J. Alloys Compd. 786: 481–488. https://doi.org/10.1016/j.jallcom.2019.01.377.

[32] Tian, Y., Chekir, N., Wang, X., Nommeots-Nomm, A., Gauvin, R. and Brochu, M. (2018). Effect of heat treatments on microstructure evolution and grain morphology of alloy 625 with 0.4 wt% boron modification fabricated by laser wire deposition. J. Alloys Compd. 764: 815–823. https://doi.org/10.1016/J.JALLCOM.2018.06.106.

[33] Sui, S., Chen, J., Li, Z., Li, H., Zhao, X. and Tan, H. (2020). Investigation of dissolution behavior of laves phase in inconel 718 fabricated by laser directed energy deposition. Addit. Manuf. 32: 101055. https://doi.org/10.1016/j.addma.2020.101055.

[34] Nie, P., Ojo, O. A. and Li, Z. (2014). Numerical modeling of microstructure evolution during laser additive manufacturing of a nickel-based superalloy. Acta Mater 77: 85–95. https://doi.org/10.1016/j.actamat.2014.05.039.

[35] Zhang, Y., Yang, L., Lu, W., Wei, D., Meng, T. and Gao, S. (2020). Microstructure and elevated temperature mechanical properties of IN718 alloy fabricated by laser metal deposition. Materials Science and Engineering: A 771: 138580.

[36] Yang, H., Meng, L., Luo, S. and Wang, Z. (2020). Microstructural evolution and mechanical performances of selective laser melting Inconel 718 from low to high laser power. J. Alloys Compd. 828: 154473. https://doi.org/10.1016/j.jallcom.2020.154473.

[37] Guo, P., Lin, X., Liu, J., Xu, J., Li, J., Zhang, Y. et al. (2020). Passive behavior of nickel-based superalloys prepared by high-deposition-rate laser solid forming additive manufacturing. Corros Sci. 177: 109036. https://doi.org/10.1016/j.corsci.2020.109036.

[38] Holland, S., Wang, X., Chen, J., Cai, W., Yan, F. and Li, L. (2019). Multiscale characterization of microstructures and mechanical properties of Inconel 718 fabricated by selective laser melting. J. Alloys Compd. 784: 182–194. https://doi.org/10.1016/j.jallcom.2018.12.380.

[39] Xie, B., Zhang, B., Yu, H., Yang, H., Liu, Q. and Ning, Y. (2020). Microstructure evolution and underlying mechanisms during the hot deformation of 718Plus superalloy. Materials Science and Engineering A 784. https://doi.org/10.1016/j.msea.2020.139334.

[40] Jiang, R., Mostafaei, A., Wu, Z., Choi, A., Guan, P. W., Chmielus, M. et al. (2020). Effect of heat treatment on microstructural evolution and hardness homogeneity in laser powder bed fusion of alloy 718. Addit. Manuf. 35. https://doi.org/10.1016/j.addma.2020.101282.

[41] Kindermann, R. M., Roy, M. J., Morana, R. and Francis, J. A. (2022). Effects of microstructural heterogeneity and structural defects on the mechanical behaviour of wire + arc additively manufactured Inconel 718 components. Materials Science and Engineering A 839: 142826. https://doi.org/10.1016/j.msea.2022.142826.

[42] Hu, Y. L., Lin, X., Li, Y. L., Wang, J., Zhang, S. Y., Lu, X. F. et al. (2019). Effect of heat treatment on the microstructural evolution and mechanical properties of GH4099 additive-manufactured by directed energy deposition. J. Alloys Compd. 800: 163–173. https://doi.org/10.1016/j.jallcom.2019.05.348.

[43] Hu, Y. L., Li, Y. L., Zhang, S. Y., Lin, X., Wang, Z. H. and Huang, W. D. (2020). Effect of solution temperature on static recrystallization and ductility of Inconel 625 superalloy fabricated by directed energy deposition. Materials Science and Engineering A 772. https://doi.org/10.1016/j.msea.2019.138711.

[44] Yu, X., Lin, X., Liu, F., Wang, L., Tang, Y., Li, J. et al. (2020). Influence of post-heat-treatment on the microstructure and fracture toughness properties of Inconel 718 fabricated with laser directed energy deposition additive manufacturing. Materials Science and Engineering A 798: 140092. https://doi.org/10.1016/j.msea.2020.140092.

[45] Wang, Y. C., Lei, L. M., Shi, L., Wan, H. Y., Liang, F. and Zhang, G. P. (2020). Scanning strategy dependent tensile properties of selective laser melted GH4169. Materials Science and Engineering A 788: 139616. https://doi.org/10.1016/j.msea.2020.139616.

[46] Galindo-Nava, E. I., Connor, L. D. and Rae, C. M. F. (2015). On the prediction of the yield stress of unimodal and multimodal γ' Nickel-base superalloys. Acta Mater 98: 377–390. https://doi.org/https://doi.org/10.1016/j.actamat.2015.07.048.

[47] Liang, L., Xu, M., Chen, Y., Zhang, T., Tong, W., Liu, H. et al. (2021). Effect of welding thermal treatment on the microstructure and mechanical properties of nickel-based superalloy fabricated by selective laser melting. Materials Science and Engineering A 819: 141507. https://doi.org/10.1016/j.msea.2021.141507.

[48] Zhang, S., Lin, X., Wang, L., Yu, X., Hu, Y., Yang, H. et al. (2021). Strengthening mechanisms in selective laser-melted Inconel718 superalloy. Materials Science and Engineering A 812: 141145. https://doi.org/10.1016/j.msea.2021.141145.

[49] Sui, S., Tan, H., Chen, J., Zhong, C., Li, Z., Fan, W. et al. (2019). The influence of Laves phases on the room temperature tensile properties of Inconel 718 fabricated by powder feeding laser additive manufacturing. Acta Mater 164: 413–427. https://doi.org/10.1016/j.actamat.2018.10.032.

[50] Cao, M., Zhang, D., Gao, Y., Chen, R., Huang, G., Feng, Z. et al. (2021). The effect of homogenization temperature on the microstructure and high temperature mechanical performance of SLM-fabricated IN718 alloy. Materials Science and Engineering A 801: 140427. https://doi.org/10.1016/j.msea.2020.140427.

[51] Ostra, T., Alonso, U., Veiga, F., Ortiz, M., Ramiro, P. and Alberdi, A. (2019). Analysis of the machining process of inconel 718 parts manufactured by laser metal deposition. Materials 12. https://doi.org/10.3390/ma12132159.

[52] Qi, H., Azer, M. and Ritter, A. (2009). Studies of standard heat treatment effects on microstructure and mechanical properties of laser net shape manufactured INCONEL 718. Metall Mater Trans A Phys Metall Mater Sci. 40: 2410–2422. https://doi.org/10.1007/s11661-009-9949-3.

[53] Soffel, F., Eisenbarth, D., Hosseini, E. and Wegener, K. (2021). Interface strength and mechanical properties of Inconel 718 processed sequentially by casting, milling, and direct metal deposition. J. Mater Process Technol. 291: 117021. https://doi.org/10.1016/j.jmatprotec.2020.117021.

[54] Zhong, C., Gasser, A., Kittel, J., Fu, J., Ding, Y. and Poprawe, R. (2016). Microstructures and tensile properties of Inconel 718 formed by high deposition-rate laser metal deposition. J. Laser Appl. 28: 022010. https://doi.org/10.2351/1.4943290.

[55] Zhu, L., Xu, Z. F., Liu, P. and Gu, Y. F. (2018). Effect of processing parameters on microstructure of laser solid forming Inconel 718 superalloy. Opt Laser Technol. 98: 409–415. https://doi.org/10.1016/j.optlastec.2017.08.027.

[56] Zhu, L., Xu, Z. F., Liu, P. and Gu, Y. F. (2021). A novel method to prevent cracking in directed energy deposition of Inconel 738 by *in-situ* doping Inconel 718. Mater Des. 197: 109214. https://doi.org/10.1016/j.matdes.2020.109214.

Chapter 8

Microstructure Characteristics and Mechanical Properties of Post Heat Treated Inconel 718 Thin-Wall

8.1 Introduction

In recent years, extensive research has been conducted on laser additive manufacturing (AM) techniques, which are considered as promising alternatives to or replacements for traditional manufacturing methods. Laser powder deposition (LPD), as a representative AM technique, offers several advantages such as high deposition efficiency, remarkable flexibility, precise forming capabilities, and the ability to repair and remanufacture metallic components with complex geometries and challenging-to-process materials like Ni-based superalloys [1]. By employing laser coaxial powder feeding, LPD can restore damaged components with minimal dilution and distortion, overcoming geometric limitations and reducing energy consumption and manufacturing costs.

Inconel 718, a commonly used Ni-based superalloy, relies on precipitation hardening phases (γ'' and γ' phases) and solid-solution hardening effects from refractory metal elements (primarily Nb and Mo) within a nickel-chromium-based γ matrix. This superalloy is renowned for its exceptional mechanical performance, excellent fatigue properties, superb weldability, high structural stability, exceptional resistance to corrosion and oxidation at elevated temperatures. As a result, it finds wide applications in industries such as aerospace, petroleum, nuclear energy, and marine sectors, particularly in demanding high-temperature service conditions [2].

Extensive research has been conducted on the laser additive manufacturing of Inconel 718 alloy [3, 4]. The as-deposited Inconel 718 superalloys undergo a series of intricate thermal cycles, including rapid heating, rapid cooling, directional heat extraction, and repeated melting. These thermal cycles result in a microstructure that is entirely distinct from that of the traditional casting and wrought processes. The as-deposited structures exhibit typical characteristics such as micro-segregation, interdendritic Laves phases, restrained strengthening precipitates, and strongly-oriented columnar structures. The uneven distribution of Nb and the presence of undesirable Laves phases negatively impact the elemental distribution, which, in turn, hinders the precipitation of γ'' and γ' strengthening phases. The hexagonal Laves phase ($(Ni,Fe,Cr)_2(Nb,Ti,Mo)$) is widely recognized as detrimental to the material's strength, ductility, stress rupture, fatigue, and creep rupture properties. This is due to its depletion of essential elements required for precipitation strengthening and its facilitation of crack initiation and propagation. Several researchers have discovered that the mechanical properties of as-deposited Inconel 718, containing discrete, fine, and granular Laves phase particles, surpass those of Inconel 718 with coarse, continuous long-chain Laves phases [5, 6].

It is widely recognized that Inconel 718 is an alloy that is mainly strengthened through precipitation. The Inconel 718 sample in its as-deposited state lacks the strengthening phases of nano γ'' and γ'. Hence, it is crucial to carry out post-heat treatments to meet the required strength levels.

DOI: 10.1201/9781003406327-8

Until now, limited research has been conducted to highlight the varying responses of the as-deposited microstructures to different heat treatments. To investigate the responses of different as-deposited Inconel 718 microstructures to heat treatments, we have chosen two representative microstructures based on our previous studies. One microstructure exhibits a coarse columnar morphology with a <001> texture and interdendritic coarse long-chain Laves phases, which was obtained through the traditional constant laser power deposition (CLP) method. The other microstructure, fabricated using the gradient laser power deposition (GLP) method, features a random texture and discrete, fine Laves phase particles.

The GLP deposition method is an innovative laser powder deposition technique that allows for the unique formation of as-deposited microstructures. It is essential to explore suitable post-heat treatments to achieve homogeneous and improved performance of GLP-built Inconel 718 superalloys. This study is motivated by the repair of thin-wall structures at the tips of aviation turbine blades. We compare the responses of the two different as-deposited microstructures to three typical types of heat treatments applied to Inconel 718. We conduct a comprehensive investigation of the microstructural evolution, hardness, tensile properties, and fracture morphologies at both room temperature and high temperature (650°C). Additionally, a comparison of the mechanical properties with those of the standard forged Inconel 718 is performed. We thoroughly discuss the intricate relationship between microstructures and mechanical properties before and after the heat treatments. The findings of this study are beneficial for the development of effective heat treatment processes tailored to specific microstructures, aiming to achieve superior performance in additively manufactured Inconel 718 alloys.

8.2 Recent research on heat treatments for laser additive manufactured Inconel 718 alloy

The main processing steps for aerospace components manufactured using additive manufacturing techniques are depicted in Figure 8.1. Post-processing is necessary to meet the geometric and metallurgical requirements of the final parts. Inconel 718 primarily gains strength from precipitation hardening phases (γ'' and γ' phases) and the solid-solution hardening effect of refractory metal elements (mainly Nb and Mo) within a nickel-chromium-based γ matrix. To mitigate the micro-segregation and regulate Laves phases in LPD, two primary approaches are employed. The first approach involves controlling the solidification conditions during the material deposition process, while the second approach relies on appropriate post-heat treatments.

It has been reported that increasing the cooling rates can significantly suppress Nb segregation and Laves phase formation [8–11]. However, the as-deposited Inconel 718 lacks strengthening phases and exhibits low strength due to the consumption of Nb by numerous Laves phases and the sluggish precipitation of γ'' precipitates during the repeated rapid-heating and rapid-cooling process of laser powder deposition (LPD). Therefore, it is essential to apply proper heat treatment to achieve the homogenization of elemental distribution, eliminate micro-segregation, regulate Laves phases, fully precipitate γ'' and γ' strengthening phases, and relieve residual stress. This is crucial to ensure that the comprehensive mechanical properties of laser powder-deposited Inconel 718 meet the required standards, both at room temperature and high temperatures.

Currently, numerous researchers have investigated the effects of various heat treatment schedules on the microstructural modification and mechanical properties of Inconel 718 fabricated through laser additive manufacturing. The commonly employed heat treatments by scholars primarily include homogenization heat treatment, solution heat treatment, and aging heat treatment.

In the study conducted by Zhao et al. [12, 13], they investigated the impact of different homogenization conditions on the relationship between microstructure and mechanical properties during subsequent aging processes for Inconel 718 produced using laser powder bed fusion. The researchers stated that when compared to the typical homogenization process performed at 1065°C, a higher homogenization temperature of 1180°C effectively eliminated residual stress, reconstructed

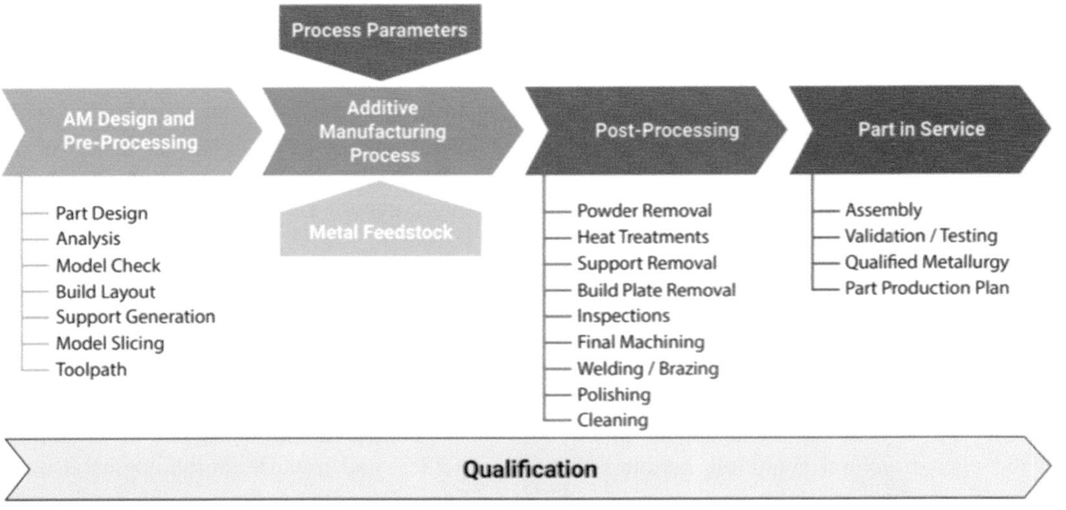

Figure 8.1. Main processing steps of additively manufactured aerospace components [7].

the microstructure, and introduced isotropic tensile properties through recrystallization. This indicates that higher homogenization temperatures can lead to improved properties in Inconel 718. Similarly, Yuan et al. [13] examined the influence of heat treatment schemes and processing parameters on the microstructural characteristics and dynamic mechanical properties of Inconel 718 samples fabricated using laser metal deposition. They discovered that a homogenization–solution–aging treatment significantly reduced the microstructural and mechanical anisotropy. As a result, the overall properties of the heat-treated Inconel 718 were found to be comparable to those of forged counterparts.

The study conducted by Li et al. [14] aimed to explore the influence of solution temperature on the microstructure and hardness of Inconel 718 produced through selective laser melting. Their findings revealed that as the solution temperature increased, a greater number of short-rod δ phases precipitated along the grain boundaries and exhibited a uniform distribution. Concurrently, the strengthening phases γ″ and γ' were formed. The presence of δ phases acted as grain boundary pins, impeding grain growth and consequently enhancing the hardness of the material. In a similar vein, Sui et al. [15] delved into the impact of heat treatments on the dissolution behavior of Laves phases. By subjecting the LPDed Inconel 718 samples to a heat treatment of 1050°C for 15 minutes, followed by double aging, notable enhancements in high-temperature stress rupture performance were observed when compared to the forged state. This improvement was attributed to the transformation of long chain-like Laves phases into granular ones, as well as the precipitation of strengthening phases in the appropriate content and size. Furthermore, Huang et al. [16] conducted a study examining the effect of different solution temperatures, time durations, and cooling rates on printed Inconel 718 samples. They conducted a meticulous analysis to establish the relationship between solution temperature and the minimum solution time, ultimately proposing a method for selecting appropriate solution treatment parameters. Their investigations confirmed a positive correlation between the minimum solution time and the dendritic arm spacing. Remarkably, the printed Inconel 718 samples with relatively fine dendrites exhibited a substantial reduction in the minimum solution time compared to Inconel 718 ingots.

For the aging heat treatment process, Zhai et al. [17] discovered that a straightforward aging method is capable of inducing the precipitation of a significant amount of γ″ and γ′ strengthening phases. This leads to an improvement in the mechanical properties of the deposited bulk Inconel 718 sample, which possesses a fine dendrite structure, minimal micro-segregation, and fine

Laves phases. Nevertheless, it is important to note that direct aging does not alter the original columnar dendrite structure or dissolve any existing Laves phases. Ozer et al. [18] examined the impact of post-fabrication aging treatment on the microstructural characteristics and mechanical properties. Following the aging process, Inconel 718 samples produced through selective laser melting exhibited a combination of columnar and equiaxed structures, deviating from the original arc-like structure. Moreover, numerous precipitates of nano-sized γ'' and γ' strengthening phases were observed. However, it should be noted that MC-type carbides and Laves phases remained present in the as-deposited sample even after undergoing aging heat treatment. As the aging time exceeded 16 hours, a slight reduction in hardness was observed. This can be attributed to the coarsening of the nano-sized γ'' and γ' strengthening phases caused by overaging, as well as the precipitation of the δ phase at the grain boundaries.

Currently, the primary focus of research on heat treatment paths for Inconel 718 manufactured by additive manufacturing (AM) is centered on examining its impact on microstructural characteristics, mechanical properties, and optimizing heat treatment paths. However, it is important to note that different laser deposition methods can significantly affect solidification conditions and thermal accumulation effects, resulting in distinct microstructures in the as-deposited state. Moreover, these microstructures will evolve differently when subjected to the same heat treatment. Sui et al. [19] conducted a study on the microstructures and stress rupture properties of Inconel 718 superalloy repaired using pulse laser after direct aging heat treatment and short time aging heat treatment at 900ºC. They observed the formation of fine and granular Laves phases through the use of pulse laser, and noted that their size further decreased after the heat treatments. Under tension, these fine and discrete Laves phases exhibited high mobility within the γ matrix, making it challenging for stress to concentrate. Consequently, the deformation behavior of Laves phases underwent a significant change, suggesting that they may serve as strengthening phases rather than detrimental ones. In another research, Li et al. [20] investigated the effect of solution treatment time on the microstructures of Inconel 718 manufactured using high-deposition-rate laser directed energy deposition (HDR-LDED). They observed that the resulting Laves phase particles were more dispersed compared to those produced by traditional laser directed energy deposition. Additionally, they found that a low solution heat treatment temperature of 1020ºC combined with a short duration of 15–45 minutes was sufficient to modify the original Laves phase particles and achieve microstructure homogeneity.

Notably, most of the intensive research conducted on evaluating the mechanical properties of Inconel 718 produced by LPD has focused on ambient temperature conditions. However, there is a lack of studies on its deformation behavior at high temperatures. While there is some existing research on the high-temperature mechanical properties of Inconel 718 manufactured using the laser powder bed method, the investigation specifically on the high-temperature tensile properties of Inconel 718 printed with the coaxial powder feeding LPD method is relatively insufficient [21–23]. Furthermore, there are currently limited comparisons available between the high-temperature performance of Inconel 718 alloy deposited by LPD and traditional processing methods. Considering that Inconel 718 is widely utilized in high-temperature conditions reaching approximately 650°C, it becomes crucial to evaluate the high-temperature mechanical properties of the as-deposited Inconel 718 alloy. This evaluation aims to verify its service performance and prevent premature failure.

8.3 Heat treatment regimens

In this chapter, we employed two methods, the GLP method and the traditional CLP method, to deposit samples and obtain different as-deposited microstructures. The GLP method involved gradually reducing the laser power from 420 W to 320 W uniformly from the 1st layer to the 51st layer (d-2 in Chapter 7). On the other hand, the CLP method utilized a constant laser power of 420 W throughout the deposition process.

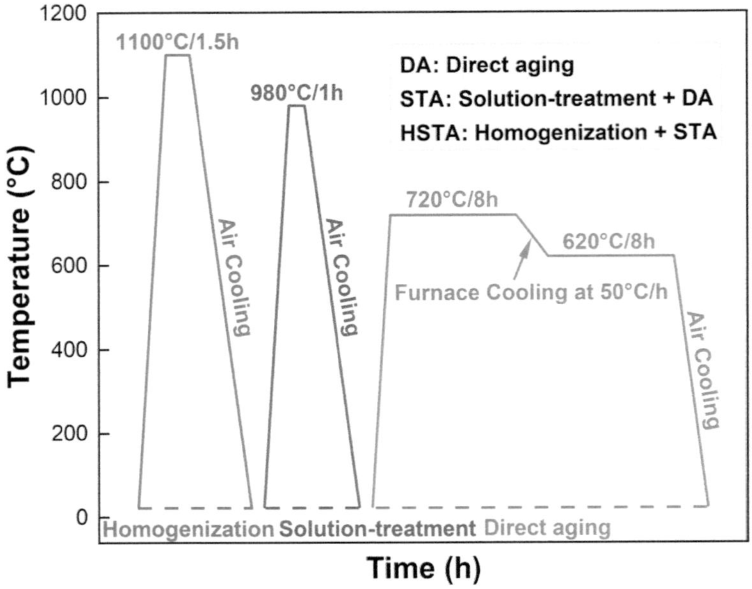

Figure 8.2. Heat treatment regimens for Inconel 718 samples.

To assess the mechanical performance and microstructural responses of the different as-deposited states under similar heat treatment conditions, we selected three typical heat treatment paths. Figure 8.2 provides a summary of the detailed steps involved in each path: direct aging (DA) heat treatment, standard solution treatment plus aging (STA), and homogenization plus STA (HSTA). We designated specific names for the GLP samples in their as-deposited states as well as for those undergoing direct aging heat treatment, standard solution treatment plus aging, and homogenization plus STA samples, namely GLP, GLP-DA, GLP-STA, and GLP-HSTA, respectively. Similarly, the samples from the CLP method were named accordingly. By comparing these various samples, we aimed to analyze the effects of different deposition methods and heat treatment paths on the mechanical properties and microstructures.

8.4 Microstructure characterization

8.4.1 As-deposited microstructures

In this study, the EDS technique was utilized to conduct a comparative analysis in a semi-quantitative manner. EDS mappings (241 μm × 181 μm) were performed at 500x magnification on unetched samples obtained through the CLP and GLP methods. The chemical compositions of the Inconel 718 alloys fabricated using both CLP and GLP methods are presented in Table 8.1. The Inconel 718 samples deposited through CLP and GLP methods demonstrate similar chemical compositions to that of the Inconel 718 powder. It is evident from the results that the GLP samples have slightly lower Nb content (4.3 wt%) compared to the CLP samples (4.7 wt%).

Table 8.1. Chemical compositions of Inconel 718 alloys fabricated by CLP and GLP methods (wt%).

Elements	Cr	Fe	Nb	Mo	Al	Ti	Mn	Ni
CLP sample	19.4	18.8	4.7	2.9	0.5	1.0	0.1	52.6
GLP sample	19.6	19.9	4.3	2.8	0.4	1.0	0.1	51.9

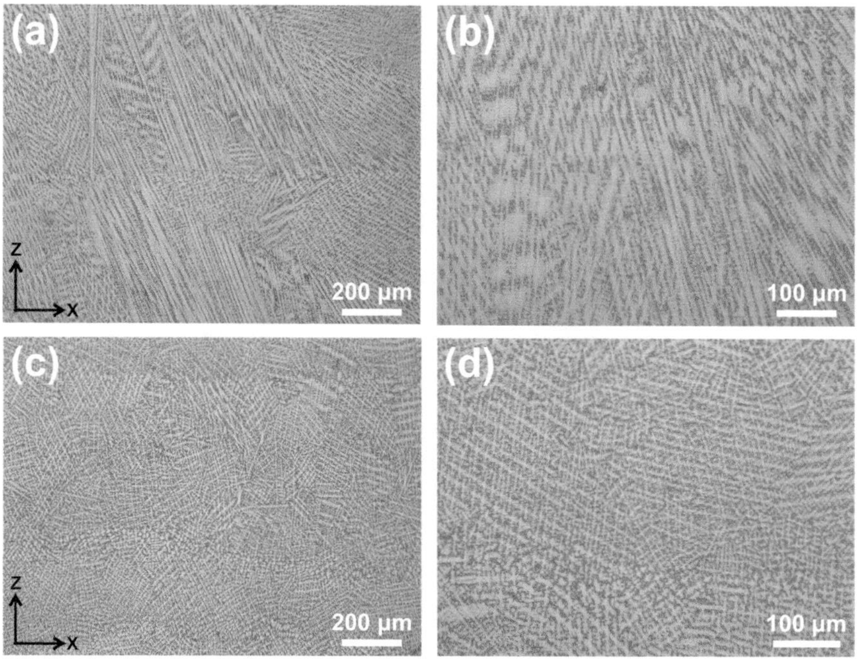

Figure 8.3. Optical micrographs of Inconel 718 thin-walls with different dendritic morphologies: (a, b) CLP sample, and (c, d) GLP sample.

Figure 8.3 illustrates the typical dendritic structures observed in as-deposited Inconel 718 samples fabricated using the CLP and GLP methods. In the case of the CLP samples, characterized by a high constant laser power, the predominant dendrites are coarse columnar structures that grow epitaxially and traverse multiple layers along the building direction. This occurrence can be attributed to the significant heat accumulation and reduced cooling rates during the process (depicted in Figure 8.3(a) and (b)). Conversely, the GLP samples exhibit a prevalence of fine columnar dendrites with a random distribution (shown in Figure 8.3(c) and (d)). The GLP approach involves gradually decreasing the laser power layer-by-layer during deposition, which enhances the stirring effect within the molten pool. As a result, the Marangoni heat flow is intensified, facilitating multi-directional heat dissipation and breaking up the dendritic structures. The remelting of the broken dendrites increases the number of heterogeneous nucleation sites, effectively impeding the epitaxial growth of grains.

Figure 8.4 illustrates the typical microstructures observed in as-deposited Inconel 718 thin-walls obtained using two different deposition methods. The presence of micro-segregation of Nb, Mo, and Ti elements leads to the formation of brittle Laves phases in the interdendritic regions. In the CLP samples, long-chain Laves phases exhibiting strong directional selectivity are observed in regions rich in these elements and located between the interdendrites (see Figure 8.4(a–c)). On the other hand, for GLP samples, discrete, fine, and granular Laves phases are randomly distributed among the dendrites (Figure 8.4(d–f)). The presence of these fine and disordered dendrites in GLP samples effectively separates the remaining liquid zone in the final stages of solidification into isolated Nb-rich regions, thus limiting the formation and growth space for Laves phases. Consequently, this promotes the formation of Laves phase particles that are evenly dispersed and fine, characterized by low content and small size. Additionally, a few MX ((Nb,Ti)(C,N)) particles are formed due to the micro-segregation of refractory elements with high concentrations and are dispersed within the γ matrix. EDS analysis has been performed on these precipitates, and the results are presented in Figure 8.4(g). Furthermore, no nanoscale strengthening phases such as γ'' and γ' are observed in the

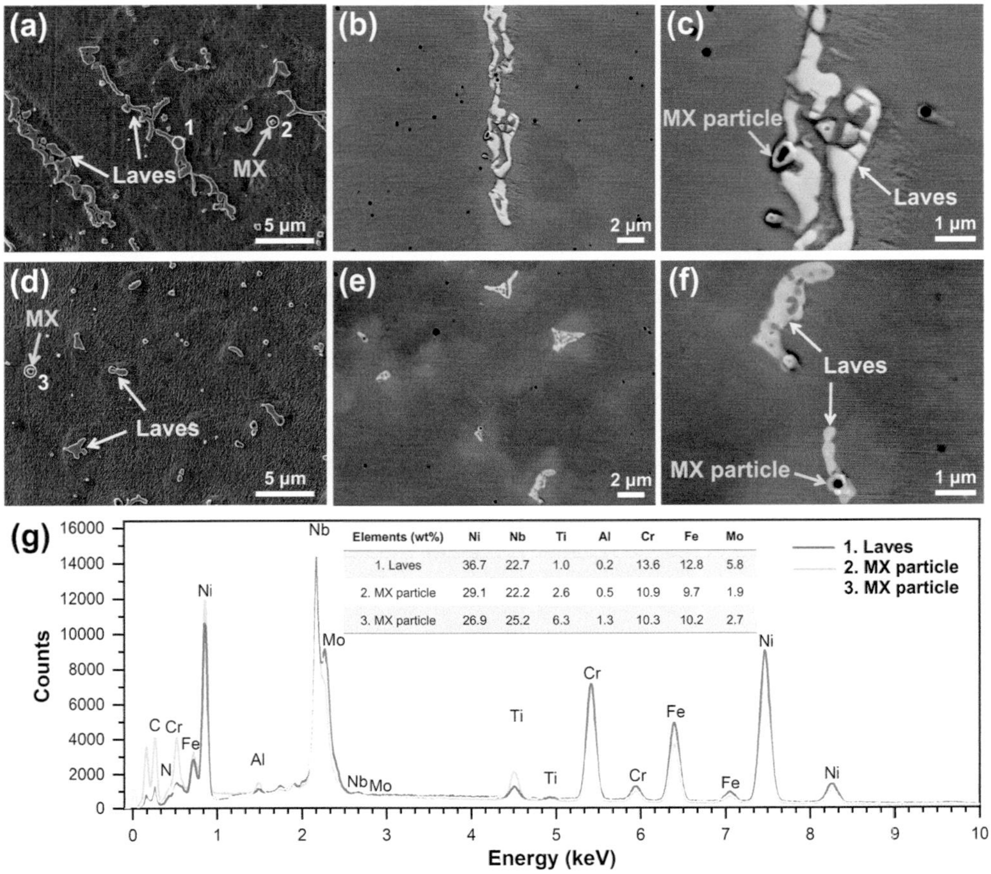

Figure 8.4. Microstructures of as-deposited Inconel 718 thin-walls: (a) SEM micrograph of CLP sample, (b, c) SEM micrographs in BSE mode of CLP sample, (d) SEM micrograph of GLP sample, (e, f) SEM micrographs in BSE mode of GLP sample, and (g) EDS analysis results.

as-deposited samples. The steep temperature gradient and rapid solidification rate associated with the LPD process effectively inhibit the sluggish precipitation of γ'' and γ' phases.

Figure 8.5(a) illustrates the FIB lamella obtained by cutting from the center of the as-deposited GLP sample along the building direction. The presence of Laves phase (Figure 8.5(b) and (c)) and MX-type precipitates (Figure 8.5(d) and (e)) in the lamella is confirmed through TEM analysis based on bright field images and corresponding selective area electron diffraction (SAED) patterns. It is evident that the Laves phases exhibit an irregular shape and possess a hard and brittle topologically close-packed structure. Moreover, the Laves phase crystal structure is hexagonal (P36/mmc), with lattice parameters of a = 0.494 nm and c = 0.787 nm. On the other hand, the refractory MX particles predominantly appear as block-shaped and exhibit relatively small sizes, measuring less than a micron in diameter. The MX particles belong to the cubic crystal structure (Fm-3m), characterized by a lattice parameter of a = 0.447 nm. The calculated lattice parameters align well with the findings reported in previous studies [24, 25].

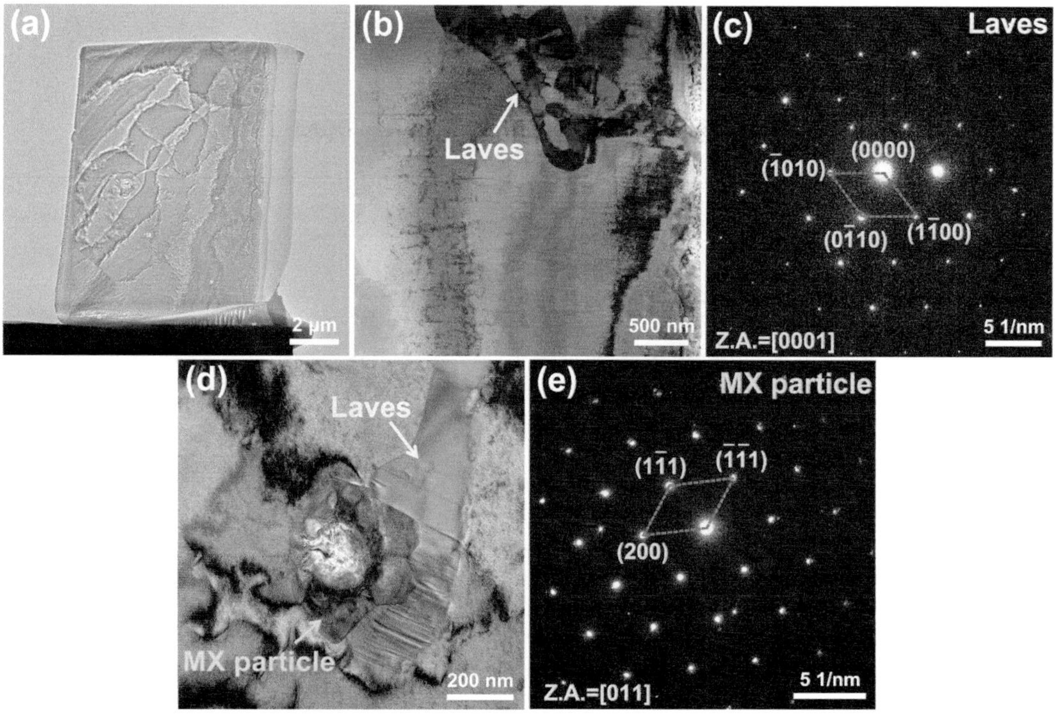

Figure 8.5. TEM analysis showing Laves phase and MX particle in as-deposited GLP sample: (a) the TEM slice after FIB processing, (b) bright field image of Laves phase, (c) corresponding SAED pattern of Laves phase shown in (b), (d) bright field image of MX particle, and (e) corresponding SAED pattern of MX particle shown in (d).

8.4.2 Heat-treated microstructures

Figure 8.6 illustrates the microstructure of etched Inconel 718 thin-wall samples in different heat-treated states. The microstructures of CLP and GLP samples after DA are similar to those in the as-deposited states. These microstructures are primarily characterized by the presence of coarse long-chain Laves phases in CLP samples and fine discrete granular Laves phases in GLP samples (Figure 8.6(a–d)). Notably, prior to the eutectic reaction during the final stage of solidification, MX carbonitrides, enriched with Nb and Ti, are formed. These particles, which exhibit a spherical or blocky shape and are embedded within the Laves phases, can act as nucleation sites (Figure 8.6(b)). Within the interdendritic region, γ'' (Ni_3Nb with a disc morphology) and γ' ($Ni_3(Al,Ti)$ with a spherical morphology) tend to cluster around the Laves phases, while their precipitation within the dendritic core is limited due to the non-uniform distribution of Nb.

The microstructures of the samples after STA heat treatment predominantly consist of composite phase particles, namely the Laves phase and the δ phase, as illustrated in Figure 8.6(e) and (g). In the case of the CLP sample, it exhibits a coarse long-chain structure composed of "Laves + δ" phase, whereas the GLP sample consists of a fine discrete arrangement of "Laves + δ" phase. Upon heat treatment, the Laves phase undergoes partial decomposition, leading to the release of Nb atoms that can diffuse into the surrounding regions. This phenomenon enhances the local driving force for the precipitation of the δ phase and promotes its formation near the Laves phase, particularly in areas rich in Nb, during the solution treatment at 980°C. The resulting δ phases exhibit acicular morphologies when observed in a 2D cross-section, consistent with the platelet-like structures reported in the literature [26]. Furthermore, during the solution heat treatment, Nb atoms can diffuse from the interdendritic region towards the dendritic core driven by the concentration difference.

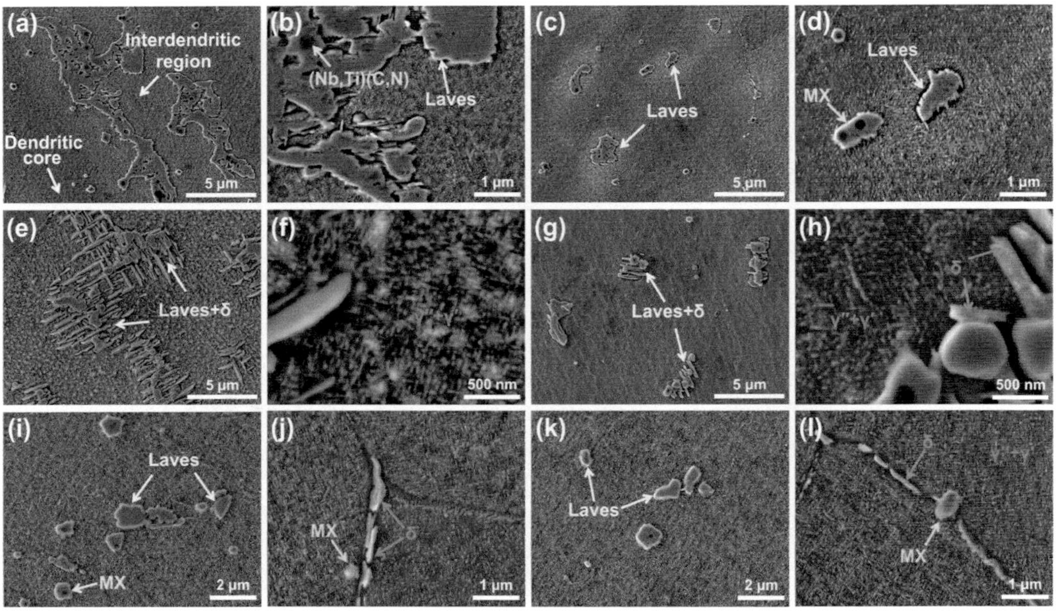

Figure 8.6. Microstructures of Inconel 718 samples following different heat treations: (a, b) CLP-DA, (c, d) GLP-DA, (e, f) CLP-STA, (g, h), GLP-STA, (i, j) CLP-HSTA, (k, l) GLP-HSTA.

Figure 8.6(f) and (h) depict the uniform distribution of γ'' and γ' phases around the "Laves + δ" phases, and these phases are also precipitated within the dendritic cores after subsequent aging.

Figure 8.6(i) and (k) depict the distribution of precipitates in the samples subjected to HSTA treatment. The Laves phases present in the HSTA-treated samples are primarily granular and sparsely distributed within the γ matrix. In the case of the conventional CLP deposition method, the long-chain Laves phase gradually dissolves during the homogenization process. Specifically, the Laves phases preferentially dissolve and form narrow connections in areas with sharp corners and grooves, where the radius of curvature is small and the solubility is high [27]. The initially sharp and uneven corners and grooves of the long-chain Laves phases gradually become smoother over time. Eventually, the Laves phases break apart and transform into granular shapes from their original long-chain structure. In contrast, the Laves phase in the as-deposited state of the GLP sample exists as discrete and fine particles, and its size becomes smaller than that of the CLP sample after undergoing HSTA heat treatment.

Figure 8.6(j) and (l) illustrate the formation of intermittent short rods and acicular δ phases along grain boundaries or at triple junctions. These locations commonly exhibit high free energies, making them favorable sites for the nucleation and growth of δ phases when the elemental compositions permit. Due to the homogenization of Nb, the γ'' and γ' phases are more uniformly distributed within the γ matrix. Notably, the MX particles are observed both within the grains and at the grain boundaries in all heat-treated states. These carbides and nitrides, present in small quantities, are relatively stable phases with high melting points and can still be retained even after undergoing high-temperature homogenization.

The occurrence of micro-segregation of refractory elements and the development of Laves phases are frequently observed in as-deposited samples of Inconel 718 during the non-equilibrium rapid solidification process in LPD process. Among the various segregated elements in Inconel 718, Nb stands out as one of the most prominently segregated elements, and its micro-segregation can significantly influence the resulting microstructure. To quantify the extent of micro-segregation, the segregation rate (*SR*) is utilized. The segregation rate of Nb can be determined using the equation provided below [28]:

$$SR = \frac{C_{id}}{C_{dc}} \tag{8.1}$$

where C_{id} represents the Nb element content in the interdendritic region, and C_{dc} is the Nb element content in the dendritic core. When $SR < 1$, the solute element tends to segregate into the dendritic core. Conversely, when $SR > 1$, the solute element exhibits a tendency to segregate towards the interdendritic regions.

Figure 8.7 presents the *SR* of Nb for different samples. The *SR* values are all greater than 1, indicating that Nb has a significant inclination to segregate towards the interdendritic regions. In comparison to the *SR* value of 5.66 for the CLP samples, the as-deposited GLP sample demonstrates a lower *SR* value of 4.23. The *SR* of the GLP sample is greatly reduced after undergoing STA heat treatment, whereas the HSTA heat treatment exhibits a strong reduction effect on the *SR* of the CLP sample. As the heat treatments progress from DA to STA and finally to HSTA, both deposition methods show a decreasing trend in *SR* values. Following the HSTA heat treatments *SR* values of the GLP (3.13) and CLP (3.12) samples are comparable.

It is widely acknowledged that severe micro-segregation leads to an uneven distribution of refractory elements and promotes the formation of harmful Laves phases. Therefore, reducing micro-segregation is a crucial aspect of optimizing the microstructure and enhancing the mechanical properties of as-deposited Inconel 718. The reduction in micro-segregation degree can be attributed to two main factors. In the as-deposited condition, it is influenced by the cooling rate and dendritic growth mode during the LPD process. In the case of heat-treated samples, it is associated with elemental diffusion and redistribution.

As mentioned earlier, the measured *SR* indicate that the GLP sample exhibits lower Nb segregation between the dendritic core and the interdendritic region compared to the CLP sample in the as-deposited condition. Our previous study has confirmed that the GLP deposition method achieves higher cooling rates and facilitates the formation of finer disordered dendrites by reducing the laser power layer by layer, as opposed to the conventional CLP LPD process that utilizes a constant laser power [29]. During the non-equilibrium solidification in the LPD process, the Nb element, having a negative distribution coefficient, is continuously rejected from the solid phase to the liquid phase. Consequently, Nb accumulates at the forefront of the solid-liquid interface, leading to the occurrence of coring effect within the γ matrix dendrites and an increase in Nb content from the dendritic core to the interdendritic region [30].

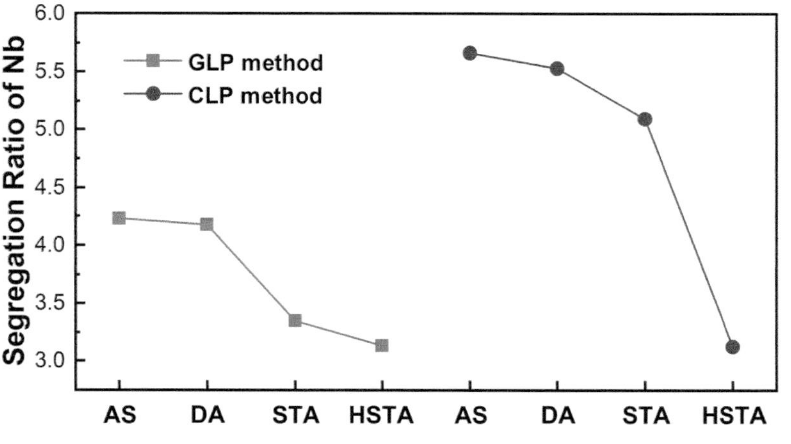

Figure 8.7. Segregation ratio of Nb in Inconel 718 samples without and with heat treatments.

It has been confirmed that there is an inverse relationship between the cooling rate and segregation ratio [31]. A higher cooling rate has the ability to reduce the solidification time of the mushy zone. Due to this rapid cooling, the Nb element has limited time for elemental redistribution and remains trapped in the γ matrix. Consequently, the solubility of Nb increases, suppressing micro-segregation and resulting in the formation of relatively fine and discrete Laves phase particles in interdendritic regions. These characteristics define the microstructure of as-deposited GLP samples.

Additionally, the dendritic growth mode significantly influences the segregation pattern observed in as-deposited samples. The CLP deposition method, characterized by a low cooling rate, produces coarse columnar dendrites. Conversely, the GLP process, which employs a high cooling rate, promotes the formation of refined columnar dendrites with disordered growth directions. This means that the solute elements have a shorter distance to travel through the dendritic core in order to compensate for the composition gradient, leading to a reduction in the *SR*.

During the rapid solidification process of LPD, certain refractory elements, such as Nb, Mo, and Ti, which have relatively larger atomic radii, tend to segregate within the interdendritic regions of the γ matrix. Subsequently, during post-heat treatments, these segregated elements diffuse to different extents in order to reduce the overall free energy of the system [32]. The diffusion and redistribution of these segregated elements are primarily influenced by various factors including the temperature and duration of the heat treatment, the size of the dendrites in the as-deposited samples, and the initial segregation ratio in the as-deposited state.

The diffusion coefficient (D_{Nb}) is dependent on temperature and can be mathematically described using the Arrhenius relation [16]:

$$D_{Nb} = D_0 \exp\left(-\frac{Q}{RT}\right) \tag{8.2}$$

where D_0 represents the diffusion coefficient, Q denotes the activation energy for diffusion, R stands for the gas constant (8.31 J/(mol·K)) and T represents the absolute temperature. In the low concentration range of Nb, the interdiffusion coefficient of Nb remains essentially constant and independent of Nb concentration [33, 34]. The interdiffusion coefficients of Nb at different temperatures exhibit a strong Arrhenius behavior. Karunaratne et al. [35] conducted a study on Ni-Nb systems in the Ni-fcc phase and reported the diffusion coefficient and activation energy for Nb interdiffusion as 8.8×10^{-5} m^2/s and 257.0 kJ/mol, respectively. By applying Eq. (8.2) for calculation, the interdiffusion coefficients of Nb in Ni at 1100°C, 980°C, 720°C, and 620°C are determined as 1.46×10^{-14} m^2/s, 1.68×10^{-15} m^2/s, 2.63×10^{-18} m^2/s, and 8.06×10^{-20} m^2/s, respectively. The diffusion coefficient of Nb significantly increases with higher heat treatment temperatures. During the HSTA heat treatment process, the elevated homogenization temperature leads to increased atomic kinetic energy. Consequently, more atoms can surpass the energy barrier and diffuse rapidly due to energy fluctuations [36]. Moreover, the concentration of vacancies also rises at elevated temperatures, promoting the diffusion of segregated components. These factors contribute to diffusion-induced grain boundary migration, resulting in smoother grain boundaries and significant grain growth, as clearly depicted in Figure 8.9(g) and (h).

The diffusion occurring between the dendritic core and the interdendritic region is a non-steady-state process that depends on the dendrite arm spacing, which is associated with the characteristic diffusion distance. When considering only solute atom diffusion, the required time (t) to reach 1% of the initial compositional segregation amplitude can be described as [37]:

$$t = 0.467\frac{(\lambda/2)^2}{D_{Nb}} \tag{8.3}$$

where λ represents the primary dendrite arm spacing (PDAS). The average PDAS of the as-deposited CLP and GLP samples are 11.03 μm and 8.37 μm based on measurements by Image-Pro-Plus software, respectively. To simplify the calculation, the interdiffusion coefficients of Nb in Nb–Ni alloy at different temperatures are applied to Eq. (8.3).

The theoretically calculated time required for the uniform diffusion of Nb within the γ matrix at different heat treatment temperatures for the CLP and GLP samples is presented in Table 8.2. The calculations reveal that the time needed for the uniform diffusion of Nb in the fine dendrites of the GLP samples is shorter compared to the coarse CLP samples at the same heat treatment temperature. Specifically, the CLP and GLP samples achieve uniform element distribution in approximately 16.26 minutes and 9.36 minutes, respectively, when subjected to a heat treatment temperature of 1100°C. However, microstructural analysis indicates the presence of residual Laves phases in both the CLP and GLP samples after undergoing HSTA heat treatment for 1.5 hours at 1100°C. The time required to achieve complete element homogeneity is significantly higher than the theoretically calculated value. This suggests that the redistribution of segregated elements is not solely governed by element diffusion but also influenced by the interface reaction-controlled mechanism during the dissolution of Laves phases. Furthermore, the interdiffusion coefficients used in the calculations were obtained by fitting and extrapolating experimental data within the temperature range of 900–1300°C (as mentioned in Ref. [35]). As a result, there is relatively large deviation in the interdiffusion coefficients of Nb at lower aging temperatures. Consequently, the calculated time required for the uniform diffusion of Nb within the γ matrix during aging heat treatment exhibits relatively large values. At lower aging temperatures such as 720°C and 620°C, aging has minimal impact on achieving element homogeneity. In such cases, the primary purpose of aging is to precipitate abundant γ'' and γ' phases for strengthening the material.

During STA and HSTA heat treatments, the disappearance of Nb-rich segregated regions and the reduction in volume percentage of the Laves phase are observed. The vanishing of Nb-rich regions can be attributed to the diffusion of segregated elements and a decrease in concentration gradient between the dendritic core and interdendritic region. And the back-diffusion of the atoms from the Laves phases into the γ matrix is responsible for the Laves phase reduction [38].

Additionally, the dissolution of Laves phases is influenced by factors such as temperature, time of heat treatments, degree of micro-segregation, content, size, morphology, and distribution of Laves phases. The dissolution of Laves phases during heat treatment can be categorized into three processes: (1) decomposition of the Laves phase, (2) short-range diffusion of solute atoms across the Laves/γ interface towards the γ matrix, and (3) long-range diffusion of solute atoms within the γ matrix [39]. The slowest among these processes determines the rate of Laves phase dissolution. The first two processes occur rapidly and are controlled by interface reactions, while the last process is governed by the long-range diffusion of solute atoms.

According to the research of Zhao et al. [40], the dissolution of Laves phase can be described by the following equation:

$$\tau = A\exp(-0.036T) \tag{8.4}$$

where τ represents the complete elimination time of Laves phase, A is a coefficient associated with the degree of micro-segregation and T denotes the temperature in degree Celsius. A reflects the element diffusion ability. The as-deposited GLP samples exhibit a low segregation rate, resulting in a smaller value of A. Consequently, at the same heat treatment temperature, the dissolution time for the Laves phase is shorter compared to the CLP sample.

Table 8.2. Theoretically calculated required time (t) for the uniform diffusion of Nb in the γ matrix at different heat treatment temperatures for the CLP and GLP samples.

	1100°C	980°C	720°C	620°C
CLP sample	16.26 min	2.34 h	1498.00 h	48942.88 h
GLP sample	9.36 min	1.35 h	862.60 h	28183.14 h

Moreover, the content, size, morphology, and distribution of the Laves phases play crucial roles in their dissolution. It has been discovered that the dissolution rate of Laves phase particles is inversely proportional to their curvature radii [37]. Regions with smaller curvature radii dissolve preferentially due to the higher concentration at the interface. Therefore, during STA and HSTA heat treatments, the coarse, elongated Laves phases in the CLP samples gradually transform into granular ones, while the discrete and granular Laves phases in the GLP samples decrease in size. Additionally, studies have reported that smaller-sized Laves phases and a smaller PDAS in the as-deposited microstructure contribute to accelerating the dissolution of Laves phases [41].

Numerous scholars have conducted extensive research on the dissolution mechanism of the Laves phase in as-deposited Inconel 718 during the heat treatment process. Sui et al. [5] employed both the Johnson-Mehl-Avrami-Kolmogorov model and the Singh-Flemings model to investigate the kinetics of Laves phase dissolution. Initially, the dissolution of the Laves phase is influenced by the long-range diffusion of Nb as well as the interfacial reaction. However, as the degree of Nb micro-segregation decreases, the dominant mechanism shifts towards the interfacial reaction. In a study by Zhang et al. [37], it was highlighted that element diffusion does not play a significant role in the dissolution of the Laves phase in L-DED-built Inconel 718. They employed a three-dimensional diffusion-controlled model (Jander equation) and a phase boundary reaction-controlled model to explore the primary mechanism and key factors controlling the dissolution of the Laves phase in the as-built Inconel 718 samples, respectively. The findings revealed that during the early stages of homogenization, Laves phase dissolution is primarily controlled by three-dimensional diffusion. However, with increasing holding time and further dissolution of the Laves phase, the chemical potential gradient at the interface and the concentration difference between the Laves phase and γ matrix gradually diminish. Consequently, the rate of interface reaction slows down, and the interface reaction process gradually becomes the dominant mechanism in the dissolution of the Laves phase.

8.4.3 Phase analysis by XRD

Figure 8.8 illustrates the XRD diffraction patterns of as-deposited samples and samples treated with various heat treatments using the CLP and GLP deposition methods. In all XRD patterns, clear diffraction peaks corresponding to the Ni-Fe-Cr rich γ matrix with a face-centered cubic (fcc) structure, including the primary (111), (200), (220), and (311) peaks, are observed. Additionally, the presence of Laves phase peaks can be detected in both the as-deposited CLP and GLP samples. Interestingly, for the CLP samples, the Laves phase peaks are only absent after undergoing the HSTA heat treatment. But it is challenging to identify the Laves phase in the GLP samples after undergoing STA or HSTA heat treatments. Following the HSTA heat treatment, the XRD patterns of both CLP and GLP samples exhibit a random crystallographic orientation similar to that of Inconel 718 powders.

From Figure 8.8, it is evident that the peaks of the DA samples shift towards higher 2θ values compared to the as-deposited samples. According to Bragg's diffraction law, this increase in the 2θ values suggests a decrease in interplanar spacings. This phenomenon is believed to be associated with elemental micro-segregation or the formation of precipitates, which can reduce the solid solubility in the γ matrix [42]. During the DA process, the elements Nb, Al, and Ti are depleted in the γ matrix, and precipitates of γ'' and γ' form in the interdendritic regions enriched with elements. This leads to an increase in the 2θ values. For STA and HSTA heat treatments, during the high-temperature solid solution and homogenization processes, micro-segregation decreases, and solid solubility increases, resulting in a decrease in the 2θ values. However, after subsequent DA, the precipitation of γ'' and γ' phases and the formation of MX carbides cause an enlargement in the 2θ values. As a result of these combined effects, the 2θ values remain nearly unchanged.

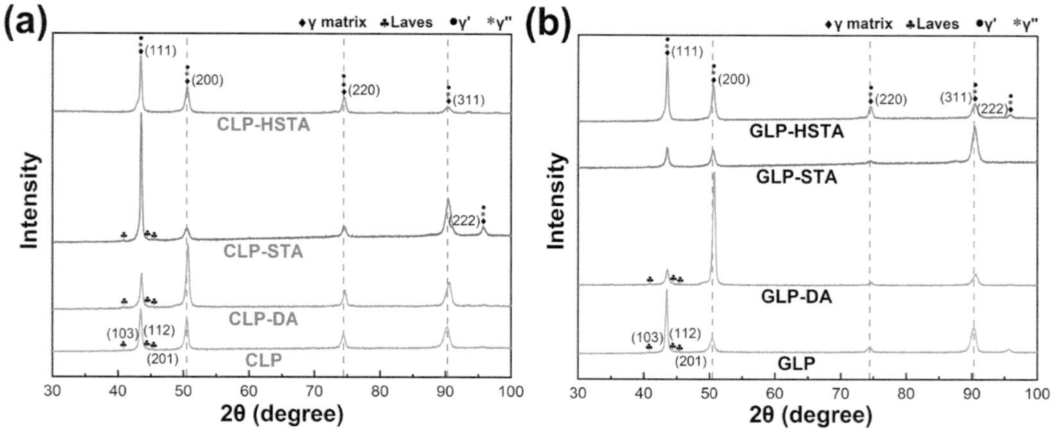

Figure 8.8. XRD patterns of as-deposited and variously heat-treated Inconel 718 samples: (a) for CLP deposition method, and (b) for GLP deposition method.

Furthermore, distinguishing the overlapping patterns of γ' and γ" phases from the γ matrix is highly challenging. This difficulty arises because γ' and γ" phases are respectively coherent and semi-coherent with the γ matrix, and they have only slight differences in their lattice parameters [43]. However, no δ phase or MX particles are detected in the XRD patterns, likely due to their low volume fractions, which are less than 3%.

8.4.4 Evolution of crystallographic texture and grain size

EBSD inverse pole figures of the γ matrix in the central region of as-deposited Inconel 718 samples, subjected to various heat-treated conditions, are depicted in Figure 8.9. The corresponding results for mean grain size (equivalent circle diameter) and grain aspect ratio can be found in Table 8.3. Figure 8.9(a) reveals that the CLP samples, deposited using constant laser powers, exhibit coarse and elongated columnar grains that epitaxially grow along the building direction (<001> texture). These grains have an average equivalent circle diameter of 67.0 μm and an aspect ratio reaching 2.8, owing to the intense heat accumulation. On the other hand, the GLP samples, as shown in Figure 8.9(b), exhibit a weak texture and consist of a mixture of equiaxed and fine columnar grains, characterized by a smaller grain size of 46.5 μm and a reduced aspect ratio of 2.2. Furthermore, Figure 8.9(c) and (d) demonstrate that the grain size and morphology exhibit minimal changes following the DA heat treatment.

Upon undergoing STA heat treatment, the CLP samples experience a reduction in the size of coarse columnar grains, accompanied by an increase in the fraction of fine equiaxed grains (Figure 8.9(e)). This suggests the occurrence of partial recrystallization. Conversely, the grains in the GLP samples exhibit growth after STA heat treatment, while a considerable number of fine equiaxed and columnar grains are still retained, as depicted in Figure 8.9(f). Moreover, Figure 8.9(g) and (h) reveal that more effective recrystallization and grain growth occur following HSTA heat treatment. The grain boundaries migrate at a higher velocity and become smoother at the elevated homogenization temperature. Compared to the CLP samples, the GLP samples exhibit noticeable grain coarsening, resulting in a final grain size larger than 84.9 μm.

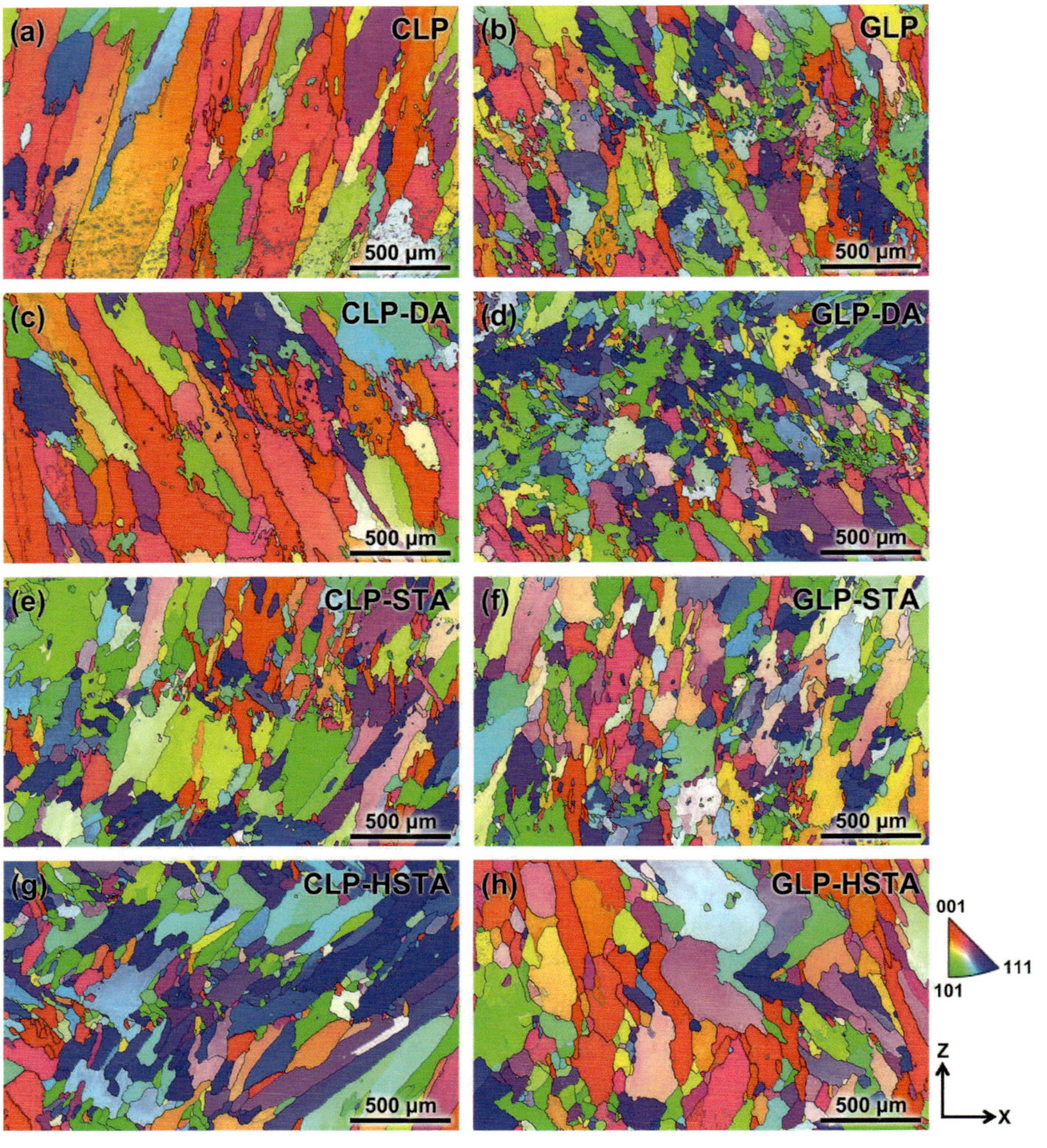

Figure 8.9. Inverse pole figures of Inconel 718 thin-wall samples: (a) CLP, (b) GLP, (c) CLP-DA, (d) GLP-DA, (e) CLP-STA, (f), GLP-STA, (g) CLP-HSTA, and (h) GLP-HSTA.

Table 8.3. Mean grain size and grain aspect ratio results of Inconel 718 samples under various heat treatment conditions.

	CLP	DA-CLP	STA-CLP	HSTA-CLP	GLP	DA-GLP	STA-GLP	HSTA-GLP
Equivalent circle diameter (μm)	67.0	62.1	55.2	66.3	46.5	42.3	55	84.9
Fitted ellipse aspect ratio	2.8	2.3	2.2	2.3	2.2	2.0	2.2	2.3

8.5 Microhardness

Figure 8.10 illustrates the average microhardness values obtained from as-deposited Inconel 718 samples subjected to various post-heat treatments. It can be observed that both the GLP samples before and after heat treatments exhibit higher hardness compared to the CLP samples. The as-deposited samples display the lowest microhardness due to the absence of strengthening precipitates. The as-deposited GLP samples experience rapid cooling rates during the LPD process, resulting in increased solubility of alloying elements, particularly Nb, within the interdendritic region of the γ phase. This phenomenon contributes to the formation of fine dendrites and discrete Laves phases, which in turn enhance the overall hardness of the material.

After undergoing heat treatment, the hardness of the material significantly increases due to the formation of γ'' and γ' phases through precipitation hardening. The hardness values of the heat-treated samples exceed the minimum requirements for the wrought Inconel 718 samples, which is approximately 355–385 HV [4].

The GLP sample itself exhibits high hardness (417.5 HV) after direct aging alone, as it has less micro-segregation. Following STA heat treatment for the GLP samples, the partially dissolved Laves phases and improved element distribution contribute to the availability of more free Nb. This facilitates the formation of strengthening phases, leading to a comparable hardness achieved after HSTA heat treatment. On the other hand, the as-deposited CLP samples contain more long-chain Laves phases and experience microstructural coarsening. Consequently, there are fewer strengthening phases, which are not uniformly distributed, resulting in a lower hardness after DA alone. Although the hardness increases after STA heat treatment, there are still residual Laves phases and precipitated δ phases, which slightly impair the hardness. Despite the grain size increase observed after HSTA heat treatment, there is a significant reduction in micro-segregation, and the element distribution becomes more uniform. A larger quantity of γ'' and γ' precipitates uniformly form, utilizing the released Nb solute from the dissolution of Laves phases. This leads to the CLP samples achieving its maximum hardness (443.3 HV). Additionally, the presence of carbides distributed along the grain boundaries contributes to the strengthening effect by impeding the migration of grain boundaries [44].

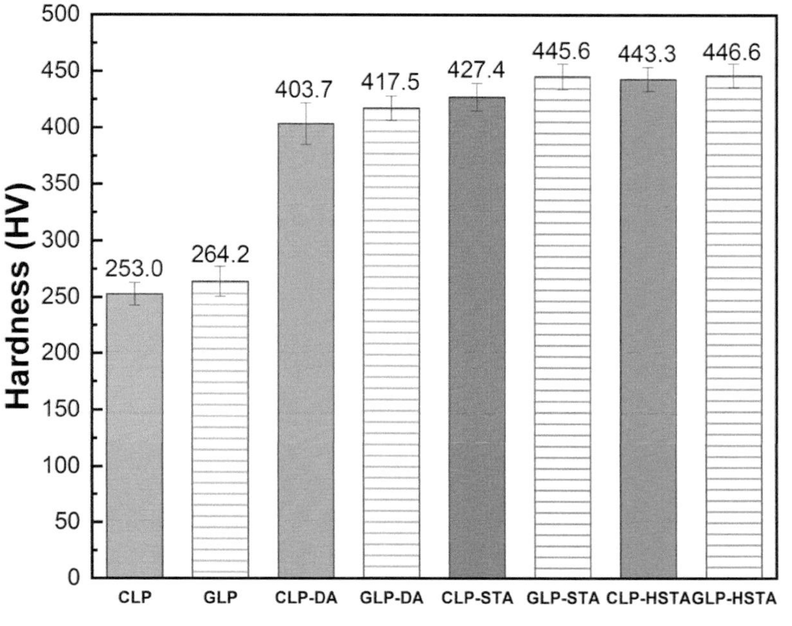

Figure 8.10. Average microhardness in Inconel 718 samples for various heat-treated conditions.

8.6 Tensile properties

8.6.1 Room-temperature results

Figure 8.11 illustrates the standard room-temperature tensile test curves of the samples and provides statistical data on yield strength (YS), ultimate strength (UTS), and elongation. The room temperature tensile properties of standard Inconel 718 forgings from the AMS5662 standard are represented by the gray dotted line: YS = 1034 MPa, UTS = 1276 MPa, and elongation = 12%. Notably, the as-deposited GLP samples exhibit superior ductility, with an elongation of up to 37.41%, compared to the as-deposited CLP samples.

Upon undergoing post-heat treatment, all samples experience a significant increase in strength, accompanied by a reduction in elongation. The GLP and CLP samples exhibit overall increasing trends in YS, UTS, and elongation as the heat treatment progresses from DA to STA and finally to HSTA. Importantly, the strength of the GLP samples after heat treatment surpasses that of the Inconel 718 forged samples and outperforms the CLP samples subjected to the same heat treatments. Among all the heat treatments, the GLP samples treated with HSTA demonstrate the highest YS (1297.9 MPa) and tensile strength (1442.5 MPa). Comparing the elongation values, the GLP samples exhibit a larger elongation after DA (7.16%) compared to the CLP samples (4.82%). However, the elongations of the samples after STA and HSTA heat treatments are comparable between the two deposition methods. It is worth noting that only the samples subjected to HSTA heat treatment, regardless of the deposition method, achieve the elongation observed in the wrought Inconel 718 alloy.

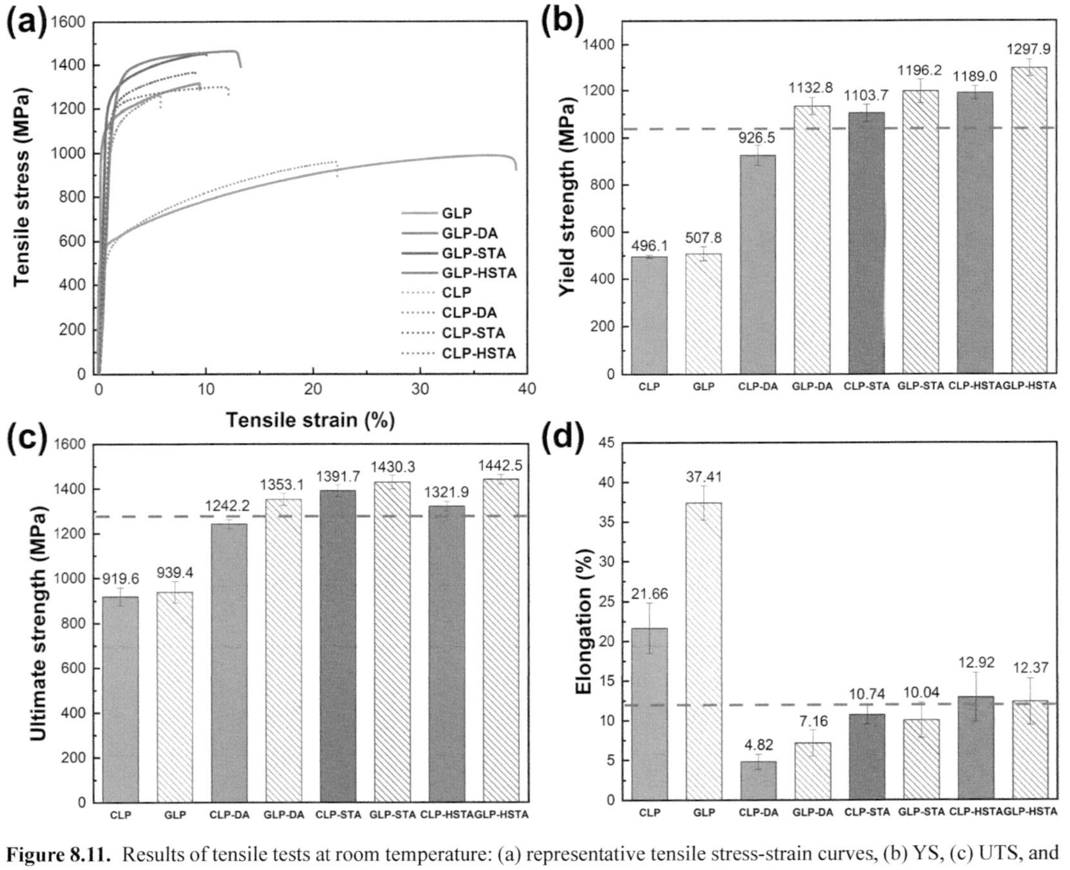

Figure 8.11. Results of tensile tests at room temperature: (a) representative tensile stress-strain curves, (b) YS, (c) UTS, and (d) elongation.

Table 8.4. Comparison of mechanical properties of Inconel 718 deposited by LPD under various heat treatment conditions.

Heat treatment	YS (MPa)	UTS (MPa)	Elongation (%)	Reference
DA	1132.8	1353.1	7.16	GLP in current work
	977	1182	7.7	[45]
	1084	1333	8.4	[46]
	1035	1294	10.5	[47]
STA	1196.2	1430.3	10.04	GLP in current work
	1013	1192	16	[20]
	1007	1221	16	[46]
	1060	1290	15.4	[47]
HSTA	1297.9	1442.5	12.37	GLP in current work
	949	1194	19.9	[46]
	1127	1303	16	[48]
	1133	1240	11	[49]
	1170	1360	18	[49]

According to the findings presented in Table 8.4, we have conducted a comprehensive analysis of the mechanical properties of GLP samples and Inconel 718 samples produced via LPD, as documented in various literature sources. These analyses were performed under different heat treatment conditions. Remarkably, when compared to the data reported in the literature, the GLP samples demonstrated exceptional strength across all three heat treatments, with the GLP samples subjected to HSTA heat treatment exhibiting the most notable improvements. Specifically, the GLP-STA sample and the GLP-HSTA sample displayed increases in YS of over 160 MPa and 200 MPa, respectively, compared to the reported results. Additionally, the UTS of the GLP-STA sample and the GLP-HSTA sample increased by more than 190 MPa and 160 MPa, respectively. In terms of elongation, the GLP-DA samples exhibited results similar to those reported in the literature. However, both the GLP-STA sample and the GLP-HSTA sample displayed slightly lower elongation values compared to the literature. This can be attributed to the fact that increased strength tends to diminish ductility to some extent. Nonetheless, considering the balance between strength and ductility, the GLP-HSTA sample still showcased outstanding mechanical properties.

As widely acknowledged, multiple factors influence the tensile mechanical properties of Inconel 718 samples produced through LPD. These factors include micro-segregation, precipitates, grain size, crystal orientation, and more. Inconel 718 is an alloy strengthened through precipitation. The formation of the strengthening phase serves as the foundation for precipitation hardening, significantly impacting its strength. The coherent γ' phase acts as an auxiliary strengthening phase, providing ordering strengthening and creating barriers to impede dislocation movement. And the semi-coherent γ'' phase, serving as the primary strengthening phase, exhibits enhanced strengthening effects due to its relatively large lattice misfit with the γ matrix and higher anti-phase boundary energy resulting from both ordering and coherency strengthening mechanisms [50].

The microstructure and distribution of Nb in the initial as-deposited states play a significant role in the precipitation of strengthening phases during subsequent heat treatment. It is evident that the as-deposited samples exhibit low strength, primarily due to the presence of a soft γ matrix and inadequate precipitation of γ'' and γ' phases. In regions within the grains where there are fewer obstacles, dislocations can easily move and propagate, making it challenging to form dislocation pile-ups and stress concentrations during plastic deformation. In terms of room temperature tensile properties (refer to Figure 8.11), the YS and UTS of samples produced using the GLP technique consistently outperform those produced using the CLP technique after undergoing the same heat treatment. This is attributed to the fact that the as-deposited GLP samples possess finer dendritic structures, reduced micro-segregation, and finer granular Laves phases compared to the CLP samples. Following the same heat treatment, the GLP samples exhibit smaller, less brittle, and hard precipitated phases. Moreover, they showcase a more homogeneous distribution of elemental constituents and a more uniform precipitation of γ'' and γ' phases, ultimately leading to higher strength.

It has also been noticed that the ductility of samples undergoes significant degradation following heat treatments. This degradation can be attributed to several factors. Firstly, the increasing presence of γ'' and γ' precipitates negatively affects ductility due to precipitation hardening. Secondly, the decline in elongation can be attributed to the formation of incoherent δ phase within the grain interior and grain boundaries, which is detrimental to ductility. As the heat treatment progresses from DA, to STA, and finally to HSTA, the differences in elongation between CLP and GLP samples gradually diminish, and the superior ductility of the GLP samples gradually diminishes as well. This observation is supported by several scholars, who have found that coarse and continuous long-chain Laves phases have a more adverse impact on ductility compared to fine and discrete ones. These long-chain Laves phases create preferred sites and paths for crack initiation and propagation [5, 51]. For CLP samples, the micro-segregation and uneven distribution of γ'' and γ' strengthening phases gradually decrease after heat treatment, along with a reduction in the size and content of the long-chain Laves phases. Similarly, the positive effect of the fine and discrete Laves phases with dispersed distribution found in the GLP samples on ductility gradually diminishes.

8.6.2 High-temperature results

Figure 8.12 illustrates the high-temperature tensile characteristics (650°C) of various types of Inconel 718 samples. The gray dotted line represents the tensile properties at 650°C of standard Inconel 718 forgings according to the AMS5662 standard: YS = 862 MPa, UTS = 1000 MPa, and elongation = 12%. As depicted in Figure 8.12, the high-temperature tensile strengths of both CLP and GLP samples show significant degradation compared to the results obtained at room temperature. This phenomenon can be explained by the thermally assisted movement of dislocations and the increased mean free path of atoms at elevated temperatures, which weakens the bonding strength [52]. Although the as-deposited samples exhibit lower strengths at high temperatures, the GLP sample maintains good elongation.

In the typical tensile stress-strain curves (Figure 8.12(a)), there is noticeable serrated flow during the yield stages for the as-deposited samples, which may be attributed to dynamic strain aging (DSA) or twin deformation in Ni-based alloys [53]. The distinctive behavior known as serrated yield is characterized by periodic drops in flow stress below the overall stress-strain curve, resulting from dislocation unlocking at high temperatures. The occurrence of this behavior is primarily attributed to DSA, which is influenced by the diffusion energy and solute-dislocation binding energy [22]. The concentration of substitutional elements in the γ matrix affects the serrated behavior. In the as-deposited samples, several substitutional elements, such as Mo and Nb, are dissolved in the γ matrix, while there is minimal precipitation of strengthening phases like γ'' and γ'. Among the substitutional atoms, Nb and Mo are considered to be the most effective in the DSA mechanism due to their significant contribution to the expansion of the Ni lattice size, leading to an increased binding energy between solute atoms and dislocations.

The heat-treated samples generally outperform the conventional standard Inconel 718 forgings in terms of strength under high-temperature tensile tests, particularly the samples subjected to STA and HSTA heat treatments. For both deposition methods, the YS and UTS are comparable in their as-deposited Inconel 718 samples and the same heat-treated Inconel 718 samples, respectively. The high-temperature tensile ductilities of the as-deposited and DA samples produced using CLP and GLP methods remain similar trends to their counterparts at room temperature. After DA, the GLP samples exhibit greater elongation at elevated temperatures compared to room temperature. It is worth noting that the high-temperature ductility of samples subjected to STA heat treatment is

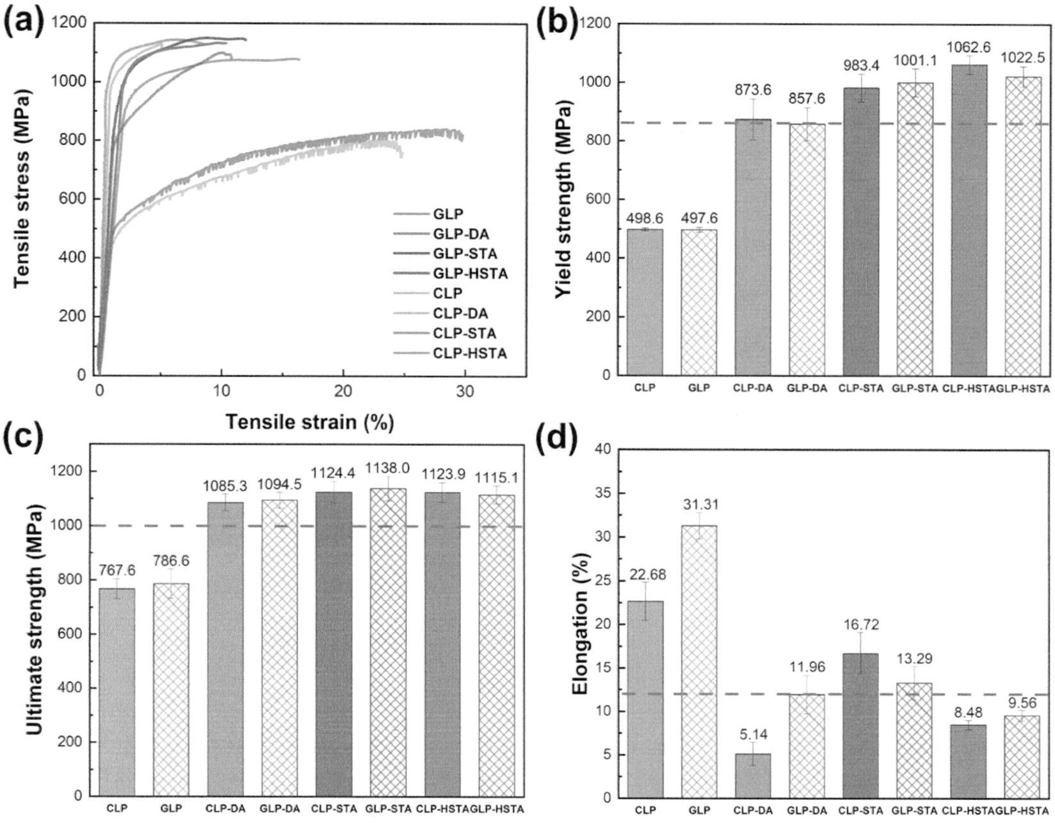

Figure 8.12. Results of tensile tests at high temperatures (650°C): (a) typical tensile stress-strain curves, (b) YS, (c) UTS, and (d) elongation.

superior. The CLP samples after the STA heat treatment, demonstrate a higher elongation (16.72%) than the GLP samples (13.29%), which surpasses the results of wrought Inconel 718 samples.

8.7 Fracture morphologies

8.7.1 Room-temperature results

To investigate the relationship between tensile characteristics and microstructures, the fracture morphologies were analyzed using a scanning electron microscope. Figure 8.13 presents the representative fracture surface morphologies of thin-wall Inconel 718 samples subjected to room-temperature tensile tests. In Figure 8.13(a) and (b), the fracture surface of conventionally CLP samples after DA reveals a dendritic pattern resembling a staircase. Within the large and deep dimples, numerous elongated Laves fragments are aligned along the direction of the columnar dendrites. Similarly, for the DA-treated GLP samples (Figure 8.13(c) and (d)), the fracture surface exhibits disordered columnar dendrites with random orientations. Additionally, discrete Laves fragments can be observed in the bottom regions of the dimples. The presence of micro-cracks resulting from the fragmentation of the Laves phase is evident in both these sample types. It is noteworthy that both CLP and GLP samples, after DA treatment, preferentially fracture along the interdendritic regions, demonstrating a transgranular ductile failure mode.

Upon subjecting the CLP sample to STA (Figure 8.13(e) and (f)), linearly distributed dimples and "Laves + δ" fragments are observed on the fracture surface, indicating that fractures occur between

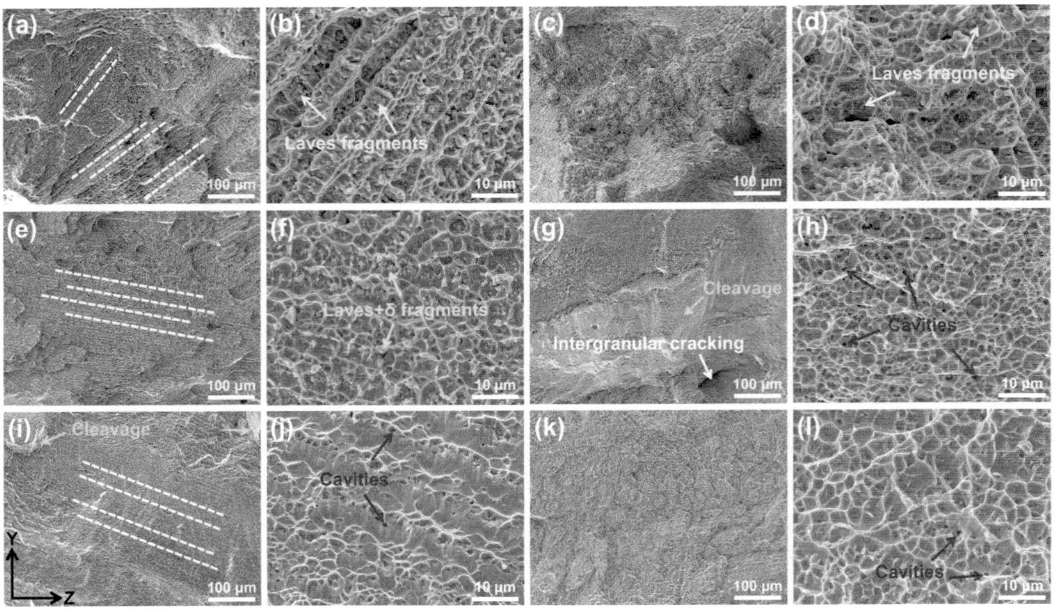

Figure 8.13. Fracture morphologies of Inconel 718 thin-wall samples after room-temperature tensile tests: (a, b) CLP-DA, (c, d) GLP-DA, (e, f) CLP-STA, (g, h), GLP-STA, (i, j) CLP-HSTA, and (k, l) GLP-HSTA.

columnar dendrites. In the case of the GLP sample after STA (Figure 8.13(g) and (h)), cleavage planes can be observed on the fracture surfaces, indicating intergranular cracking. Furthermore, a significant number of micro-cavities are visible within the fine and shallow dimples of the STA-treated GLP sample, resulting from the interfacial debonding of granular "Laves + δ" phases from the γ matrix. Although the presence of "Laves + δ" fragments is noticeable, the primary mode of failure in STA-treated GLP samples is attributed to the separation between the "Laves + δ" phase and the γ matrix. This behavior can be ascribed to the surrounding δ phase playing a role in the pinning effect, which inhibits the cracking and fragmentation of the Laves phase.

In the case of CLP samples treated with HSTA, the distribution of dimples becomes more uniform and shallow, and the cleavage plane can be observed on the fractured surface, indicating a combination of transgranular and intergranular fracture modes (see Figure 8.13(i) and (j)). Close to the dimples, several micro-cavities are present, which are believed to be a result of the separation of fine and globular Laves phases, carbides, nitrides, and other constituents from the γ matrix. The fracture surface characteristics of HSTA-treated GLP samples are similar to those of CLP samples after undergoing HSTA. However, the dimples in the HSTA-treated GLP samples appear larger and deeper (see Figure 8.13(k) and (l)). It is worth noting that the dendritic structures become increasingly indistinct as the heat treatment temperature rises. As the heat treatment progresses from DA to STA and finally to HSTA, the fracture mode transitions from ductile fracture through micro-void coalescence to a mixture of ductile and brittle fracture.

8.7.2 High-temperature results

Figure 8.14 illustrates the SEM fractography of the fracture surface observed after high-temperature tensile tests conducted at 650°C. The fracture surfaces of all samples after DA still retain their dendrite characteristics, indicating a ductile transgranular fracture mode. This fracture mode is attributed to the micro-void coalescence resulting from the presence of secondary phase particles. Detailed observations of the fracture surfaces are presented in Figure 8.14(a) and (b) for the CLP samples after DA. In these images, distinct fracture steps can be observed, and the dimples are

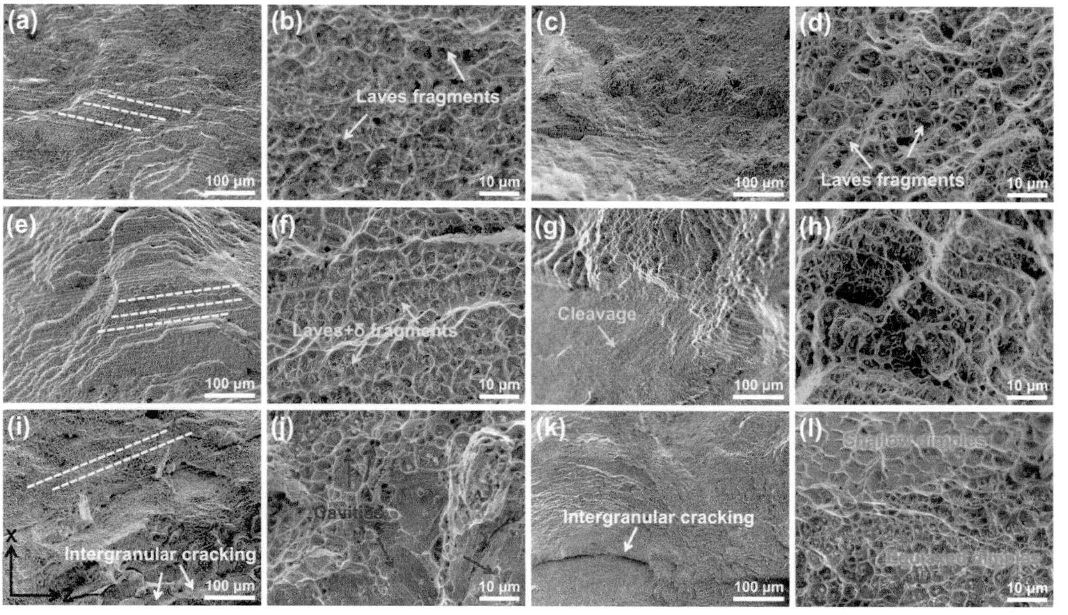

Figure 8.14. Fracture morphologies of Inconel 718 thin-wall samples after high-temperature tensile tests: (a, b) CLP-DA, (c, d) GLP-DA, (e, f) CLP-STA, (g, h), GLP-STA, (i, j) CLP-HSTA, and (k, l) GLP-HSTA.

aligned along the building direction (Z direction). In contrast, the fracture surfaces of the GLP samples after DA, as depicted in Figure 8.14(c) and (d), exhibit a disordered arrangement of dimples without any discernible directionality. Notably, both sets of samples display cracking of the Laves phase, with remnants of fragments observed within the dimples.

Upon subjecting the CLP samples to STA heat treatment, the fracture mode shifts to a clearly evident ductile mode with significantly increased ductility. The fracture morphologies and characteristics of these STA-treated CLP samples closely resemble those observed in room temperature tests, as shown in Figure 8.14(e) and (f). In contrast to the GLP samples, the CLP samples exhibit larger dimples due to the presence of coarser "Laves + δ" fragments. Furthermore, the GLP samples subjected to STA heat treatment display a combination of both transgranular and intergranular fracture modes, with deeper dimples compared to the room temperature tests (Figure 8.14(g) and (h)).

After undergoing HSTA heat treatment, both the CLP and GLP samples exhibit deteriorated dendritic structures and display a combination of mixed fracture modes. In the case of the CLP sample, micro-cavities can be observed within the dimples resulting from transgranular fracture, as well as cleavage planes from intergranular fracture, as shown in Figure 8.14(j). This phenomenon occurs due to the separation of small residual secondary phase particles from the γ matrix. On the other hand, the GLP samples reveal a mixed distribution of shallow dimples and deep equiaxed dimples, as depicted in Figure 8.14(l).

Figure 8.15 illustrates the lateral morphologies of the fractured surfaces near the fracture zone at the Y-Z plane following high-temperature tensile tests. The yellow arrows indicate the fragmentation of second-phase particles, while the green arrows represent the debonding of second-phase particles from the γ matrix. Due to the disparate thermal expansion coefficients and elasticity modulus of the second-phase particles and the γ matrix, localized stress is generated around the second-phase particles, leading to stress concentrations and eventual failure [54]. In the DA sample, both Laves phase fragmentation and Laves phase debonding from the γ matrix can be observed. For the DA-CLP sample (Figure 8.15(a) and (b)), the dominant phenomenon during the tensile process

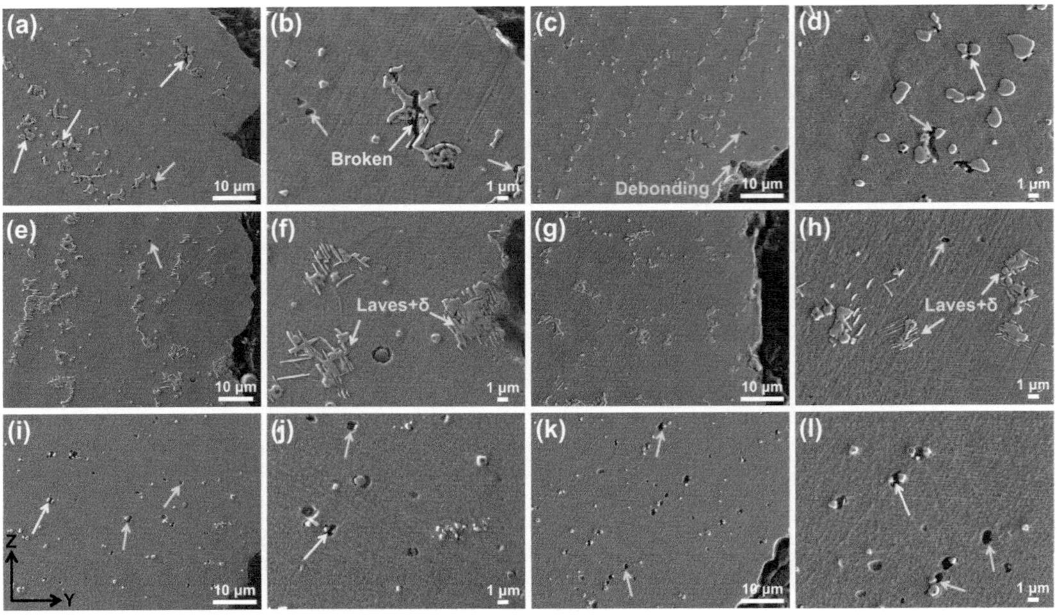

Figure 8.15. SEM micrographs of the fractures' longitudinal sections after high-temperature tensile tests: (a, b) CLP-DA, (c, d) GLP-DA, (e, f) CLP-STA, (g, h), GLP-STA, (i, j) CLP-HSTA, and (k, l) GLP-HSTA.

is the fragmentation of long-chain Laves phases, which promotes the nucleation of micro-voids and subsequently results in premature fracture at high temperatures. Granular and discrete Laves phases are dispersed in the DA-CLP sample, with only a few instances of Laves phase fractures observed (Figure 8.15(d)). When the Laves phases are fine and discrete, they are able to move with the γ matrix during plastic deformation without easily breaking apart. In this case, particle-matrix decohesion plays a predominant role.

The microstructure of the sample primarily consists of the "Laves + δ" phases following the STA heat treatment. The CLP samples subjected to STA heat treatment exhibit the presence of long-chain "Laves + δ" phases, while the GLP samples shows discrete "Laves + δ" phases. When observing the longitudinal section of fractures, it becomes apparent that the majority of the "Laves + δ" phases do not fracture or detach from the γ matrix even after subjecting the sample to high-temperature tensile tests. Instead, these phases persist within the γ matrix (refer to Figure 8.15(e–h)). This phenomenon can be attributed to the Smith-Zener pinning effect exerted by the surrounding δ phase, which hinders the fragmentation and separation of the Laves phases from the γ matrix. Furthermore, the δ phase also demonstrates its ability to impede dislocation motion and hinder grain boundary migration.

It has been reported that the presence of δ phases can negatively affect the mechanical properties of materials. Li et al. [20] highlighted that the combination of "Laves + δ" phases appears to have a greater detrimental impact on mechanical qualities compared to individual Laves phases. This is because dislocations tend to accumulate around the "Laves + δ" phase more easily during room-temperature tensile processes. Liu et al. [27] suggested that incoherent and needle-shaped δ phases can facilitate the initiation and propagation of cleavage cracks due to their sharp edges, which act as stress concentrators. Furthermore, the formation of the δ phase involves the depletion of Nb, which is a critical constituent element in strengthening phases such as γ" and γ'. However, the investigation of high-temperature tensile properties reveals inconsistencies with these claims. In our study, we discovered that the presence of "Laves + δ" phases does not significantly impair the mechanical properties of the material. Instead, these phases play an excellent role in enhancing high-temperature ductility. This can be attributed to the vacancy diffusion and the thermal activation energy provided by the elevated temperatures, which enable stacking dislocations to overcome short-range obstacles more easily [55]. The δ phases are softer and more ductile compared to γ'' and γ' phases. The elevated temperature, up to 650°C, promotes coordinated deformation between the "Laves + δ" phases and the γ matrix, resulting in improved ductility.

The HSTA samples (Figure 8.15(i–l)) exhibit only a small amount of particle fractures. However, fine Laves phases and MX particles are mainly detached from the γ matrix, indicating the prevalence of interfacial decohesion. The interfaces between the particles and the matrix serve as favorable sites for the initiation of micro-voids. Upon closer inspection in the magnified images (Figure 8.15(j) and (l)), numerous micro-voids can be observed following the decohesion between the particles and the matrix.

Furthermore, it has been observed that the HSTA heat-treated samples exhibit a decrease in ductility when subjected to high-temperature tensile tests, as depicted in Figure 8.12(d). The enlarged fracture image of the CLP-HSTA sample after the high-temperature tensile tests, along with the corresponding EDS mappings (Figure 8.16), provides evidence of the presence of Al and O enriched oxides within the dendrites, indicated by white circles. The GLP samples, followed by HSTA heat treatment, also display the existence of oxides. This occurrence can be attributed to the high-temperature homogenization process, which facilitates the accumulation of O, S, and C elements in the interdendrites, grain boundaries, and other regions. Consequently, this leads to premature failure, a reduction in ductility, and the occurrence of intermediate temperature embrittlement [56].

Based on the results and analysis provided in Section 8.6 and Section 8.7, it becomes evident that the failure modes of heat-treated samples are significantly influenced by the heat treatment paths, as compared to the current test temperatures and deposition methods. In DA samples, the distribution of γ″ and γ′ nano-strengthening phases within the γ matrix is non-uniform. Additionally, interdendritic retained Laves phases tend to create stress concentration, fragment into small pieces, and detach from the γ matrix, thus promoting the formation of micro-voids during the tensile process. Consequently, DA samples exhibit low intra-grain strength and failure occurs through a ductile transgranular mode.

For STA or HSTA heat treatments, the distribution of γ″ and γ′ phases becomes more uniform, but the grain boundaries weaken. Due to their resistance to plastic deformation, the grain boundaries act as barriers for dislocations. The presence of MX particles and δ phase at the grain boundaries facilitates a transition from a transgranular fracture mode to an intergranular fracture mode, as these particles act as crack initiators. When these particles are closely dispersed along the grain boundaries, microcracks easily propagate and promote intergranular fracture.

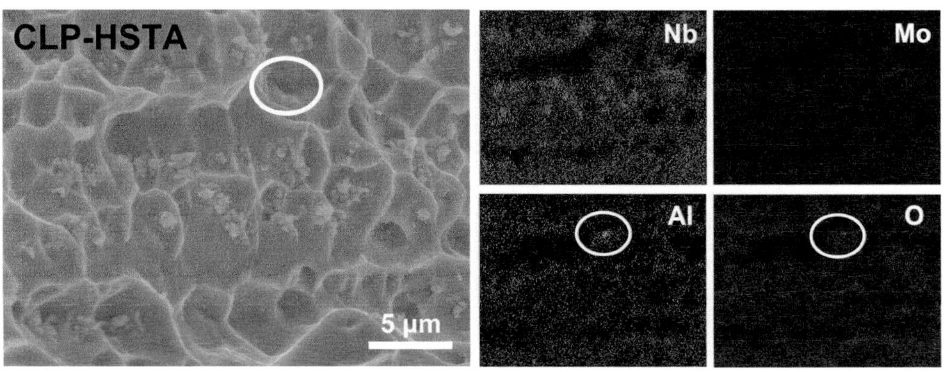

Figure 8.16. SEM fracture surface of the CLP-HSTA sample after high-temperature tensile tests and corresponding EDS mappings.

8.8 Summary

Post-heat treatments are crucial in the production of laser powder deposition-based nickel-based superalloys for aviation industry applications, aiming to attain exceptional performance levels. These heat treatments play an indispensable role in enhancing the material properties and optimizing the alloy's microstructure. In this chapter, we summarized the current research on heat treatments for laser additive manufacturing Inconel 718. A systematic study on the responses of Inconel 718 deposited by CLP and GLP methods to three typical heat treatments was conducted. We extensively investigated the microstructural evolution, hardness, tensile properties, and fracture morphologies at both room temperature and high temperature.

The microstructure and degree of element segregation in the as-deposited state have a significant influence on the microstructure evolution after heat treatments. The microstructure of as-deposited GLP samples primarily consists of fine discrete Laves phases within fine disordered dendrites and a low *SR*. After STA heat treatment, we can achieve a relatively uniform element distribution along with precipitation of γ'' and γ' phases. In the case of CLP samples, they exhibit coarse and continuous long-chain Laves phases within coarse dendrites and a high *SR*, necessitating HSTA heat treatment with a high homogenization temperature to achieve a uniform element distribution.

It is found that the dissolution of Laves phases is faster when they are smaller in size and when the PDAS is smaller in the as-deposited microstructure. Under the same heat treatment, GLP samples with a uniform distribution of elements and smaller Laves phases dissolve more quickly compared to CLP samples.

After undergoing the same heat treatments, GLP samples exhibit superior microhardness and room-temperature tensile strength compared to CLP samples, and they also outperform forged Inconel 718 in these aspects. However, the excellent ductility observed in the as-deposited GLP samples is reduced. The positive effect of finely dispersed Laves phases in the GLP samples on ductility gradually diminishes.

Following STA heat treatment, the δ phase precipitates around the Laves phase, exerting a pinning effect that inhibits Laves phase fragmentation and separation from the matrix. At high temperatures up to 650°C, the coordination deformation between "Laves + δ" phases and the γ matrix promotes high ductility during high-temperature tensile testing.

Both GLP and CLP samples exhibit similar fracture mechanisms during room-temperature and high-temperature tensile tests. As the heat treatments transition from DA, STA to HSTA, the fracture mode shifts from a ductile transgranular fracture with micro-void coalescence to a mixed mode involving both ductile transgranular and brittle intergranular fractures.

The findings presented in this chapter enhance our understanding of how different as-deposited microstructures and segregation ratios influence the evolution of microstructures and mechanical characteristics following heat treatments. It emphasizes the need to employ specific heat treatments based on the characteristics of the as-deposited microstructure. This knowledge provides a new perspective to optimize the post-heat treatment of Inconel 718 fabricated using laser additive manufacturing by considering the microstructure-defined parameters.

References

[1] Zhang, W., Zheng, Y., Liu, F., Wang, D., Liu, F., Huang, C. et al. (2021). Effect of solution temperature on the microstructure and mechanical properties of Hastelloy X superalloy fabricated by laser directed energy deposition. Materials Science and Engineering A 820: 141537. https://doi.org/10.1016/j.msea.2021.141537.

[2] Jiang, R., Mostafaei, A., Wu, Z., Choi, A., Guan, P.-W., Chmielus, M. et al. (2020). Effect of heat treatment on microstructural evolution and hardness homogeneity in laser powder bed fusion of alloy 718. Addit. Manuf. 35. https://doi.org/10.1016/j.addma.2020.101282.

[3] Kok, Y., Tan, X. P., Wang, P., Nai, M. L. S., Loh, N. H., Liu, E. et al. (2018). Anisotropy and heterogeneity of microstructure and mechanical properties in metal additive manufacturing: A critical review. Mater Des. 139: 565–586.

[4] Hosseini, E. and Popovich, V. A. (2019). A review of mechanical properties of additively manufactured Inconel 718. Addit. Manuf. 30: 100877. https://doi.org/10.1016/j.addma.2019.100877.

[5] Sui, S., Chen, J., Li, Z., Li, H., Zhao, X. and Tan, H. (2020). Investigation of dissolution behavior of laves phase in inconel 718 fabricated by laser directed energy deposition. Addit. Manuf. 32: 101055. https://doi.org/10.1016/j.addma.2020.101055.

[6] Sui, S., Tan, H., Chen, J., Zhong, C., Li, Z., Fan, W. et al. (2019). The influence of Laves phases on the room temperature tensile properties of Inconel 718 fabricated by powder feeding laser additive manufacturing. Acta Mater 164: 413–427. https://doi.org/10.1016/j.actamat.2018.10.032.

[7] Gradl, P., Tinker, D. C., Park, A., Mireles, O. R., Garcia, M., Wilkerson, R. et al. (2022). Robust metal additive manufacturing process selection and development for aerospace components. Journal of Materials Engineering and Performance 31(8): 6013–6044. https://doi.org/10.1007/S11665-022-06850-0.

[8] Liu, F., Cheng, H., Yu, X., Yang, G., Huang, C., Lin, X. et al. (2018). Control of microstructure and mechanical properties of laser solid formed Inconel 718 superalloy by electromagnetic stirring. Opt Laser Technol. 99: 342–350. https://doi.org/10.1016/j.optlastec.2017.09.022.

[9] Wang, H., Hu, Y., Ning, F. and Cong, W. (2020). Ultrasonic vibration-assisted laser engineered net shaping of Inconel 718 parts: Effects of ultrasonic frequency on microstructural and mechanical properties. J. Mater Process Technol. 276: 116395. https://doi.org/10.1016/j.jmatprotec.2019.116395.

[10] Chen, Y., Guo, Y., Xu, M., Ma, C., Zhang, Q.,Wang, L. et al. (2019). Study on the element segregation and Laves phase formation in the laser metal deposited IN718 superalloy by flat top laser and gaussian distribution laser. Materials Science and Engineering A 754: 339–347. https://doi.org/10.1016/j.msea.2019.03.096.

[11] Xiao, H., Xie, P., Cheng, M. and Song, L. (2020). Enhancing mechanical properties of quasi-continuous-wave laser additive manufactured Inconel 718 through controlling the niobium-rich precipitates. Addit. Manuf. 34: 101278. https://doi.org/10.1016/j.addma.2020.101278.

[12] Zhao, Y., Meng, F., Liu, C., Tan, S. and Xiong, W. (2021). Impact of homogenization on microstructure-property relationships of Inconel 718 alloy prepared by laser powder bed fusion. Materials Science and Engineering A 826: 141973. https://doi.org/10.1016/j.msea.2021.141973.

[13] Yuan, K., Guo, W., Li, P., Wang, J., Su, Y., Lin, X. et al. (2018). Influence of process parameters and heat treatments on the microstructures and dynamic mechanical behaviors of Inconel 718 superalloy manufactured by laser metal deposition. Materials Science and Engineering A 721: 215–225. https://doi.org/10.1016/j.msea.2018.02.014.

[14] Li, J., Zhao, Z., Bai, P., Qu, H., Liu, B., Li, L. et al. (2019). Microstructural evolution and mechanical properties of IN718 alloy fabricated by selective laser melting following different heat treatments. J. Alloys Compd. 772: 861–870. https://doi.org/10.1016/j.jallcom.2018.09.200.

[15] Sui, S., Li, Z., Zhong, C., Zhang, Q., Gasser, A., Chen, J. et al. (2021). Laves phase tuning for enhancing high temperature mechanical property improvement in laser directed energy deposited Inconel 718. Compos B Eng. 215: 108819. https://doi.org/10.1016/j.compositesb.2021.108819.

[16] Huang, W., Yang, J., Yang, H., Jing, G., Wang, Z. and Zeng, X. (2019). Heat treatment of Inconel 718 produced by selective laser melting: Microstructure and mechanical properties. Materials Science and Engineering A 750: 98–107. https://doi.org/10.1016/j.msea.2019.02.046.

[17] Zhai, Y., Lados, D. A., Brown, E. J. and Vigilante, G. N. (2019). Understanding the microstructure and mechanical properties of Ti-6Al-4V and Inconel 718 alloys manufactured by Laser Engineered Net Shaping. Addit. Manuf. 27: 334–344. https://doi.org/10.1016/j.addma.2019.02.017.

[18] Ozer, S., Bilgin, G. M., Davut, K., Esen, Z. and Dericioglu, A. F. (2022). Effect of post fabrication aging treatment on the microstructure, crystallographic texture and elevated temperature mechanical properties of IN718 alloy fabricated by selective laser melting. J. Mater Process Technol. 306: 117622.

[19] Sui, S., Chen, J., Ma, L., Fan, W., Tan, H., Liu, F. et al. (2019). Microstructures and stress rupture properties of pulse laser repaired Inconel 718 superalloy after different heat treatments. J. Alloys Compd. 770: 125–135. https://doi.org/10.1016/j.jallcom.2018.08.063.

[20] Li, Z., Ma, X., Zhong, C., Sui, S., Gasser, A. and Chen, J. (2022). Microstructure homogeneity and mechanical property improvement of Inconel 718 alloy fabricated by high-deposition-rate laser directed energy deposition. Materials Science and Engineering A 832. https://doi.org/10.1016/j.msea.2021.142430.

[21] Li, Y., Dlouhý, J., Koukolíková, M., Kirana, A., Vavřík, J. and Džugan, J. (2022). Effect of deposit thickness on microstructure and mechanical properties at ambient and elevated temperatures for Inconel 718 superalloy fabricated by directed energy deposition. J. Alloys Compd. 908. https://doi.org/10.1016/j.jallcom.2022.164723.

[22] Bhowmik, S., McWilliams, B. A. and Knezevic, M. (2022). Effect of powder reuse on tensile, compressive, and creep strength of Inconel 718 fabricated via laser powder bed fusion. Mater Charact. 190: 112023. https://doi.org/10.1016/j.matchar.2022.112023.

[23] Zhao, J. R., Hung, F. Y. and Lui, T. S. (2020). Microstructure and tensile fracture behavior of three-stage heat treated inconel 718 alloy produced via laser powder bed fusion process. Journal of Materials Research and Technology 9: 3357–3367. https://doi.org/10.1016/j.jmrt.2020.01.030.

[24] Li, X., Shi, J. J., Wang, C. H., Gao, G. H., Russell, A. M., Zhou, Z. J. et al. (2018). Effect of heat treatment on microstructure evolution of Inconel 718 alloy fabricated by selective laser melting. J. Alloys Compd. 764: 639–649. https://doi.org/10.1016/j.jallcom.2018.06.112.

[25] Ferreri, N. C., Vogel, S. C. and Knezevic, M. (2020). Determining volume fractions of γ, γ', γ'', δ, and MC-carbide phases in Inconel 718 as a function of its processing history using an advanced neutron diffraction procedure. Materials Science and Engineering A 781: 139228. https://doi.org/10.1016/j.msea.2020.139228.
[26] Kumara, C., Segerstark, A., Hanning, F., Dixit, N., Joshi, S., Moverare, J. et al. (2019). Microstructure modelling of laser metal powder directed energy deposition of alloy 718. Addit. Manuf. 25: 357–364. https://doi.org/10.1016/j.addma.2018.11.024.
[27] Liu, F., Lyu, F., Liu, F., Lin, X. and Huang, C. (2020). Laves phase control of inconel 718 superalloy fabricated by laser direct energy deposition via δ aging and solution treatment. Journal of Materials Research and Technology 9: 9753–9765. https://doi.org/10.1016/j.jmrt.2020.06.061.
[28] Sun, R., Shi, Y., Yang, Y., Wang, X. and Zhou, X. (2021). Microstructure, element segregation and performance of Inconel 625 metal layer deposited by laser assisted ultra-high frequency induction deposition. Surf. Coat Technol. 405: 126715.
[29] Xu, L., Chai, Z., Chen, H., Zhang, X., Xie, J. and Chen, X. (2021). Tailoring Laves phase and mechanical properties of directed energy deposited Inconel 718 thin-wall via a gradient laser power method. Materials Science and Engineering A 824: 141822. https://doi.org/10.1016/j.msea.2021.141822.
[30] Wang, Y., Shi, J. and Liu, Y. (2019). Competitive grain growth and dendrite morphology evolution in selective laser melting of Inconel 718 superalloy. J. Cryst. Growth 521: 15–29. https://doi.org/10.1016/j.jcrysgro.2019.05.027.
[31] Manikandan, S. G. K., Sivakumar, D., Rao, K. P. and Kamaraj, M. (2014). Effect of weld cooling rate on Laves phase formation in Inconel 718 fusion zone. J. Mater Process Technol. 214: 358–364. https://doi.org/10.1016/j.jmatprotec.2013.09.006.
[32] Cao, M., Zhang, D., Gao, Y., Chen, R., Huang, G., Feng, Z. et al. (2021). The effect of homogenization temperature on the microstructure and high temperature mechanical performance of SLM-fabricated IN718 alloy. Materials Science and Engineering A 801: 140427. https://doi.org/10.1016/j.msea.2020.140427.
[33] Patil, R. V. and Kale, G. B. (1996). Chemical diffusion of niobium in nickel. Journal of Nuclear Materials 230: 57–60. https://doi.org/10.1016/0022-3115(96)80010-9.
[34] Liu, X. J., Hu, H. H., Han, J. J., Lu, Y. and Wang, C. P. (2012). Assessment of the diffusional mobilities in fcc Ni-Nb and fcc Ni-Mo alloys. CALPHAD 38: 140–145. https://doi.org/10.1016/j.calphad.2012.05.003.
[35] Karunaratne, M. S. A. and Reed, R. C. (2005). Interdiffusion of niobium and molybdenum in nickel between 900–1300°C. Defect and Diffusion Forum 237–240: 420–425. https://doi.org/10.4028/www.scientific.net/ddf.237-240.420.
[36] Tucho, W. M., Cuvillier, P., Sjolyst-Kverneland, A. and Hansen, V. (2017). Microstructure and hardness studies of Inconel 718 manufactured by selective laser melting before and after solution heat treatment. Materials Science and Engineering: A 689: 220–232.
[37] Zhang, S., Wang, L., Lin, X., Yang, H. and Huang, W. (2022). The formation and dissolution mechanisms of Laves phase in Inconel 718 fabricated by selective laser melting compared to directed energy deposition and cast. Compos B Eng. 109994.
[38] Sohrabi, M. J., Mirzadeh, H. and Rafiei, M. (2018). Solidification behavior and Laves phase dissolution during homogenization heat treatment of Inconel 718 superalloy. Vacuum 154: 235–243.
[39] Hu, Y. L., Lin, X., Zhang, S. Y., Jiang, Y. M., Lu, X. F., Yang, H. O. et al. (2018). Effect of solution heat treatment on the microstructure and mechanical properties of Inconel 625 superalloy fabricated by laser solid forming. J. Alloys Compd. 767: 330–344.
[40] Zhao, L., Tan, Y., Wang, Y., You, X., Chang, K., Zhuang, X. et al. (2021). Homogenization behavior of IN 718 superalloy prepared by electron beam layered solidification technology. Journal of Materials Research and Technology 13: 1567–1575.
[41] Deng, D., Peng, R. L., Söderberg, H. and Moverare, J. (2018). On the formation of microstructural gradients in a nickel-base superalloy during electron beam melting. Mater Des. 160: 251–261.
[42] Jiang, R., Mostafaei, A., Pauza, J., Kantzos, C. and Rollett, A. D. (2019). Varied heat treatments and properties of laser powder bed printed Inconel 718. Materials Science and Engineering: A 755: 170–180.
[43] Zhao, Z., Li, L., Yang, W., Zeng, Y., Lian, Y. and Yue, Z. (2022). A comprehensive study of the anisotropic tensile properties of laser additive manufactured Ni-based superalloy after heat treatment. Int. J. Plast. 148: 103147.
[44] Holland, S., Wang, X., Chen, J., Cai, W., Yan, F. and Li, L. (2019). Multiscale characterization of microstructures and mechanical properties of Inconel 718 fabricated by selective laser melting. J. Alloys Compd. 784: 182–194. https://doi.org/10.1016/j.jallcom.2018.12.380.
[45] Sui, S., Chen, J., Zhang, R., Ming, X., Liu, F. and Xin, L. (2017). The tensile deformation behavior of laser repaired Inconel 718 with a non-uniform microstructure. Materials Science and Engineering A 688: 480–487. https://doi.org/10.1016/j.msea.2017.01.110.
[46] Qi, H., Azer, M. and Ritter, A. (2009). Studies of standard heat treatment effects on microstructure and mechanical properties of laser net shape manufactured INCONEL 718. Metall Mater Trans A Phys Metall Mater Sci. 40: 2410–2422. https://doi.org/10.1007/s11661-009-9949-3.
[47] Yu, X., Lin, X., Liu, F., Wang, L., Tang, Y., Li, J. et al. (2020). Influence of post-heat-treatment on the microstructure and fracture toughness properties of Inconel 718 fabricated with laser directed energy deposition additive manufacturing. Materials Science and Engineering A 798: 140092. https://doi.org/10.1016/j.msea.2020.140092.

[48] Zhong, C., Gasser, A., Kittel, J., Fu, J., Ding, Y. and Poprawe, R. (2016). Microstructures and tensile properties of Inconel 718 formed by high deposition-rate laser metal deposition. J. Laser Appl. 28: 022010. https://doi.org/10.2351/1.4943290.

[49] Zhao, X., Chen, J., Lin, X. and Huang, W. (2008). Study on microstructure and mechanical properties of laser rapid forming Inconel 718. Materials Science and Engineering A 478: 119–124. https://doi.org/10.1016/j.msea.2007.05.079.

[50] Zhong, C., Gasser, A., Kittel, J., Wissenbach, K. and Poprawe, R. (2016). Improvement of material performance of Inconel 718 formed by high deposition-rate laser metal deposition. Mater Des. 98: 128–134.

[51] Liang, L., Xu, M., Chen, Y., Zhang, T., Tong, W., Liu, H. et al. (2021). Effect of welding thermal treatment on the microstructure and mechanical properties of nickel-based superalloy fabricated by selective laser melting. Materials Science and Engineering A 819: 141507. https://doi.org/10.1016/j.msea.2021.141507.

[52] Firoz, R., Basantia, S. K., Khutia, N., Bar, H. N., Sivaprasad, S. and Murthy, G. V. S. (2020). Effect of microstructural constituents on mechanical properties and fracture toughness of Inconel 718 with anomalous deformation behavior at 650°C. J. Alloys Compd. 845: 156276.

[53] Zhang, X., Chai, Z., Chen, H., Xu, J., Xu, L., Lu, H. et al. (2021). A novel method to prevent cracking in directed energy deposition of Inconel 738 by *in-situ* doping Inconel 718. Mater Des. 197: 109214. https://doi.org/10.1016/j.matdes.2020.109214.

[54] Zhao, Y., Guan, K., Yang, Z., Hu, Z., Qian, Z., Wang, H. et al. (2020). The effect of subsequent heat treatment on the evolution behavior of second phase particles and mechanical properties of the Inconel 718 superalloy manufactured by selective laser melting. Materials Science and Engineering: A 794: 139931.

[55] Zhang, Y., Yang, L., Lu, W., Wei, D., Meng, T. and Gao, S. (2020). Microstructure and elevated temperature mechanical properties of IN718 alloy fabricated by laser metal deposition. Materials Science and Engineering: A 771: 138580.

[56] Sang, L., Lu, J., Wang, J., Ullah, R., Sun, X., Zhang, Y. et al. (2021). *In-situ* SEM study of temperature-dependent tensile behavior of Inconel 718 superalloy. J. Mater Sci. 56: 16097–16112.

Chapter 9

Ultrasonic Vibration Assisted Deposition of Inconel 718 Bulk Alloys

Microstructure, Deposition Process, and Mechanical Property

9.1 Introduction

Additive manufacturing technology is widely employed in the fabrication of metallic materials. Laser-based additive manufacturing has emerged as a crucial technique for precision manufacturing, owing to its benefits such as small laser spot radius, high energy density, and minimal forming errors [1].

Presently, laser additive manufacturing methods find extensive usage in the production of Inconel 718, a material widely employed in the aerospace industry due to its remarkable strength at elevated temperatures and excellent resistance to fatigue, wear, thermal corrosion, and favorable weldability. Despite the numerous advantages of Directed Energy Deposition (DED) technology, several challenging issues remain to be addressed [2]. The fast heating and cooling characteristics of the DED process lead to the formation of large grain sizes, precipitated Laves phases, inhomogeneous grain shapes, and porosity, accompanied by the effects of thermal stresses, ultimately resulting in inferior mechanical properties.

Utilizing an applied energy field during laser-directed deposition can significantly enhance material properties. Ultrasonic vibration, for instance, induces periodic positive and negative pressures, leading to two fundamental non-linear effects: acoustic flow and cavitation. Acoustic flow involves the absorption of acoustic oscillations in a liquid, driven by momentum transfer to establish a steady flow. On the other hand, cavitation is a dynamic phenomenon that generates tiny bubbles or cavities within a liquid material, followed by their growth, pulsation, and collapse [3]. Both these effects promote the movement of the liquid material. The formation and collapse of acoustic cavitation exert a dramatic influence on the instantaneous pressure inside the melt pool, while acoustic flow and mechanical effects generated by ultrasound within the melt pool accelerate the pool flow. The combined impact of ultrasonic vibration results in a significant organizational and property control effect [4, 5]. As depicted in Figure 9.1, during the solidification and crystallization of molten metal materials, ultrasonic vibration facilitates microstructure refinement, chemical composition homogenization, particle distribution arrangement, and reduction of porosity, grain size, and segregation [6].

Consequently, in numerous studies [7, 8], ultrasonic vibration-assisted laser deposition is utilized to fabricate high-quality laser additive parts and achieve robust metallurgical bonding. For example, Ning et al. [9] compared ultrasonic vibration-assisted laser deposition with and without ultrasonic vibration-assisted laser deposition parts, and it was observed that ultrasonic vibration

DOI: 10.1201/9781003406327-9

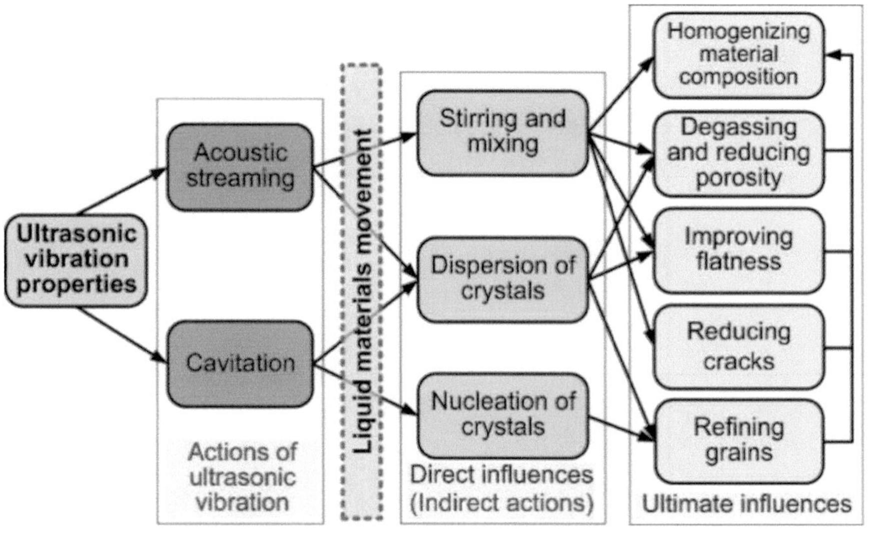

Figure 9.1. Effect of ultrasonic vibration on solidification of UV-A melted metals [6].

homogenized the temperature field in the local area, resulting in grain refinement and facilitating the dissolution of the Laves phase. Additionally, it reduced porosity, leading to increased yield strength, tensile strength, and microhardness of the part. Similarly, Srivastava et al. [10] found that during ultrasound-assisted laser additive processes, ultrasonic vibration caused grain and boundary fragmentation, thereby generating more grains. Furthermore, Cong et al. [7] suggested that the acoustic flow and cavitation induced by ultrasonic vibration can break up the porosity and homogenize the melt pool with reduced residual stress. This led to the elimination or reduction of deposition defects, including pores, cavities, and microcracks, and improved the bond strength of the deposited part to the substrate, ultimately enhancing the tensile properties and hardness of the prepared part.

In this chapter, we employ ultrasonic vibration during the DED process to fabricate Inconel 718 alloy and compare the organization and properties of samples fabricated using ultrasonic vibration-assisted DED (UA-DED) and conventional DED (DED) through various characterization methods. Additionally, we investigate the changes in melt pool flow and heat transfer during the UA-DED printing process through simulations and explore the intrinsic mechanism of the effect of ultrasonic vibration on the structure and properties of the deposited samples in-depth.

9.2 Microstructure and mechanical properties analysis of Ultrasonic vibration-assisted deposition of Inconel 718

9.2.1 Typical features of the microstructure

Figures 9.2, 9.3, and 9.5 present a comparative analysis of the microstructure of Direct Energy Deposition (DED) printed Inconel 718 and Ultrasonic Vibration-Assisted DED (UA-DED) printed Inconel 718 specimens in three different directions: the Z-X plane, Z-Y plane, and X-Y plane, respectively.

Figure 9.2 illustrates the comparison of the microstructure organization in the Z-X plane between DED and UA-DED.

As shown in Figure 9.2(a) and (b), the DED melt pool appears deeper with more curvature at the boundary, while the UA-DED melt pool is shallower and has less curvature. Figure 9.2(c) to Figure 9.2(f) demonstrate the direction and morphology of dendrite growth within the melt pool. DED printing results in columnar dendrites mainly draped over the melt pool edge towards the

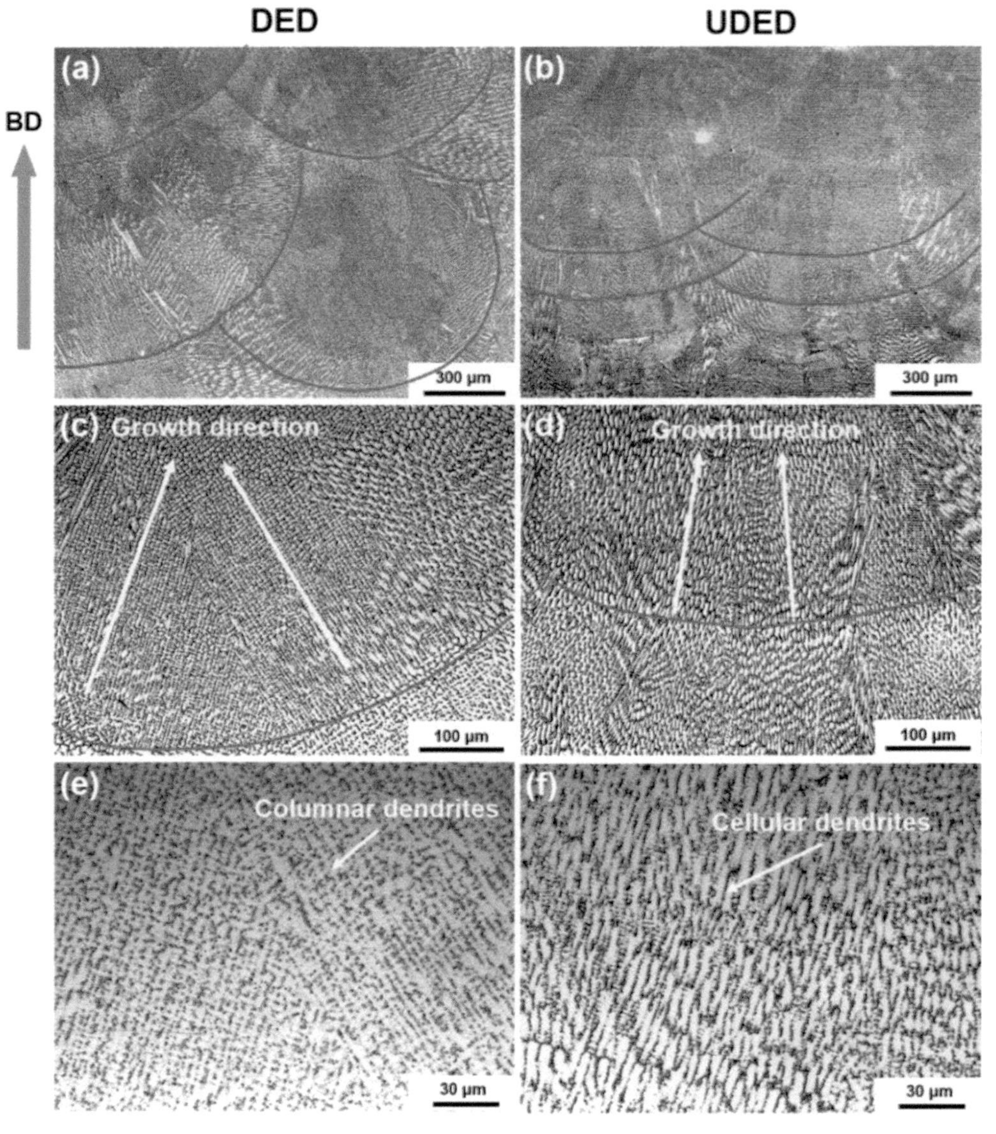

Figure 9.2. Z-X structure of DED and UA-DED (a) (c) (e) DED; (b) (d) (f) UA-DED.

center. In contrast, UA-DED printing yields cellular dendrites growing perpendicular to the melt pool edge, with the direction of cellular dendrite growth almost parallel to the printing direction.

Figure 9.3 shows the comparison of the organization of DED and UA-DED in the Z-Y plane. As observed in Figure 9.3(a) to Figure 9.3(d), dendrite growth is influenced by the direction of heat flow during DED printing. The direction of heat flow during laser printing is skewed according to the laser scanning direction, causing dendrites to grow in the opposite direction of the heat flow. In multilayer deposition, the newly deposited layer's dendrite tissue grows epitaxially on the main rod of the previous deposited layer, primarily at an angle of about 45° to the scanning direction, forming columnar dendrites, as shown in Figure 9.3(e).

In UA-DED printing, the shallow melt depth and small curvature of the melt pool edges enable better distribution of liquid metal within the pool, leading to less inclination compared to DED printing. Consequently, the liquid metal remains at a higher temperature at the end of the melt pool, reducing heat transfer in the opposite scan direction. The direction of heat dissipation largely aligns

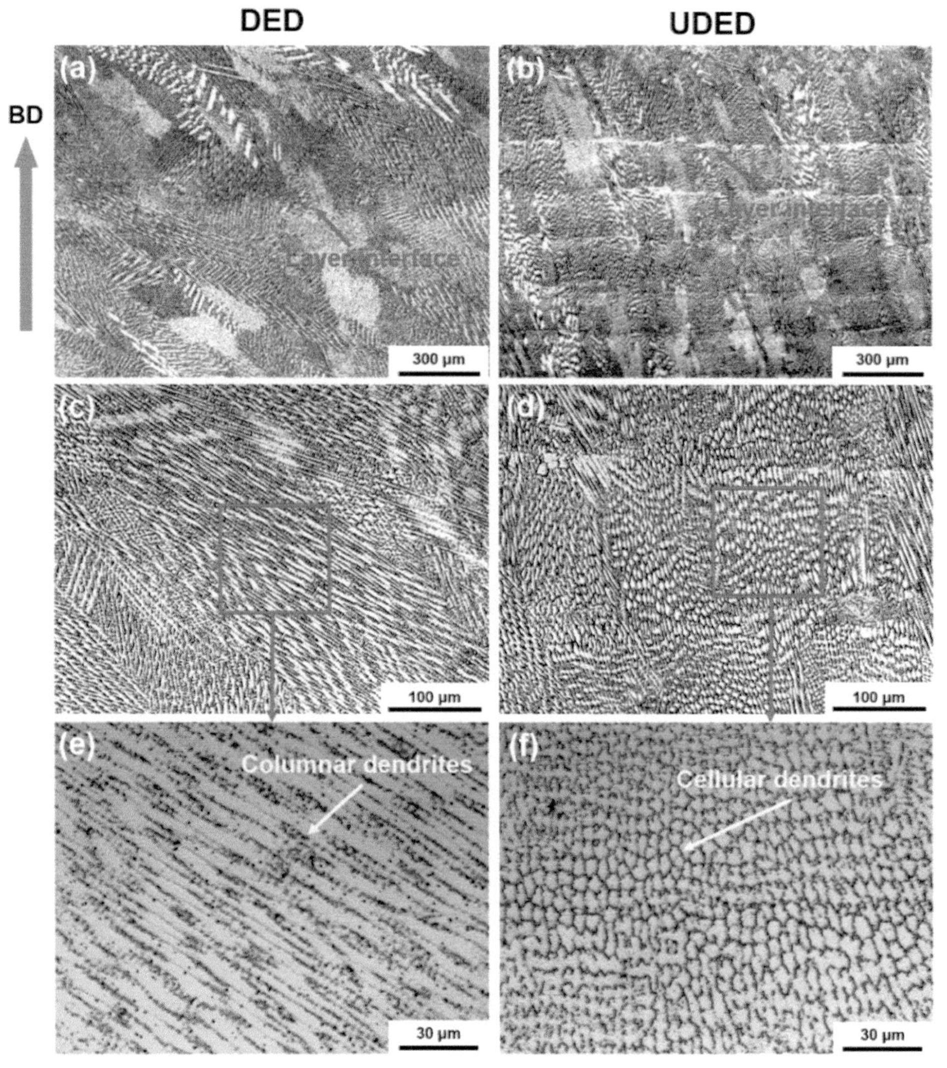

Figure 9.3. Z-Y structure of DED and UA-DED (a) (c) (e) DED; (b) (d) (f) UA-DED.

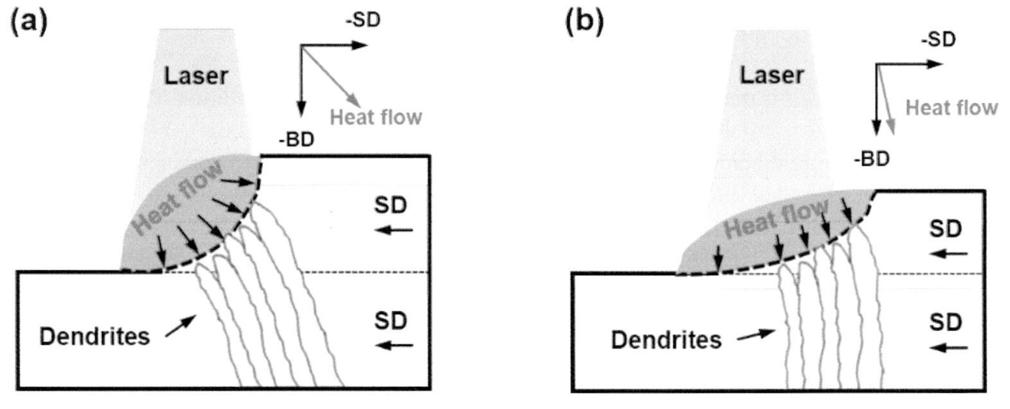

Figure 9.4. Schematic diagram of dendrite growth (a) DED; (b) Ultrasound assisted DED.

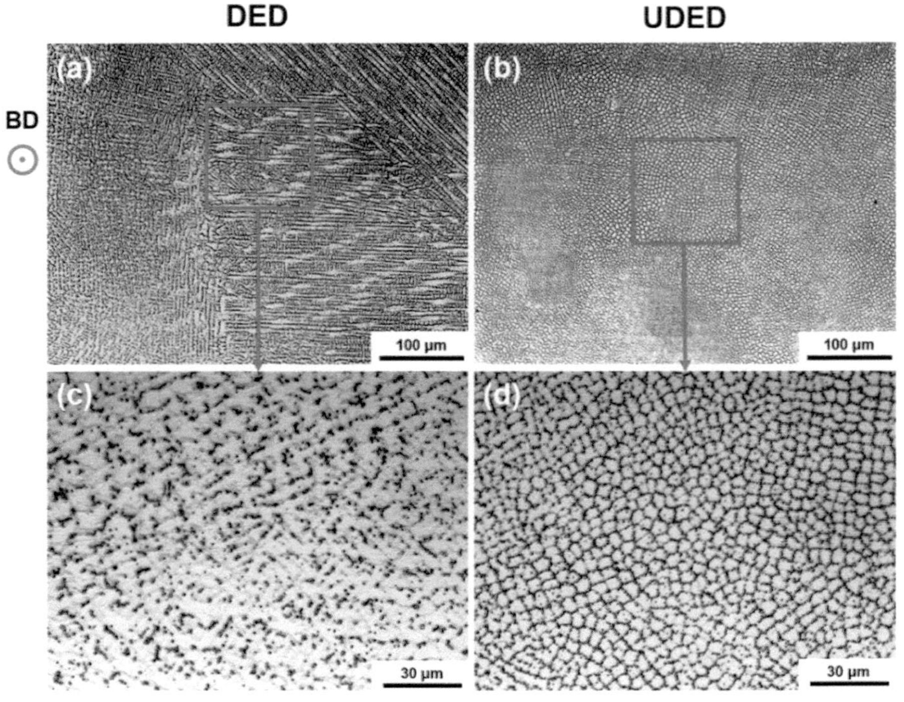

Figure 9.5. X-Y structure of DED and UA-DED (a) (c) DED; (b) (d) UA-DED.

with the substrate's normal direction, causing dendrites to grow in the opposite direction to the heat flow (Figure 9.4(b)). As a result, the dendrites tend to grow vertically, essentially aligned with the deposition direction, forming cellular dendrites (Figure 9.3(f)).

Figure 9.5 presents a comparison of the cross-sectional organization of DED and UA-DED in the X-Y plane. In DED printing, dendrites grow mainly in the opposite direction of the heat flow, resulting in irregularly shaped cross-sectional organization (Figure 9.5(c)). In UA-DED, the direction of heat flow in the melt pool changes due to ultrasonic vibration, leading to vertically growing dendrites with a neatly arranged honeycomb shape in their cross-section (Figure 9.5(d)).

9.2.2 The morphology and distribution of the Laves phase

Figure 9.6 depicts the Laves phase morphology and distribution in the Z-X, Z-Y, and X-Y planes after DED and UA-DED printing, respectively.

Figure 9.6(a) and Figure 9.6(b) show the Laves phase morphology in the Z-X plane. DED-printed specimens exhibit mainly long chains of Laves phases with precipitation at the boundaries of columnar dendrites. In the Z-Y plane (Figure 9.6(c) and Figure 9.6(d)), UA-DED-printed specimens demonstrate discrete granular Laves phases around the cytosolic dendrites, which grow predominantly in a vertical direction.

As shown in Figure 9.6(e) and Figure 9.6(f), the X-Y plane mainly contains the Laves phase of the dendrite cross-section. DED printing results in an irregular cross-sectional Laves distribution due to the difference in heat flow direction. On the other hand, UA-DED printed specimens, with their vertically growing dendrites, exhibit a neatly arranged honeycomb-shaped Laves phase distribution at the boundaries of the dendrites.

For a more detailed analysis of the effect of ultrasonic vibration on the Laves phase, the results of the Energy Dispersive Spectroscopy (EDS) analysis in the Z-X and Z-Y planes are shown in Figure 9.7 and Figure 9.8, respectively.

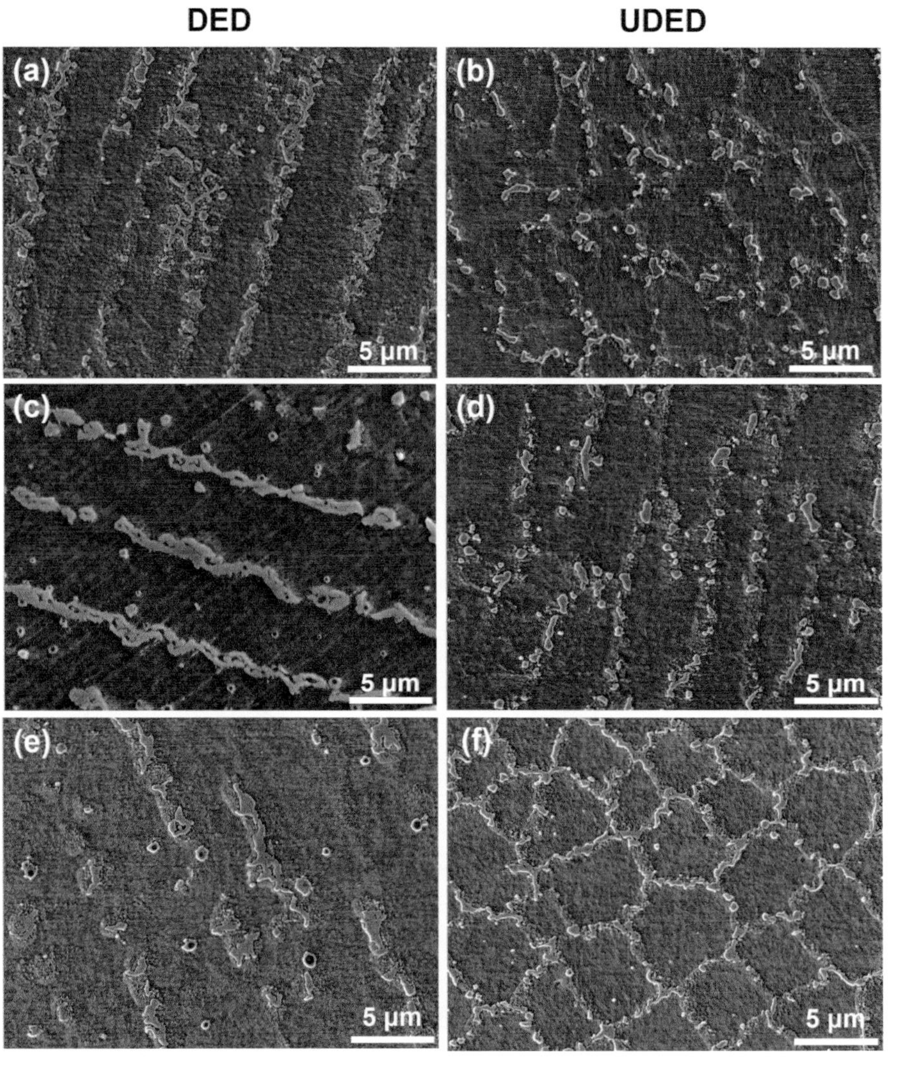

Figure 9.6. Laves phase appearance of DED and UA-DED (a) (c) (e) DED; (b) (d) (f) UA-DED.

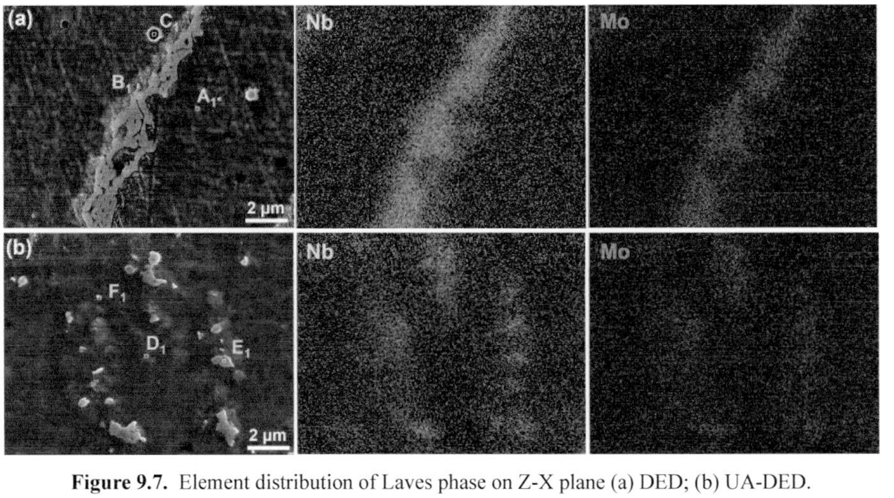

Figure 9.7. Element distribution of Laves phase on Z-X plane (a) DED; (b) UA-DED.

As seen in Figure 9.7, the conventionally DED printed specimens demonstrate long chains of Laves phases between the dendrites, while UA-DED printed specimens show granular and discrete Laves phases between the dendrites. Moreover, the Laves phases formed during solidification exhibit significant micro-partitioning of Nb and Mo alloying elements. The Laves phase is enriched in Nb and Mo but depleted in Ni, Cr, Fe, and Ti compared to the interlayer eutectic γ. The most significant absolute compositional differences between the γ matrix and the Laves phase are observed in the elements Ni and Nb.

Deposition (DED) and Ultrasonic Vibration-Assisted DED (UA-DED) is presented in Table 9.1. The DED printed alloy consists mainly of a γ matrix (A1), Laves phase (B1), and MC particles of titanium-rich nitride (C1). On the other hand, UA-DED printed Inconel 718 alloys consist mainly of a γ matrix (D1), Laves phase (E1), and Nb-rich carbide (F1).

Figure 9.8 illustrates the morphology and elemental distribution of the Laves phase in the Z-Y plane. The Nb elements are enriched at the edges of the dendrites, and the Laves phase precipitates at the boundaries of the dendrites. In DED printing, the Laves phase is primarily precipitated at the edges of the columnar dendrites, forming long chains. Conversely, in UA-DED printing, the Laves phase precipitates at the edges of the cytosol, with particles of the Laves phase being small and diffusely distributed at the cytosol's edges. Additionally, Nb and Mo elements precipitate both in and around the Laves phase.

Table 9.1. EDS analysis results of the chemical compositions of the γ matrix, Laves and MC particles in the deposited samples (in wt%).

Position	Ni	Nb	Ti	Al	O	Cr	Fe	Co	Mo	C
$_{A1}$ (Matrix)	49.58	2.49	0.67	0.49	0.83	19.67	19.12	0.15	2.54	3.98
$_{B1}$ (Laves)	41.79	18.04	1.57	0.36	0.97	14.18	12.28	0.13	4.14	5.87
$_{C1}$ (MC)	38.11	2.94	1.24	11.46	10.05	15.54	14.96	0.18	1.98	3.37
$_{D1}$ (Matrix)	49.87	1.91	0.68	0.44	0.96	19.32	20.17	0.35	2.12	3.32
$_{E1}$ (Laves)	45.33	7.94	1.54	0.49	1.1	17.69	17.44	0.19	2.3	5.51
$_{F1}$ (MC)	37.06	17.84	3.3	0.43	1.01	15.08	14.32	0.02	2.05	8.08

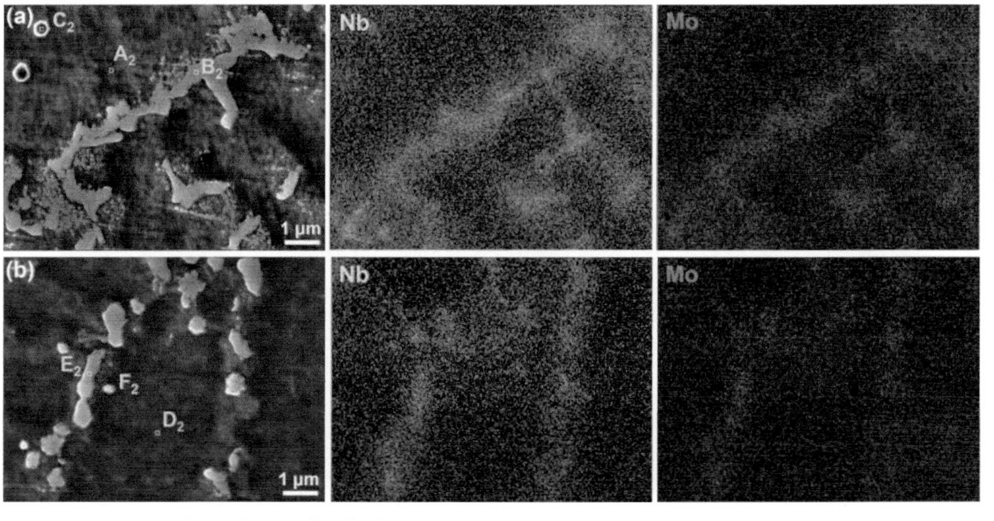

Figure 9.8. Element distribution of Laves phase on Z-Y plane (a) DED; (b) UA-DED.

Table 9.2. EDS analysis results of the chemical compositions of the γ matrix, Laves and MC particles in the deposited samples (in wt%).

Position	Ni	Nb	Ti	Al	O	Cr	Fe	Co	Mo	C
A2 (Matrix)	49.05	3.78	0.95	0.55	0.75	18.85	17.79	0.17	2.84	4.58
B2 (Laves)	30.58	27.13	2.88	0.31	1.00	12.88	10.83	0.12	2.84	10.84
C2 (MC)	35.39	9.66	3.1	7.75	8.53	14.58	12.91	0.05	2.42	5.03
D2 (Matrix)	49.18	1.89	0.69	0.44	0.84	19.41	20.78	0.22	2.19	3.97
E2 (Laves)	44.3	10.7	1.34	0.48	1.36	16.27	15.49	0.14	3.4	5.84
F2 (MC)	44.67	5.8	1.91	0.85	1.78	17.83	17.84	0.2	2.34	6.11

The results in Table 9.2 show that the DED printed Inconel 718 alloy primarily comprises a γ matrix (A2), Laves phase (B2), and MC particles of titanium-rich nitride (C2). On the other hand, UA-DED printed Inconel 718 alloys consist mainly of a γ matrix (D2), Laves phase (E2), and Nb-rich carbide (F2).

To investigate the micro-precipitation of elements and its relation to the morphology of the Laves phases, a statistical comparison of the elemental distribution was conducted. Specifically, the γ matrix (A1 and A2) and Laves phases (B1 and B2) of the long chain-like Laves phases and granular Laves phases in the Z-X and Z-Y planes were compared, as depicted in Figure 9.9.

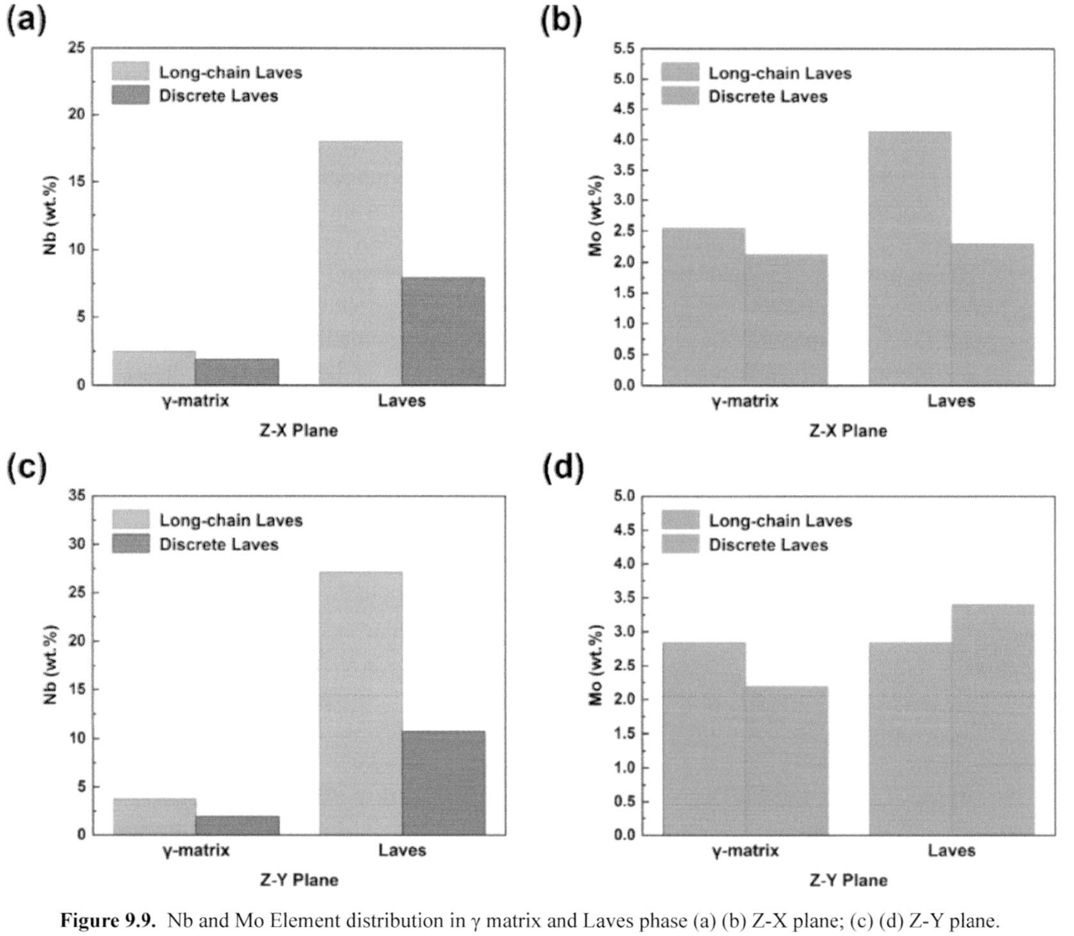

Figure 9.9. Nb and Mo Element distribution in γ matrix and Laves phase (a) (b) Z-X plane; (c) (d) Z-Y plane.

In Figure 9.9(a) and Figure 9.9(b), it can be observed that the concentration of Nb in the γ matrix of the Z-X facets ranges from 1.9 to 2.5 wt%, while the concentration of Nb in the Laves phase can reach 8–18 wt%, significantly higher than the nominal chemical composition of 5 wt% for the Inconel 718 alloy. In comparison, the degree of micro-bias in Mo is less severe due to its lower diffusivity and higher density. The concentrations of Mo in the γ matrix and the long-chain Laves phase are 2.54 wt% and 4.14 wt%, respectively, both higher than the concentrations of Mo in the γ matrix and granular Laves phase, which are 2.12 wt% and 2.3 wt%, respectively. Furthermore, the concentrations of Nb in the γ matrix of the Z-Y surface range from 1.9 to 3.8 wt%, while the concentrations of Nb in the Laves phase can reach 10 to 27 wt%. The enrichment of Nb in the long-chain Laves phase and its matrix is higher than that in the granular Laves phase. Compared to conventionally DED-printed samples, the UA-DED-printed samples show higher levels of dissolved Nb and Mo in the element-rich regions around the γ matrix and granular Laves phase, which leads to the reduction of Nb and Mo element concentrations around the granular Laves phase and the matrix.

The solidification process of Inconel 718 alloy during laser directed deposition commences with a liquid γ matrix, and as the process advances, the liquid metal between the dendrites becomes enriched with elements like Nb and Mo until a eutectic reaction concludes the solidification [11]. However, due to the poor diffusion rate of Nb atoms, the Laves phase is challenging to dissolve in the γ matrix. With the introduction of ultrasonic vibration, the heat transfer from the melt pool temperature increases, leading to a higher solubility of Nb and Mo in the matrix phase. The ultrasonic vibration also causes the Laves phase to break up and dissolve, resulting in a more uniform distribution of the Laves phase in the matrix, forming small particle shapes.

9.2.3 Grain growth and grain size

Figure 9.10 illustrates the polar and antipodal plots of the Z-X plane of specimens for DED printing and UA-DED printing.

In the inverse pole diagram shown in Figure 9.10(a), the grains exhibit random orientation in the case of DED printing, with a <001>/BD weave inside the melt pool and a <111>/BD weave at the edge of the melt pool, due to grain growth from the edge of the melt pool towards its center. However, with the assistance of ultrasonic vibration, as shown in Figure 9.10(b), a weave of <001>//BD mainly appears, and the grain growth direction is almost identical to the deposition direction. The polar plots depicted in Figure 9.10(c) and Figure 9.10(d) show that both DED printed and UA-DED printed specimens have a strong <001> weave, but the location of the maximum polar density is different. Ultrasonic vibration assistance leads to the maximum pole density appearing almost at the same location as the deposition direction, with the grain growth direction tending to be vertical, as shown in Figure 9.10(d). In contrast, the position of the maximum polar density for DED printing is more dispersed, as shown in Figure 9.10(c), which is related to the grain growth perpendicular to the edge of the melt pool towards the center.

Figure 9.11 presents the polar and antipodal plots for the Z-Y plane of DED printed and UA-DED printed specimens. The preferred growth orientation for face-centered cubic crystal systems is <001>, so the direction of dendrite growth is usually in one of the <001> crystal directions close to the heat flow direction. In conventionally DED printed specimens, the dendrite growth direction is about 45° to the deposition direction (BD), resulting in a <101>//BD weave, as shown in Figure 9.11(a). However, with the aid of ultrasonic vibration, the dendrite growth direction becomes essentially parallel to the deposition direction (BD), leading to a <001>/BD weave, as shown in Figure 9.11(b).

In Figure 9.11(d), the DED-printed sample shows a weaker <001> weave compared to the ultrasonic vibration sample, with a maximum pole density of 9.61 and a maximum pole density position at 45° to the deposition direction (BD), where the dendrites grow epitaxially. The polar diagram in Figure 9.11(c) shows that the microstructure of the UA-DED-printed specimens has a strong <001> weave with a maximum polar density of 15.0. The position of the maximum polar

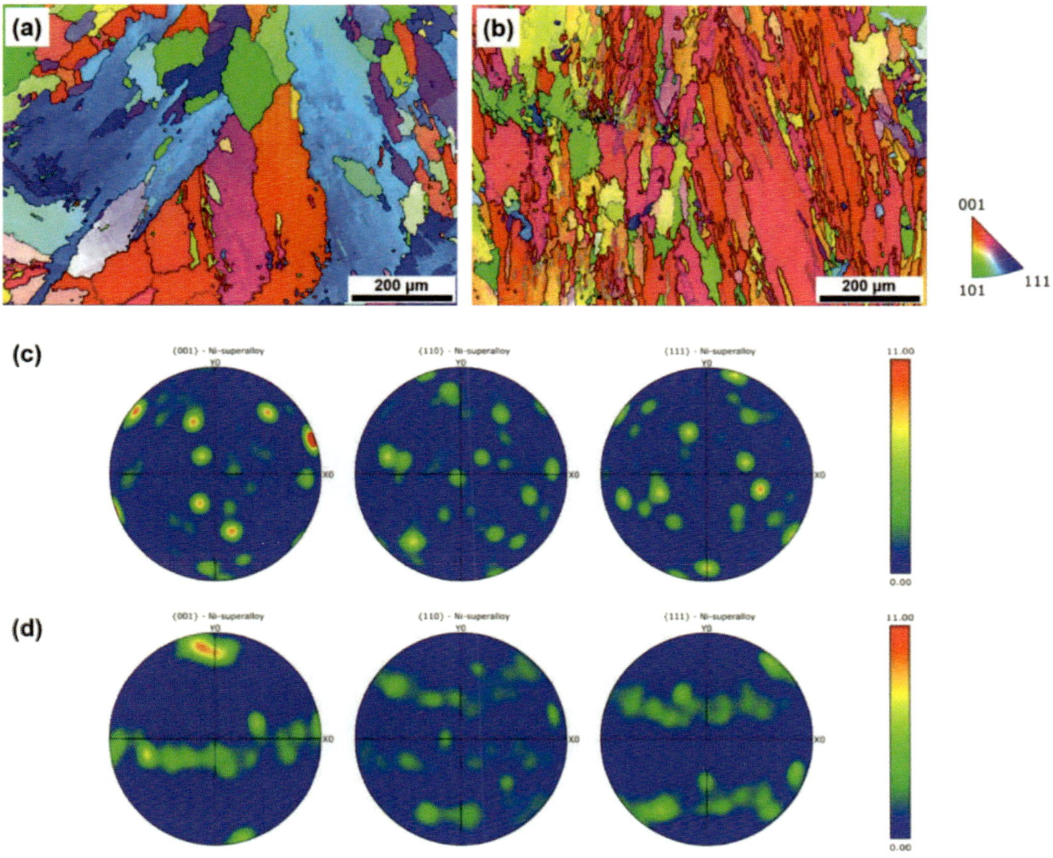

Figure 9.10. EBSD pole diagram and reverse pole diagram of Z-X deposited sample (a) The reverse pole diagram of the sample is printed by DED; (b) Reverse pole diagram of UA-DED sample; (c) The pole diagram of the sample is printed by DED; (d) Polar diagram of UA-DED.

density also tends to be close to the deposition direction (BD), indicating that the dendrites exhibit vertical growth. This suggests that ultrasonic vibration enhances the <001> orientation of the grains.

The grain size was further measured by EBSD, and the average fitted circle diameter was used to assess the grain size. The results, shown in Figure 9.12 and Figure 9.13, indicate that over 70% of the grains of the ultrasonic vibration-assisted DED printed specimens are concentrated in the 0–25 μm interval, while around 60% of the grains of the conventionally DED printed specimens are distributed in the same interval. Therefore, the grain size of the UA-DED printed samples tends to be finer. Additionally, the average grain diameter and the number of grains per unit area (mm^2) on the Z-X and Z-Y surfaces were counted in Figure 9.13, revealing that the ultrasonic vibration-assisted samples have smaller grain sizes and twice the number of grains compared to the DED printed samples.

The average grain diameter and the number of grains per unit area (mm^2) for the Z-X and Z-Y surfaces are quantified in Figure 9.13. As seen in Figure 9.13(a), the grain size is smaller for UA-DED printing compared to DED printed specimens, with the average grain diameter reducing from 31.4 μm to 21.6 μm on the Z-X surface, and 29.2 μm for DED printing and 22 μm for UA-DED printed specimens on the Z-Y surface. The number of grains per unit area (mm^2) also indicates that the ultrasonic vibration-assisted samples have twice the number of grains compared to the DED printed samples, as shown in Figure 9.13(b).

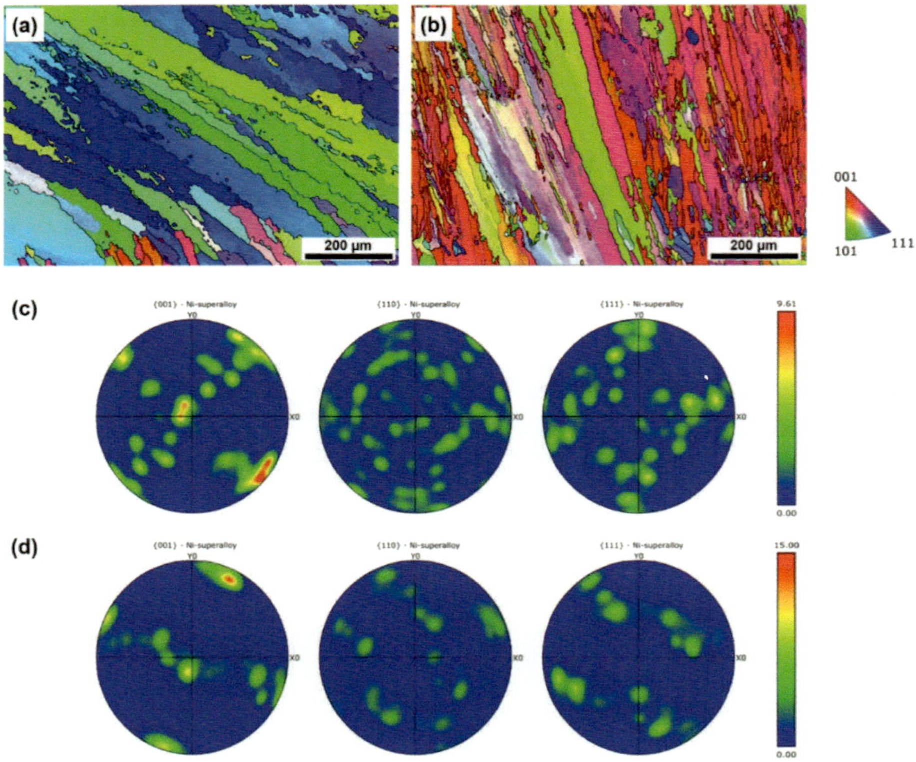

Figure 9.11. EBSD pole diagram and reverse pole diagram of Z-Y deposited sample (a) The reverse pole diagram of the sample is printed by DED; (b) Reverse pole diagram of UA-DED sample; (c) The pole diagram of the sample is printed by DED; (d) Polar diagram of UA-DED.

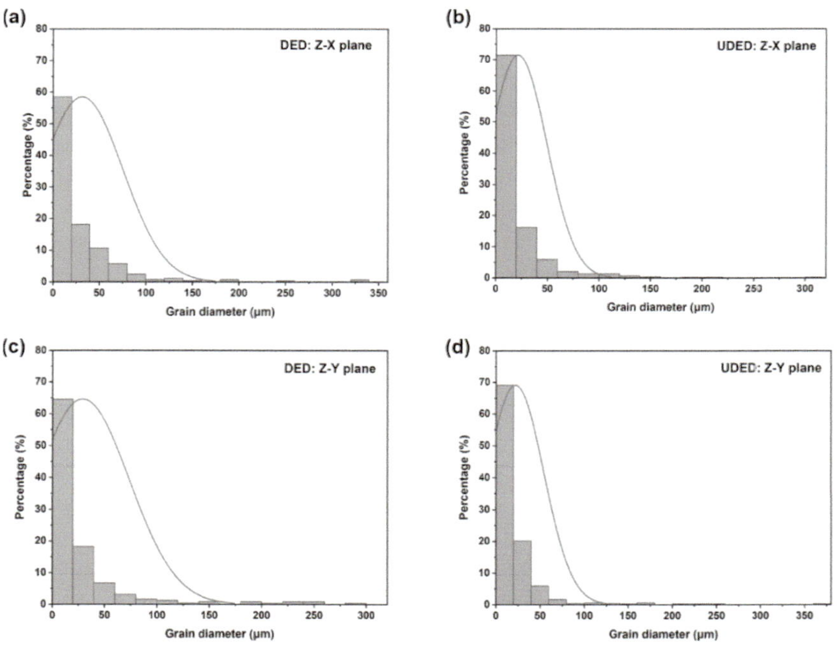

Figure 9.12. Grain size distribution (a) Z-X side DED sample; (b) Z-X UA-DED sample; (c) Z-Y side DED sample; (d) Z-Y UA-DED sample.

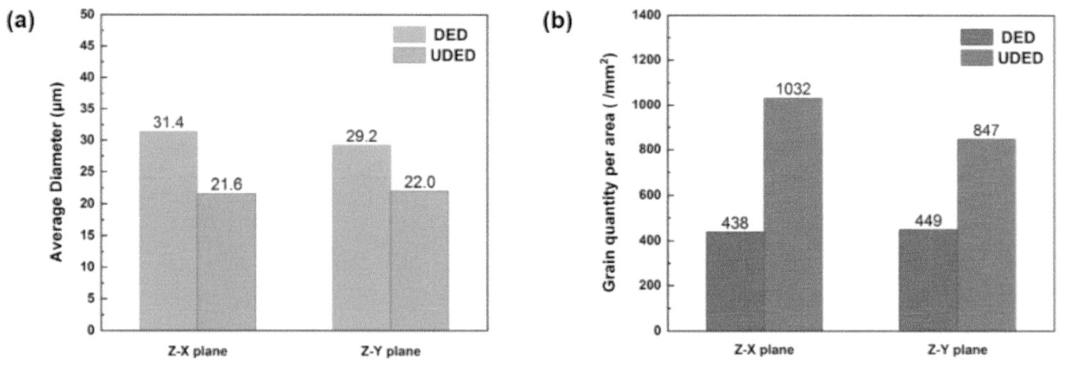

Figure 9.13. Statistics of grain characteristics of deposited samples (a) average grain size; (b) Number of grains per unit area.

9.2.4 *Defects*

The process of laser directional deposition is characterized by rapid heating and cooling, which makes it susceptible to two types of defects, namely porosity and unfused defects. Pores are small, round holes, typically 10–50 μm in diameter, caused by air bubbles inside the melt pool that cannot be expelled before solidification [12, 13]. Unfused defects result from the presence of unmelted powder during the laser directional deposition process, leading to incomplete fusion of the material in an irregular shape, usually in the range of 50–500 μm [13].

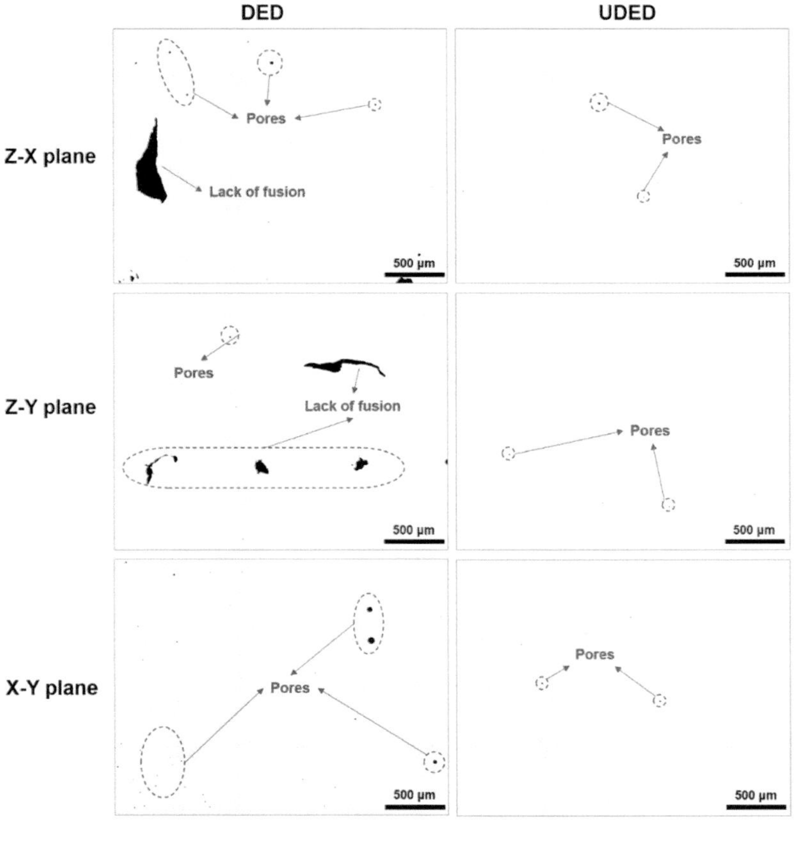

Figure 9.14. Internal defects of DED printing and UA-DED printing samples.

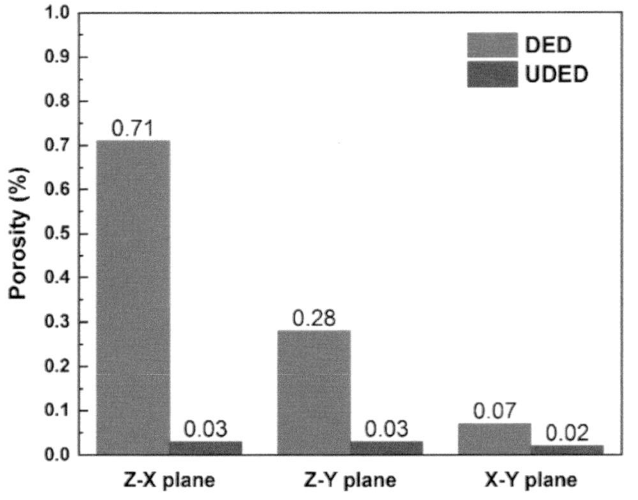

Figure 9.15. Porosity statistics of DED and UA-DED samples.

Defects like lack of fusion and porosity are the primary concerns in laser DED printing. Figure 9.14 depicts the distribution of defects on three different surfaces for conventional and UA-DED printed specimens. As evident, the UA-DED printed specimens exhibit no unfused defects and show fewer instances of porosity.

Figure 9.15 presents the porosity statistics for the three different surface faces of the conventionally DED printed and UA-DED printed specimens. In this case, the porosity statistics include the percentage of both porosity and unfused defects. As observed from the graph, conventional printing has lower porosity due to unfused and porous defects on the Z-X and Z-Y surfaces, whereas ultrasound-assisted DED printing only has porous defects and no unfused defects, resulting in lower porosity for ultrasound-assisted DED printed specimens. Overall, the porosity of ultrasound-assisted printing was approximately 0.02%, which is lower than the porosity reported by Ning et al. [14] in their study of ultrasonic vibration-assisted laser printing of Inconel 718, where they obtained a lower porosity result of 0.09%.

Figure 9.16 illustrates the porosity of DEDs and UA-DEDs separately and shows that the difference in porosity between DEDs and UA-DEDs in the Z-X plane is not significant, at 0.04%

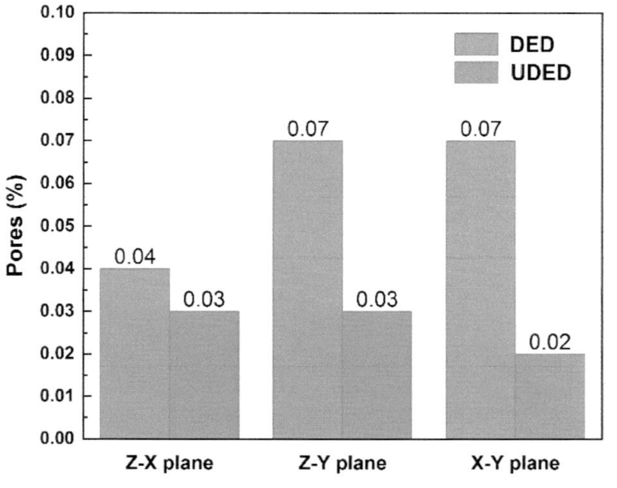

Figure 9.16. Pores statistics of DED and UA-DED samples.

and 0.03%, respectively. In the Z-Y and X-Y planes, the porosity of UA-DED printing was only 50% of that of DED printing.

9.2.5 Mechanical properties

9.2.5.1 Microhardness

Figure 9.17 presents the average hardness values for samples produced using DED printing and UA-DED printing in the Z-X, Z-Y, and X-Y planes. As observed from the graphs, the introduction of ultrasonic vibration resulted in higher average hardness values in all three directions for the laser directionally deposited samples compared to conventionally printed samples.

The DED-printed sample exhibited an average hardness of 264 HV on the Z-X surface, while the ultrasound-assisted print showed an average hardness of 319 HV, indicating an increase of 55 HV. On the Z-Y surface, the average hardness of the ultrasound-assisted print increased by 42 HV from 272 HV of the DED print, reaching 314 HV. Similarly, the average hardness on the X-Y surface with DED printing was 285 HV, whereas the ultrasound-assisted print showed a significant increase to 322 HV, marking a gain of 37 HV.

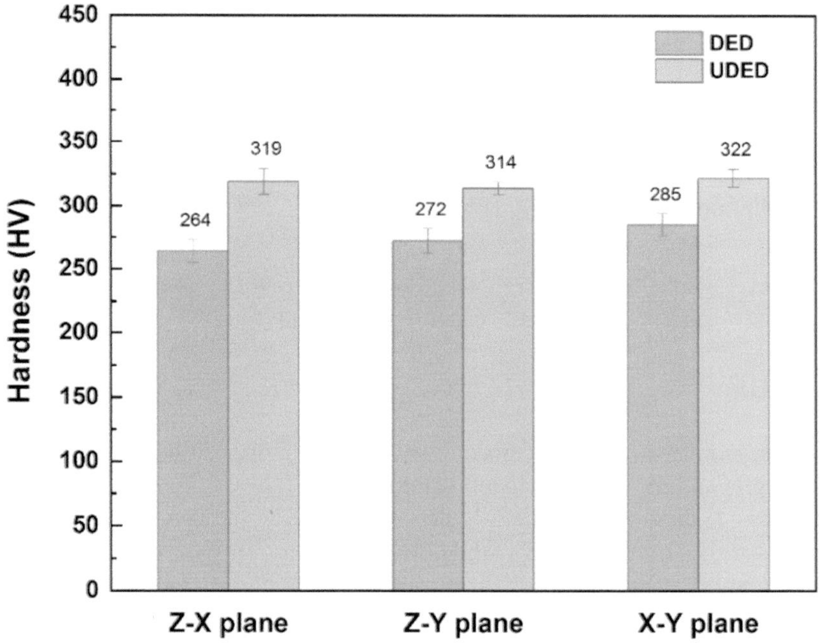

Figure 9.17. Average hardness of conventional and UA-DED samples.

9.2.5.2 Microscopic deformation

During laser directional deposition, higher thermal gradients and repetitive local heat transfer typically lead to thermal strain buildup and residual stresses in the deposited sample. Residual stresses in alloys occur at three scales: macroscopic scale as type I stress, grain scale as type II or microscale stress, and nanoscale as type III stress [15].

The Grain Orientation Spread (GOS) diagram obtained from EBSD analysis reflects the magnitude of plastic strain and residual stresses, specifically the type II stress distribution at the grain size level. The GOS map enables analysis of the microscopic deformation characteristics of

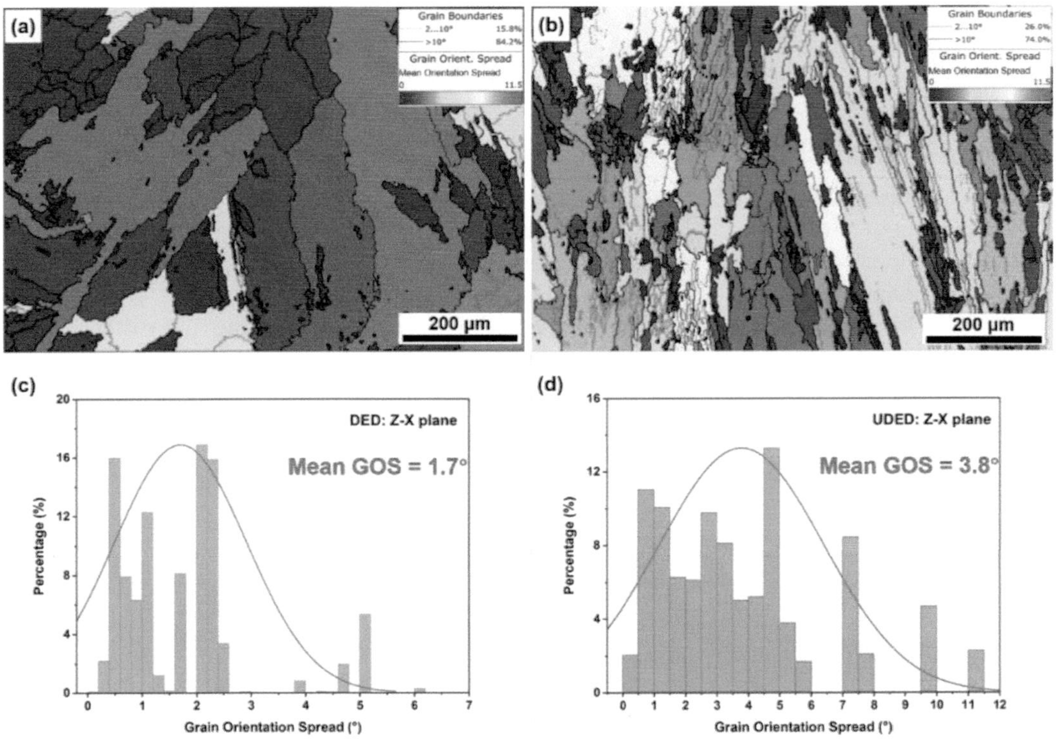

Figure 9.18. GOS maps of Z-X plane deposited sample (a) (c) DED sample; (b) (d) UA-DED sample.

the grains. The average GOS value is calculated by considering the orientation variation between each pixel and the mean grain orientation, defined as follows [16] :

$$GOS(i) = \frac{1}{J(i)} \sum_{j} \omega_{ij} \tag{9.1}$$

Where $J(i)$ is the number of pixels of grain i, and ω_{ij} is the orientation error angle between the orientation of pixel j and the mean orientation of grain i.

Based on the GOS distribution in the Z-X plane (Figure 9.18), the average orientation differences within the grains range from 0 to 11.5°, while the introduction of ultrasonic vibration leads to an increase in the number of grains experiencing larger deformations. The average GOS value for conventionally DED printed samples was 1.7°, while for UA-DED printed samples, it rose to 3.8°.

Figure 9.19 illustrates the GOS distribution in the Z-Y plane. As observed from the graph, ultrasonic vibration enhances the degree of grain deformation, resulting in more grains undergoing large deformations. The average GOS value for the DED-printed samples was 2.2°, while for the UA-DED printed samples, it increased to 3.7°.

In summary, the UA-DED printed samples exhibit the highest average GOS values, indicating a non-uniform distribution of plastic strain and residual stress. The GOS plots reveal grain-scale differences in geometric dislocations, residual plastic strain, and the degree of grain deformation [110]. The introduction of ultrasonic vibration causes molten material to experience ultrasonic shock oscillations, cavitation, and acoustic flow effects, leading to violent movements of the liquid metal. As a consequence, the samples show high average GOS values due to the local thermal effects caused by ultrasonic vibration and the severe thermal expansion and contraction during cooling.

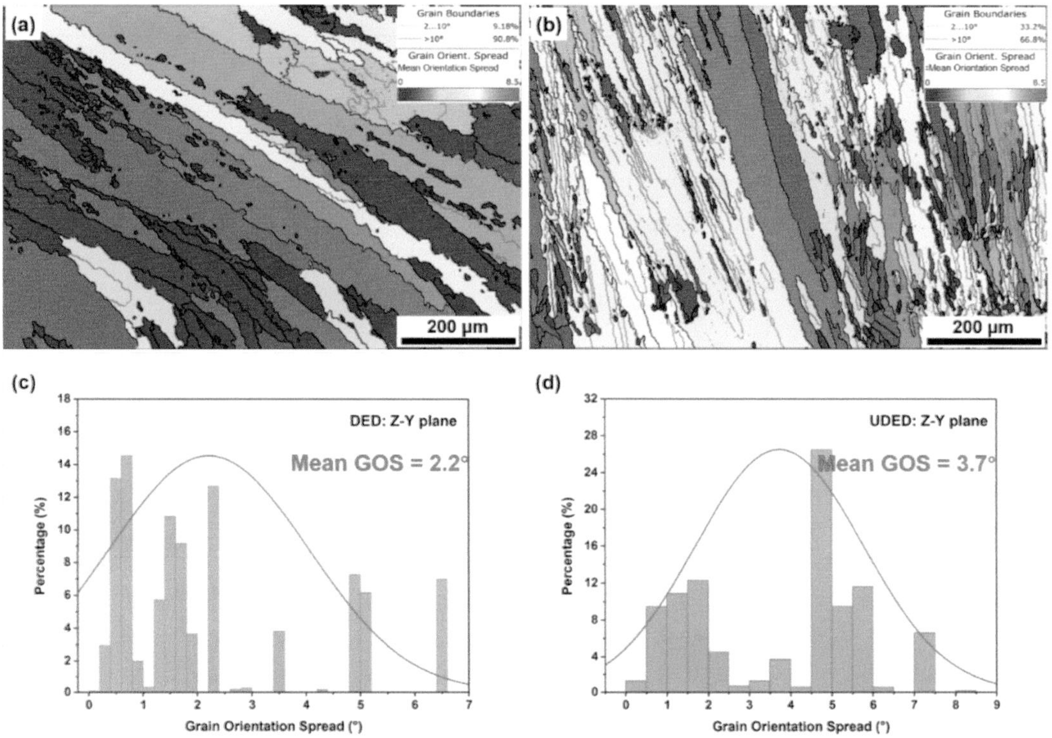

Figure 9.19. GOS maps of Z-Y plane deposited sample (a) (c) DED sample; (b) (d) UA-DED sample.

9.2.5.3 Tensile properties and fracture analysis

To investigate the effect of ultrasonic vibration on the tensile properties of the material, the samples were tested in tension along the scanning direction parallel (0°) and perpendicular (90°) to the scanning direction, and the results are shown in Figure 9.20. Figure 9.20(a) presents the stress-strain curve for tensile testing, while the ultimate tensile strength (UTS), yield strength (YS), and elongation (EL) of the samples are summarized in Figure 9.20(b), (c), and (d), respectively.

For the 0° tensile specimens, the UA-DED printed samples exhibited higher ultimate tensile strength and elongation compared to the DED printed samples. The ultrasound-assisted printed samples showed an ultimate tensile strength of 879 MPa, a yield strength of 563 MPa, and an elongation of 23.5%, whereas the DED-printed samples had an ultimate tensile strength of 713 MPa, a yield strength of 480 MPa, and an elongation of only 11.8%. The introduction of ultrasonic vibration increased the ultimate tensile and yield strengths of the material, while the UA-DED-printed samples exhibited significantly higher elongation and improved plasticity. This can be attributed to the effect of ultrasonic vibration, which refines the grain and reduces the tissue's porosity, creating a granular Laves phase that promotes a more homogeneous distribution of elements, resulting in increased yield strength and UTS.

Regarding the 90° tensile specimen, the ultimate tensile and yield strengths of the UA-DED printed samples were similar to those of the DED printed samples. The DED-printed samples had an ultimate tensile strength of 688 MPa and a yield strength of 483 MPa, whereas the ultrasound-assisted printed samples had an ultimate tensile strength of 722 MPa and a yield strength of 527 MPa. However, the elongation of the UA-DED printed samples was slightly reduced, from 9.1% to 7.3%, in comparison to the conventionally printed samples. Thus, ultrasonic vibration also increased the ultimate tensile and yield strengths of the deposited samples, but resulted in a slight reduction in their plasticity. The ultimate tensile strength of the 90° specimen is also a reflection of the degree of channel-to-channel bonding of the deposited layer, as the tensile test is performed

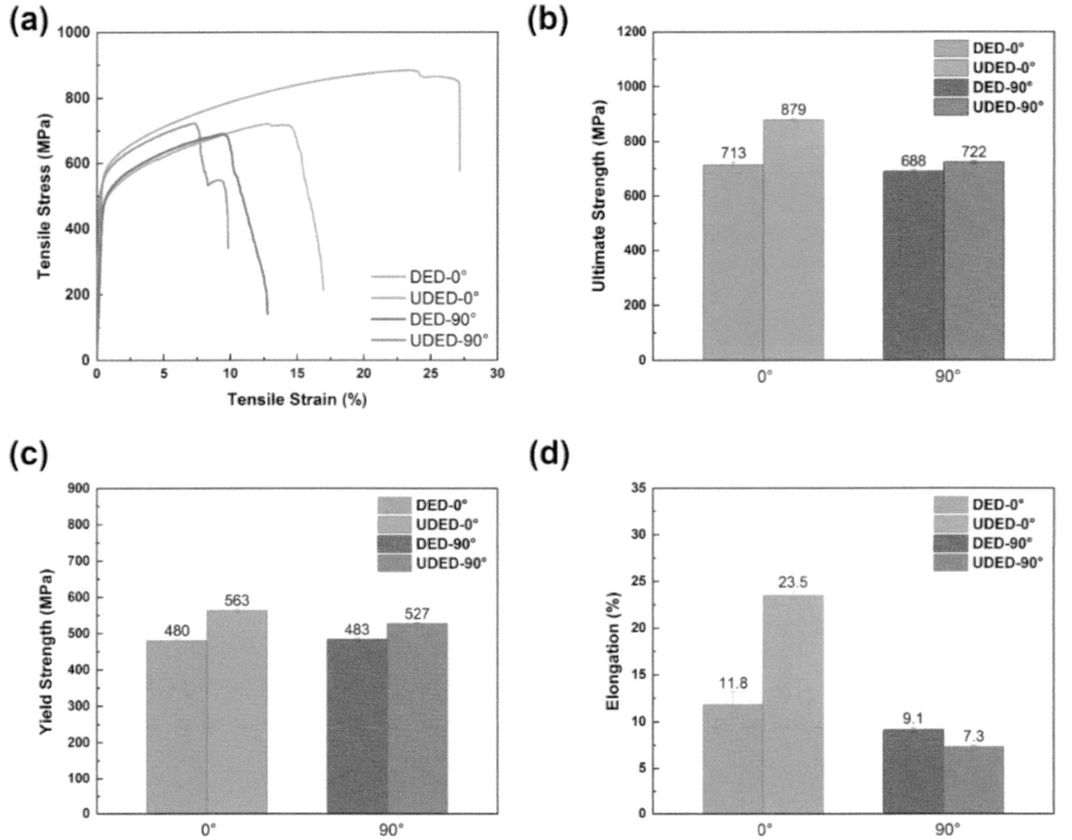

Figure 9.20. Tensile properties of deposited samples (a) stress-strain curve; (b) Ultimate tensile strength; (c) yield strength; (d) Elongation.

perpendicular to the scanning direction. Ultrasonic vibration enhances the flow and heat transfer within the melt pool, leading to better fusion of the molten material, a reduction in porosity, and grain refinement, resulting in increased UTS and yield strength of the deposited samples.

The fractures of the 0° tensile specimens are morphologically analyzed in Figure 9.21. As shown in the figure, the DED printed tensile fractures exhibit shallow tough nests and deconfined plateaus indicating combined ductile and brittle fracture. The fracture occurs along the columnar crystals, which are distributed parallel to the fracture surface, and internal fractures occur in the long chain-like Laves phase distributed at the boundary of the dendrites. Due to its large size, the long chain-like Laves phase is difficult to deform plastically with the γ matrix, leading to a tendency to slip and disintegrate during deformation, which increases the risk of crack extension [17]. Consequently, a large number of broken Laves phases and cavities remain in the crater after fragmentation of the long-chain Laves phase and separation from the γ matrix, as shown in Figure 9.21(c).

The tensile fracture of the UA-DED samples exhibits a large number of deep tough nests, indicating ductile fracture. At the same time, the granular Laves phase is dislodged at the interface, creating numerous cavities. The fine, discrete, and granular Laves phases favor ductility and have high coordination capacity with the γ matrix, retaining their original form or debonding from the γ matrix without breaking. The critical stress required for the discrete Laves phase to break is higher than that of its long-chain counterpart.

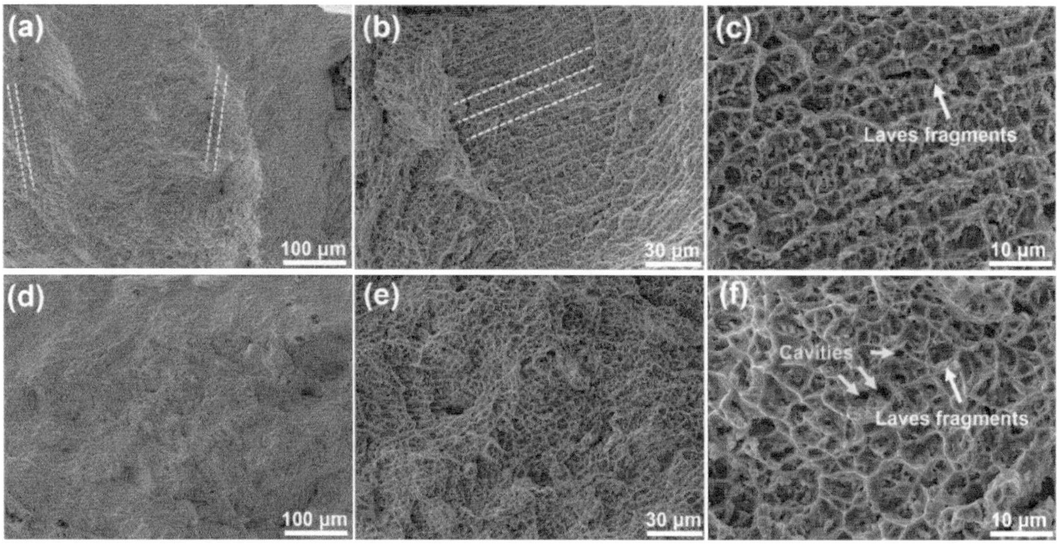

Figure 9.21. Tensile sample fracture morphology (a) (b) (c) DED; (d) (e) (f) Ultrasound-assisted DED.

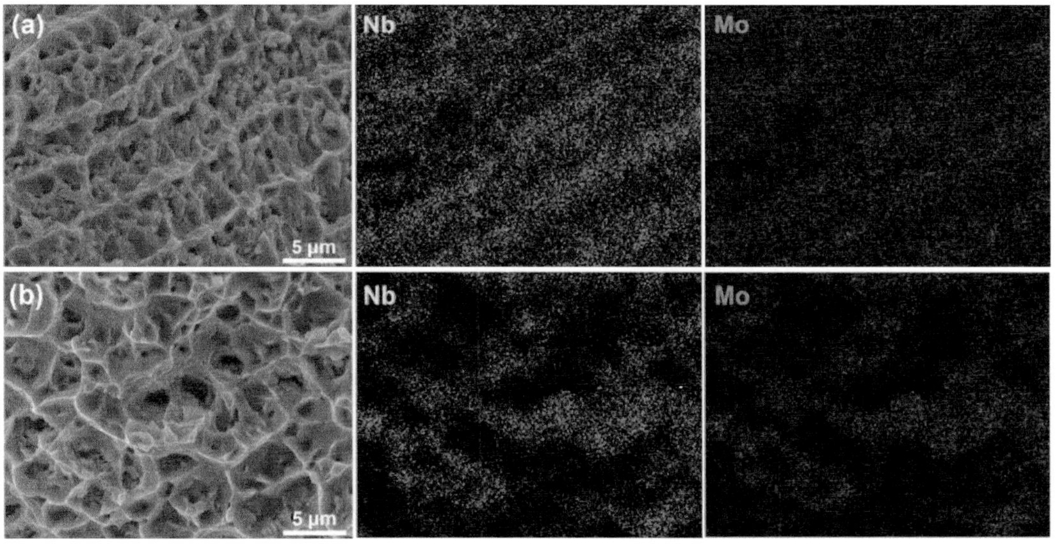

Figure 9.22. Distribution of elements at the fracture of tensile samples. (a) DED; (b) Ultrasound-assisted DED.

Figure 9.22 shows the elemental distribution on the fracture surface. In Figure 9.22(a), fracture occurs along the columnar dendrites with Nb and Mo elements enriched at the columnar dendrite boundaries where the Laves phase fractures. Figure 9.22(b) illustrates the tensile fracture organization of the UA-DED sample, revealing a deeper tough fossa with Nb and Mo elements enriched around it and cavities resulting from the dislodgment of the Laves phase. The UA-DED samples exhibit high strength and excellent ductility compared to the DED samples. The Laves phase in the tissue is brittle and hard, with poor plastic deformability, and the Laves phase and γ interface are weak points due to incoherent bonding. Many of the Laves phases precipitate in the interdendritic region and act as barriers to the movement of dislocations, leading to stress concentrations [18]. When the sample is subjected to tensile loading, microvoids and microcracks easily develop at the location of the Laves phase, gradually spreading and interconnecting, resulting in eventual failure. The small, granular Laves phase is less prone to breakage and exhibits better ductility. The granular Laves phase also generally experiences interfacial exfoliation under load, while irregularly shaped particles of long-chain-like Laves phase often fail due to internal fracture.

9.3 Mechanisms for the effect of ultrasonic vibration on microstructure and mechanical properties

9.3.1 Effect of ultrasonic vibration on the melt pool

9.3.1.1 Effect of ultrasonic vibration on melt pool morphology

Figure 9.23 presents the shape of the melt pool obtained from the simulation, with the yellow line representing the isotherm of the melting point of Inconel 718 alloy. The figure indicates that the melt pool for DED printing is deeper, approximately 1 mm, and wider, about 3.4 mm, whereas the melt pool for UA-DED printing is shallower, approximately 0.6 mm, and wider, about 3.6 mm. Consequently, ultrasonic vibration can alter the shape of the melt pool, making it less deep and slightly wider.

Among the various forces acting on the melt pool during laser directional deposition, surface tension plays a significant role [19]. It determines the shape and size of the melt pool to a large extent. The driving force behind the pool flow, caused by surface tension, is known as Marangoni flow, which significantly influences the depth of the resulting molten metal pool [20].

Marangoni flow arises due to the surface tension gradient, which propels the melt flow inside the pool. Temperature variations lead to a surface tension gradient, driving molten material from the center to the periphery of the pool. The equation governing the surface tension-induced Marangoni flow inside the laser melt pool is as follows:

$$\mu\frac{\partial u}{\partial z} = -\frac{\partial T}{\partial x}\frac{\partial \gamma}{\partial T} \tag{9.2}$$

Where μ is the dynamic viscosity of the liquid, and $\frac{\partial \gamma}{\partial T}$ is the surface tension temperature coefficient, and $\frac{\partial T}{\partial x}$ is the temperature gradient.

For nickel-based high-temperature alloys, the surface tension temperature coefficient is negative, indicating that the surface tension decreases as the temperature increases. Ultrasonic vibration enhances the Marangoni flow by carrying heat from the center to the edges of the melt pool, affecting its cooling rate and altering its shape. As a result, heat build-up at the bottom of the melt pool is less severe, resulting in a shallower melt depth.

The cavitation effect of ultrasound induces localized high temperatures and pressures in the melt pool, promoting heat transfer within it. The ultrasonic waves impact the free surface area of the melt pool, leading to increased heat transfer from the center to the periphery and faster liquid metal flow, resulting in a wider melt pool. The surface tension gradient causes the surface layer of

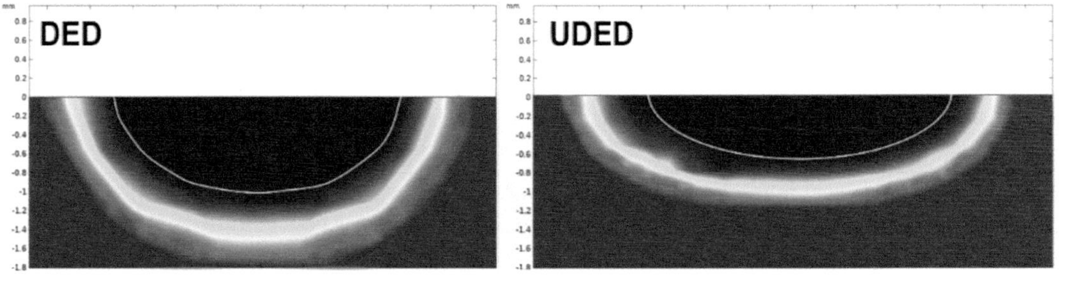

Figure 9.23. Molten pool shape simulation results.

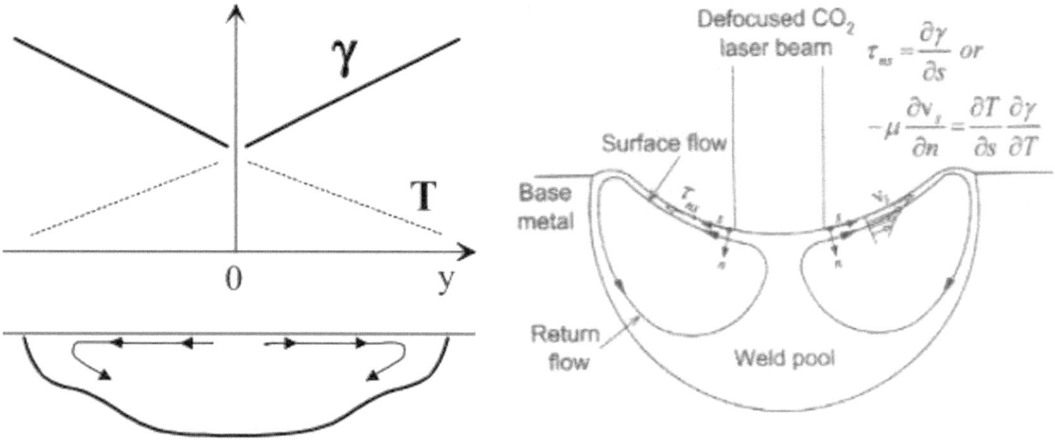

Figure 9.24. Schematic diagram of the Marangoni flow [21].

liquid metal to flow from the center to the periphery, causing the center of the melt pool to depress downwards. Gravity then causes the metal at the periphery of the melt pool to reflow along the bottom towards the center, forming a circular convection pattern (as shown in Figure 9.24).

9.3.1.2 *Effect of ultrasonic vibration on heat transfer in the melt pool*

Figure 9.25 illustrates the results of the temperature field for simulated conventional and ultrasound-assisted printing. The laser beam's heat source acts on the material's surface, and when the temperature reaches the melting point of Inconel 718 high-temperature alloy, the substrate and powder material melt, forming a molten pool of metal. The line in the diagram represents the isotherm of the melting point of Inconel 718 high-temperature alloy, with the area above the isotherm indicating the molten pool and the part below indicating the unmelted substrate material. The graph shows that the temperature is highest at the center of the melt pool, where the laser heat source directly acts, and decreases progressively away from the center towards the edge of the pool. For conventional printing, the maximum temperature at the center of the melt pool is 1820 K, while for ultrasound-assisted DED printing, it is 1950 K.

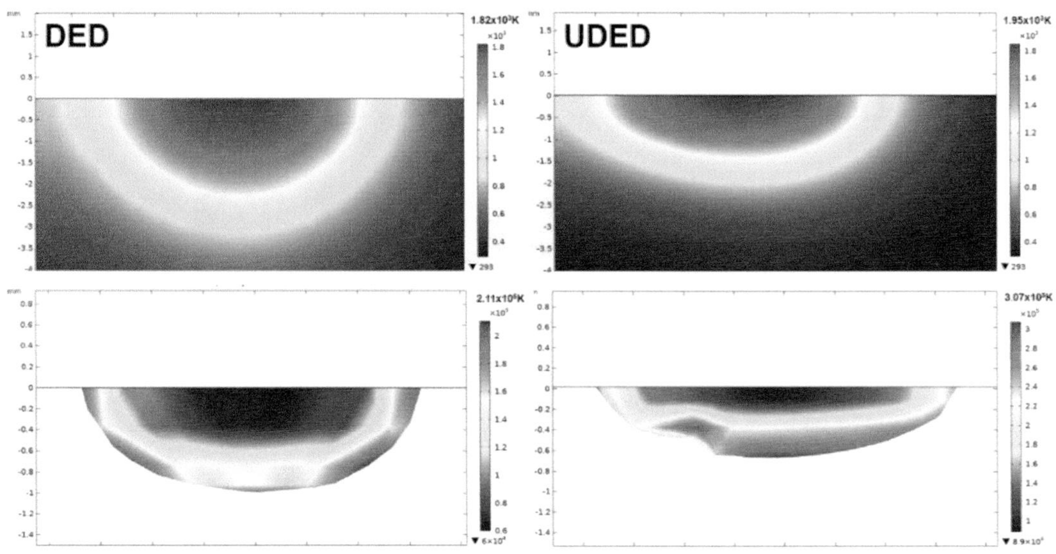

Figure 9.25. Distribution of the temperature field inside the molten pool.

The temperature gradient of the melt pool, as shown in Figure 9.25, is highest at the edge and decreases from the edge towards the center, with the lowest temperature gradient at the center of the melt pool. The highest temperature gradient for conventional printing is 2.11×10^5, which is lower than the maximum temperature gradient of 3.07×10^5 for ultrasound-assisted printing. This difference can be attributed to the introduction of ultrasonic vibration energy, which causes tiny bubbles or stretched cavities within the melt pool to experience periodic alternating acoustic pressure, leading them to continuously expand and oscillate until they collapse. This collapse generates a significant temperature gradient, increasing the surface tension gradient between the center and edge of the melt pool, thereby contributing to the Marangoni effect.

9.3.1.3 Effect of ultrasonic vibration on melt pool flow

Figure 9.26 illustrates the simulated flow field diagram, depicting the liquid metal flow from the center to the edge of the melt pool, followed by re-flow of the liquid metal at the pool's edge along the bottom towards the center. This creates a ring flow pattern within the melt pool.

In the flow vector diagram, the arrows represent the flow, and their color and length indicate the magnitude of the flow velocity. The flow velocity is higher near the free surface area close to the melt pool, where the liquid flows fastest due to Marangoni flow caused by surface tension. For ultrasound-assisted DED printing, the maximum melt pool flow velocity is 0.08, which is twice the velocity of conventional DED printing. The introduction of ultrasonic vibration enhances the Marangoni effect inside the melt pool, increasing convective intensity and flow velocity at the melt pool's surface area. Ultrasonic vibration persists throughout the laser printing process, resulting in more convective cycles, intensifying the agitation of the melt pool.

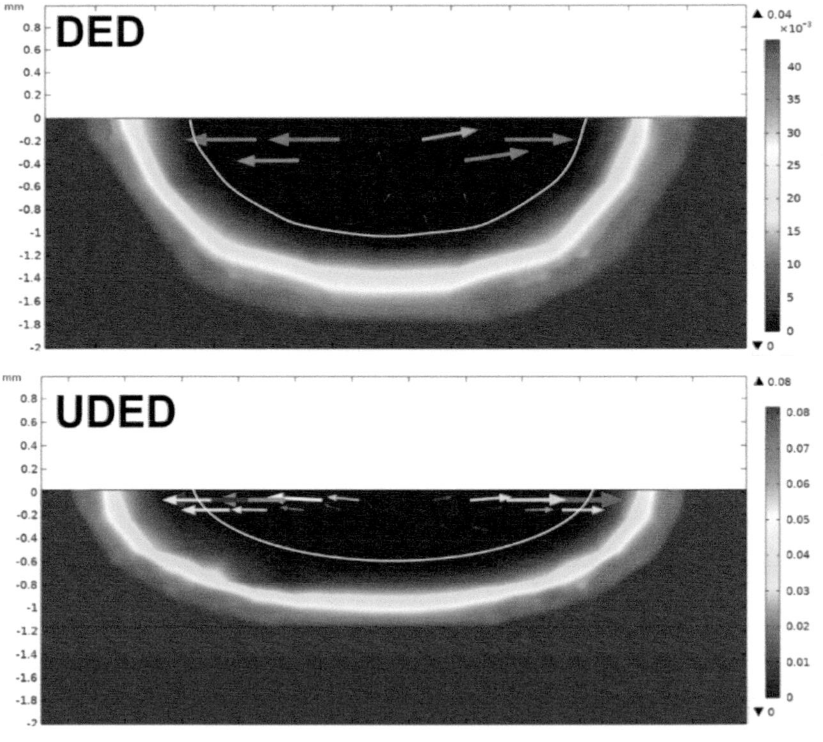

Figure 9.26. Distribution diagram of the flow field inside the molten pool.

9.3.2 Effect of ultrasonic vibration on microstructure

9.3.2.1 Effect of ultrasonic vibration on dendrite growth direction

Grain growth is determined by two key thermal factors: the temperature gradient (which determines the direction of grain growth) and the cooling rate (which determines the size of grain growth). According to crystallisation theory, each crystal structure has its own unique preferred crystal orientation where grain growth is fastest and the temperature gradient plays an important role in the growth rate. Grain growth can only be sustained if the preferred crystal orientation and the direction of the maximum temperature gradient (always perpendicular to the edge of the melt pool) remain constant and the time for grain growth is sufficient.

The nickel-based high temperature alloy Inconel 718 has a face-centred cubic structure with a face-centred cubic crystal system with a meritocratic production orientation of <001> and dendritic growth usually in a certain <001> crystal direction close to the direction of heat flow. As can be seen in Figure 9.27, the boundary line of the melt pool becomes flat and the direction of heat dissipation changes due to the intense marangoni flow caused by ultrasonic action. This results in the dendrite growth direction for ultrasound assisted DED printing being essentially parallel to the deposition direction, with dendrites growing upwards. In contrast, with DED printing of the metal melt pool, the arc of the melt pool boundary is larger and the dendrites grow from the edge of the melt pool towards the centre of the melt pool.

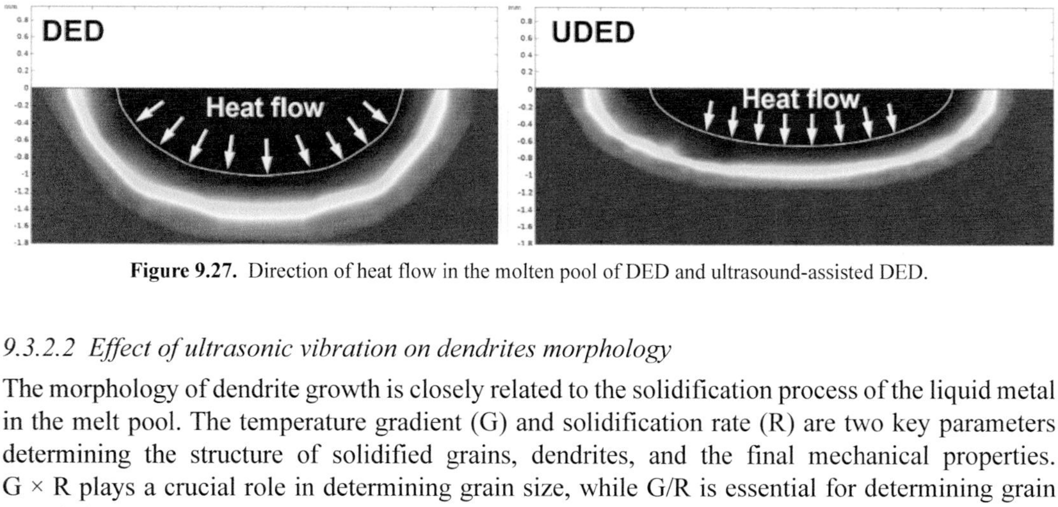

Figure 9.27. Direction of heat flow in the molten pool of DED and ultrasound-assisted DED.

9.3.2.2 Effect of ultrasonic vibration on dendrites morphology

The morphology of dendrite growth is closely related to the solidification process of the liquid metal in the melt pool. The temperature gradient (G) and solidification rate (R) are two key parameters determining the structure of solidified grains, dendrites, and the final mechanical properties. $G \times R$ plays a crucial role in determining grain size, while G/R is essential for determining grain morphology. The equation for these parameters is as follows:

$$G = \nabla T \cdot \overrightarrow{n_0} \tag{9.3}$$

$$R = V_s \cdot \vec{i} \cdot \overrightarrow{n_0} \tag{9.4}$$

Where $\overrightarrow{n_0}$ is the unit normal vector of the solidification front, and $\vec{i}$ is the unit vector in the scanning direction, and V_s is the scanning speed.

During laser deposition, different temperature gradients and grain growth rates are generated due to multiple laser heat inputs and heat accumulation. Solidification of high temperature alloys produces a region of subcooling which, due to the complex thermal history of the fabrication process, may propagate steadily as a planar front or unsteadily as a cellular or dendritic structure. kurz and fisher [22] developed a solid-liquid interface model for the instability criterion during solidification, where the grain morphology may undergo a transition from cellular to columnar to

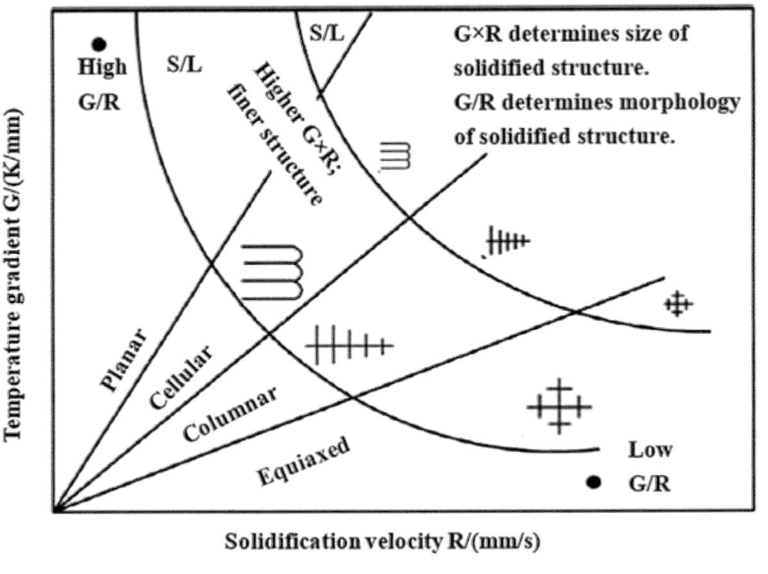

Figure 9.28. The effects of G and R on the morphology and the scale of the solidified structure [24].

equiaxed dendrites as the parameter (G/R) decreases due to severe structural subcooling. Subcooling occurs when the following conditions are met [23], the solidification front will no longer be stable:

$$\frac{G}{R} < \frac{\Delta T}{D_L} \tag{9.5}$$

Where ΔT is the temperature difference between the two sides of the solid-liquid interface (K), and D_L is the diffusivity of the solute in the liquid (m^2/s).

As shown in Figure 9.28 [24], when the G/R ratio is high, the solidification front remains planar, facilitating the formation of planar dendrites. As the G/R ratio decreases, the cellular dendrites and columnar dendrites sequentially emerge. When the G/R ratio is low, the dendritic morphology transitions to an equiaxed form. Additionally, it indicates that a higher G×R value yields finer crystalline structures, whereas a lower G×R value tends to produce coarse grains. In the DED process, as shown in Figure 9.29(a), the temperature gradient (G) and solidification rate (R) lead to dendritic dendrites growing epitaxially along the solidification front. The melt pool's deeper depth in DED printing results in a larger angle, causing heat flux to dissipate downwards during solidification and generating a higher temperature gradient at the bottom. Additionally, the lower cooling rate of the melt pool in DED printing facilitates the growth of columnar dendrites [25]. The temperature gradient distribution in the simulation shows an increasing heat build-up from the edge towards the center of the melt pool, resulting in a decreasing temperature gradient with the lowest

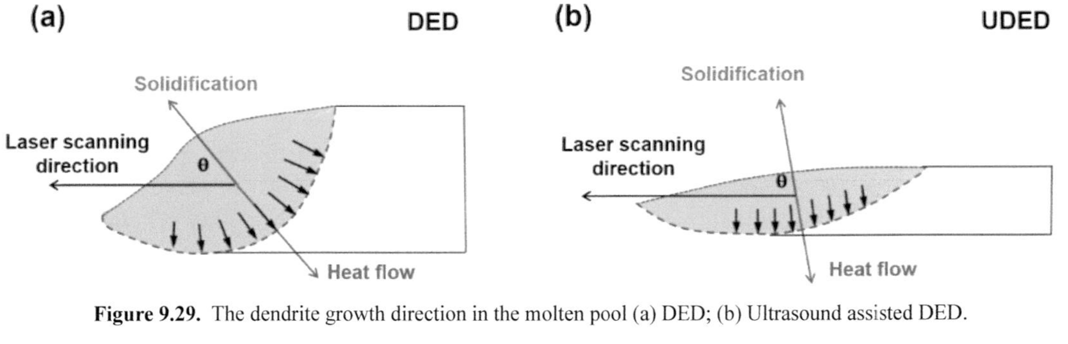

Figure 9.29. The dendrite growth direction in the molten pool (a) DED; (b) Ultrasound assisted DED.

gradient at the center. A lower G/R ratio leads to more severe structural subcooling and promotes a change in grain morphology from columnar to equiaxed dendrites, leading to a CET transition.

In Figure 9.29(b), the heat dissipation direction of ultrasound-assisted DED printing is nearly parallel to the deposition direction. The larger angle results in a smaller solidification speed (R) when the scanning speed (Vs) is the same. The ultrasonic energy converting to thermal energy inside the melt affects the temperature gradient of the melt pool. Therefore, the temperature gradient (G) at the edge of the melt pool for ultrasound-assisted DED printing is greater than that of conventional printing, and the G/R ratio for ultrasound-assisted DED printing is greater than that of DED printing. This tendency leads to a cellular dendrite organization at the solidification front when the liquid metal solidifies.

9.3.2.3 Effect of ultrasonic vibration on the morphology of the Laves phase

The morphology and size of Laves phase particles are influenced by the ratio of cooling rate to dendrite growth rate (G/R) [26, 27]. Laves phase typically precipitates at dendrite edges, where Nb elements are enriched. High cooling rates and high G/R ratios in high-temperature alloys lead to discrete Laves phase particles, while low cooling rates result in continuous long chains of Laves phase particles.

In DED printing, the continuous laser deposition causes dendrites to grow in their previous direction, forming continuous coarse Laves with strong directional selectivity. At low cooling rates, the Laves phase particles form long chains. The introduction of ultrasonic vibration increases melt pool mobility, breaking up the Laves phase and producing fine spherical Laves phases at dendrite boundaries. At high cooling rates, particles are finer and more discrete. Columnar dendrite growth separates the liquid between adjacent dendrites, leading to long chain morphology, while cellular dendrites promote the formation of discrete Laves phase particles. During UA-DED printing, ultrasonic energy promotes Marangoni flow, transforming Laves phase particles from long chains to fine discrete ones, enhancing the mechanical properties of IN 718 alloy.

9.3.2.4 Effect of ultrasonic vibration on grain size

Simulation results demonstrate that ultrasonic vibration increases melt flow velocity, promoting diffusion and uniform distribution of broken dendrites in the melt pool. The ultrasonic energy transmitted directly into the melt pool induces cavitation and acoustic flow, refining the grains.

The reduction in grain size is related to the formation of nuclei. From classical nucleation theory it is clear that two conditions are required for the formation of a nucleus, one is that the maximum size of the embryo r_{max} is greater than the critical nucleus size r_k and the other is that the energy fluctuations in the melt pool are sufficient to provide the energy required for the formation of the embryo. The expressions for the critical nucleus r_k and the critical energy of formation are as follows

$$r_k = \frac{2\sigma}{\Delta G} \tag{9.6}$$

$$\Delta G_c = \frac{16\pi\sigma^3}{3\Delta G^2 f(\theta)} \tag{9.7}$$

$$\Delta G = \Delta H_s + T\Delta S_s + \Delta V_s \Delta P \tag{9.8}$$

Of these, the σ is the Gibbs-Thomson effect, and $f(\theta)$ is the non-uniform nucleation factor ($0 < f(\theta) < 1$). The introduction of ultrasound causes the ΔP increase, which causes the free energy to ΔG increases and the critical nucleation r_k and the critical nucleation free energy decreases, so that ultrasonic vibration-assisted printing can increase the nucleation rate and reduce grain size compared to conventional printing.

9.3.3 Effect of ultrasonic vibration on defects

The two types of defects in laser deposition are porosity and unfused defects. Pores are usually round, spherical voids formed by entrapped gas during the melting process and are more frequently found inside the melt pool. Unfused defects, irregular in shape, occur due to incomplete material fusion during manufacturing.

Simulation results reveal that ultrasonic vibration enhances Marangoni flow, increasing the flow rate of liquid metal on the melt pool's surface, creating strong mixing and agitation that facilitates rapid bubble escape. Ultrasonic vibration also affects heat transfer within the melt pool, leading to the collapse of cavitation bubbles and an increase in the melt pool's peak temperature, aiding in the melting of powder and base material. Ultrasound promotes Marangoni flow, altering the melt pool's morphology, reducing its depth, and minimizing the contact area between channels.

9.3.4 Effect of ultrasonic vibration on mechanical properties

9.3.4.1 Effect of ultrasonic vibration on microhardness

Microhardness is influenced by mass and heat transfer, Marangoni convection, and solidification behavior within the melt pool during laser-directed deposition (UA-DED printing). The simulation results indicate that UA-DED printing exhibits a greater temperature change during cooling, resulting in a higher cooling rate of the melt pool. This increased cooling rate leads to grain refinement, enhancing microhardness due to the effects of cavitation and acoustic flow caused by ultrasonic energy. Additionally, enhanced Marangoni convection in the shallower melt pool of UA-DED printing allows air bubbles to discharge before solidification, reducing porosity within the clad layer and further enhancing microhardness. Moreover, the increased cooling rate and the ratio of temperature gradient to solidification front movement rate (G/R) generate finely fragmented Laves phases. Accompanied by ultrasonic energy, these phases disperse finely at the dendrite boundaries, suppressing the formation of long chain-like Laves phases and thereby enhancing microhardness.

9.3.4.2 Effect of ultrasonic vibration on tensile properties

The mechanical properties of Inconel 718 high temperature alloys, manufactured by laser directional deposition, are closely related to their microstructure. During this process, the material is melted by the laser heat source to form a molten pool where heat transfer and Marangoni flow lead to diverse solidification structures. The grain size, number of Laves phase particles, and presence of pores in the printed part influence deformation behavior, fracture stress, strain, and ductility.

Grain size significantly affects the fracture behavior of microscale materials, as demonstrated by numerous studies [28–31], which have shown that fracture stresses and strains decrease with increasing grain size. Grain boundaries hinder dislocation movement during deformation, leading to stress concentration in the grain boundary region. Ultrasonic vibration refines the grains and weakens grain boundaries, thereby improving the mechanical properties of the sample.

In addition to grain size, porosity also affects the mechanical properties of the material. Prior research [32, 33] has experimentally and through modeling methods demonstrated that tensile properties are adversely affected by porosity. Zhou et al. [34] quantitatively investigated the effect of porosity on tensile properties using analytical (i.e., critical local strain model) and empirical methods based on the projected area of porosity on the fracture surface. They found that both ultimate tensile strength (UTS) and elongation (EL) decreased with increasing porosity levels. The influence of porosity on tensile properties can be further described by the empirical correlations [35] :

$$EL = EL_0 \ (1 - f)^a \tag{9.9}$$

Where EL_0 and a are empirical constants. The parameter EL_0 can be interpreted as the EL of a non-porous material and the parameter a is the defect sensitivity factor of the EL to porosity changes.

Similarly, the ultimate tensile strength (UTS) can be described by the dependence of the following equation on the change in porosity [35] :

$$UTS = UTS_0 (1 - f)^b \tag{9.10}$$

Where UTS_0 and b are empirical constants. The parameter UTS_0 can be interpreted as UTS for non-porous materials and parameter b is the defect sensitivity factor of UTS to changes in porosity.

As observed from the simulation results, ultrasonic vibration energy enhances the Marangoni convective circulation within the melt pool, leading to a reduction in porosity and defect rate of the sample. This, in turn, increases the final fracture stress and strain of the material.

The presence of the Laves phase significantly influences the yield strength of Inconel 718 alloy. There are two main mechanisms by which the precipitated phase resists dislocation movement [36]: the Orowan mechanism, where dislocations bypass the precipitates, and the shear mechanism, where dislocations cut through the precipitates. The reinforcement increment of dispersed Laves phase particles can be calculated using the Orowan-Ashby relationship [37] :

$$\Delta\sigma_{OA} = \frac{0.84MGb}{2\pi(1-v)^{0.5}(\lambda - 2r_L)} \ln\left(\frac{r_L}{b}\right) \tag{9.11}$$

Where M is the Taylor factor for random grain orientation, G is the shear modulus, b is the Burgers vector and v is the Poisson's ratio. λ is the mean particle spacing with a value equal to $r_L\sqrt{\frac{2\pi}{3f_L}}$. r_L and f_L denotes the radius and volume fraction of the Laves phase, with the subscript "L" representing the Laves phase.

During deformation, dislocations bypass the Laves phase, which is disjoint from the matrix. The morphology and size of the Laves phase play a vital role in determining the yield strength. Ultrasonic vibration changes the morphology of the Laves phase, causing it to be finely dispersed at the cellular dendrite boundaries, thereby promoting the precipitation strengthening of the material and increasing the yield strength of the sample.

The plastic deformation behavior and ductility of metals are closely related to their microstructure. The influence of the brittle phase on the tensile ductility of nickel-based high-temperature alloys has been investigated [38], and the results show that ductility degradation is mainly influenced by brittle orientation, morphology, size, and brittleness distribution. Normal orientation of the brittle phase results in a greater reduction in strength and ductility than parallel orientation. In laser-deposited samples, multiple Laves phases are distributed in dendritic regions, and the Laves phase is a topologically compact packed (TCP) phase with high hardness and brittleness. The Laves phase impedes dislocation movement during the stretching process, leading to dislocation accumulation and stress concentrations. As stress increases, cracks develop around the Laves phase. The relationship between material ductility and the size of the Laves phase can be expressed as [39]:

$$\varepsilon_p = Cf(n, p) \tag{9.12}$$

$$C = \frac{1}{\tilde{\varepsilon}_n(\theta)} \cdot \frac{\varepsilon_c}{2} \tag{9.13}$$

$$f(n, p) = \left[\frac{I}{0.405\pi h}\right]^{\frac{1}{n+1}} * p^{\frac{1}{n+1}} \tag{9.14}$$

$$p = \frac{\lambda_c}{2r_c} - 1 \tag{9.15}$$

Where θ is the angle in polar coordinates, $\widetilde{\varepsilon_n}(\theta)$ is the angle when $\theta = 0$ a constant, λ_c and r_c are the spacing and radius of the Laves phase, respectively. ε_c is the critical value of the microplastic strain, I and h are functions of n, which is the strain hardening index.

It can be observed that plastic deformation of the material increases as the size of the Laves phase decreases. The plate-like brittle phase induces embrittlement to a significantly higher extent than the spherical brittle phase. The finely dispersed distribution of the brittle phase has a smaller influence on plasticity than the continuous distribution of the brittle phase. Ultrasonic vibration, by changing the cooling rate inside the melt pool and through the accompanying cavitation effect of ultrasonic energy, results in a finer dispersion of the Laves phase in the printed Inconel 718 alloy sample, which facilitates the improvement of material ductility.

9.4 Summary

In this chapter, we have developed a system platform for Ultrasonic Assisted Laser Directed Energy Deposition (UA-DED) manufacturing to enhance the organization and mechanical properties of Inconel 718 alloy printed using laser directional deposition. To achieve this, we incorporated ultrasonic vibration assistance into the laser additive process by directly contacting the ultrasonic vibration bar with the printed workpiece to minimize ultrasonic loss.

To investigate the effects of ultrasonic vibration on the microstructure and properties, we conducted numerical simulations of the melt pool temperature field and flow field distribution during the UA-DED process using COMSOL simulation software. This simulation approach effectively transformed ultrasonic vibration energy into a dynamic load applied to the bottom of the melt, taking into account various factors such as laser heat source, dissipative radiation, melt flow, Marangoni effect, surface tension, and solid-liquid phase change. By considering the interaction of multiple physical fields, we gained precise insights into the mechanism behind the improvement of material structure and mechanical properties due to ultrasonic vibration.

Ultrasonic vibration has a significant impact on marangoni flow and mass and heat transfer in the melt pool, thereby altering the pool's shape and heat dissipation direction. Specifically, the melt pool in conventional DED printing appeared deeper, while the melt pool in ultrasound-assisted DED printing was wider and shallower. This change in melt pool shape influenced the direction and morphology of dendrites. In conventional DED-printed samples, columnar dendrites were predominant, growing at a 45° angle from the deposition direction. On the other hand, ultrasound-assisted DED samples exhibited mainly cellular dendrites, with dendrites tending to grow vertically. Moreover, ultrasonic vibration played a role in refining the grains and affecting the shape of the Laves phase. In ultrasound-assisted DED-printed samples, the Laves phase appeared granular and discrete at the boundaries of the cytosolic dendrites, whereas conventionally DED-printed samples displayed long chain-like Laves phases regularly distributed at the columnar dendrite boundaries. The improved organization through ultrasonic vibration resulted in increased hardness, strength, toughness, and reduced porosity in the UA-DED samples.

The introduction of ultrasonic vibration into the laser deposition process demonstrates its potential to improve the mechanical properties of materials by enhancing their microstructure. In future research, it will be important to explore the effects of adjusting processing parameters on material organization and properties. This will help us better understand the relationship between ultrasonic parameters and laser printing parameters concerning microstructure and properties, thus allowing us to flexibly adjust the microstructure and mechanical properties of printed samples.

References

[1] Thivillon, L., Bertrand, P., Laget, B. and Smurov, I. (2009). Potential of direct metal deposition technology for manufacturing thick functionally graded coatings and parts for reactors components[J]. Journal of Nuclear Materials 385(2): 236–241.

[2] DebRoy, T., Wei, H. L., Zuback, J. S., Mukherjee, T., Elmer, J. W., Milewski, J. O. et al. (2018). Additive manufacturing of metallic components–process, structure and properties[J]. Progress in Materials Science 92: 112–224.

[3] Komarov, S. V., Kuwabara, M., Abramov, O. V. (2005). High power ultrasonics in pyrometallurgy: current status and recent development[J]. ISIJ International 45(12): 1765–1782.

[4] Ju, Z., Li, X. and Jiang, R. (2012). Numerical simulation of the stirring action of ultrasound in the process of semi-continuous casting aluminum alloy[J]. Tezhong Zuzao Ji Youse Hejin (Special Casting & Nonferrous Alloys) 32(11): 1045–1048.

[5] Nagira, T., Nakatsuka, N., Yasuda, H., Uesugi, K., Takeuchi, A. and Suzuki, Y. (2015). Impact of melt convection induced by ultrasonic wave on dendrite growth in Sn–Bi alloys[J]. Materials Letters 150: 135–138.

[6] Abramov, O. V. (1987). Action of high intensity ultrasound on solidifying metal[J]. Ultrasonics 25(2): 73–82.

[7] Cong, W. and Ning, F. (2017). A fundamental investigation on ultrasonic vibration-assisted laser engineered net shaping of stainless steel[J]. International Journal of Machine Tools and Manufacture 121: 61–69.

[8] Gorunov, A. I. (2020). Additive manufacturing of Ti6Al4V parts using ultrasonic assisted direct energy deposition[J]. Journal of Manufacturing Processes 59: 545–556.

[9] Ning, F., Hu, Y., Liu, Z., Wang, X., Li, Y. and Cong, W. (2018). Ultrasonic vibration-assisted laser engineered net shaping of inconel 718 parts: Microstructural and mechanical characterization[J]. Journal of Manufacturing Science and Engineering 140(6).

[10] Srivastava, N., Chaudhari, G. P. and Qian, M. (2017). Grain refinement of binary Al-Si, Al-Cu and Al-Ni alloys by ultrasonication[J]. Journal of Materials Processing Technology 249: 367–378.

[11] Hongxi, L., Zifeng, L., Xiaowei, Z., Hai, S. and Yehua, J. (2017). Design of steady-state magnetic field and effect of current intensity on microstructure of Fe55 alloy coating by laser cladding[J]. 红外与激光工程 46(4): 406001–0406001 (7).

[12] Choo, H., Sham, K. L., Bohling, J., Ngo, A., Xiao, X., Ren, Y. et al. (2018). Effect of laser power on defect, texture, and microstructure of a laser powder bed fusion processed 316L stainless steel[J]. Materials & Design.

[13] Everton, S. K., Hirsch, M., Stravroulakis, P., Leach, R. K. and Clare, A. T. (2016). Review of *in-situ* process monitoring and *in-situ* metrology for metal additive manufacturing[J]. Materials & Design 95(Apr.): 431–445.

[14] Ning, F., Hu, Y., Liu, Z., Wang, X., Li, Y. and Cong, W. (2018). Ultrasonic vibration-assisted laser engineered net shaping of inconel 718 parts: microstructural and mechanical characterization[J]. Journal of Manufacturing Science and Engineering 140(6).

[15] HengChen, ZebangHe and LinLu. (2020). Correlation of surface features with corrosion behaviors of interstitial free steel processed by temper rolling[J]. 材料科学技术:英文版 2020(1): 37–44.

[16] Zhao, Y., Li, K., Gargani, M. and Xiong, W. (2020). A comparative analysis of Inconel 718 made by additive manufacturing and suction casting: microstructure evolution in homogenization[J]. Additive Manufacturing 2020: 101404.

[17] Sui, S., Chen, J., Zhang, R., Ming, X., Liu, F. and Lin, X. (2017). The tensile deformation behavior of Laser Repaired Inconel 718 with a non-uniform microstructure[J]. Materials Science and Engineering A 688(Complete): 480–487.

[18] Wang, K., Liu, Y., Sun, Z., Lin, J., Lv, Y. and Xu, B. (2019). Microstructural evolution and mechanical properties of Inconel 718 superalloy thin wall fabricated by pulsed plasma arc additive manufacturing[J]. Journal of Alloys and Compounds 819: 152936.

[19] Anthony, T. R. and Cline, H. E. (1977). Surface rippling induced by surface-tension gradients during laser surface melting and alloying[J]. Journal of Applied Physics 48(9): 3888–3894.

[20] Limmaneevichitr, C. and Kou, S. (2000). Visualization of marangoni convection in simulated weld pools: marangoni convection resembling that in weld pools is revealed by flow visualization[J]. Welding Journal 2000(5): 79.

[21] Drezet, J. M., Pellerin, S., Bezenon, C. and Mokadem, S. (2004). Modelling Marangoni convection in laser heat treatment[J]. Journal de Physique IV (Proceedings) 120(12): 299–306.

[22] Kurz, W. and Fisher, D. J. (1992). Fundamentals of Solidification.

[23] Tiller, W. A., Jackson, K. A., Rutter, J. W. and Chalmers, B. (1953). The redistribution of solute atoms during the solidification of metals[J]. Acta Metallurgica 1(4): 428–437.

[24] Kou, S. (2003). Welding Metallurgy, Second Edition[M].

[25] Farshidianfar, M. H., Khajepour, A. and Gerlich, A. P. (2016). Effect of real-time cooling rate on microstructure in Laser Additive Manufacturing[J]. Journal of Materials Processing Technology 231: 468–478.

[26] Nie, P., Ojo, O. A. and Li, Z. (2014). Numerical modeling of microstructure evolution during laser additive manufacturing of a nickel-based superalloy[J]. Acta Materialia 77: 85–95.

[27] Xiao, H., Li, S. M., Xiao, W. J., Li, Y. Q., Cha, L. M., Mazumder, J. et al. (2017). Effects of laser modes on Nb segregation and Laves phase formation during laser additive manufacturing of nickel-based superalloy[J]. Materials Letters 188: 260–262.

[28] Petch, N. J. (1953). The cleavage strength of polycrystals[J]. Journal of the Iron and Steel institute 174: 25–28.

[29] Fan, Z. (1995). The grain size dependence of ductile fracture toughness of polycrystalline metals and alloys[J]. Materials Science and Engineering: A 191(1-2): 73–83.

[30] Raulea, L. V., Goijaerts, A. M., Govaert, L. E. and Baaijens, F. P. T. (2001). Size effects in the processing of thin metal sheets[J]. Journal of Materials Processing Technology 115(1): 44–48.

[31] Liu, B. B., Han, J. Q., Zhao, R., Liu, W. and Wan, M. (2016). Grain size effect on fracture behavior of the axis-tensile test of Inconel 718 sheet[J]. High Temperature Materials and Processes 35(10): 989–998.

[32] Lee, C. D. and Shin, K. S. (2007). Effect of microporosity on the tensile properties of AZ91 magnesium alloy[J]. Acta Materialia 55(13): 4293–4303.

[33] Lee, C. D. (2007). Effects of microporosity on tensile properties of A356 aluminum alloy[J]. Materials Science & Engineering A 464(1-2): 249–254.

[34] Zhou, B., Meng, D. H., Wu, D., Tang, J. F., Chen, R. S., Li, P. J. et al. (2019). Characterization of porosity and its effect on the tensile properties of Mg-6Gd-3Y-0.5Zr alloy[J]. Materials Characterization 152: 204–212.

[35] Lee, C., Youn, J. and Kim, Y. (2017). Effect of strain rate on the defect susceptibility of tensile properties to porosity variation[J]. Materials Science & Engineering, A. Structural Materials: Properties, Misrostructure and Processing 683: 135–142.

[36] Kini, A. R., Maischner, D., Weisheit, A., Ponge, D., Gault, B., Jägle, E. A. et al. (2020). *In-situ* synthesis via laser metal deposition of a lean Cu–3.4 Cr–0.6 Nb (at%) conductive alloy hardened by Cr nano-scale precipitates and by Laves phase micro-particles[J]. Acta Materialia 197: 330–340.

[37] Anderson, K. R. and Groza, J. R. (2001). Microstructural size effects in high-strength high-conductivity Cu-Cr-Nb alloys[J]. Metallurgical and Materials Transactions A 32(5): 1211–1224.

[38] Sui, S., Tan, H., Chen, J., Zhong, C., Li, Z., Fan, W. et al. (2019). The influence of Laves phases on the room temperature tensile properties of Inconel 718 fabricated by powder feeding laser additive manufacturing[J]. Acta Materialia 164: 413–427.

[39] Hu, Y. L., Li, Y. L., Zhang, S. Y., Lin, X., Wang, Z. H. and Huang, W. E. (2020). Effect of solution temperature on static recrystallization and ductility of Inconel 625 superalloy fabricated by directed energy deposition[J]. Materials Science and Engineering: A 772: 138711.

Chapter 10

Cracking Mechanisms and Suppression of the γ' Strengthen Nickel-based Superalloy During Additive Manufacturing

10.1 Introduction

Nickel-based superalloys are alloys with comprehensive properties such as high strength and certain resistance to oxidation and corrosion at high temperatures of 650 to 1000°C. They are widely used in aircraft, ships, industrial gas turbines and other fields, especially in the need to withstand high temperatures and corrosive environments. For example, the nickel-based superalloy content in the GE CF6 turbine engine of Boeing 787 aircraft reached approximately 47%, as illustrated in Figure 10.1, with predominant distribution within the combustion chamber, high-pressure compressor, and turbine blade [1]. According to the main strengthening methods of alloys, nickel-based alloys can be divided into solid solution strengthened nickel-based superalloys and precipitation strengthened nickel-based superalloys. The solid solution strengthened superalloys undergo solid solution treatment to achieve a homogeneous composition and obtain an appropriate grain size, and then are fabricated into components for application. Such alloys are typically utilized as the flame tube materials in combustion chambers, such as Inconel 625, GH 3030, GH 3039, etc. Precipitation strengthened superalloys are generally subjected to solid solution treatment and aging treatment, causing the precipitation strengthening phase or the like to uniformly and dispersedly precipitate, hindering dislocation movement and significantly enhancing the strength of the superalloys. These alloys are mainly employed as turbine blade and guide vane materials. The precipitation of these superalloys can be γ' or γ''.

In recent years, additive manufacturing (AM) has breaked through traditional preparation methods and opening up new possibilities for nickel-based superalloys in terms of multi-component alloying, structure size, and complexity. This benefited from the unique characteristics of additive manufacturing, including ultra-high temperature, strong convection, and supernormal metallurgical conditions such as instantaneous formation of small molten pools and ultra-high temperature gradients [2–4]. These conditions enable non-equilibrium rapid solidification with a fast cooling rate, However, the high ratio between metals and the requirement for high alloying to confer desired properties result in a wide solidification temperature interval (between liquidus and solid phase line), which poses significant challenges in AM of γ' phase strengthened nickel-based superalloys. During rapid solidification of nickel-based superalloy, local concentration of Al and Ti in the liquid phase occurs, leading to microscopic segregation and the formation of solute-rich liquid between dendrites. Differential shrinkage rates of the liquid film within the molten pool during cooling can induce tensile stress and cracking. Below the melting point, stress can be adjusted through

DOI: 10.1201/9781003406327-10

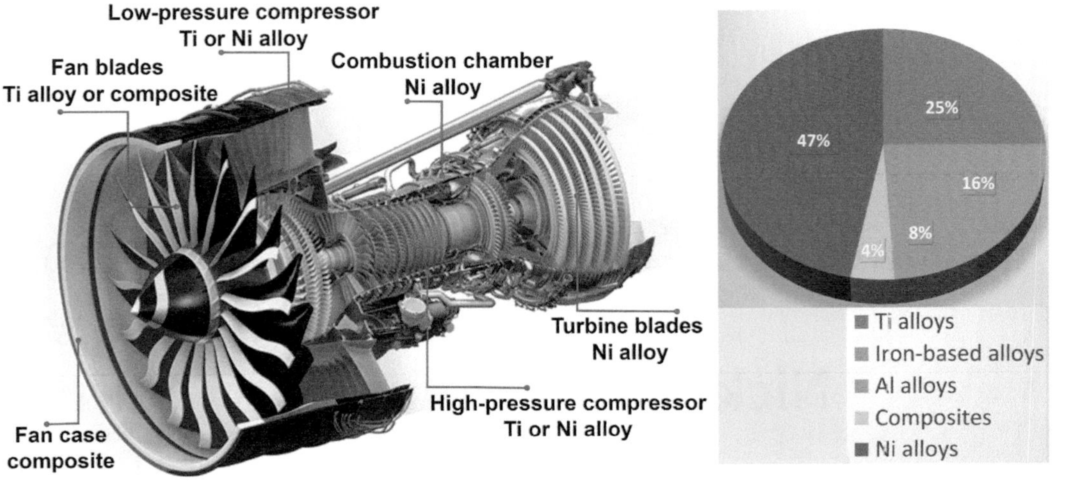

Figure 10.1. Material distribution in the GE CF6 turbine engine of Boeing 787 aircraft [1].

solid-phase plastic deformation. However, for high-volume fraction γ' phase nickel-based superalloys, the window between the thermodynamically stable temperature of the strengthening phase and the solidus temperature is narrow, leading to high crack sensitivity during the late solidification and in the solid state [5].

In general, the content of Al element in the nickel-based superalloys for hot forming is relatively low, and most of them are hot-deformed superalloys and cast superalloys with γ'' phase as the main strengthening phase. However, the content of Al in the difficult-to-form alloys is higher, and they are mostly cast alloys or even single-crystal superalloys with γ' phase as the main strengthening phase. [6, 7]. In order to improve hot formability, the types and contents of strengthening elements need to be strictly controlled, but a low alloying level is bound to sacrifice some mechanical properties, i.e., there is a contradiction between the processing compatibility of AM and mechanical properties of difficult-to-form nickel-based superalloys. The cracking susceptibility of high-strength nickel-based superalloys increases with the increase of the content of γ' phase. Although nickel-based superalloys with low γ' phase content can be fabricated without cracks, their strength and high-temperature resistance fail to meet the application requirements for high strength and high-temperature stability. In this chapter, we provide a comprehensive review of the cracking mechanisms and suppression strategies for γ'-strengthened nickel-based superalloys during AM, including but not limited to laser directed energy deposition.

10.2 Cracking mechanism of γ' strengthened superalloys

Most nickel-based superalloys especially γ' strengthened superalloys are susceptible to crack defects when prepared by AM. These defects can be categorized as solidification cracks, liquation cracks, strain-age cracks, ductility-dip cracks, and cold cracks, as depicted in Figure 10.2 [8]. The first two types are liquid cracks, which necessitate the presence of liquid films, while the latter three are solid cracks. Due to the intricate thermal history during AM, distinguishing between liquation and solidification cracks based on liquid films is relatively straightforward, however, discriminating ductility-dip cracks from strain-age cracks poses challenges since both occur in a solid state.

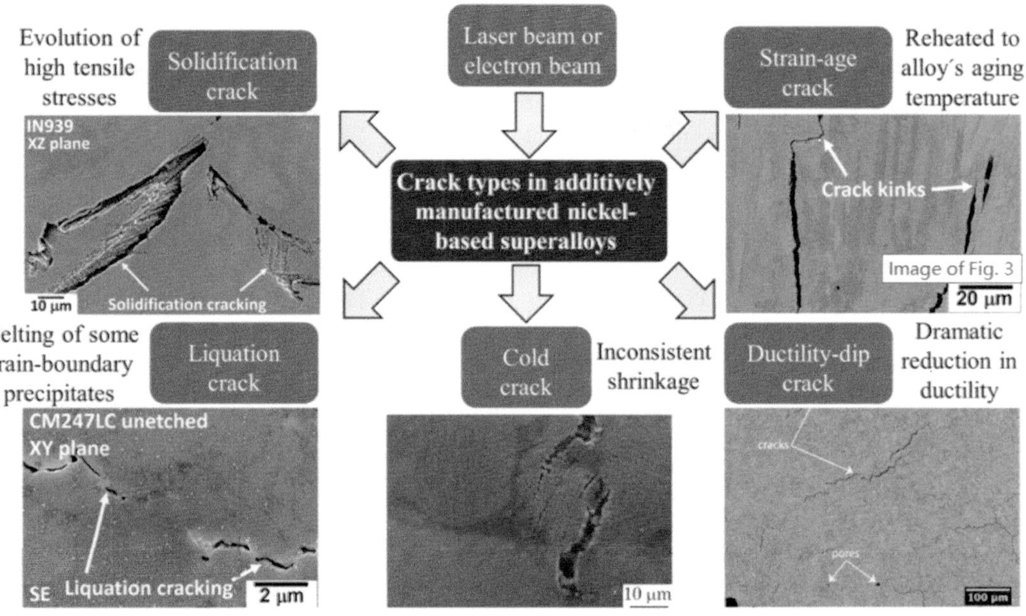

Figure 10.2. Main crack types in AM-ed nickel-based superalloys: Solidification cracks, [8] liquation cracks, strain-age cracks, ductility-dip cracks, and cold cracks [8].

10.2.1 Hot cracks related to liquid films

Hot cracks related to liquid films occur in the liquid-solid two-phase zone during the solidification or reheating process, including solidification cracks and liquation cracks, both of which are related to the intergranular liquid films. The susceptibility to liquid film cracking is higher at large-angle grain boundaries. Shrinkage porosity between dendrites is the origin of solidification cracking. In the early stage of solidification, the brittle and hard high-melting carbides formed in the inter-dendritic region are pinned in the liquid-filled channels and impede the liquid flow during the filling process, causing the liquid films to be prone to tearing and thereby initiating solidification cracks. In the final stage of solidification, when the mushy liquid-solid two-phase zone in the molten pool is subjected to tensile thermal stress and has a high solid fraction (0.7–0.9), the flow of liquid metal is hindered by large-sized dendrites and it is difficult to backfill the long liquid channels. The insufficient feeding of liquid between dendrites leads to solidification cracks. The solidification cracks in γ′ strengthened nickel-based superalloys are related to trace elements or impurities. For instance, although the addition of Zr can stabilize grain boundaries and enhance creep resistance, the low-melting liquid film formed by Zr segregation at the grain boundaries and covering dendrites in the liquid phase embrittles the grain boundaries. This implies that the grain boundaries cannot transfer the residual tensile stresses or the contraction caused by cooling the melt, facilitating the separation of adjacent grains and the formation of solidification cracks. Additionally, elements such as Zr, B, Si, and P all lower $T_{solidus}$ and are key factors that need to be focused on when regulating the susceptibility to solidification cracking [9, 10]. The commonly used evaluation index for the susceptibility to solidification cracking is ΔT. The larger the span of ΔT, the poorer the feeding capacity of the liquid phase at the end of solidification, and the more likely it is to form shrinkage pores in the dendrite-encapsulated regions that cannot be filled by the liquid phase, which then evolve into solidification cracks under the effect of thermal stress. Therefore, a small ΔT is beneficial for alloys to rapidly pass through microstructures with high crack susceptibility [9, 11, 12].

Liquation cracks mainly occur during the temperature peak stage in the thermal cycling process, encompassing two types: constitute liquation and direct melting of the low-melting point phase. Specially, constitute liquation typically emerges below the equilibrium $T_{solidus}$ of the alloy. During rapid heating, the solute concentration near the precipitate-matrix interface rises with the dissolution of the precipitate phase. If the solute concentration exceeds the alloy composition at a temperature not lower than the equilibrium reaction temperature of the precipitate-matrix eutectic, a metastable liquid film rich in solute will form at the interface. Adjacent liquefied regions gradually connect to form a large-scale film. For the direct melting of the low-melting point phase, the liquid films are deduced to direct melting of the secondary phases at the grain boundaries, such as the γ-γ' eutectic phase, γ-Laves eutectic phase, carbides, etc., those phases generally with lower eutectic point and melting point. Ultimately, both types of liquid films will fracture under the action of residual tensile stress to form liquation cracks, and the crack propagation path is rather tortuous, featuring obvious intergranular cracking characteristics [13–16]. In the γ′ strengthened nickel-based superalloys during welding, the γ-γ′ eutectic phase is prevalently present in the inter-dendritic region of solidification dendrites, and the solidus temperature of the eutectic phase is lower than the overall $T_{solidus}$ of the alloy, making it highly prone to liquation cracking. In LDED of the γ′ strengthened nickel-based superalloys, the liquation of γ-γ′ eutectics also observed in our previous research [17].

However, in the LPBF, the high cooling rate results in the very limited γ-γ′ eutectic phase, and the deposited microstructure is closer to an oversaturated solid solution. Therefore, some studies suggest that the γ-γ′ eutectic phase is not the main cause of liquation cracking in the γ′ phase-strengthened nickel-based superalloys. The liquation is most likely the segregation-induced liquation. During the thermal cycling process, the partially dissolved intermetallic carbides (such as MC and M_6C) and second phases such as the Laves phase and σ phase lower the melting point, increasing the cracking susceptibility in the heat-affected zone [18–19]. The liquation susceptibility of nickel-based superalloys can be reflected by the phase transformation during the heating-cooling process. The distinct $T_{solidus}$ value indicates that even within the cells/dendrites in the final solidification region, there is a relatively uniform composition. The significantly reduced nano-segregation and the absence of grain boundary carbides and the γ-γ′ eutectic phase suggest that the alloy has a lower ability to form liquid films during cyclic heating [13].

10.2.2 Solid-state cracks

Solid-state cracks mainly occur in the solid phase and are unrelated to the liquid film thus the fracture does not possess the exposed dendritic morphologies. They can be further classified into high-temperature ductility dip cracking (DDC) and strain aging cracking (SAC). Subsequent thermal cycles subject the previously deposited layers to a high-temperature stage, during which the plasticity of grain boundaries significantly decreases, and the grain boundaries slide under the action of residual stress and phase transformation shrinkage stress to form intergranular cracks. DDC is mainly located at the positions of the triple junction grain boundaries or the bottoms of liquation cracks, and the crack tips are wedge-shaped. DDC is considered to be generated due to the "ductility exhaustion" of the material, causing cracks to pass through the grain boundaries via a creep-like mechanism. During the rapid cooling process of AM, the ductility of the alloy varies in a fluctuating manner rather than monotonically decreasing with the decrease in temperature. The first trough appears between the peak heating temperature ($\approx T_{liquidus}$) and the ductility recovery temperature ($\approx T_{solidus}$), which is known as the "brittle temperature range". In this time, the alloy is in a "pasty" state, and a large amount of molten metal is prone to causing solidification cracking. The second trough is positioned between $T_{solidus}$ and $T_{liquidus}/2$, known as the "high-temperature ductility loss temperature range", within which the ductility of the alloy further declines and DDC emerges [20]. The loss of ductility in γ′ strengthened nickel-based superalloys typically occurs within the critical temperature range of 0.5 $T_{solidus}$ to 0.9 $T_{solidus}$. As the triangular connection area at the grain boundary is more prone to stress concentration, the columnar grain structure obtained through AM makes the alloy more susceptible to DDC.

SAC is a unique cracking phenomenon in γ' strengthened nickel-based superalloys. Subsequent thermal cycling makes the first deposited layer in the aging temperature range, sometimes, after deposition the γ' is precipitated continuously by the heat treatment, and then the strength of the alloy is increased but the ductility is decreased simultaneously. When the strain caused by the superposition of the phase transformation shrinkage stress and the unreleased residual stress exceeds the plasticity of the alloy, intergranular cracks are formed, which is known as SAC. The crack surfaces of SAC are relatively smooth, the paths are tortuous and accompanied by sharp kinks. Meanwhile, SAC penetrates multiple deposited layers, extends for hundreds of micrometers, and the expansion direction is almost consistent with the building direction and the grain growth direction [21]. Unlike γ' strengthened alloys, γ'' strengthened nickel-based superalloys have a relatively slow precipitation and dissolution process, ensuring a larger residual stress relaxation window, therefore, SAC rarely occurs in the AM of γ'' strengthened nickel-based superalloys. Essentially, the γ' forming elements dominate SAC, and alloys with a higher total content of Al + Ti are more sensitive to SAC than those with a lower total content of Al + Ti [22].

10.2.3 Humping-induced cracking

The origin of cracks discussed above primarily stems from the perspective of microstructure and metallurgical defects. Besides, the condition of stress and strain within the deposited layers constitutes another crucial factor contributing to cracks. However, due to the inherent technical challenges associated with real-time monitoring of stress and strain during the additive manufacturing process, there is a scarcity of literature addressing this aspect. Recently, Tristan G. Fleming et al. [23] employed the inline coherent imaging (ICI) to optically monitor surface topology and detect cracking in situ during the DED of CM247LC, while the synchrotron X-ray imaging is used to observe sub-surface crack healing and growth. As shown in Figure 10.3, in their study, 2–3 mm tall thin walls were pre-built on the substrate, the ICI signatures were captured from the layers deposited on the pre-built thin walls. Surface waves are induced into the pre-built layers by modulating the laser on and off during DED. These waves change the distribution of residual stress in the deposited layer, resulting in higher residual stress in valleys or pits. Similar to a Charpy V-notch test, cracking is then consistently localized in the valleys and occurs within the subsequent 5–10 layers deposited on the pre-existing thin wall. This enables to visualize cracking using synchrotron beamline imaging. The progression of layers deposited during beamtime is illustrated in Figure 10.4. Particularly, radiographic images were taken before, during (operando), and after depositing layers 1(a) and 9(b) onto a thin-walled structure prior to construction. As commonly seen in such structures before construction begins, there was an observable presence of a narrow crack measuring approximately 7 μm wide prior to deposition (a, pre). However, it was only after deposition that crack growth became noticeable while applying the first layer onto the structure (pre-build +1; a, post). Upon reaching layer nine (pre-build+9), it was observed that crack growth

	No. layers	z-increment (μm)
Beamtime	5	100
Pre-build	5	100
	4 (laser on/off)	50
	20	100

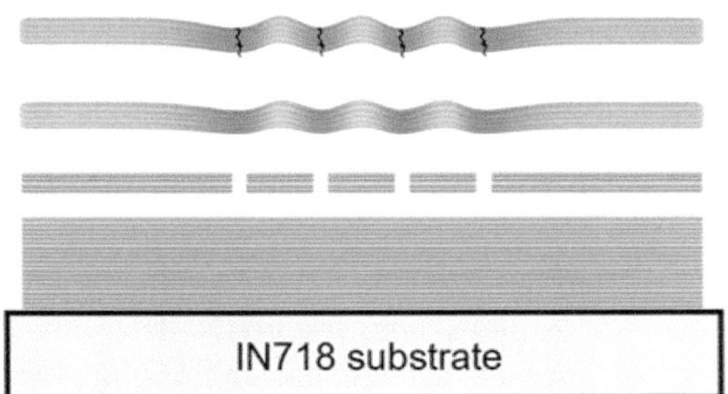

Figure 10.3. The preparation principle and process diagram of DED sample containing hump [23].

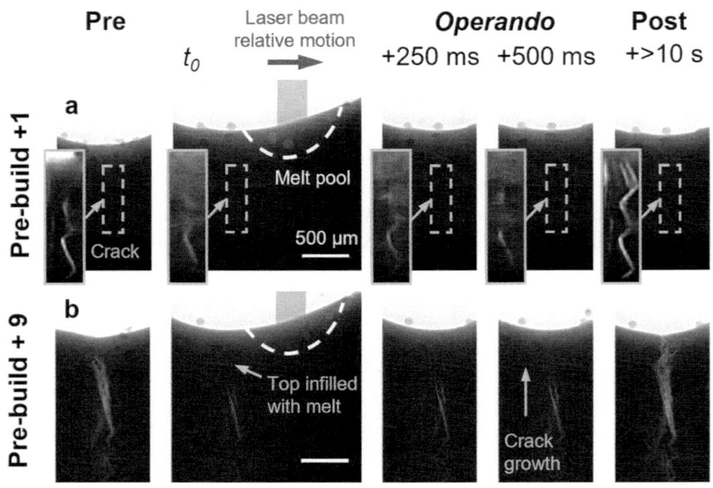

Figure 10.4. Sub-surface crack infill and growth captured with operando X-ray imaging [23].

occurred within just 500 ms following each laser pass (b; operando), indicating a reduction in time required for cracks to develop and spread towards the surface as residual stress increased. Despite utilizing identical process parameters for constructing the thin-wall pre-builds, variations in valley depths were observed, resulting in different levels of residual stress and cracking prior to beamtime.

Preferential formation of cracks was observed in depressed areas or low points of slender walls exhibiting seeded humping, which resulted from localized elevation of residual stress. X-ray imaging revealed that each layer experienced partial healing of the upper crack section as it passed through the molten pool, subsequently reopening and propagating towards the surface of the slender wall. Eventually, these cracks opened at the surface where they became detectable using ICI, indicating accumulation of adequate residual stress. The expansion and subsequent emergence of cracks on the surface were witnessed for a duration ranging from 1–10 seconds, surpassing the time required for solidification. This emphasizes how residual stress at a component level plays a significant role along with detecting defects between layers. In cases where there are greater temperature differences, even quicker crack propagation and surfacing (within less than 10 milliseconds after passing through the laser beam) have been documented during Powder Bed Fusion (PBF) processes involving CM247LC material [24]. By incorporating an appropriate trailing offset technique, it is possible to observe these instances of crack formation in real-time using ICI technology. This enables accurate determination of when cracks propagate relative to each pass made by the laser.

10.3 Cracking susceptibility indicators

10.3.1 Criteria for hot tear cracks

Researchers have developed a set of criteria to quantify the susceptibility of hot tear cracks during solidification, with a focus on understanding their formation mechanism. The evaluation model proposed in the casting literature [25] and later applied in the Ni system utilizes the freezing range (FR) as a key parameter for assessing crack proneness [26]. It is defined as follows:

$$FR = T_{Liquidus} - T_{Solidus} \tag{10.1}$$

where the solidus temperature is determined when 0.99 mole fraction of solid is reached. Compositions that result in a wide freezing range and low melting point solidification products such as carbides, borides, and $\gamma+\gamma'$ eutectics are considered. The FR has been widely used in the designs of novel Ni based superalloys for AM [27].

Clyne and Davies introduced the crack susceptibility coefficient (CSC), which quantifies the duration of the mushy zone in a critical solidification window based on the solid fraction in the semisolid state. The initial stage of solidification is considered as a period for relieving stress and facilitating mass and liquid feeding, while the final stage of solidification is identified as a vulnerable phase characterized by the presence of a continuous liquid film with limited permeability that hinders sufficient liquid feeding. Consequently, the CSC criterion can be defined as the ratio between the time an alloy remains susceptible to hot tearing and the available time for stress relaxation. The expression is as follows:

$$CSC = \frac{t_{(fs=0.99)} - t_{(fs=0.90)}}{t_{(fs=0.99)} - t_{(fs=0.40)}} \approx \frac{T_{(fs=0.99)} - T_{(fs=0.90)}}{T_{(fs=0.99)} - T_{(fs=0.40)}} \tag{10.2}$$

where f_s is the solid fraction in the semisolids, and t and T correspond to the time and temperature at a certain solid fraction, respectively. Based on the observed rapid cooling rate in AM processing, it is reasonable to infer a linear relationship between time and temperature. Consequently, the CSC can be approximately expressed as a function of temperature corresponding to their respective regions [28].

Kou proposed a simplified approach to solidification, wherein two dendritic grains grow concurrently within the mushy zone. The spatial separation between these grains induces tension that leads to cracking, while their mutual growth promotes bonding and inhibits cracking. Hence,

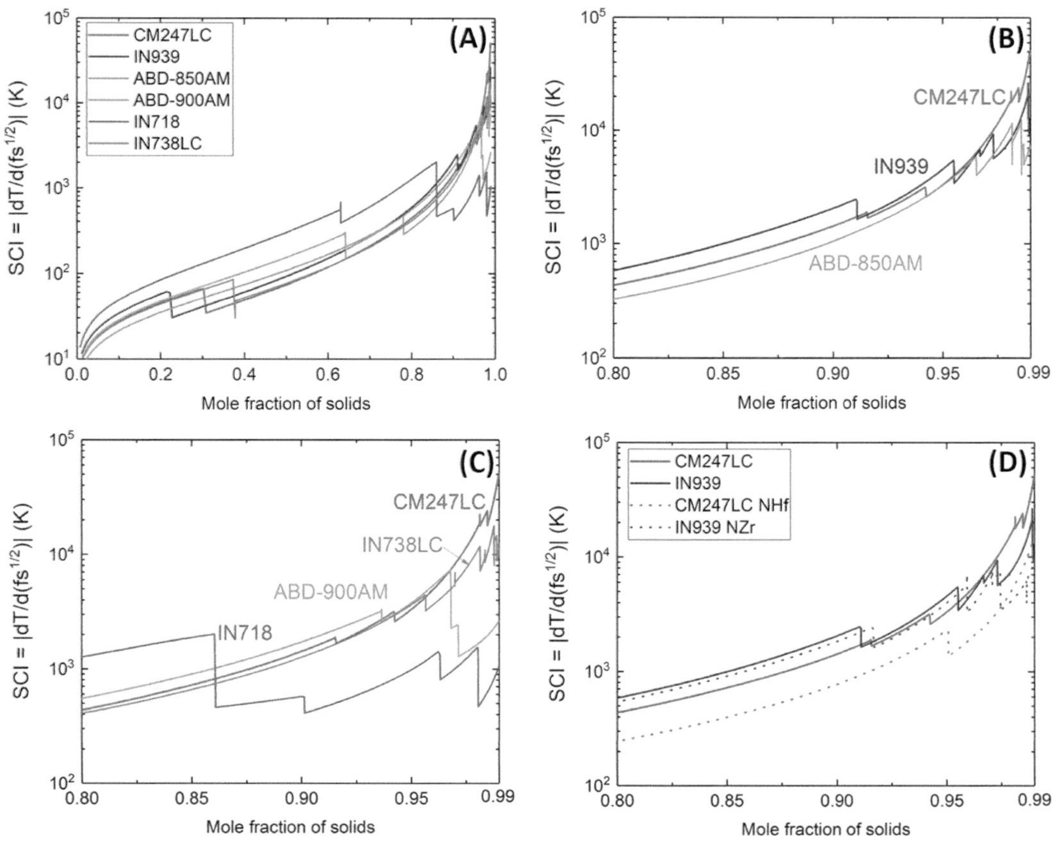

Figure 10.5. Solidification cracking index (SCI) of various alloys based upon solidification curves calculated using Thermo-Calc with TTNi8 database. (a) SCI value over the full range of solid fraction, (b, c & d) SCI value of the last stage solidification in the range of 0.8–0.99 of CM247LC, IN939, ABD-850AM, ABD-900AM, IN718, IN738LC, CM247LC Hf free and IN939 Zr free [13].

it is imperative to consider liquid feeding along the grain boundary as a means of suppressing crack formation. The Kou criterion pertains to the steepness of the solid fraction-temperature curve during terminal solidification. The solidification cracking index (*SCI*) is proposed to quantify the sensitivity of solidification cracks, the expression of *SCI* is as follows [29].

$$SCI = \left| \frac{dT}{d(f_s^{1/2})} \right| \tag{10.3}$$

Where f_s is the solid fraction, T is the temperature. It has been clarified by Kou that the use of one specific temperature interval, generally, the range f_s is defined between 0.87 and 0.94.

A high value of the *SCI* indicates a sluggish lateral growth rate, wherein adjacent columnar grains exhibit slow growth towards each other for bonding. Simultaneously, this slow lateral growth allows the columnar grains to attain considerable length without bridging. Consequently, the intergranular liquid channel can become significantly elongated, impeding fluid flow and resulting in the formation of a pronounced notch that facilitates crack initiation. Taking *SCI* into consideration as a crucial reference factor, Yuanbo T. Tang et al. [13] successfully developed a range of novel nickel-based superalloys strengthened by γ′ for additive manufacturing, such as ABD-850AM and ABD-900AM. Figure 10.5 shows the *SCI* of various alloys based upon solidification curves calculated using Thermo-Calc with TTNi8 database [13].

10.3.2 Criteria for strain-age cracks

Strain-age cracks (SAC) in Ni superalloys typically occur during thermal cycling or subsequent heat treatment due to the combined effect of residual stress and additional stress induced by the precipitation of γ′ from solid solution. To evaluate the susceptibility of SAC, an empirical measure has been proposed by Thompson and further modified by Reed [13]. This measure is based on the assumption that weight percents of certain elements (Al, Ti, Nb, and Ta) compromise local ductility.

$$M_{SAC} = [Al] + 0.5[Ti] + 0.3[Nb] + 0.15[Ta] \tag{10.4}$$

The empirical strain age cracking index provides only a first approximation to a very difficult problem. However, it has been demonstrated that M_{SAC} has a strong positive correlation with the precipitation driving force of γ′.

Besides, Hao Yu et al. [28] proposed a different index to characterize the sensitivity of SAC, the definition is as follows:

$$SAC = \frac{dV_f^{\gamma}}{dT}, T \in [T_{\gamma}^{*}, T_{solidus}] \tag{10.5}$$

Where V_f^{γ} means the volume fraction of γ matrix, T_{γ}^{*} can be defined as the critical temperature when the volume percent of γ is about 80%, or $0.7 * T_{solidus}$. A high SAC value indicates a high crack sensitivity brought by the volume change of precipitations.

10.4 Cracking suppression of γ' strengthened superalloys

10.4.1 Process optimization

Based on the mechanism of cracking, it is evident that cracking is intricately linked to heat, mass, and momentum transfer, as well as the stress state during the AM process. Key parameters of the AM, including laser power, scanning speed, and scanning strategy, directly influence these transfer processes. Consequently, optimizing process parameters is the most effective method to control cracking.

1) Reduce energy input in the AM process

   The study of Yang et al. [30] indicated that crack sensitivity is directly associated with heat input. When the body energy density was decreased from 827 J/mm^3 to 429 J/mm^3 in Rene 104 alloy specimens, the quantity and density of cracks on the specimens dropped. Nevertheless, it was impossible to fabricate crack-free and high-density Rene 104 components merely by optimizing process parameters. Wang et al. [31] summarized the process window for attaining crack-free and high-density (relative density approximately 99.76%) formation of IN738LC. Chen et al. [32] also reported that reducing the power density and line energy can significantly reduce the number of cracks in the Inconel 738LC cladding layers. However, there are also some conflicting conclusions. For example, Canadian Idowu [33] found that in the laser welding study of ATI Allvac 718Plus high-temperature alloy, more HAZ grain boundary liquidation cracks appeared at lower laser heat input conditions, while more HAZ grain boundary liquidation cracks appeared with higher heat input, but no cracks appeared. Osoba [34] studied the effect of laser welding on the cracking of Haynes282 high-temperature alloy and found that reducing the heat input would lead to an increase in cracking. It is evident that altering the laser process parameters not only modifies the stress state during the additive manufacturing process but also affects the microstructure of the material within the brittle-sensitive temperature range. Consequently, the impact of heat input on crack formation during the additive manufacturing of nickel-based superalloys is inconsistent, necessitating a detailed investigation to identify and summarize to the specific alloy compositions and process parameters.

2) Optimize the laser scanning strategy

   The scanning strategy determines the heat input and heat accumulation, which has a significant impact on the solidification process and stresses states, thus playing an important role in the control of grain structure and crack behavior. Common scanning strategies include unidirectional, bidirectional, orthogonal, and checkerboard strategies. In addition, rotating one angle (0°, 90°, 67°) between adjacent deposition layers can also have a positive effect on obtaining equiaxed grains and eliminating cracks. The local equiaxed grains generated by dynamic recrystallization can release residual stresses, and the refined grains and reduced residual stresses can effectively inhibit crack initiation and propagation [35, 36]. However, due to variations in laser spot size and the method by which drives the scanning of the laser beam, the freedom degree of adjustment of the laser scanning strategy in LDED is relatively limited compared to the LPBF.

3) Reheat the substrate

   Given that temperature gradient is a key factor affecting residual stress, reducing temperature gradient has become a reliable method for eliminating cracks. Preheating methods usually include the use of laser low-energy rapid scanning substrate preheating, resistance heating table preheating substrate and electromagnetic induction coil local preheating. However, the overall preheating of the substrate always sacrifices flexibility and is not suitable for practical applications of DED, such as manufacturing large parts where the heating effect decreases sharply with the increase in the number of layers, and laser repair where full substrate preheating changes the microscopic structure of body parts. Therefore, simultaneous heating was proposed. H. Liu et al. [37] proved that adding synchronized local preheating using a defocusing laser beam can significantly reduce the temperature gradient around the molten pool in the DED process, thereby helping to reduce thermal stress. However, as a surface heat source, the additional laser beam can only preheat the near-surface region and has little effect on the temperature gradient in areas away from the molten pool. Therefore, a more flexible and applicable method is the hybrid approach, namely synchronous induction-assisted laser deposition (SILD), which combines laser and induction heating technologies. Due to the complementary properties of induction heating and laser heat sources, the entire temperature field can be controlled *in situ*

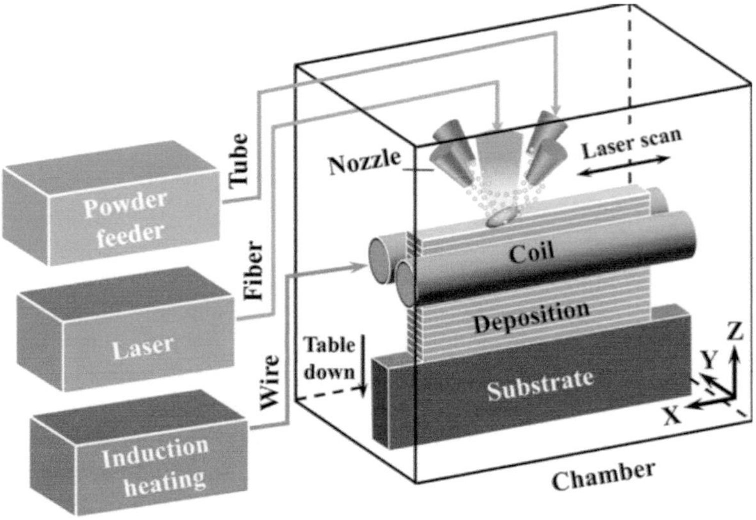

Figure 10.6. Schematic diagrams of synchronous induction-assisted laser deposition [40].

by adjusting the hybrid laser-induction parameters [38, 39]. Consequently, SILD significantly reduces residual stress and eliminates cracks.

By constructing a 3D finite element model, Fan W. et al. [40] analyzed the stress evolution and the impact of a synchronous additional heat source on thermal stress during SILD. The Schematic diagrams of synchronous induction-assisted laser deposition are shown in Figure 10.6. The results indicate that integrating synchronous induction heating reduces thermal stress by approximately 80% during deposition. Additionally, the distribution of residual stress is altered, with the location of maximum stress shifting from the bottom corner of the deposition to the central region of the substrate.

4) Hot isostatic pressing

The processing characteristics of the hot isostatic pressing (HIP) process encompass high temperatures (up to 2000°C) and high pressures (up to 200 MPa). Inert gas serves as the pressure transmission medium, ensuring uniform application of pressure from all directions onto the surface of the components. This unique treatment environment facilitates diffusion and creep phenomena, thereby optimizing the microstructure of the material and enhancing its overall properties.

M. Vilanova et al. [41] found HIP treatment was significantly reducing cracks below 6 μm but had limited effect on wider cracks. By the selection of proper laser power and rotation strategy, which in conjunction with hot isostatic pressing (HIP) treatment, the defects that arise during the manufacturing of IN738LC samples via LPBF can be substantially minimized. Moreover, during the HIP of IN738LC, a substantial amount of γ' phase precipitated, thereby significantly enhancing the material's hardness. Xiaoming Zhao et al. [42] found HIP treatment can effectively heal cracks in LSF Rene88DT, especially short cracks. By HIP diffusion bonding, short cracks can be completely healed without obvious traces. In addition, HIP treatment temperatures exceeding the saturated temperature of Rene88DT (1130°C) may induce recrystallization, which can potentially enhance the mechanical properties. Moreover, the precipitate phase formed after HIP exhibits a coarser, yet more uniformly distributed microstructure.

From the current study, it is evident that HIP is effective in healing short and narrow internal micro-cracks. However, once the crack size increases or extends to the sample surfaces, HIP is unable to effectively repair these defects. Meanwhile, HIP can change the microstructure of the materials, thus may affect the properties of the samples.

10.4.2 Chemical composition optimization

It is relatively straightforward and feasible to suppress cracks in the additive manufacturing process through process optimization for the nickel-based superalloys that exhibit a certain degree of crack susceptibility. However, the effect is limited for Nickel-based superalloys with very high crack sensitivity. Therefore, the most fundamental way to suppress cracks is to optimize the chemical composition or even novel compositions that minimize crack sensitivity while maintaining the required mechanical properties. To shorten the development cycle and reduce costs, numerous developers have investigated modifying existing commercial nickel-based alloy compositions by adding or removing specific chemical elements to achieve crack-free additive manufacturing. Jinghao Xu et al. [43] developed a novel nickel-based superalloy, MAD542, by incorporating optimal levels of interdendritic segregating elements, specifically 5 wt.% molybdenum (Mo) and 2 wt.% niobium (Nb). These elements segregated to the interdendritic regions, effectively mitigating both solidification and liquation cracking during processing. Additionally, Mo and Nb functioned as efficient stacking fault energy (SFE) reducers, promoting the formation of annealing twins. This, in turn, reduces the susceptibility to SAC and DDC during post-processing treatments.

Yong HU et al. [44] conducted selective laser melting experiments using two commercial IN738LC alloy powders with varying Zr contents (0.024 wt.% and 0.12 wt.%). As seen in Figure 10.7, the crack density in the XOY plane increased from 0.15% to 0.87%, and in the

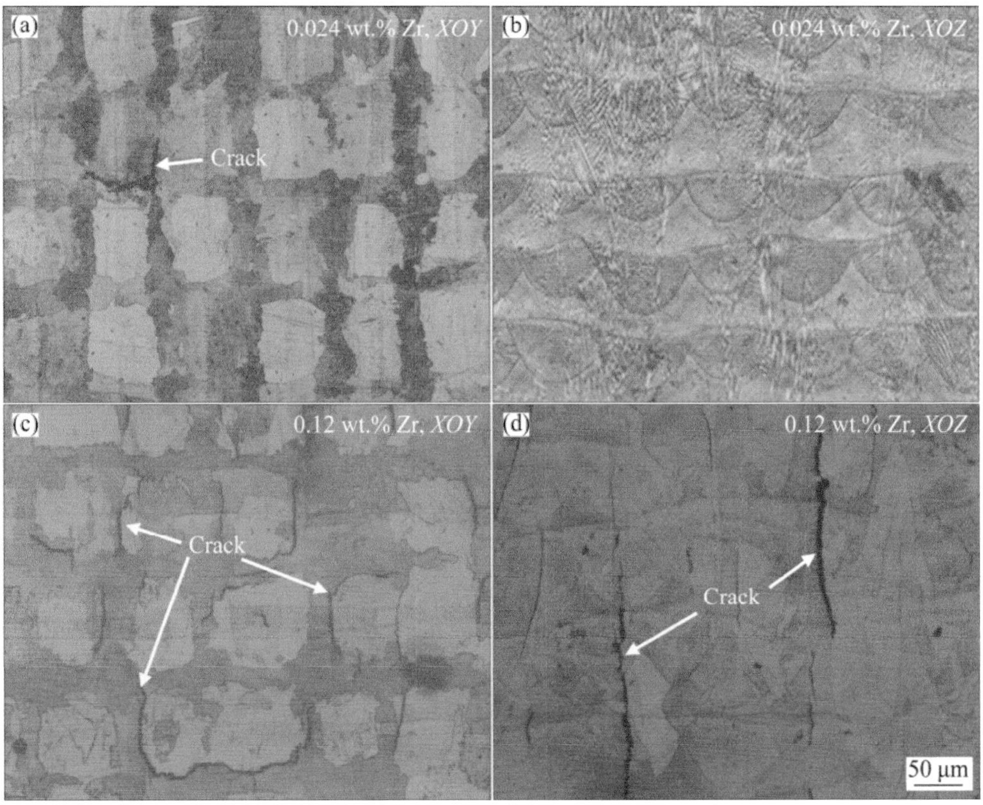

Figure 10.7. Optical micrographs of Sample 1 (a, b) and Sample 2 (c, d) showing crack formation location [44].

XOZ plane from 0.21% to 1.81%, as the Zr content in the powder increased from 0.024 wt.% to 0.12 wt.%. They concluded that while excessive Zr content significantly increases crack sensitivity during casting and welding, an appropriate level of Zr can mitigate grain boundary defects, enhance grain boundary cohesion, and reduce grain boundary diffusion rates.

Arthur Després et al. [45] investigated the hot cracking susceptibility and creep resistance of three powder batches of the γ'-hardened AD730® alloy with different contents of boron, carbon and zirconium fabricated by laser powder bed fusion. They found alloys containing boron, carbon, and zirconium (B+C+Zr) exhibit significant thermal cracking in their untreated state, whereas the zirconium-free alloy (B+C) and the boron- and zirconium-free alloy (C) remain crack-free (Figure 10.8). After heat treatment, the zirconium-free alloy (B+C) demonstrates superior creep resistance with a rupture time exceeding 2000 hours; the B+C+Zr alloy shows moderate creep life; and the boron- and zirconium-free alloy (C) exhibits negligible creep resistance despite the absence of cracks. SEM and TEM observations reveal that boron, carbon, and zirconium are enriched at high-angle grain boundaries in the untreated state. Following heat treatment, the grain boundary enrichment of boron and carbon significantly increases, while zirconium is predominantly distributed within the γ' precipitates. The distribution of boron, carbon, and zirconium at grain boundaries plays a crucial role in determining thermal crack sensitivity and creep resistance. Specifically, grain boundary enrichment of boron and carbon enhances creep resistance, whereas the presence of zirconium may increase thermal crack sensitivity.

KenHee Ryou et al. [46] studied the hot cracking of gas-atomized IN738 superalloy during direct laser deposition by adding Hf element, which is helpful to improve the application performance. The nominal Hf concentrations are 0, 0.5, 1.5, and 2.5 wt%, the results show with the increase of Hf content, the columnar grain width and length decrease, and the grain orientation divergence decreases, forming micron η-Ni_3Ti phase islands. Through theoretical calculation and phase field simulation, they found the calculated thermal strain and crack sensitivity decrease with the addition of Hf. Specifically, DSC analysis and associated model calculations revealed that the critical temperature range for thermal strain accumulation (ΔT_{CTR}) was reduced from 135°C (0 wt% Hf) to 71°C (2.5 wt% Hf). Additionally, the thermal strain and crack sensitivity were reduced by approximately 40% due to the incorporation of Hf. These findings are consistent with observed microstructural changes, where Hf promotes the formation of complex dendritic structures and discontinuous interdendritic regions, thereby mitigating thermal strain and crack sensitivity, the schematic illustrations of the mechanisms are shown in Figure 10.9.

However, Seong-Jin Lee et al. [47] reached contrasting conclusions regarding the impact of Hf on hot cracking during their investigation of the weldability of 247LC superalloy. As seen in Figure 10.10, they found the absence of Hf significantly reduced the solidification brittle temperature

Figure 10.8. SEM-SE contrast micrographs of the as-built microstructure of (a) the (B+C+Zr) alloy, (b) the (B+C) alloy, (c) the (C) alloy. An enlarged view of the crack surface is shown in (a). The enlarged view was acquired with an in-lens detector [45].

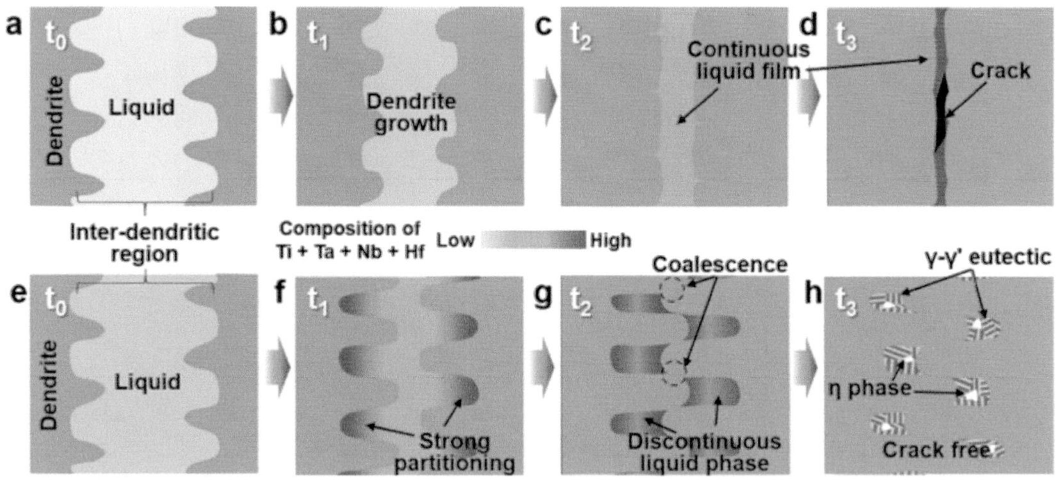

Figure 10.9. Schematic illustrations of the mechanisms of hot cracking and prevention in 0Hf and 2.5Hf: (a)–(d) morphology and solidification process of 0Hf. (e)–(h) morphology and solidification process of 2.5Hf [46].

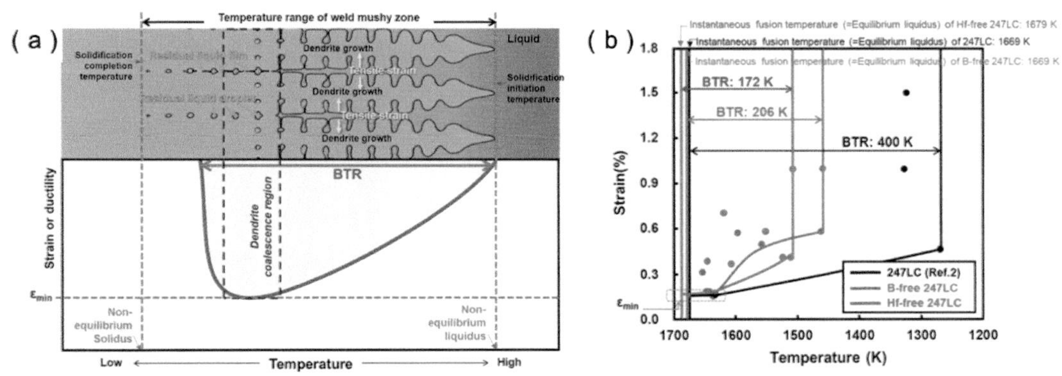

Figure 10.10. Schematic of solidification brittle temperature range (a) and the calculated value in the B- and Hf-free 247LC alloys [47].

range (BTR) value from 400 K (Hf-free) to 172 K (1.4%wt Hf), primarily due to the suppression of carbide formation. Specifically, the carbide area fraction in the Hf-free 247LC alloy is only 0.15%, markedly lower than the 4.72% observed in the original 247LC alloy (1.4%wt Hf). Additionally, the Hf-free 247LC alloy exhibits a narrower weld paste zone range (170 K), which is another critical factor contributing to the substantial reduction in BTR.

Wenzhe Zhou et al. [48] examines the impact of carbon content on the microstructure, tensile properties, and crack susceptibility of IN738 superalloy processed by laser powder bed fusion (LPBF). A moderate increase in carbon content from 0.11 wt% (alloy IN738LC) to 0.3 wt% (alloy IN738-0.3C) and 0.6 wt% (alloy IN738-0.6C) enables the fabrication of crack-free LPBF samples within a broad range of processing parameters. As seen in Figure 10.11, with the carbon content rises, more fine single carbides (not exceeding 200 nm) form quasi-continuous or continuous networks at the boundaries of the cellular structure, while the low-melting-point phases (containing boron and γ/γ′ eutectics) are significantly reduced. Due to the decreased level of elemental segregation, elimination of the liquid film initiation source, and reduction in local strain concentration, the susceptibilities to both solid-state and liquid-state cracks are substantially reduced. LPBF IN738-0.3C and IN738-0.6C demonstrate outstanding tensile strength (1320 MPa and 1598 MPa, respectively) and superior total elongation (14.7% and 9.0%, respectively).

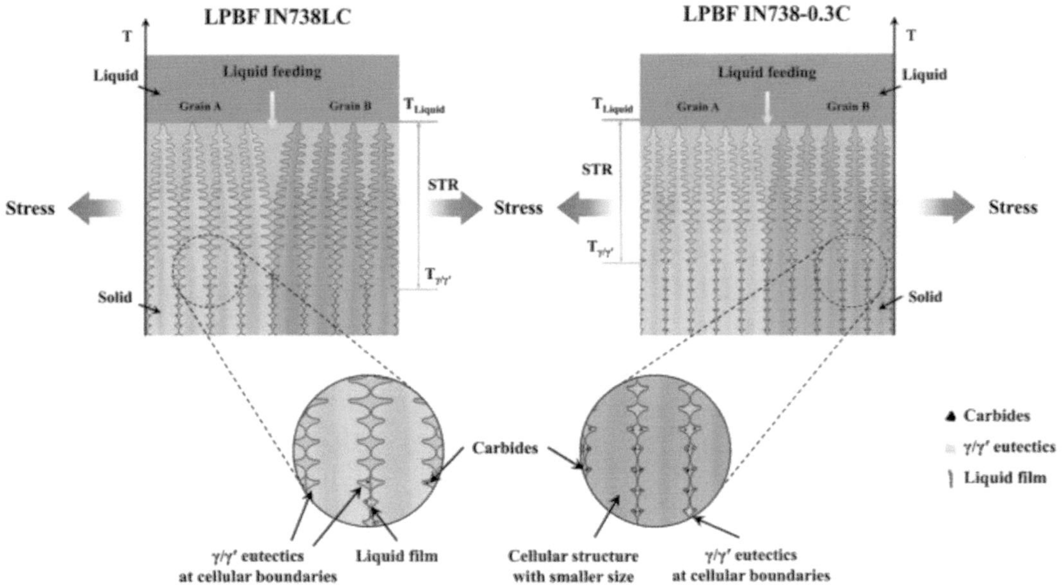

Figure 10.11. Schematic diagram of the inhibition effect of carbon on the cracking of LPBF IN738 [48].

Tang et al. [13] introduced the concept of "Design-Driven Alloy Development" (ABD). Utilizing a high-throughput integrated computational tool, they identified a novel superalloy design framework that balances cracking sensitivity and creep properties. This was achieved by comprehensively considering factors such as ΔT, SAC index, yield strength, and creep resistance, based on the non-equilibrium solidification Scheil model. Consequently, γ' strengthened ABD-850AM and ABD-900AM alloys designed for laser additive manufacturing (LAM) have been developed. Compared to IN939 and CM247LC alloys fabricated via laser powder bed fusion (LPBF), these new alloys exhibit significantly reduced cracking, with no observed solidification, liquation or solid-state cracks. This is attributed to their narrow ΔT, absence of low-melting eutectic phases, and sufficient high-temperature ductility. Ghoussoub et al. [49] further advanced this approach by adjusting the atomic ratio of Nb+Ta to Al in CM247LC alloy and employing a heat treatment strategy with a solution temperature lower than that of γ'. This resulted in a new type of nickel-based superalloy with excellent oxidation resistance and comprehensive mechanical properties, including extended creep life.

10.4.3 Adding the second phase particles

Introducing particles with high modulus and excellent thermal stability into nickel superalloys is considered an effective method to improve performance and reduce cracks. The basic principle is to promote the formation of equiaxed grains to change the magnitude and distribution of residual stress at the grain boundary positions, and to strengthen the grain boundaries or the matrix, thereby achieving effective inhibition of cracks. Aiming at the problem that Inconel 738LC is prone to cracks during the laser powder bed fusion process, Xiaoqi Chen et al. [50] proposed a method of improving its printability and crack resistance by adding micron-sized TiC particles (1 wt%, 6~24 μm). The research results show that the addition of TiC particles significantly reduces cracks and defects (as seen in Figure 10.12), refines the microstructure, promotes the precipitation of nano-sized carbides, improves the grain shape and grain boundary strengthening, thereby effectively suppressing the generation of cracks.

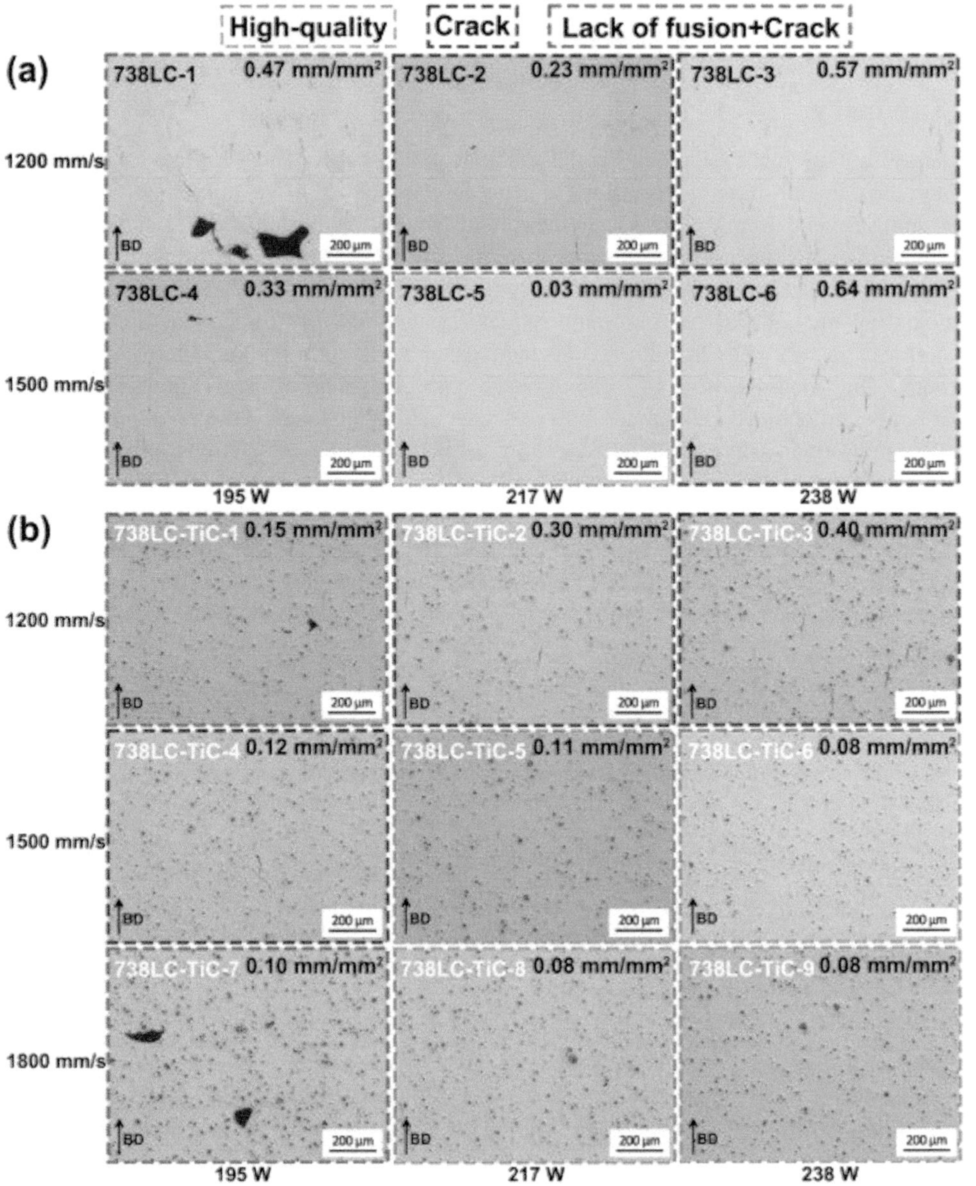

Figure 10.12. Optical microscope images of the unetched surfaces of the PBF-LBed Inconel 738LC samples: (a) without TiC and (b) containing 1 wt% TiC illustrating the defects under various laser powers and scanning speeds along with the corresponding average crack density [50].

Similarly, Chang Shu et al. [51] mixed 2 wt% nano-TiC with nickel superalloy IN738LC. Through the synergistic effect of fine grain strengthening, load-bearing strengthening, and Orowan strengthening mechanisms, the nano-sized TiC particles, effectively enhanced the comprehensive performance of IN738LC and achieved crack-free printing. Chen Z. et al. [52] fabricated the GNPs-reinforced K418 nickel-based superalloy composites by the L-PBF. It was found that the addition of GNPs (0.1 wt%) significantly refined the grains, reduced the size of columnar grains, and increased the total grain boundary area, thereby enhancing the strength and toughness of the material. The main strengthening mechanisms of GNPs include grain refinement, load transfer, forest dislocation strengthening, and Orowan strengthening. In addition, the uniform dispersion of

GNPs and the good interface bonding with the matrix also contribute to improving the toughness of the material, thereby suppressing the generation of cracks.

10.5 Summary

This chapter comprehensively reviews the cracking mechanisms and suppression strategies for γ' strengthened nickel-based superalloys during laser-based additive manufacturing. It begins by examining several primary cracking mechanisms, including hot cracks associated with liquid films (solidification cracks, liquation cracks), solid-state cracks (strain-age cracking, ductility-dip cracking), and cracks induced by humping. Subsequently, the chapter discusses the significance and calculation methods of some common cracking susceptibility indicators. Finally, it details crack suppression strategies from three perspectives: process optimization, chemical composition adjustment, and incorporation of second-phase particles. The review highlights significant advancements and achievements made by researchers in understanding the cracking mechanisms and developing suppression methods for additive manufacturing of γ'-strengthened superalloys. Looking forward, with the advancement of artificial intelligence, computational materials science, and high-throughput research methodologies, there is a promising opportunity to develop universal material design approaches for new superalloys tailored specifically for additive manufacturing, ensuring a balance between processability and desired performances.

References

[1] Tan, C., Weng, F., Sui, S., Chew, Y. and Bi, G. (2021). Progress and perspectives in laser additive manufacturing of key aeroengine materials[J]. International Journal of Machine Tools and Manufacture 170: 103804.

[2] Mostafaei, A., Ghiaasiaan, R., Ho, I. T., Strayer, S., Chang, K. C., Shamsaei, N. et al. (2023). Additive manufacturing of nickel-based superalloys: a state-of-the-art review on processstructure-defect-property relationship[J]. Progress in Materials Science 136: 101108.

[3] Ghoussoub, J. N., Klupś, P., Dick-Cleland, W. J., Rankin, K. E., Utada, S., Bagot, P. A. et al. (2022). A new class of alumina-forming superalloy for 3D printing[J]. Additive Manufacturing 52: 102608.

[4] Behera, A., Sahoo, A. K. and Mahapatra, S. S. (2023). Application of Ni-based superalloy in aero turbine blade: A review[J]. Proceedings of the Institution of Mechanical Engineers, Part E: Journal of Process Mechanical Engineering, 2023: 09544089231219104.

[5] Murray, S. P., Pusch, K. M., Polonsky, A. T., Torbet, C. J., Seward, G. G. E., Zhou, N. et al. (2020). A defect-resistant Co-Ni superalloy for 3D printing[J]. Nature Communications 11: 4975.

[6] Wang, H., Zhang, X., Wang, G. B., Shen, J., Zhang, G. Q., Li, Y. P. et al. (2019). Selective laser melting of the hard-to-weld IN738LC superalloy: efforts to mitigate defects and the resultant microstructural and mechanical properties[J]. Journal of Alloys and Compounds 807: 151662.

[7] Wang, Y., Roy, S., Choi, H. and Rimon, T. (2022). Cracking suppression in additive manufacturing of hard-to-weld nickel-based superalloy through layer-wise ultrasonic impact peening[J]. Journal of Manufacturing Processes 80: 320–327.

[8] Zhou, W., Tian, Y., Tan, Q., Qiao, S., Luo, H., Zhu, G. et al. (2022). Effect of carbon content on the microstructure, tensile propertiesandcrackingsusceptibilityofIN738superalloyprocessedbylaserpowderbedfusion[J].AdditiveManufacturing 58: 103016.

[9] Chauvet, E., Kontis, P., Jägle, E. A., Gault, B., Raabe, D., Tassin, C. et al. (2018). Hot cracking mechanism affecting a non-weldable Ni-based superalloy produced by selective electron beam melting[J]. Acta Materialia 142: 82–94.

[10] Cloots, M., Uggowitzer, P. J. and Wegener, K. (2016). Investigations on the microstructure and crack formation of IN738LC samples processed by selective laser melting using Gaussian and doughnut profiles[J]. Materials & Design 89: 770–784.

[11] Enrique, P. D., Minasyan, T. and Toyserkani, E. (2025). Laser powder bed fusion of difficult-to-print γ' Ni-based superalloys: a review of processing approaches, properties, and remaining challenges[J]. Additive Manufacturing 106: 104811.

[12] Shi, S., Li, Z., Yang, C., Zeng, Z., Cheng, X., Tang, H. et al. (2024). Research progress on crack control and composition design of γ'' phase strengthened nickel-based superalloys suitable for laser additive manufacturing (Invited) [J]. Chinese Journal of Lasers 51(10): 1002302.

[13] Tang, Y. T., Panwisawas, C., Ghoussoub, J. N., Gong, Y., Clark, J. W. G., Nemeth, A. A. N. et al. (2021). Alloys-bydesign:application to new superalloys for additive manufacturing[J]. Acta Materialia 202: 417–436.

[14] Xu, J., Lin, X., Zhao, Y., Guo, P., Wen, X., Li, Q. et al. (2018). HAZ liquation cracking mechanism of IN-738LC superalloy prepared by laser solid forming[J]. Metallurgical and Materials Transactions A 49(10): 5118–5136.

[15] Wang, Y., Guo, W., Zheng, H., Xie, Y., Zhang, X., Li, H. et al. (2023). Microstructure, crack formation and improvement on nickel-based superalloy fabricated by powder bed fusion[J]. Journal of Alloys and Compounds 962: 171151.
[16] Jeong, S. G., Ahn, S. Y., Kim, E. S., Kang, S. H., Yoo, S. H., Ryu, J. Y. et al. (2023). Liquation cracking in laser powder bed fusion-fabricated Inconel718 of as-built, stressrelieved, and hot isostatic pressed conditions[J]. Materials Science and Engineering: A 888: 145797.
[17] Zhang, X., Chen, H., Xu, L., Xu, J., Ren, X. and Chen, X. (2019). Cracking mechanism and susceptibility of laser melting deposited Inconel 738 superalloy[J]. Materials & Design 183: 108105.
[18] Wei, B., Liu, Z., Nong, B., Cao, B., Lv, X., Ren, Y. et al. (2021). Microstructure, cracking behavior and mechanical properties of René104 superalloy fabricated by selective laser melting[J]. Journal of Alloys and Compounds 867: 158377.
[19] Kontis, P., Chauvet, E., Peng, Z., He, J., Silva, A, K, D., Raabe, D. et al. (2019). Atomic-scale grain boundary engineering to overcome hot-cracking in additively manufactured superalloys[J]. Acta Materialia 177: 209–221.
[20] Guo, C., Li, G., Li, S., Hu, X., Li, X., Xu, Z. et al. (2023). Additive manufacturing of Ni-based superalloys:residual stress, mechanisms of crack formation and strategies for crack inhibition[J]. Nano Materials Science 5(1): 53–77.
[21] Xu, J., Kontis, P., Peng, R. and Moverare, J. (2022). Modelling of additive manufacturability of nickel-based superalloys for laser powder bed fusion[J]. Acta Materialia 240: 118307.
[22] Yu, L. and Cao, R. (2021). Welding crack of Ni-based alloys: a review[J]. Acta Metallurgica Sinica 57(1): 16–28.
[23] Fleming, T. G., Rees, D. T., Marussi, S., Connolley. T., Atwood, R. C., Jones, M. A. et al. (2023). *In situ* correlative observation of humping-induced cracking in directed energy deposition of nickel-based superalloys[J]. Additive Manufacturing 71: 103579.
[24] Ghasemi-Tabasi, H., de Formanoir, C., Van Petegem, S., Jhabvala, J., Hocine, S., Boillat, E. et al. (2022). Direct observation of crack formation mechanisms with operando Laser Powder Bed Fusion X-ray imaging[J]. Additive Manufacturing 51: 102619.
[25] Singer, A. R. E. and Jennings, P. H. (1946). Hot-shortness of the aluminium-1043 silicon alloys of commercial purity[J]. J. Inst. Metals 72: 197–211.
[26] Lippold, J. C. (2014). Welding Metallurgy and Weldability[M]. John Wiley & Sons.
[27] Liu, Z., Han, Q., Zhang, Z., Wang, L., Ma, T., Gao, Z. et al. (2023). Design of a novel crack-free precipitation-strengthened nickel-based superalloy and composites for laser powder bed fusion[J]. Virtual and Physical Prototyping 18(1): e2224769.
[28] Yu, H., Liang, J., Bi, Z., Li, J. and Xu, W. (2022). Computational design of novel Ni superalloys with low crack susceptibility for additive manufacturing[J]. Metallurgical and Materials Transactions A 53(6): 1945–1954.
[29] Kou, S. (2015). A criterion for cracking during solidification[J]. Acta Materialia 88: 366–374.
[30] Yang, J., Li, F., Wang, Z. and Zeng, X. (2015). Cracking behavior and control of Rene 104 superalloy produced by direct laser fabrication[J]. Journal of Materials Processing Technology 225: 229–239.
[31] Wang, H., Zhang, X., Wang, G., Shen, J., Zhang, G., Li, Y. et al. (2019). Selective laser melting of the hard-to-weld IN738LC superalloy: efforts to mitigate defects and the resultant microstructural and mechanical properties[J]. Journal of Alloys and Compounds 807: 151662.
[32] Chen, Z., Zhang, Q., Lou, C. and Yao, J. (2013). Methods of crack control for Inconel 738 laser cladding layer[J]. Appl. Laser 33(1): 7–13.
[33] Idowu, O. A., Ojo, O. A. and Chaturvedi, M. C. (2007). Effect of heat input on heat affected zone cracking in laser welded ATI Allvac 718Plus superalloy[J]. Materials Science and Engineering: A 454: 389–397.
[34] Osoba, L. O. and Ojo, O. A. (2012). Influence of laser welding heat input on HAZ cracking in newly developed Haynes 282 superalloy[J]. Materials Science and Technology 28(4): 431–436.
[35] Xu, J., Ding, Y., Gao, Y., Hao, W., Hu, Y. and Zhang, D. (2021). Grain refinement and crack inhibition of hard-to-weld Inconel 738 alloy by altering the scanning strategy during selective laser melting[J]. Materials & Design 209: 109940.
[36] Liu, X., Hu, R., Zou, H., Yang, C., Luo, X. and Ma, R. (2023). Investigation of cracking mechanism and yield strength associated with scanning strategy for an additively manufactured nickel-based superalloy[J]. Journal of Alloys and Compounds 938: 168532.
[37] Liu, H., Du, X., Guo, H., Liu, J., Chen, P., Yang, H. et al. (2021). Finite element analysis of effects of dynamic preheating on thermal behavior of multi-track and multi-layer laser cladding[J]. Optik 228: 166194.
[38] Fan, W., Tan, H., Lin, X. and Huang, W. (2020). Thermal analysis of synchronous induction-assisted laser deposition of Ti-6Al-4V using different laser-induction interaction modes[J]. Additive Manufacturing 35: 101267.
[39] Huang, Y. and Zeng, X. (2010). Investigation on cracking behavior of Ni-based coating by laser-induction hybrid cladding[J]. Applied Surface Science 256(20): 5985–5992.
[40] Fan, W., Tan, H., Zhang, F., Feng, Z., Wang, Y., Hu, Y. et al. (2023). Effect of synchronous induction heating on residual stress for laser-based directed energy deposition of thin-walled structures[J]. Materials Today Communications 35: 105702.
[41] Vilanova, M., Garciandia, F., Sainz, S., Badiola, D. J., Guraya, T. and Sebastian, M. S. (2022). The limit of hot isostatic pressing for healing cracks present in an additively manufactured nickel superalloy[J]. Journal of Materials Processing Technology 300: 117398.

[42] Zhao, X., Lin, X., Chen, J., Xue, L. and Huang, W. (2009). The effect of hot isostatic pressing on crack healing, microstructure, mechanical properties of Rene88DT superalloy prepared by laser solid forming[J]. Materials Science and Engineering: A 504(1-2): 129–134.

[43] Xu, J., Gruber, H., Peng, R. L. and Moverare, J. (2020). A novel γ′-strengthened nickel-based superalloy for laser powder bed fusion[J]. Materials 13(21): 4930.

[44] Hu, Y., Yang, X., Kang, W., Ding, Y., Xu, J. and Zhang H. (2021). Effect of Zr content on crack formation and mechanical properties of IN738LC processed by selective laser melting[J]. Transactions of Nonferrous Metals Society of China 31(5): 1350–1362.

[45] Després, A., Antonov, S., Mayer, C., Tassin, C., Veron, M., Blandin, J. J. et al. (2021). On the role of boron, carbon and zirconium on hot cracking and creep resistance of an additively manufactured polycrystalline superalloy[J]. Materialia 19: 101193.

[46] Ryou, K. H., Im, H. J., Park, J. and Choi, P. (2023). Microstructural evolution and hot cracking prevention in direct-laser-deposited Ni-based superalloy through Hf addition[J]. Materials & Design 234: 112298.

[47] Lee, S. J., Seo, S. M. and Chun, E. J. (2024). Effects of Hf and carbides formation behavior on the weld solidification cracking susceptibility of 247LC superalloy[J]. Materials Chemistry and Physics 316: 129147.

[48] Zhou, W., Tian, Y., Tan, Q., Qiao, S., Luo, H., Zhu, G. et al. (2022). Effect of carbon content on the microstructure, tensile properties and cracking susceptibility of IN738 superalloy processed by laser powder bed fusion[J]. Additive Manufacturing 58: 103016.

[49] Ghoussoub, J. N., Klupś, P., Dick-Cleland, W. J. B., Rankin, K., Utada, S., Bagot, P. A. J. et al. (2022). A new class of alumina-forming superalloy for 3D printing[J]. Additive Manufacturing 52: 102608.

[50] Xu, L., Zhou, W., Zhai, W., Luu, D. N., Nai, S. M. L. and Chen, X. (2024). Cracking mitigation of additively manufactured Inconel 738LC through addition of micro-TiC particles[J]. Journal of Manufacturing Processes 131: 641–658.

[51] Shu, C., Chen, S., Zheng, Z., Lu, X., Li, W., Lisi, M. D. et al. (2023). Fabrication and strengthening mechanism of crack-free nano-TiC reinforced IN738LC with enhanced mechanical properties by laser powder bed fusion[J]. Journal of Materials Research and Technology 27: 3835–3848.

[52] Chen, Z., Wei, P., Zhang, S., Lu, B., Zhang, L., Yang, X. et al. (2020). Graphene reinforced nickel-based superalloy composites fabricated by additive manufacturing[J]. Materials Science and Engineering: A 769: 138484.

Chapter 11

Depositing Inconel 738LC

Microstructure Characteristics and Cracking

11.1 Introduction

Inconel 738LC alloy is a high-aluminum, high-titanium nickel-based superalloy developed in the United States, characterized by an aluminum content of approximately 3.2 to 3.7 wt% and a titanium content of about 3.0 to 3.5 wt%. This alloy exhibits excellent resistance to thermal corrosion, with a maximum service temperature around 900°C, making it primarily suitable for applications in industrial gas turbine blades, guide vanes, as well as components and corrosion-resistant parts in aerospace engine turbines. Due to its elevated Al and Ti contents, Inconel 738LC has poor weldability and is prone to severe cracking phenomena. Ojo et al. [1–2] conducted extensive research on welding cracks in Inconel 738LC and identified numerous $\gamma + \gamma'$ eutectic liquation cracks and γ' constitution liquation cracks within the heat-affected zone.

In recent years, additive manufacturing of Inconel 738LC alloy has garnered increasing attention from both domestic and international scholars, leading to several studies on microstructural characteristics and crack formation mechanisms. Chen et al. [3] suggested that during the Directed Energy Deposition (DED) process of Inconel 738LC, rapid cooling rates prevent γ' particles from precipitating out from the matrix; thus necessitating subsequent aging treatments to ensure adequate precipitation of γ', thereby meeting the high-temperature mechanical performance requirements for deposited components. Xu et al. [4] discovered that significant amounts of approximately 100 nm-sized γ' particles precipitate within the laser re-melted structure of Inconel 738LC while also noting liquation cracks induced by either $\gamma + \gamma'$ eutectic structures or borides in the heat-affected zone.

Cloots et al. [5] posited that cracks formed during Selective Laser Melting (SLM) processes involving Inconel 738LC are classified as solidification cracks with zirconium playing a significant role in their initiation. However, Ramakrishnan et al. [6] indicated that both liquation cracks and solidification cracks arise during DED processing of Inconel 738LC. They attributed crack formation to factors related to $\gamma + \gamma'$ eutectic structures along with carbides and borides.

In this chapter, we conduct DED additive manufacturing of a single-layer thin wall made from Inconel 738LC and systematically study the dendrite growth, precipitated phase precipitation, and crack formation mechanisms in the additive process. This research provides an experimental and theoretical foundation for performance control in additive manufacturing of Inconel 738LC.

11.2 Microstructure characteristics

11.2.1 Grain growth characteristics

The solidification of liquid metal is essentially a process involving nucleation and growth of the solid phase within the liquid phase, both homogeneous and heterogeneous nucleation occur in this

DOI: 10.1201/9781003406327-11

process. In laser-directed deposition manufacturing, a high positive temperature gradient exists at the leading edge of the solid-liquid interface in the laser melt pool, promoting epitaxial growth of grains. Figure 11.1 illustrates the morphology of epitaxial growth for Inconel 738LC on a substrate. Specifically, Figure 11.1(a) displays dendritic structures at a lower magnification, with BD (build-up direction) representing the deposition height direction and SD (scanning direction) representing the laser scanning direction. Upon further magnification, Figure 11.1(b) reveals detailed morphology of these bottom dendrites. Epitaxial growth is essentially a form of heterogeneous nucleation, which does not necessitate super-cooling or other driving forces. Once the temperature falls below the liquidus temperature, the molten pool liquid can solidify on the substrate grain. As depicted in Figure 11.1(b), the substrate grains exhibit fine equiaxed grains, resulting in relatively random grain orientation. During this stage, initial epitaxial grain growth will maintain the orientation of the substrate grain and select the preferred orientation closest to the direction of maximum temperature gradient for further growth, as illustrated in Figure 11.2. The primary phase of Inconel 738LC possesses an FCC structure; hence, it typically exhibits a preferred growth orientation along <100> directions [7, 8]. With continued grain growth, those grains with significantly different orientations from that of heat flow gradually diminish and eventually give rise to a columnar crystal exhibiting more consistent orientation.

As depicted in Figure 11.1(a), the dendrites present in the lowermost deposition layer exhibit a diminutive size, indicating rapid cooling of the liquid melt pool near the substrate. As illustrated in Figure 11.1(b), the average primary dendrite spacing within this layer measures approximately

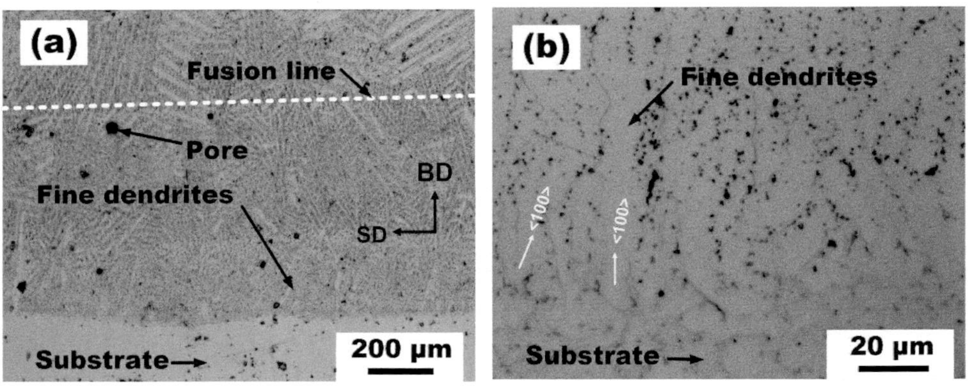

Figure 11.1. Epitaxial growth of the bottom cladding layer: (a) macro morphology, (b) micro morphology.

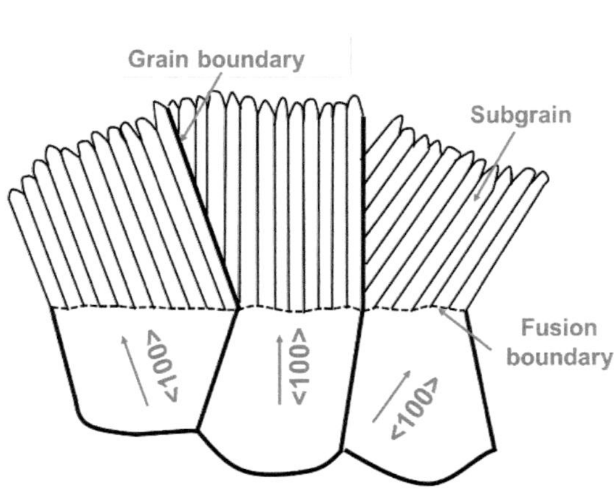

Figure 11.2. Schematic diagram of epitaxial growth of bottom dendrites.

8 μm. According to empirical formula [9] correlating cooling rate of nickel-based alloy and primary dendrite spacing:

$$\lambda = 60.6\,(G\,R)^{-0.31} \qquad (11.1)$$

it can be inferred that the cooling rate of the bottom deposition layer is estimated to be around 687 K/s.

The morphology of dendrites within the central deposition layers is presented in Figure 11.3, revealing distinct growth orientations under different technological conditions. Specifically, Figure 11.3(a)–(c) respectively display dendritic morphologies formed by reciprocating bidirectional scanning at fixed scanning speed (5 mm/s) and powder delivery speed (7.6 g/min). Notably, when laser power amounts to 320 W (Figure 11.3(a)), adjacent layers exhibit relatively uniform dendritic angles measuring approximately 90° each other; whereas with an increase in laser power to 410 W (Figure 11.3(b)), these angles fluctuate between roughly 100° and 130°. Furthermore, upon further increase of laser power up to 500 W (Figure 11.3(c)), the angles surpass 150°, with certain dendrites even maintaining their angle while crossing fusion lines. Thus, it becomes evident that as laser power increases, the likelihood for deflection of dendrites from one layer to another diminishes, and epitaxial growth phenomena become increasingly pronounced. In case unidirectional scanning mode is employed (as shown in Figure 11.3(d)) (laser power is 320 W), dendrites that form between layers also demonstrate favorable epitaxial growth, and their direction tilts about 66° along the scanning direction.

A similar phenomenon of dendrite growth has been reported in the DED of Inconel 718 alloy [10]. It can be seen that the above phenomenon should to the general rule of the DED of Nickel-based

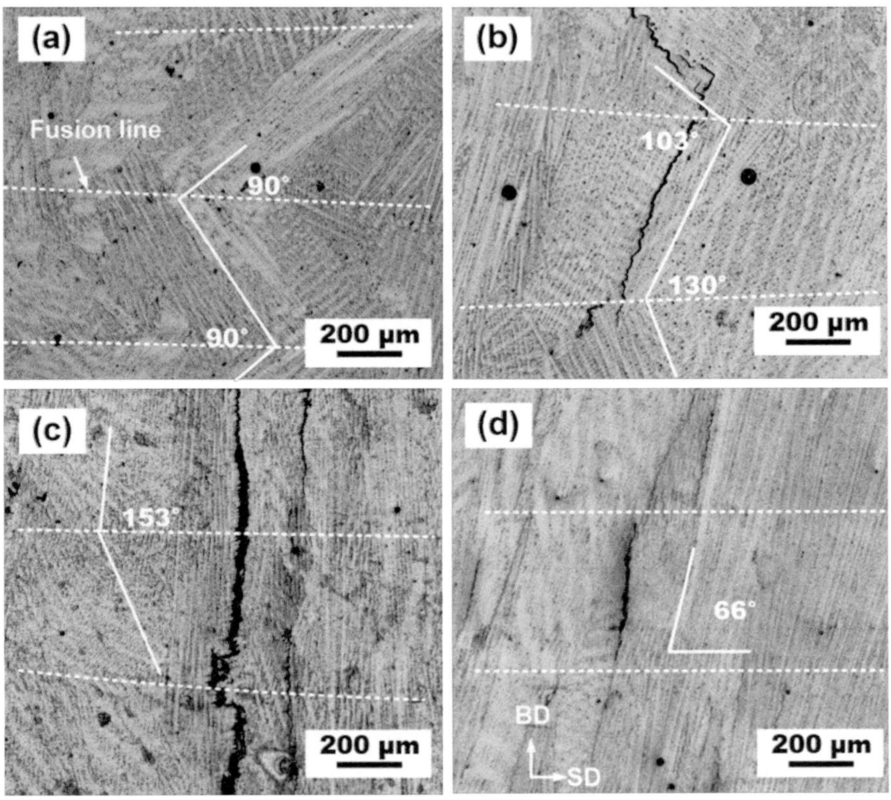

Figure 11.3. Dendrites morphologies of the central deposited layers under different processes: (a) bidirectional scanning under 320 W, (b) bidirectional scanning under 410 W, (c) bidirectional scanning under 410 W, (d) unidirectional scanning under 320 W.

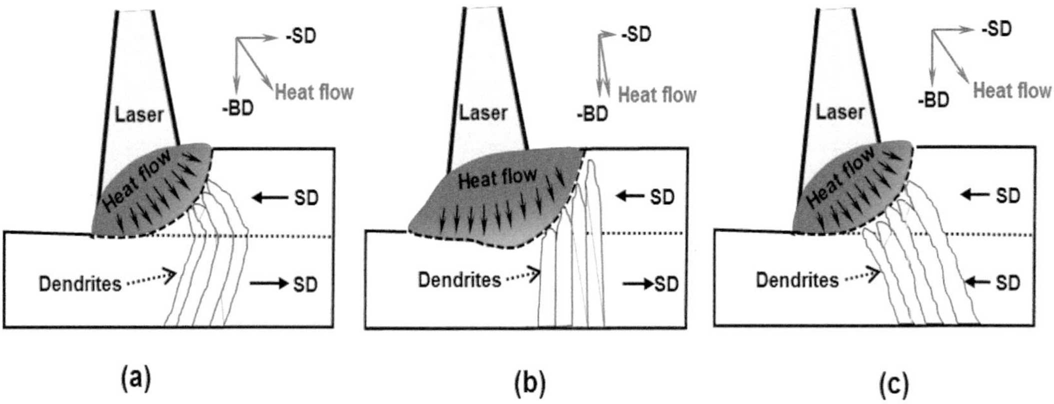

Figure 11.4. Schematic growth of dendrite during DED, (a) bidirectional scanning with low heat input, (b) bidirectional scanning with high heat input, (c) unidirectional scanning with low heat input.

alloy, and its mechanism is shown in Figure 11.4. When the heat input is small (Figure 11.4(a), (c)), When the heat input is low (Figure 11.4(a), (c)), the molten pool size and melting depth are small, resulting in lower temperature and poor fluidity. Under the combined effects of surface tension and gravity, the liquid molten pool tilts opposite to the laser scan direction. Heat is conducted backward towards the laser scanning area while also being conducted towards the substrate, causing a tilted total heat flow direction opposite to the laser scanning direction. The dendrite growth direction also exhibits a similar tilt due to heat flow influence. In this case, if reciprocating bidirectional scanning is employed as shown in Figure 11.4(a), current deposition layer dendrites can epitaxially grow on secondary dendrite arms from previous layers at an approximate angle of 90° between adjacent layers' dendrites. However, with unidirectional scanning, current deposition layer dendrites will continue their epitaxial growth on previous layer's primary dendritic rod with an approximately 60°~70° angle relative to the scanning direction (Figure 11.4(c)). When high heat input is applied as illustrated in Figure 11.4(b), larger molten pool size and deeper melting depth result in higher temperature and improved liquid metal spreading ability, leading to reduced tilt degree of molten pool inclination. At this stage, newly deposited metal at the tail end of molten pool remains at a high temperature state which inhibits heat conduction along directions opposite to scanning; thus aligning total heat flow closer to base normal direction. However, in the case of adopting unidirectional scanning, the dendrites of the current deposition layer will continue to epitaxially grow on the dendrite master rod of the previous layer. The direction of dendrite growth is approximately opposite to that of heat flow, with an angle ranging from 60° to 70° from the scanning direction (Figure 11.4(c)). When a large amount of heat input is applied, as shown in Figure 11.4(b), the molten pool size increases along with deeper melting depth and higher temperature, resulting in improved liquid metal spreading and reduced tilt degree of the molten pool. Consequently, newly deposited metal at the tail end of the molten pool remains at a high temperature state which inhibits heat conduction along the opposite scanning direction, causing total heat flow to align closer to that of base normal direction. As a result, dendrites tend to grow vertically. It can be observed that employing a process strategy involving unidirectional laser scanning with high thermal input promotes grain epitaxy growth and facilitates obtaining crystalline structures similar to directional solidification. Such microstructures are highly advantageous for nickel-based alloys used under high temperatures. However, as depicted in Figure 11.3, Inconel 738LC deposited under these process conditions exhibits severe cracking issues which hinder achieving high-quality Inconel 738LC additives.

The dendrite morphology of the uppermost deposition layer is illustrated in Figure 11.5. Figure 11.5(a) depicts the longitudinal section of the dendrite morphology, while Figure 11.5(b) presents the transverse section. It can be observed that the top deposition layer exhibits two distinct

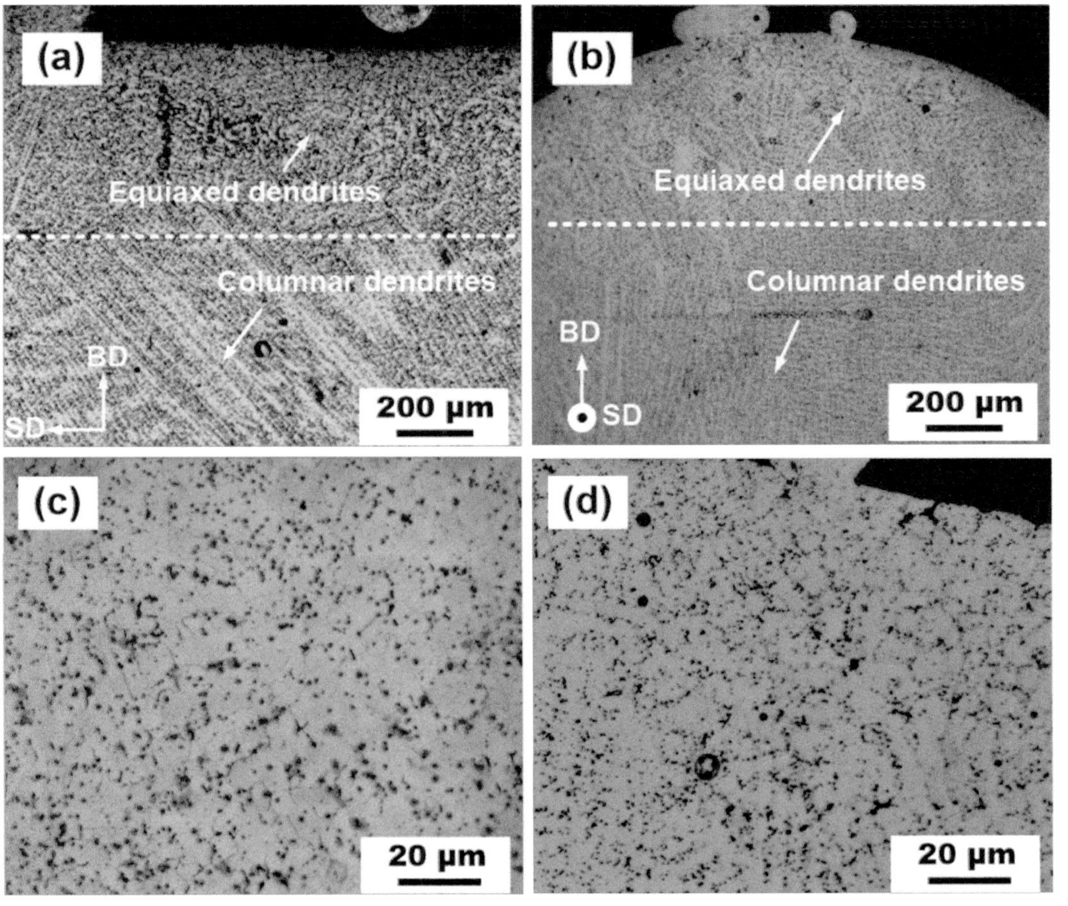

Figure 11.5. Dendrites morphologies of the topmost deposited layer, (a) dendrite morphology transformation in longitudinal section, (b) dendrite morphology transformation in cross section, (c) equiaxed grains in longitudinal section, (d) equiaxed grains in cross section.

dendritic morphologies: slender columnar grains dominate at the bottom, whereas a mixed crystal region is present at the top layer. The mixed zone predominantly consists of equiaxed dendrites. Figure 11.5(c) and (d) display high-magnification views of longitudinal and cross-sectional equiaxed dendrites, respectively.

Taking CMSX-4 alloy as an example, Gaumann et al. [11] investigated the mechanism behind columnar to equiaxial crystal transition (CET) during laser remelting of nickel-based alloys. Their findings indicate that the dendritic morphology resulting from solidification in a laser melting pool is influenced by temperature gradient and solidification rate, as depicted in Figure 11.6. The rectangular inserted window represents the parameter range for laser forming processes. Notably, higher temperature gradients and slower solidification rates favor columnar crystal growth. In nickel-based alloy laser melt pools, most regions fall within this range conducive to columnar crystal growth; however, towards the top of these melt pools, there is a decrease in temperature gradient accompanied by an increase in solidification rate leading to CET phenomenon.

Since the above research was carried out based on laser remelting, no powder was delivered to the molten pool, thus neglecting the influence of powder delivery on the resulting CET mechanism. However, in DED of Inconel 738LC, we find that the powder feeding at the top of the molten pool also has an important influence on CET. During the DED process, continuous transportation of metal powder leads to incomplete melting at lower temperature positions such as the tail or edge of

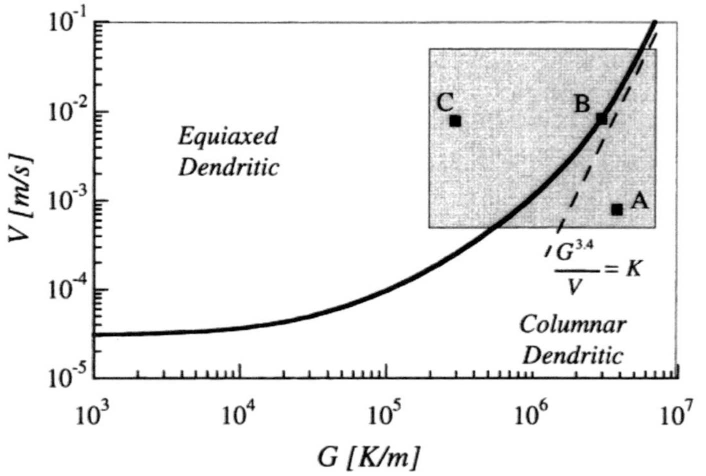

Figure 11.6. Dendrite transformation during the solidification of CMSX-4 alloy [11].

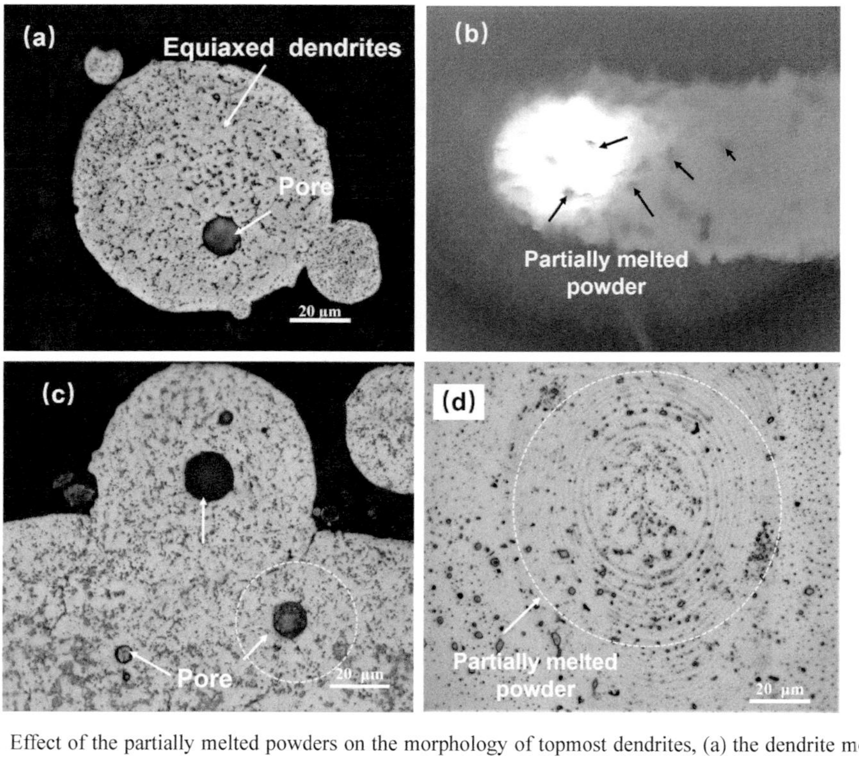

Figure 11.7. Effect of the partially melted powders on the morphology of topmost dendrites, (a) the dendrite morphology of powder, (b) the CCD image of molten pool during DED, (c) partially melted powders which are attached to the topmost layer, (d) a partially melted powder which is wrapped in deposited layers.

the molten pool (as shown in Figure 11.7(b)). Consequently, partially melted powders are present on and near the surface of the molten pool. Some partially melted particles adhere to the deposit layer's surface (Figure 11.7(c)), while others are encapsulated by liquid metal from the molten pool and solidify on top of it (Figure 11.7(d)). This partial melting retains equiaxed grains within their core and serves as nucleation sites for liquid molten pools, resulting in an abundance of fine equiaxed grains atop the deposited layer. Furthermore, it should be noted that if these partially melted powders possess hollow structures like those illustrated in Figure 11.7(a), they can introduce additional porosity into deposition layers (Figure 11.7(c)). Increasing heat input proves effective for reducing porosity since it promotes complete melting of powders and extends retention time within liquid molten pools so as to extend the time for gas to overflow the liquid molten pool.

11.2.2 Precipitation and distribution of carbides

The XRD pattern of INCONEL 738LCin the as-deposited state is presented in Figure 11.8, revealing a predominant composition of γ, γ′, and MC carbides. Other potential precipitated phases are challenging to directly detect via XRD due to their limited abundance. Furthermore, both γ (Fm-3m) and γ′ (Pm-3m) belong to the same cubic crystal system with highly similar lattice constants (approximately 3.524 A), resulting in overlapping diffraction peaks that hinder their differentiation.

The MC carbides can be easily identified through transmission electron microscopy (TEM). Figure 11.9(a) shows the bright field image of MC under TEM, while Figure 11.9(b) displays the selected electron diffraction pattern of MC carbide. Additionally, Figure 11.9(c) presents the selected electron diffraction pattern of the matrix (γ and γ′) surrounding MC, and Figure 11.9(d) exhibits the selected electron diffraction at the interface between the MC particle and the matrix. It is evident that the morphology of MC particles is blocky, with an orientation relationship $[001]_{\gamma}//[\bar{1}12]_{MC}$ to γ matrix. The crystal structure of MC belongs to point group Fm-3m (225), with lattice parameters a=b=c=11.427 Å and α=β=γ=90°.

In addition, as depicted in Figure 11.9(a), certain MC carbides exhibit minute heterogeneous cores. The high-resolution investigation of these cores is presented in Figure 11.10. Among them, Figure 11.10(a) displays the high-resolution image, Figure 11.10(b) showcases the selected electron diffraction pattern between the core and MC particles, and Figure 11.10(c) presents the compositional analysis results of the core. The diffraction pattern and EDS composition test outcomes reveal that the heterogeneous core comprises Al_2O_3 particles with a rhombidal crystal system belonging to space point group R-3C:R, featuring a lattice constant of approximately 5.121 Å and α=β=γ=55.28° angles. It can be observed that Al_2O_3 particles serve as effective nucleation sites for MC carbide precipitation during INCONEL 738LC deposition by reducing the required nucleation work for MC solidification within the liquid phase of molten pool. Similarly, in IN100 casting formation, researchers have also identified aluminum-rich particles within the cores of MC particles [12]. This indicates that this phenomenon of heterogeneous nucleation is common during large particle

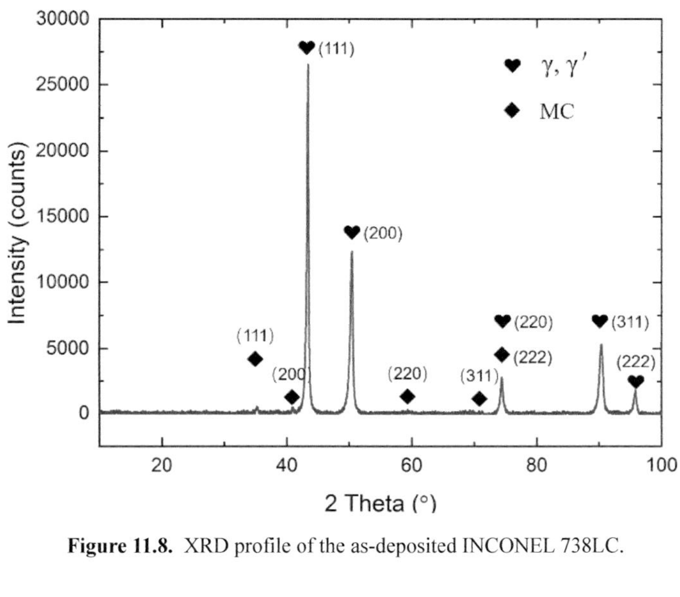

Figure 11.8. XRD profile of the as-deposited INCONEL 738LC.

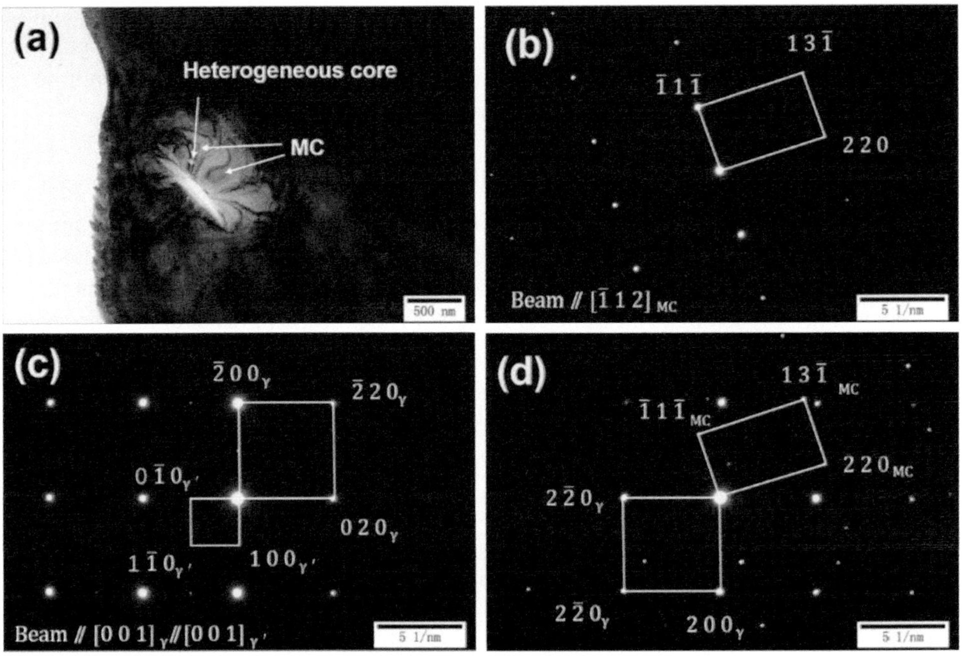

Figure 11.9. TEM bright-field image and SAED patterns of MC carbide and the surrounding matrix, (a) bright-field image, (b) SAED patterns of MC, (c) SAED patterns of the surrounding matrix, (d) SAED patterns of both MC and matrix.

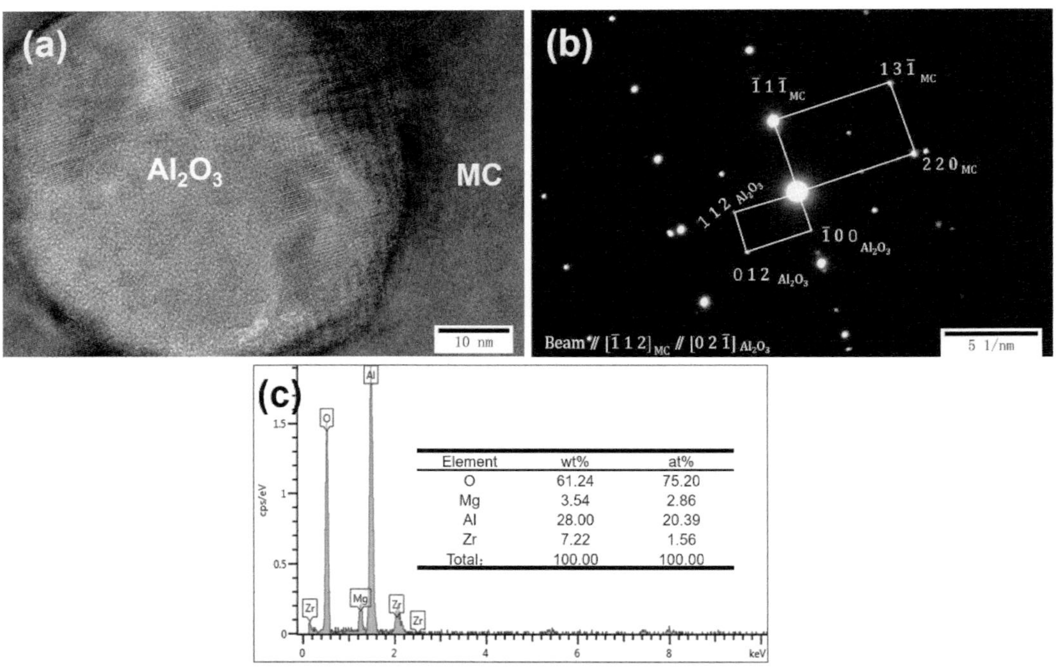

Figure 11.10. TEM analysis of the heterogeneous core in MC, (a) high-resolution image, (b) SAED patterns of MC and Al_2O_3 cores, (c) constitution of Al_2O_3 core by EDS.

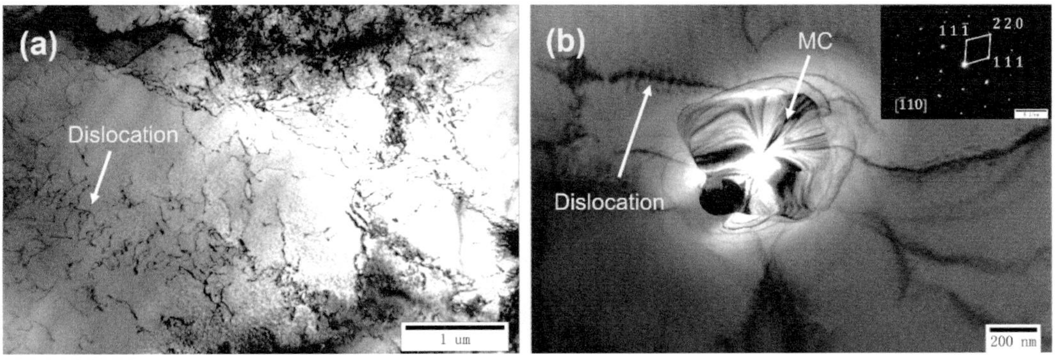

Figure 11.11. The dislocation in as-deposited INCONEL 738LC, (a) distribution of dislocations, (b) pinning effect of MC on dislocations, the insert is SAED patterns of MC.

precipitation of MC. Apart from aluminum-rich oxides, other oxides, nitrides or carbonitride particles may also act as potential substrates for initiating MC nucleation [13].

In the DED of Inconel 738LC, the rapid thermal cycling and precipitation of a precipitated phase result in significant internal stress within the obtained structure. Consequently, a large number of dislocations are present in the deposited microstructure, as depicted in Figure 11.11(a). The distribution of these dislocations is non-uniform, suggesting that the denser regions likely correspond to interdendritic areas. This observation can be attributed to potential variations in orientation between adjacent dendrites during solidification. To mitigate such differences post-solidification, it becomes necessary to generate an adequate number of geometrically necessary dislocations (GNDs) within the interdendritic region. As illustrated in Figure 11.11(b), MC carbides effectively pin these dislocations by forming a surrounding dislocation ring.

The distribution of MC particles in the intergranular region of columnar dendrites is illustrated in Figure 11.12(a). The morphologies of these particles under SEM are depicted in Figure 11.12(b). It is evident that, apart from the abundant blocky MC, there exist rod-like particles as well. EDS constitutionanalysis revealed similar elemental composition (at %) between bulk and rod-shaped particles, with high concentrations of Ti, Nb, Ta, Mo, and Zr. This confirms that the rod-shaped particles also belong to the MC phase. In comparison to blocky MCs, the content of C, Ti, and Ta decreases while Al, Cr, and Zr increase in rod-shaped MC. By calculating the ratio of element content between dendrite stems and dendrites themselves, it becomes possible to estimate the solute partition coefficient k for major elements in Inconel 738LC. The results are presented in Table 11.1 [14]. Elements with $k > 1$ exhibit a reverse segregation phenomenon, and the magnitude of this phenomenon increases with higher values of k. Conversely, elements with $k < 1$ display positive segregation, and the severity of segregation intensifies as the value decreases. It is evident that the primary MC forming elements Ti, Nb, Ta, Mo, and Zr all exhibit positive segregation phenomena and tend to concentrate near grain boundaries or interdendritic regions. On the other hand, Cr, Co, and W demonstrate a reverse segregation phenomenon and preferentially distribute within dendrite frameworks [15, 16]. This element segregation leads to MC carbides primarily being distributed in intercrystalline regions between dendrites. Similar segregation patterns also exist in INCONEL 738LC tissues formed through DED; however, due to rapid cooling of liquid metal in the molten pool during DED processing, dendrites predominantly form as columnar grains while secondary dendrite growth is effectively suppressed resulting in a more regular distribution of MC carbides. Additionally, Chen et al. [17] discovered that during directional solidification, the growth rate of MC, the content of required elements, and the shape of surrounding solid phase determine the final morphology of MC. The growth process for MC involves stepwise expansion on {111} surfaces. At low-speed solidification, the leading edge MC can fully develop into massive large particles. However, during high-speed solidification, growing MC particles become entrapped by rapidly

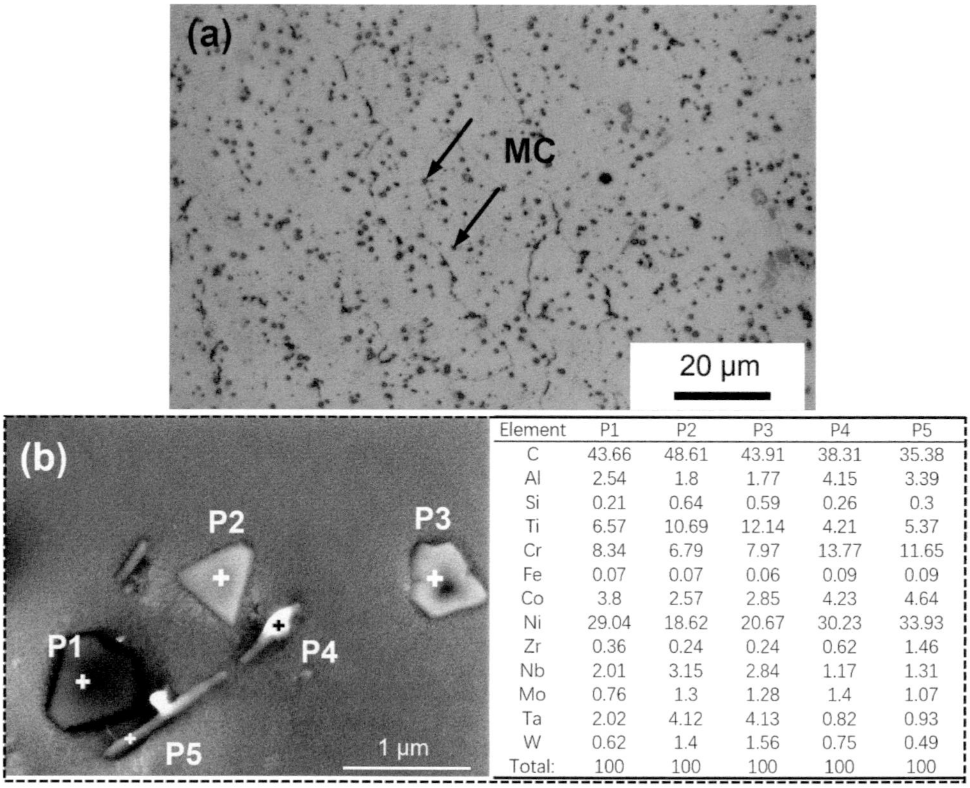

Element	P1	P2	P3	P4	P5
C	43.66	48.61	43.91	38.31	35.38
Al	2.54	1.8	1.77	4.15	3.39
Si	0.21	0.64	0.59	0.26	0.3
Ti	6.57	10.69	12.14	4.21	5.37
Cr	8.34	6.79	7.97	13.77	11.65
Fe	0.07	0.07	0.06	0.09	0.09
Co	3.8	2.57	2.85	4.23	4.64
Ni	29.04	18.62	20.67	30.23	33.93
Zr	0.36	0.24	0.24	0.62	1.46
Nb	2.01	3.15	2.84	1.17	1.31
Mo	0.76	1.3	1.28	1.4	1.07
Ta	2.02	4.12	4.13	0.82	0.93
W	0.62	1.4	1.56	0.75	0.49
Total:	100	100	100	100	100

Figure 11.12. Distribution (a) of MC carbides and their morphology (b).

Table 11.1. Distribution coefficients of the elements in INCONEL 738LC [18].

Element	Ni	Cr	Co	Al	Ti	Ta	Nb	Mo	W	Zr	C	B
k	1.03	0.98	1.09	0.92	0.69	0.74	0.52	0.87	1.13	0.13	0.3	0.008

solidifying γ dendrites, resulting in rod-shaped or even Chinese font-like morphologies. This theory can also explain variations observed in deposition structures composed from Inconel 738LCs. It can be inferred that bulk precipitation preferentially occurs during molten pool solidification; however, towards end-stage solidification, the liquid phase proportion becomes minimal, resulting in increased solidification rates which potentially lead to rod-like MC precipitation between dendrites.

11.2.3 Precipitation of γ′

The TEM image in Figure 11.13 illustrates the matrix structure of Inconel 738LCin as-deposited state. Specifically, Figure 11.13(a) represents a dark field image, while Figures 11.13(b) and (c) correspond to bright field images associated with 11.13(a) and selected electron diffraction spots respectively. The dark field phase reveals the dense precipitation of nano-sized particles within the matrix. Furthermore, the selected area electron diffraction pattern confirms that these particles belong to the γ′ phase – an ordered phase capable of unveiling superlattice spots during electron diffraction analysis (e.g., (100) spots). Notably, both γ′ and γ phases exhibit highly similar lattice structures. Additionally, it is challenging for bright field imaging to accurately depict the morphology of the γ′ phase. As depicted in Figure 11.13(a), deposition γ′ appears as spherical granules with two distinct sizes: smaller particles measuring approximately 20 nm in diameter and larger ones

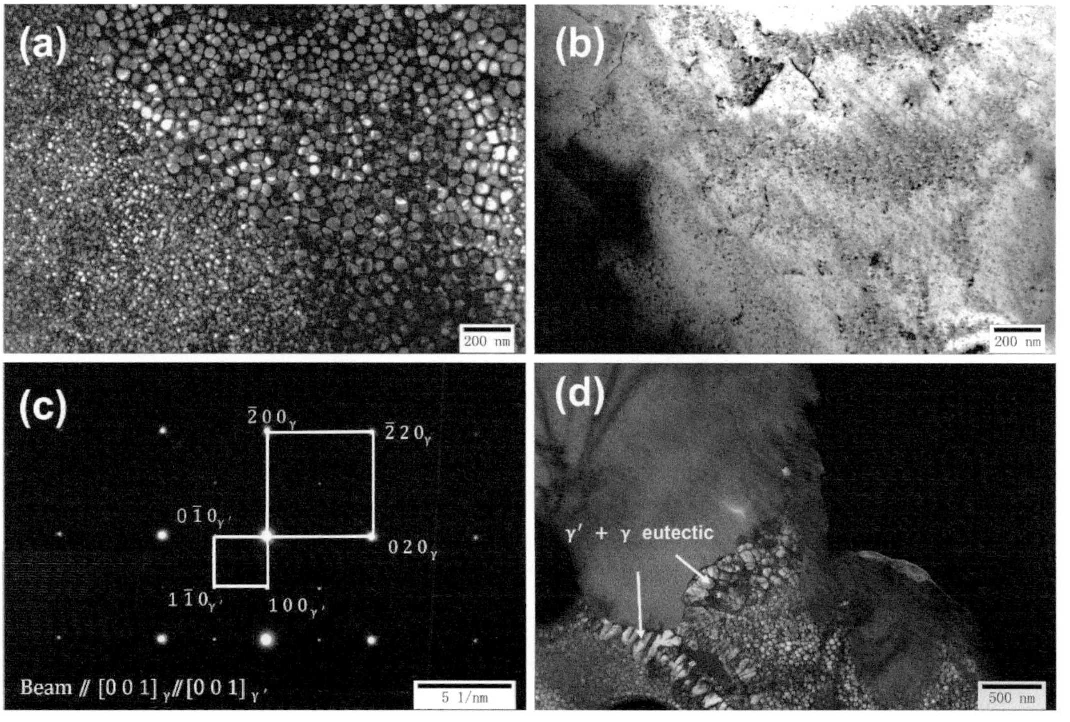

Figure 11.13. TEM images of the as-deposited INCONEL 738LCmatrix, (a) dark-field image, (b) and (c) are the bright-field image and SAED patterns of (a), (d) dark-field image of γ+γ′ eutectic.

transitioning from spherical to cuboidal shapes with a diameter around 50 nm. These spheroidal nanoscale precipitates of γ′ represent solid-phase transformation products derived from the γ phase.

Due to the high Al and Ti content in Inconel 738LC (Al+Ti > 6% wt), γ′ can still precipitate densely within the γ phase even under rapid cooling rates during laser additive manufacturing. Extensive studies have reported on the nucleation and growth of γ′ phase in nickel-based alloys [15–19]. The findings reveal that the interdendritic region is enriched with Ti, Al, Mo, and Nb, while the size of γ′ particles tends to be larger than those present in dendrite stems. Decreasing cooling rates result in an increase in both interdendritic and dendrite stem regions, leading to a morphological transformation from a sphere to cube and further to cube-cluster structure. Notably, the {111} face represents the six facets of the γ′ cube due to its minimal interfacial energy preference for growth within the γ phase.

In addition to the presence of small γ′ particles, the dark field image in Figure 11.13(d) also reveals the existence of a γ+γ′ eutectic structure. These eutectic structures primarily consist of larger strips of isolated γ′ particles separated by thin sheets of γ′. However, TEM images have limited sample size and do not accurately reflect the distribution of γ′ and γ+γ′ eutectic structures. Therefore, high-resolution SEM analysis (as shown in Figure 11.14) is employed to investigate their distribution further. It can be observed that these eutectic structures are predominantly located at dendrite boundaries (Figure 11.14(a)) or around large particles precipitated from the liquid phase (Figure 11.14(b)). The majority of these large particles are MC particles, but they may also include trace amounts of low melting point particles such as Ni_7Zr_2 or (Cr, Mo)B particles discussed later on. Additionally, micro-cracks or voids are often found near the γ+γ′ eutectic structure in Figure 11.14, which can lead to severe cracking under internal stress. Henceforth, it is crucial to conduct further analysis and discussion on the mechanism behind γ+γ′ eutectic precipitation while separately studying its associated cracking mechanism in subsequent sections.

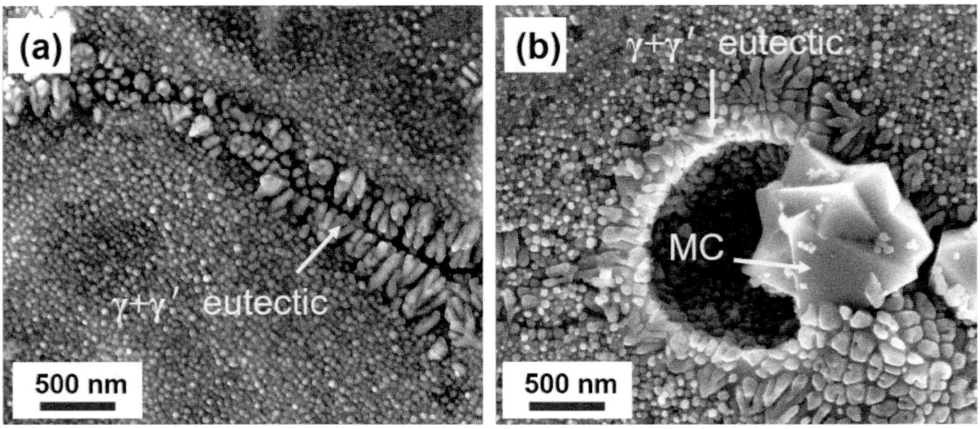

Figure 11.14. Distribution of γ+γ′ eutectic under SEM, (a) eutectic at the dendrites boundary, (b) eutectic around the MC carbide.

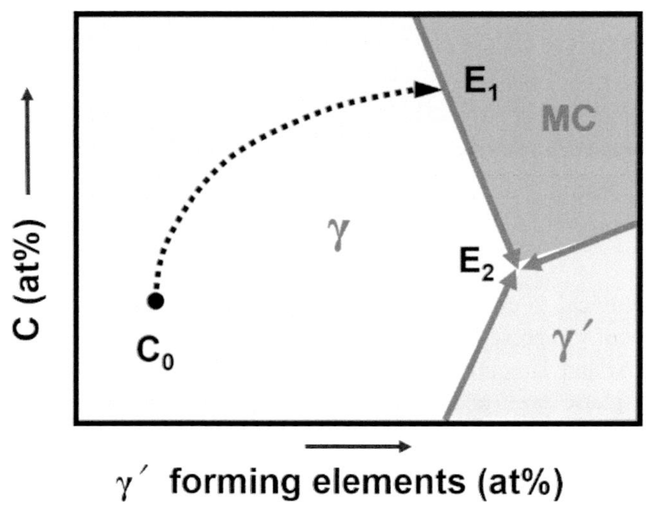

Figure 11.15. Liquidus projection of the pseudo ternary γ-MC-γ′ of Inconel 738LC [14].

According to the pseudo-ternary γ-MC-γ′ liquidus projection diagram of Inconel 738LC (Figure 11.15), it can be observed that during solidification, C elements and γ′ forming elements (Ti, Al, Ta, etc.) gradually segregate towards the residual liquid phase. The solidification route is indicated by dotted lines in Figure 11.15. Consequently, the residual liquid phase cools and solidifies within the interdendrite region. Initially, this leads to the formation of large particles primarily composed of MC carbides; however, there may also be trace amounts of Ni_7Zr_2 particles or (Cr, Mo) B particles present. Subsequently, further cooling and solidification occur along the EE_3 liquidus until reaching the terminal solidification region where reactions such as γ+γ′ or even γ+MC+γ′ eutectic reactions take place around dendritic boundaries or near large particles. The basic order of solidification can be expressed using the following formula:

$$\mathrm{L} \rightarrow \mathrm{L}' + \gamma \tag{11.2}$$

$$\mathrm{L}' \rightarrow \mathrm{MC} + \gamma' + \mathrm{L}'' \tag{11.3}$$

$$\mathrm{L}'' \rightarrow (\gamma' + \gamma)_{\mathrm{eutectic}} \tag{11.4}$$

The symbol L represents the initial molten pool liquid phase, while L′ and L″ represent the residual liquid phase in different stages.

11.2.4 Other trace precipitates

In addition to the MC carbides, γ′ precipitated phase, and γ+γ′ eutectic structure mentioned above, there are also trace precipitates present in the deposited INCONEL 738LCstructure. These phases typically consist of low melting point precipitates formed during the final stages of solidification, which may not exhibit distinct characteristics under light microscopy or scanning electron microscopy (SEM), making it challenging to differentiate them from the MC phase. Therefore, it becomes necessary to employ transmission electron microscopy (TEM) for precise identification of these phases.

The bright field image in Figure 11.16(a) reveals the presence of Ni_7Zr_2 particles. Selected electron diffraction and EDS components (Figure 11.16(b) and (c)) correspond to the Ni_7Zr_2 particles observed in Figure 11.16(a). Additionally, an irregularly shaped Ni_7Zr_2 particle at the grain boundary is shown in Figure 11.16(d). Based on diffraction spots, the lattice parameters for the Ni_7Zr_2 phase are as follows: a=11.70 Å, b=8.23 Å, c=12.19 Å, β=95.8°, α=γ=90°. In addition to Zr and Ni elements, there is also a minor presence of Cr, Co, and Ti.

TEM analysis of the M_3B_2 is presented in Figure 11.17. The bright field image (Figure 11.17(a)) displays the M_3B_2 phase while Figures 11.17(b) and (c) show selected electron diffraction patterns and corresponding EDS components respectively for this region of interest. It can be determined that M_3B_2 belongs to P4/m space point group with lattice parameters: a=b=5.83 Å, c=3.12 Å, α=β=γ=90°. The EDS constitution indicates that M mainly consists of Cr and Mo. Notably, B being a light element poses challenges for accurate measurement using EDS; hence B content was not included here.

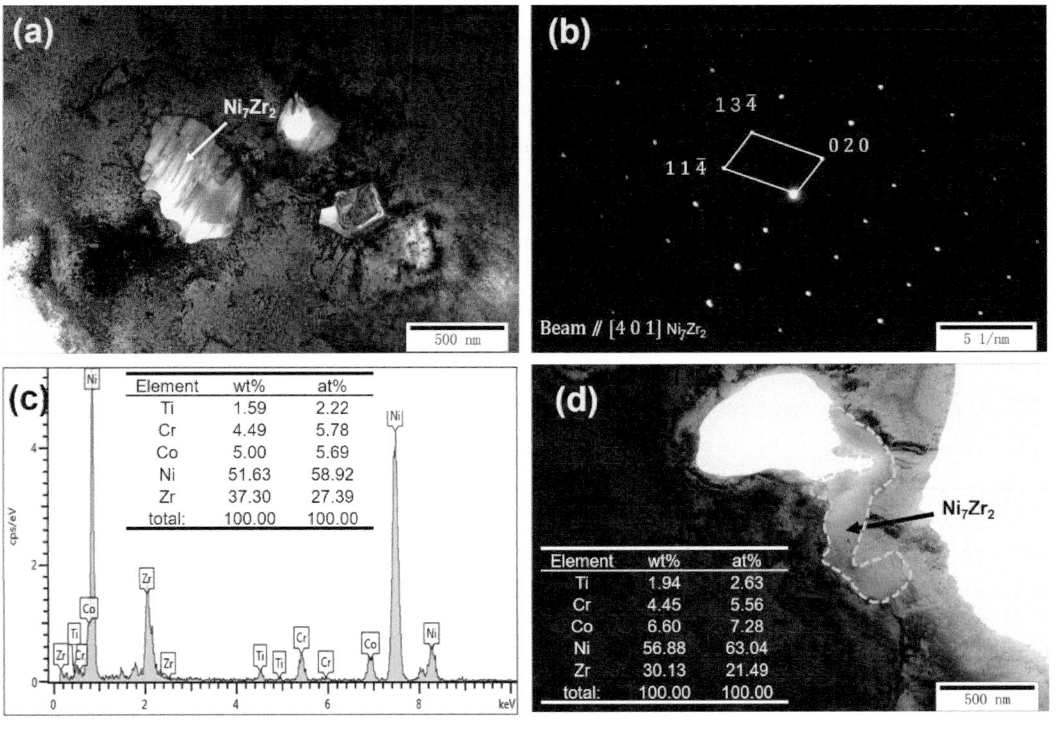

Figure 11.16. TEM analysis of Ni_7Zr_2 particles: (a) bright-field image containing Ni_7Zr_2 a particle, (b) and (c) are the SAED patterns and the EDS tested components of the Ni_7Zr_2 particle in (a), (d) another Ni_7Zr_2 particle at grain boundary.

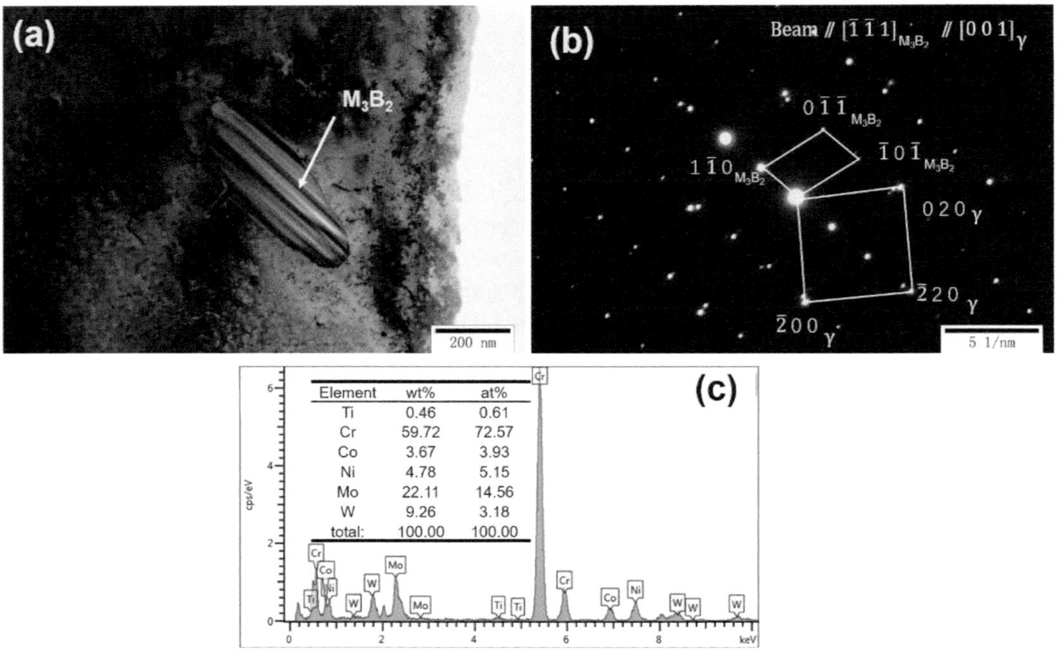

Figure 11.17. TEM analysis of M_3B_2 phase, (a) bright-field image, (b) SAED patterns of M_3B_2 and the neighbor matrix, (c) EDS analysis of M_3B_2.

It can be observed from Table 11.1 that the partition coefficient (k) of Zr between dendrite stem and dendrite is merely 0.13, whereas the corresponding k value for element B is even lower, at only 0.008. This clearly indicates that Zr and B are prone to segregation into the residual liquid phase, leading to the formation of Ni_7Zr_2 phase and M_3B_2 phase during late solidification processes. Similarly, O.A. Ojo et al. reported the presence of Ni_7Zr_2 phase and M_3B_2 phase in TIG welding welds of Inconel 738LC [20], suggesting that these two phases possess a relatively low melting point (approximately 1154°C~1184°C), which makes them susceptible to heat affected zone liquid cracking during post-welding heat treatment. Michael Cloots et al., utilizing atom probe tomography (APT) technology [21], investigated the SLM structure of INCONEL 738LC and discovered that the Zr content at grain boundaries was approximately twenty times higher than its average content in the alloy, while the B content at grain boundaries also exceeded its average level significantly. Based on this observation, it can be inferred that aggregation of Zr and B at grain boundaries could readily induce solidification cracks.

11.3 Crack formation mechanism

Inconel 738LC is prone to cracking in additive manufacturing. This phenomenon has been reported in many studies, and extensive research has shown that cracking is related to the $\gamma+\gamma'$ eutectic microstructure, Zr-Ni intermetallic compounds, and borides, among other final solidification products. This section will conduct a systematic study on the cracking mechanism in the DED process based on the as-deposited microstructure of Inconel 738LC.

11.3.1 Solidification cracking

Solidification cracking thus creates an opening dendrite structure to the cracking surface. Solidification cracking could be observed in any kind of fusion-based AM superalloy, especially the non-weldable Nickel-based superalloys. The susceptibility of Nickel-based superalloys to solidification crack is closely related to the segregation and formation of dendrites within the

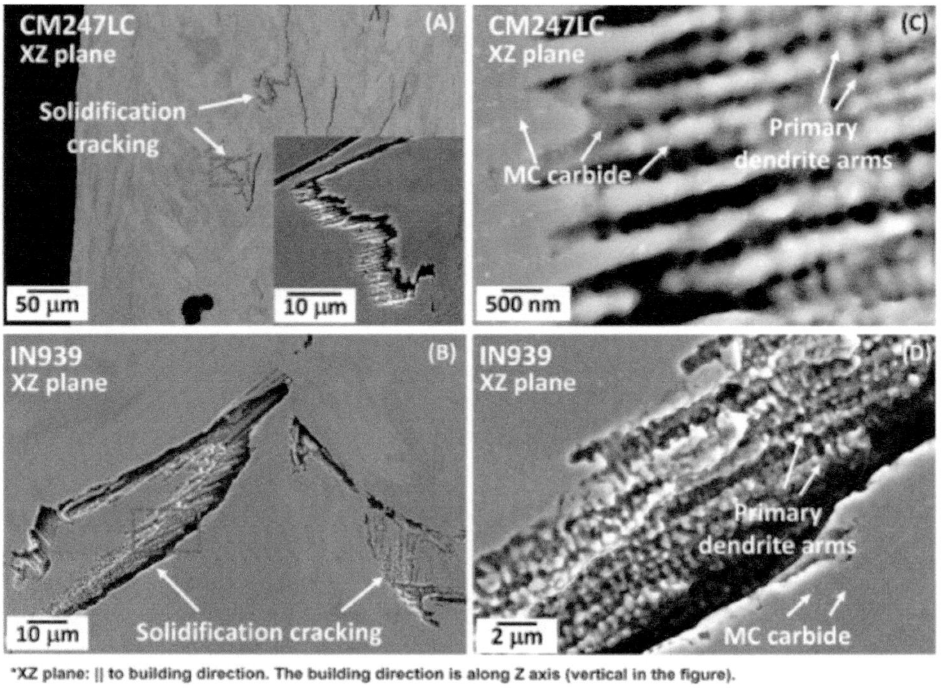

Figure 11.18. Solidification cracking observed in both CM247LC and IN939 processed by L-PBF. The jagged cracks were identified as dendrite arms were separated in the final stage of solidification [35].

material. Figure 11.18 presents the typical morphologies of solidification cracking in AM processed CM247LC [35].

Solidification cracking commonly occurs during the welding process of nickel-based alloys. In the additive manufacturing of Inconel 738LC, Cloots et al. [21] identified the cracks in the depositedstructure as solidification cracks, while Ramakrishnan et al. [6] suggested that the cracks in the deposition structure encompassed both liquid and solidification cracks. These studies primarily rely on structural characteristics near the crack to determine its properties. However, both solidification and liquid cracks are associated with liquid film between dendrites, leading to similar microstructure characteristics. Therefore, accurately determining crack properties solely based on microstructure characteristics near the crack is challenging. Henceforth, a comprehensive analysis considering factors such as time, location, and mode of crack generation will be conducted to systematically elucidate the cracking mechanism of Inconel 738LC in DED manufacturing.

The typical crack morphology in the Inconel 738LC deposited microstructure is illustrated in Figure 11.19. Among these, the microcracks depicted in Figure 11.19(a) can be considered as the prototypical cracks. It can be observed that these cracks are distributed within the solidified dendrites and exhibit narrow widths with occasional discontinuities. Upon closer examination, it becomes evident that such micro-cracks are primarily concentrated in the middle and upper regions of the deposition layer, while no such micro-cracks are detected near the top deposition layer or close to its base. Figure 11.19(b) show cases the expansion morphology of a microcrack, revealing distinct extensions towards both its upper and lower ends based on variations in crack opening width. Additionally, Figure 11.19(c) and (d) present macroscopic cracks formed after expansion, with maximum widths reaching up to 210 μm as depicted in the Figures themselves; these cracks may either close near their tops (Figure 11.19(c)) or directly penetrate through multiple layers from top to bottom (Figure 11.19(d)). Specifically, solidification cracks occur at the end of solidification and are typically found in the mushy region behind the solid-liquid interface. The mechanism

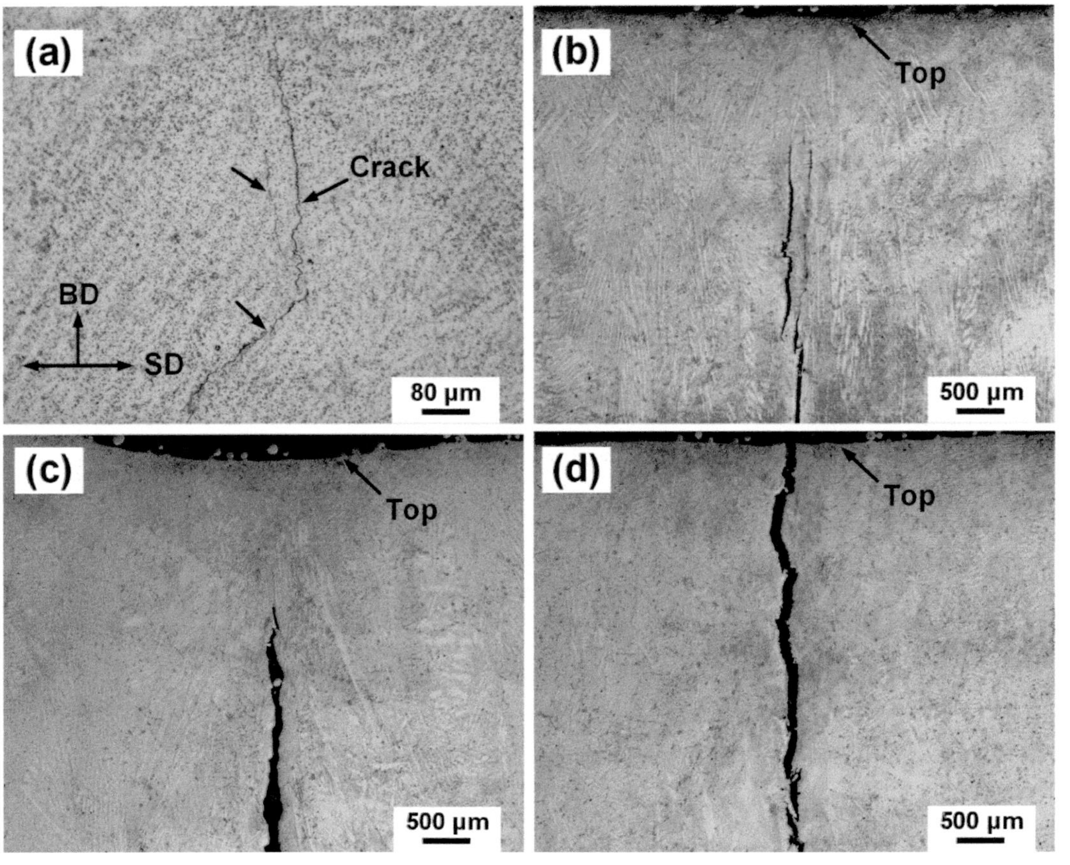

Figure 11.19. Typical morphology of cracks in the plane formed by the building-up direction (BD) and the scanning direction (SD), (a) incipient micro-cracks which generally formed in middle zone, (b) macro-cracks which grow from micro-cracks, (c) macro-crack which closed at top zone, (d) macro-crack which unclosed at top zone.

behind their formation involves constitution segregation during the final stage of solidification, resulting in continuous liquid films between solidified grains or dendrites. If these liquid films cannot accommodate the cooling contraction strain of the solid phase, they lead to separation of the solidification boundary and subsequent crack formation. The solidification crack typically exhibits a considerable width, accompanied by prominently exposed dendritic arms on both sides of the crack. In our experiment, we found that the number of such cracks is small.

11.3.2 Liquation cracking

In addition to solidification cracks, there are also HAZ liquation cracking related to liquid film. The heat affected zone (HAZ) in DED exhibits unique characteristics. Any region beneath the current sediment layer experiencing microstructural changes due to thermal influence can be considered as part of the HAZ. As depicted in Figures 11.5 and 11.3, the uppermost portion of the present deposition layer typically consists of equiaxed grains; however, with continued deposition, this layer undergoes re-melting and transforms into columnar grains. To further investigate this re-melting phenomenon, a DSC heating curve analysis was conducted on the deposited microstructure, and the results are presented in Figure 11.20.

The DSC heating curve reveals the presence of two smaller endothermic peaks preceding the corresponding matrix gamma melting peak. In accordance with the solidification sequence described for Inconel 738LC, each endothermic peak represents a distinct phase transition. Specifically, at

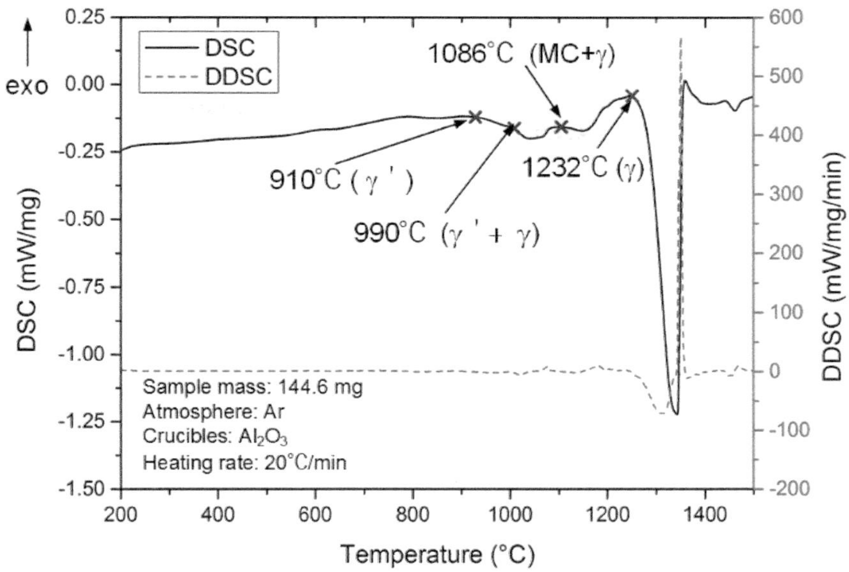

Figure 11.20. DSC heating curve and its differential (DDSC) of the as-deposited sample.

910°C, the heating curve exhibits a significant downward trend indicating the initiation of γ′ to γ phase transformation. At 990°C, the first smaller endothermic peak signifies the onset of γ+γ′ eutectic structure melting. Subsequently, at 1086°C, a second smaller endothermic peak indicates MC carbide melting initiation. Finally, at 1232°C, the main endothermic peak signifies matrix γ melting commencement.

The temperature field during DED was measured using an infrared thermal phase meter (laser power: 410 W, scanning speed: 5 mm/s, powder delivery speed: 7.6 g/min, planned layer height: 0.42 mm). The results are presented in Figure 11.21. Among them, Figure 11.21(a) depicts the infrared thermal image of point A deposition; (b) represents the corresponding thermal cycle curve for point A with the dashed line indicating the initial phase transition temperature measured by DSC; and (c) shows the final additive obtained. As shown in Figure 11.21(c), a prominent macroscopic crack is observed in the middle of the additive. Point A within this crack-sensitive region was selected to observe its thermal cycle curve (Figure 11.21(b)). It can be observed that during point A deposition, the molten pool temperature reaches approximately 1500°C, exceeding the melting interval of γ phase as determined by DSC heating curve measurements (T(γ)~T, range from 1232°C to 1353L°C). In subsequent six thermal cycles, point A consistently exceeds the initial temperature of γ+γ′ eutectic melting as measured by DSC heating curves (T(γ+γ′),990°C). The solidification of point A can lead to the subsequent remelting of the γ+γ′ eutectic structure, thereby re-establishing a liquid film between the dendrites. However, due to the high stress levels in the solidified structure at this stage, there is an increased risk of forming liquation cracks in the heat-affected zone. It should be noted that the heating speed during DSC testing (0.33°C/s) is significantly lower than that in laser-based Directed Energy Deposition (DED) processes (approximately 390°C/s). Consequently, element diffusion and heat absorption time within the sample are more abundant during DSC testing, resulting in lower measured phase change temperatures compared to those observed during laser thermal cycling.

Currently, there exist two well-established liquation cracking mechanisms, namely the segregation mechanism and the penetration mechanism. The segregation mechanism can account for the occurrence of liquation cracking in most single-phase alloys. According to this mechanism, impurity elements (primarily S, P, B) within the single-phase alloy and certain other positively segregating alloy elements tend to accumulate near grain boundaries, thereby reducing the actual

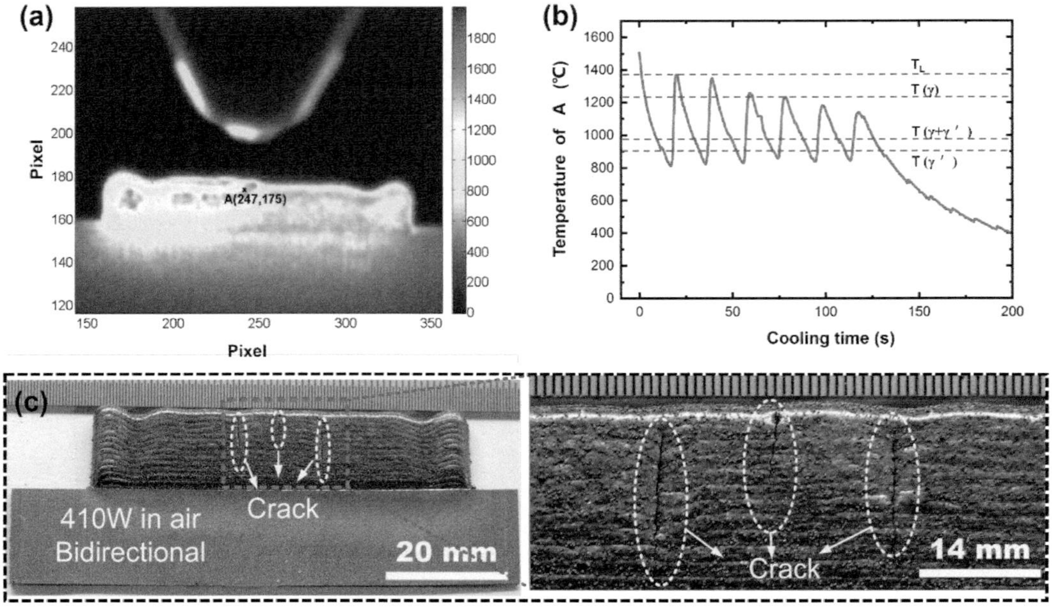

Figure 11.21. The temperature ranges and cracking during DED: (a) the infrared image when point A was deposited, (b) thermal cycles of A during DED, (c) the cracks of the as-deposited part.

melting temperature at these interfaces. Consequently, under elevated temperatures, these grain boundaries can undergo effective liquation and form continuous liquid films along them which eventually lead to the development of liquation cracks under specific stress conditions. On the other hand, the penetration mechanism explains liquation phenomena involving eutectic or secondary phases. When heating rates in heat-affected zones are rapid or when eutectic components exceed their solubility limits within an alloy matrix, residual eutectic structures may not dissolve promptly into the matrix phase. Once temperatures surpass eutectic values, direct liquification of eutectic substances occurs resulting in liquid film formation along grain boundaries that experience local stress and strain levels conducive to crack initiation. In such cases, while secondary phases themselves do not melt entirely; transition regions between secondary phases and matrices often possess lower melting points making them susceptible to liquification-induced cracking or void formations due to localized stress-strain interactions.

Figure 11.22 illustrates the occurrence of liquation in the deposited microstructure of Inconel 738LC, demonstrating conformity to the permeability mechanism. The liquation phenomena can be attributed to two factors: (1) liquation of $\gamma+\gamma'$ eutectic at grain boundaries (Figure 11.22(a), (b)), and (2) liquation of MC phase or $N_{i7}Zr_2$, M_3B_2 composition (Figure 11.22(c), (d)). Eutectic liquation at grain boundaries directly leads to the formation of discontinuous liquation cracks, while constituent liquation primarily results in constituent liquation holes near the grain boundary. These holes exhibit smooth boundaries and are characterized by distributed $\gamma+\gamma'$ eutectic with evident signs of liquification.

11.3.3 Solid-state cracking

The liquation cracking of Inconel 738LC has been previously reported in the fields of welding and additive manufacturing. However, we have observed a distinct crisp cracking sound during the DED of Inconel 738LC, which is indicative of solid-state cracking. This suggests that besides simple liquation cracking, there also exists solid phase cracking phenomenon during the additive

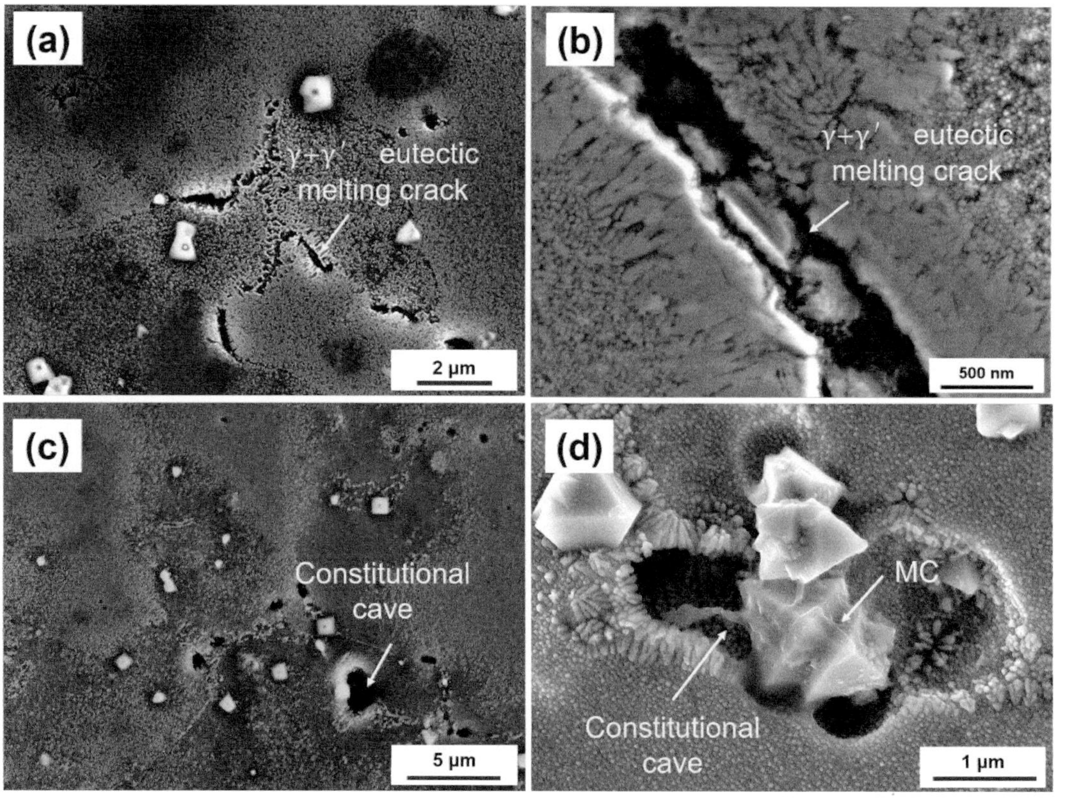

Figure 11.22. The liquation signs in the as-deposited texture, (a) and (b) are the $\gamma+\gamma'$ eutectic liquation cracking, (c) and (d) are the cracking caused by constitutional liquation of MC.

manufacturing process. Surprisingly, there are limited studies on solid phase cracking in the additive manufacturing of Inconel 738LC. Therefore, this study aims to investigate the mechanism behind this solid-state cracking phenomenon in subsequent sections.

γ' precipitation-strengthened nickel-based alloys typically exhibit two modes of solid phase cracking, namely ductility-dip cracking (DDC) and strain aging cracking (SAC). The mechanisms responsible for DDC formation include grain-boundary sliding induced cracking, brittle cracking of grain boundaries due to segregation of impurity elements (mainly S, P, H), and cracking caused by grain boundary precipitates [22–24]. SAC formation is associated with the precipitation [25–27] of the γ' phase, primarily occurring during the heat treatment stage after welding as a result of rapid accumulation of strain during aging. Deposited INCONEL 738LC structures display cracks consistent with the DDC mechanism. Figure 11.23 illustrates the morphology and formation mechanism of INCONEL 738LC. In Figure 11.23(a), it can be observed that carbides visibly detach or separate from the matrix, leading to corresponding pits in the matrix. Unlike rounded potholes resulting from liquation, these potholes are angular and bright, and have clear signs of the separation and shedding of carbide particles in solid state. By analyzing the relative position of the residual carbides and their corresponding pits, it is possible to infer the local deformation direction of the grain where the carbide is located, as depicted by the yellow arrow in Figure 11.23(a). It can be inferred that when two adjacent grains have opposite strain directions, significant relative sliding can occur at their grain boundary. This sliding may lead to fragmentation of grain boundaries parallel to slip direction, resulting in DDC cracking (Figure 11.23(b)). The straight nature of grain boundary between two neighboring grains along with strain concentration areas caused by precipitates or voids can induce relative sliding under tangential stress, ultimately leading to DDC cracks formation. Figure 11.23(c)

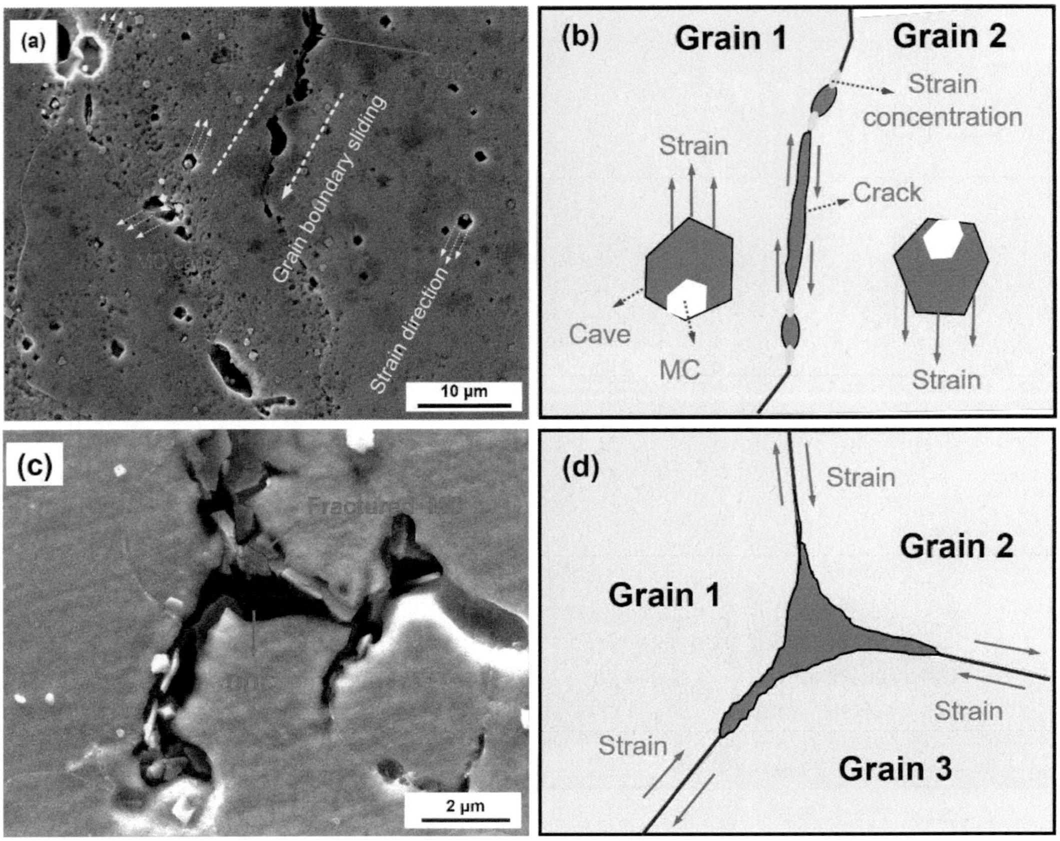

Figure 11.23. Morphology and mechanism of crack occurred in solid state.

illustrates DDC topography at three intersection points of a grain boundary, revealing numerous flaky MC particles present at these intersections which exhibit brittle fracture characteristics indicating high stress concentration. Correspondingly, Figure 11.23(d) shows a diagram depicting DDC cracking occurring at these triple intersections of grain boundaries due to substantial strain accumulation resulting from relative sliding among three different grains.

According to the afore-mentioned cracking mechanism, it is evident that the formation of DDC cracks is closely associated with the stress level in proximity to the grain boundary. During the DED process, rapid thermal cycling subjects the material to significant thermal stress. Furthermore, a substantial amount of γ' particles and MC particles precipitate during the DED process of Inconel 738LC, leading to an enhancement in overall internal stress levels due to phase transition stress induced by lattice mismatch. Importantly, γ' phase rapidly precipitates during high-temperature stages following liquid phase solidification, impeding sufficient plastic deformation for complete release of thermal stress and resulting in continuous accumulation of stress levels. In welding field, the solid-phase cracking caused by precipitation of γ' phase typically occurs during post-welding heat treatment stage known as SAC cracking. During DED of INCONEL 738LC, γ' phase the precipitate at a high temperature, and the precipitation increases the inner stresses, in this process, yield strength within grains is higher than grain boundaries, plastic deformation primarily concentrates near grain boundaries, thus promoting mutual slip between grain boundaries and ultimately forming cracks. This aligns with both SAC cracking mechanism and DDC cracking. Therefore, the SAC and DDC cracks can occur at the same location (grain boundaries) and within the same temperature range (γ' phase precipitation temperature interval), This kind of hot crack unrelated to liquid film, and can be collectively referred to as solid-state crack.

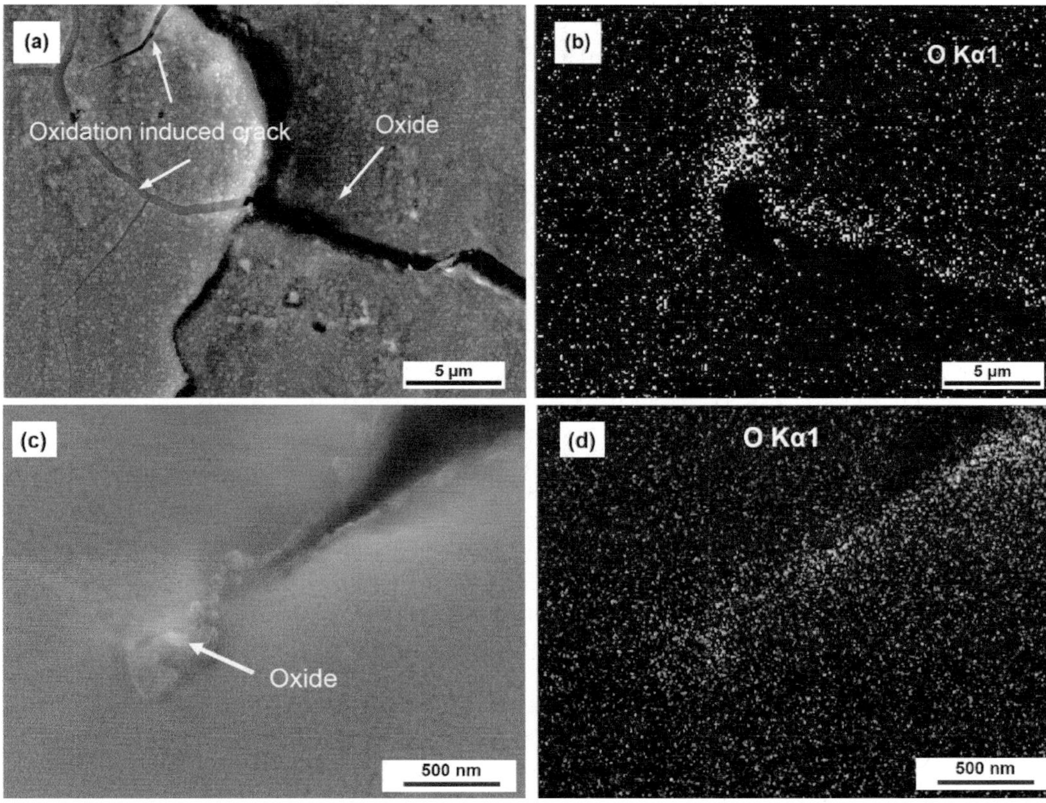

Figure 11.24. Effects of oxidation on cracking, (a) and (c) is the oxides nearby cracks, (b) and (d) are the EDS maps of (a) and (c) in respectively.

Additionally, in regions exhibiting severe cracking, conspicuous oxidation is also observed. As depicted in Figure 11.24(a), the severely oxidized area even displays distinct sharp solid-state cracks. The corresponding EDS constitution analysis (Figure 11.24(b)) confirms the presence of oxidation near the crack site. The influence of oxygen on high-temperature cracking of nickel-based alloys primarily arises from two mechanisms: dynamic embrittlement (DE) and stress-assisted grain boundary oxidation [28–31] (SAGBO) [32–34]. According to the theory of dynamic embrittlement, at elevated temperatures, oxygen elements undergo short-range diffusion at the crack tip, thereby accelerating intergranular crack propagation. Consequently, as cracks dynamically expand, oxides such as Cr_2O_3, TiO_2, NiO or Al_2O_3 are formed on both sides of these cracks. On the other hand, based on the stress-assisted grain boundary oxidation theory, under stress and high temperature conditions, oxygen can propagate over long distances towards the crack tip and form oxides at its leading edge; subsequently allowing for further expansion through rupture or separation from the substrate by these oxide layers themselves. The observed cracks in Figure 11.24(c) align with this stress-assisted grain boundary oxidation theory since oxides are detected at both sides and tips of these cracks; corroborated by an EDS distribution diagram illustrating oxygen elements shown in Figure 11.24(d). Considering that DED-induced oxidation diminishes grain boundary plasticity while promoting intergranular crack propagation; it can be considered a contributing factor inducing solid-state cracking.

Generally, the liquation cracking and solid-state cracking during DED of INCONEL 738LC don't happen entirely in isolation. Liquation cracks at grain boundaries or liquation voids surrounding large particles induce significant stress concentration phenomena. These stress concentration regions experience preferential sliding along grain boundaries, thereby promoting the formation of solid-state cracking during the subsequent cooling process. Subsequent accumulation of stresses can rapidly propagate those initial micro-cracks, ultimately leading to the development of macro-cracks on a larger scale.

11.4 Summary

In this chapter, the microstructure characteristics of Inconel 738LC and the cracking mechanism during deposition manufacturing are investigated. The main conclusions are as follows:

1) The deposited layer primarily consists of columnar crystals, with potential stray crystal regions at the top and fusion area between layers. The growth direction of columnar crystals is influenced by heat flow direction and crystal's preferred orientation, which can be affected by process parameters such as laser power, scanning speed, and scanning path during deposition. Two main factors contribute to the formation of top stray crystal: firstly, a decrease in temperature gradient at the molten pool's top leads to an increased solidification rate triggering a transition from columnar to stray crystals; secondly, incomplete melted metal powder remaining at the top promotes fine stray crystal formation.
2) The precipitated phase of INCONEL 738LC primarily consists of discretely distributed MC carbides, γ+γ' eutectics, and densely precipitated γ' nanoparticles. The MC type carbides are typically enriched with Ti, Nb, Ta, Mo, Zr and may exist in various forms such as rods or blocks. Moreover, the MC core often contains nano-sized aluminum-rich oxides. The γ' phase particles generally occur in two distinct sizes: inner-dendritic γ' particles have a diameter of approximately 20 nm while inter-dendritic γ' particles have a diameter of around 50 nm. Additionally, there is a small number of Ni_7Zr_2-type Zr-rich particles and M_3B_2-type particles.
3) The cracking during DED of INCONEL 738LCmainly attributed to liquation cracking and solid-state cracking. Liquation cracking can be categorized into two types: one involves the liquation of γ+γ' eutectic at grain boundaries, while the other pertains to constitution liquation surrounding MC or Ni_7Zr_2 and M_3B_2 particles. Solid-state cracking is associated with grain boundary sliding induced by internal stress, intensive precipitation of γ' particles, and oxidation along grain boundaries. The internal stress within the structure encompasses not only thermal stress resulting from rapid thermal cycling but also phase transition stress caused by the precipitation of γ', MC, and other particles.

References

[1] Ojo, O. A., Richards, N. L. and Chaturvedi, M. C. (2004). Contribution of constitutional liquation of gamma prime precipitate to weld HAZ cracking of cast Inconel 738 superalloy [J]. Scripta Materialia 50(5): 641–646.

[2] Ola, O. T., Ojo, O. A. and Chaturvedi, M. C. (2013). Laser arc hybrid weld microstructure in nickel based IN738 superalloy [J]. Materials Science and Technology (United Kingdom) 29(4): 426–438.

[3] Chen, J. and Xue, L. (2010). Process-induced microstructural characteristics of laser consolidated IN-738 superalloy [J]. Materials Science and Engineering: A 527(27): 7318–7328.

[4] Xu. J., Lin, X., Zhao, Y., Guo, P., Wen, X., Li, Q. et al. (2018). HAZ liquation cracking mechanism of IN-738LC superalloy prepared by laser solid forming [J]. Metallurgical & Materials Transactions A 49(10): 5118–5136.

[5] Cloots, M., Uggowitzer, P. J. and Wegener, K. (2016). Investigations on the microstructure and crack formation of IN738LC samples processed by selective laser melting using Gaussian and doughnut profiles [J]. Materials & Design 89: 770–784.

[6] Ramakrishnan, A. and Dinda, G. P. (2019). Direct laser metal deposition of Inconel 738 [J]. Materials Science and Engineering A 740-741: 1–13.

[7] Wei, H. L., Mazumder, J. and Debroy, T. (2015). Evolution of solidification texture during additive manufacturing [J]. Scientific Reports 5: 16446.

[8] Zhou, Y. Z., Volek, A. and Green, N. R. (2012). Mechanism of competitive grain growth in directional solidification of a nickel-base superalloy [J]. Acta Materialia 56(11): 2631–2637.

[9] Blecher, J. J., Palmer, T. A. and Debroy, T. (2014). Solidification map of a nickel-base alloy [J]. Metallurgical & Materials Transactions A 45(4): 2142–2151.

[10] Parimi, L. L., Ravi, G. A., Clark, D. and Attallah, M. M. (2014). Microstructural and texture development in direct laser fabricated IN718 [J]. Materials Characterization 89: 102–111.
[11] Gäumann, M., Bezençon, C., Canalis, P. and Kurz, W. (2001). Single-crystal laser deposition of superalloys: processing–microstructure maps [J]. Acta Materialia 49(6): 1051–1062.
[12] Fernandez, R., Lecomte, J. C. and Kattamis, T. Z. (1978). Effect of solidification parameters on the growth geometry of MC carbide in IN-100 dendritic monocrystals [J]. Metallurgical Transactions A 9(10): 1381–1386.
[13] Chen, J., Huang, B. and Lee, J. H. (1999). Carbide formation process in directionally solidified MAR-M247 LC superalloy [J]. Journal of Materials Ence & Technology 15(1): 48–52.
[14] Ojo, O. A., Richards, N. L. and Chaturvedi, M. C. (2006). Study of the fusion zone and heat-affected zone microstructures in tungsten inert gas-welded INCONEL 738LC superalloy [J]. Metallurgical & Materials Transactions A 37(2): 421–433.
[15] Rosenthal, R. and West, D. R. F. (1999). Continuous γ' precipitation in directionally solidified IN738 LC alloy [J]. Metal Ence Journal 15(12): 1387–1394.
[16] Sidhu, R. K., Ojo, O. A., Chaturvedi, M. C. (2007). Microstructural analysis of laser-beam-welded directionally solidified INCONEL 738 [J]. Metallurgical & Materials Transactions A 38(4): 858–870.
[17] Chen, J., Huang, B., Lee, J. H., Choe, S. J. and Lee, Y. T. (1999). Carbide formation process in directionally solidified MAR-M247 LC superalloy [J]. Journal of Materials Sciences and Technology 15(1): 48–52.
[18] Singh, A. R. P., Nag, S., Chattopadhyay, S., Ren, Y., Tiley, J., Viswanathan, G. B. et al. (2013). Mechanisms related to different generations of γ' precipitation during continuous cooling of a nickel base superalloy [J]. Acta Materialia 61(1): 280–293.
[19] Behrouzghaemi, S. and Mitchell, R. J. (2008). Morphological changes of γ' precipitates in superalloy IN738LC at various cooling rates [J]. Materials Ence & Engineering A 498(1-2): 266–271.
[20] Ojo, O. A., Richards, N. L. and Chaturvedi, M. C. (2004). Microstructural study of weld fusion zone of TIG welded IN 738LC nickel-based superalloy [J]. Scripta Materialia 51(7): 683–688.
[21] Cloots, M., Uggowitzer, P. J. and Wegener, K. (2016). Investigations on the microstructure and crack formation of IN738LC samples processed by selective laser melting using Gaussian and doughnut profiles [J]. Materials & Design 89(JAN.): 770–784.
[22] Ramirez, A. J. and Lippold, J. C. (2004). High temperature behavior of Ni-base weld metal: Part II – Insight into the mechanism for ductility dip cracking [J]. Materials Ence & Engineering A 380(1-2): 245–258.
[23] Ramirez, A. J. and Lippold, J. C. (2004). High temperature behavior of Ni-base weld metal Part I. Ductility and microstructural characterization [J]. Materials Science & Engineering A Structural Materials A380(1/2): 259–271.
[24] Torres, E. A., Montoro, F., Righetto, R. D. and Ramirez, A. J. (2014). Development of high-temperature strain instrumentation for *in situ* SEM evaluation of ductility dip cracking [J]. J. Microsc 254(3): 157–165.
[25] Kazempour-Liasi, H., Tajally, M. and Abdollah-Pour, H. (2019). Effects of pre- and post-weld heat treatment cycles on the liquation and strain-age cracking of IN939 superalloy [J]. Engineering Research Express 1(2).
[26] Rowe, M. D. (2006). Ranking the resistance of wrought superalloys to strain-age cracking [J]. Welding Journal 85(2): 27s–34s.
[27] Lippold, J. C. (2014). Solid-State Cracking [M]. John Wiley & Sons, Ltd.
[28] Németh, A. A. N., Crudden, D. J., Armstrong, D. E. J., Collins, D. M., Li, K., Wilkinson, A. J. et al. (2017). Environmentally-assisted grain boundary attack as a mechanism of embrittlement in a nickel-based superalloy [J]. Acta Materialia 126: 361–371.
[29] Jiang, R., Proprentner, D., Callisti, M., Shollock, B., Hu, X. T., Song, Y. D. et al. (2018). Role of oxygen in enhanced fatigue cracking in a PM Ni-based superalloy: Stress assisted grain boundary oxidation or dynamic embrittlment? [J]. Corrosion Science 139: 141–154.
[30] Kitaguchi, H. S., Li, H. Y., Evans, H. E., Ding, R. G., Jones, I. P., Baxter, G. et al. (2013). Oxidation ahead of a crack tip in an advanced Ni-based superalloy [J]. Acta Materialia 61(6): 1968–1981.
[31] Krupp, U., Wackermann, K., Christ, H.-J., Colliander, M.-H. and Stiller, K. (2017). Intergranular oxidation effects during dwell-time fatigue of high-strength superalloys [J]. Oxidation of Metals 88(1): 3–14.
[32] Jiang, R., Gao, N. and Reed, P. A. S. (2015). Influence of orientation-dependent grain boundary oxidation on fatigue cracking behaviour in an advanced Ni-based superalloy [J]. Journal of Materials Science 50(12): 4379–4386.
[33] Evans, H. E., Li, H. Y. and Bowen, P. (2013). A mechanism for stress-aided grain boundary oxidation ahead of cracks [J]. Scripta Materialia 69(2): 179–182.
[34] Karabela, A., Zhao, L. G., Tong, J., Simms, N. J., Nicholls, J. R. and Hardy, M. C. (2011). Effects of cyclic stress and temperature on oxidation damage of a nickel-based superalloy [J]. Materials Science & Engineering A 528(19): 6194–6202.
[35] Tang, Y. T., Panwisawas, C., Ghoussoub, J. N., Gong, Y., Clark, J. W., Németh, A. A. et al. (2020). Alloys-By-Design: Application to new superalloys for additive manufacturing[J]. Acta Materialia.

Chapter 12

Effects of Laser Deposition Processes on Crack Sensitivity for Depositing Inconel 738LC

12.1 Introduction

The microstructure characteristics and cracking mechanisms of DEDed IN738LC were primarily investigated in Chapter 11. The results demonstrate the propensity for liquation cracks and solid-state cracks to form during the DED process of IN738LC. It is evident that suppressing crack formation is crucial for ensuring the quality of IN738LC additive. Therefore, this chapter aims to explore the influence of laser deposition processes on crack sensitivity when depositing Inconel 738LC. These processes encompass laser power, deposition rate, scanning path, substrate temperature, and environmental atmosphere within the DED system. Comprehensive analysis utilizing infrared thermal imaging, digital image correlation (DIC) method, and electron backscatter diffraction (EBSD) characterization will be employed to investigate the underlying correlations between these processes and crack sensitivity from perspectives including temperature field gradient, macroscopic strain, and microscopic strain.

12.2 Effects of laser power on crack sensitivity

12.2.1 Cracking under different laser power

The influence of laser power on cracking during the DED process is significant. As shown in Figure 12.1, the macroscopic cracks of the samples were observed at laser powers of 250 W, 320 W, and 410 W respectively. The influence of laser power on cracking during the DED process is significant. Figure 12.1 shows the macroscopic cracks of the samples which were deposited at laser powers of 250 W, 320 W, and 410 W respectively, the other process parameters kept constant: the scanning speed is 5 mm/s, the powder feeding speed is 7.6 g/min, and the designed layer height is 0.42 mm. It can be seen that at a laser power of 250 W, no macroscopic cracks (at the milimeter level) are observed in the obtained sample. However, when the laser power is increased to 320 W, three narrow macro-cracks emerge within the sample. Moreover, with further increase in laser power to 410 W, both narrow and macro-cracks are evident in the specimen. Figure 12.1(d) provides an enlarged view of cracks within the dotted box shown in Figure 12.1(c). It can be observed that these cracks predominantly propagate along the build-up direction (BD), traversing multiple deposition layers. The deposited metal adjacent to narrow cracks exhibits relatively flat surfaces; however, invagination is clearly visible around wider cracks. This phenomenon primarily arises from progressive widening of earlier-formed cracks during the additive process. When the laser spot passes through a wide crack, it leads to an actual defocusing effect on a microscale zone which

DOI: 10.1201/9781003406327-12

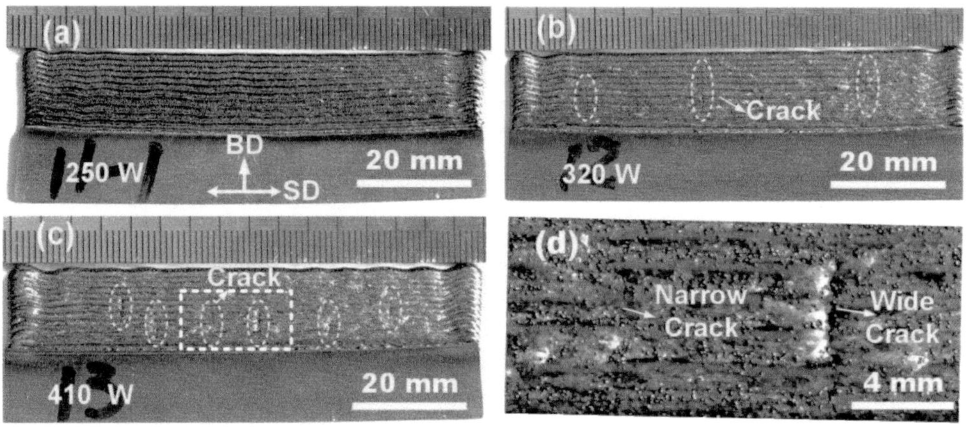

Figure 12.1. Macro-crack morphology in the deposited parts with different power, (a) 250 W, (b) 320 W, (c) 410 W, (d) is the local zoom of (c) showing the amplified cracks.

reduces local laser energy density and ultimately diminishes metal powder deposition efficiency while causing collapsible occurrences.

In summary, the crack sensitivity increases with the rise in laser power. This finding aligns with the majority of research studies on laser processing of nickel-based alloys [1, 2]. The literature suggests two main reasons for the increased occurrence of cracks: firstly, higher power and heat input result in greater thermal deformation and subsequent internal stress; secondly, elevated heat input prolongs high-temperature residence time and intensifies element segregation between dendrites, thereby raising the risk of solidification or liquation cracks formation. However, Idowu et al. [3] discovered that liquation cracks in the heat-affected zone (HAZ) significantly increased at low heat input during their investigation on laser welding of IN718Plus (an alloy derived from Inconel 718 with increased Co, Al and Ti content). They found that at high heat input levels where obvious liquation occurred within HAZ, however liquation cracks were rarely observed. This phenomenon can be attributed to a thick liquid film forming between dendrites under high heat input conditions which effectively releases stress between dendrites and reduces the likelihood of liquation cracking. Moreover, this thicker liquid film at high heat input also fills existed liquation cracks thus minimizing their occurrence. It is evident that the influence mechanism of laser power on liquation cracks may exhibit variations. Therefore, in the subsequent paragraphs, we will conduct a comprehensive analysis of the impact of laser power on IN738 DED cracking, considering factors such as temperature gradient, macro strain, micro strain, and grain characteristics.

12.2.2 Effect of laser power on temperature gradient

Generally, it is believed that an increase in laser power will result in elevated heat input and thermal deformation, consequently leading to heightened stress within the heat affected zone. However, it should be noted that the internal stress induced by thermal strain is not solely dependent on strain magnitude but also closely associated with the constraints imposed by the object. In unconstrained condition, even if thermal expansion and contraction occur due to high temperatures, the level of internal stress remains relatively low, as long as the material can withstand this degree of strain, cracks will not occur. Conversely, under strong constraint conditions, thermal expansion at elevated temperatures can cause irreversible inelastic deformation within the metal structure, resulting in macroscopic internal stresses distributed throughout the entire component. In additive manufacturing or welding processes, the confinement of the heat affected zone is closely associated with the non-uniformity of thermal deformation in different regions, which in turn depends on the temperature gradient in space and cooling rate over time. Therefore, to accurately assess internal stress levels, it is essential to investigate the distribution and dynamic changes of the temperature field. Currently, finite element analysis and actual temperature measurements are commonly employed methods for studying temperature gradient and cooling rate. However, when using finite element to calculate additive manufacturing's temperature field, numerous simplifications and assumptions regarding laser energy absorption rates, boundary conditions for heat dissipation, material heat transfer parameters, etc., need to be made; improper settings often lead to significant calculation errors. On the other hand, actual temperature measurement methods typically employ tools such as thermocouples, infrared thermometers and infrared thermal imagers. Thermocouples and infrared thermometers usually provide single point measurements that do not facilitate spatial distribution analysis of the temperatures. Henceforth in this study section we utilize an infrared thermal imager to examine the temperature field during additive manufacturing process aiming at exploring how laser power affects it under real conditions. It should be noted that temperatures obtained through infrared measurement are influenced by material emissivity towards infrared spectrum; thus in this chapter we have calibrated emissivity values using a thermocouple.

To achieve the *in-situ* analysis of deformation during the DED process, the Digital Image Correlation Method (DIC) was employed for strain measurement. The hardware components of the DIC system encompass the CMOS high-speed camera of the German Optronis brand and the laser-assisted light source with synchronous triggering functionality. The maximum frame rate of the CMOS camera can reach 125 fps, while the actual frame rate utilized in this paper is 25 fps, with a resolution of 1280 × 1024 pixels. The wavelength of the auxiliary light source is 900–910 nm, and the maximum power is 35 W. Concurrently, an infrared thermal imager was also utilized in this paper for in-situ measurement of the temperature field during the DED process. The model of the thermal imager is AVIO500EX-PRO, with a wavelength of 8–14 μm, a resolution of 640 × 480 pixels, and a temperature measurement range of –40 to 2000°C. The sampling frame rate is 3 fps. Figure 12.2 presents the device diagram and example results of the DIC strain measurement and infrared temperature measurement.

Figure 12.3(a), (b), and (c) respectively show the temperature distribution of the substrate, the temperature distribution in the scanning direction (SD), and the temperature distribution in the build-up direction (BD). These results were obtained below the molten pool when the first four layers of IN738LC were deposited under the 250 W laser power. As can be seen that the laser molten pool has an obvious heating phenomenon on the tail metal, and after four layers of deposition, the temperature of the base metal has been raised to about 300°C. At the front of the laser molten pool,

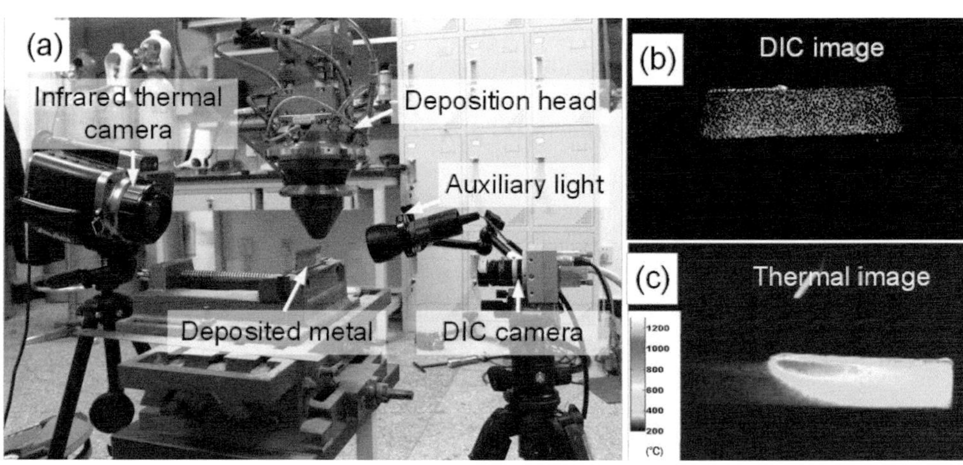

Figure 12.2. Strain measurement by DIC camera and temperature measurement by infrared thermal camera during DED, (a) measurement setup, (b) DIC image example, (c) thermal image example.

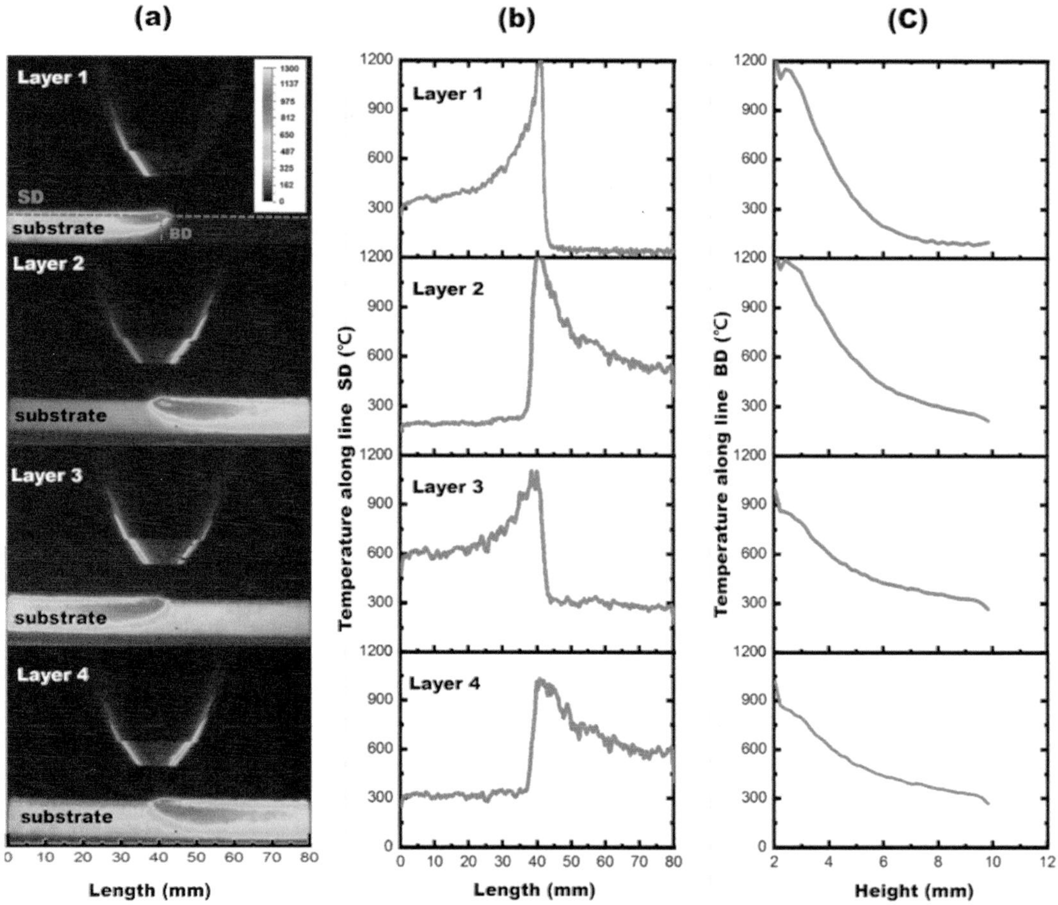

Figure 12.3. Temperature field of the substrate during the deposition with 250 W power, (a) infrared temperature map for the first 4 layers, (b) and (c) is the temperature distribution along the scanning direction (SD) and build-up direction (BD) in respectively.

the temperature drops sharply, forming a large temperature gradient in this area, while at the rear of the molten pool, the temperature changes gently and the temperature gradient is small.

Figure 12.4 presents quantitative statistics of the temperature gradient at the leading edge of the molten pool during laser power of 250 W. Specifically, for the initial layer deposition, the maximum temperature gradient in the SD reaches to 364°C/mm, while in the BD it is 295°C/mm. For the second layer, the temperature gradient in the SD and decreased to 318°C/mm and 224°C/mm (height direction). For the third, highest temperature gradients were observed as 277°C/mm in SD and 229°C/mm in BD, and for the fourth layers, they decreased to 173°C/mm in SD and 154°C/mm in BD. Evidently, with an increase in deposition layers, heat accumulation leads to a gradual decrease in the maximum temperature gradients. The movement of heat source results in larger temperature differences along the SD to BD, hence resulting in higher temperature gradients at molten pool front.

Figure 12.5 shows the temperature gradient maximum at the front of the molten pool varies with laser power. When laser power is increased, the temperature gradient along the SD rises significantly. At a laser power of 410 W, the first layer's temperature gradient can reach 500°C/mm, which is about 1.4 times that of a power of 250 W. However, in terms of BD, increasing laser power leads to a decrease in temperature gradient; specifically, at a power level of 410 W it is approximately only 0.7 times that observed at a level of 250 W. Combined with macroscopic

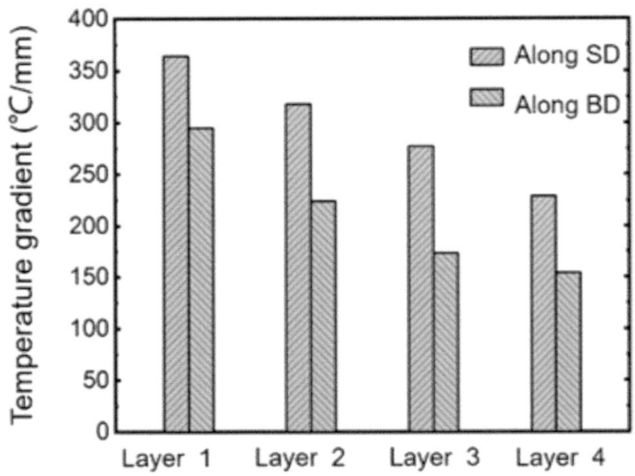

Figure 12.4. Temperature gradient at the front edge of the molten pool with 250 W power.

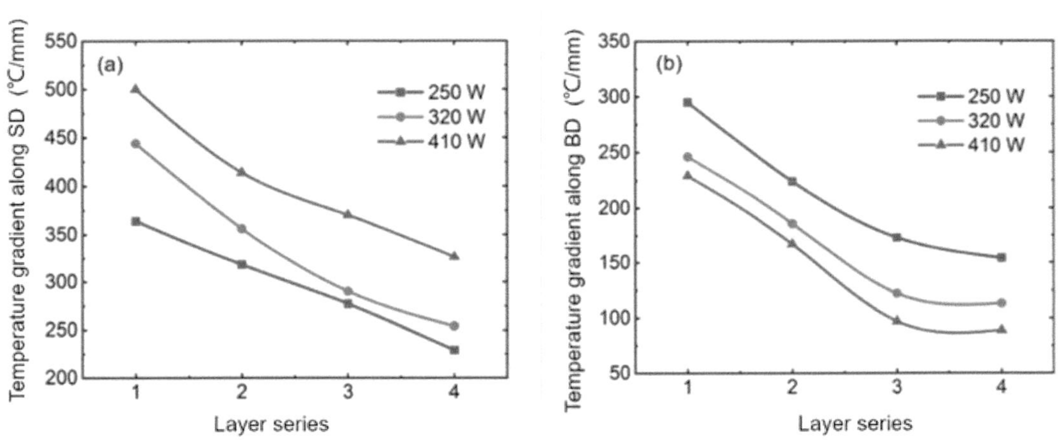

Figure 12.5. Effects of laser power on the spatial temperature gradient, (a) the SD directional gradient in front of molten pool, (b) the BD directional gradient below the molten pool.

cracking observations shown in Figure 12.1, it can be inferred that increasing laser power enhances sensitivity to cracking by elevating thermal gradients in SD.

The infrared thermal imager can also capture real-time crack formation processes. As depicted in Figure 12.6, noticeable color variations between cracks and continuous metal are observed in the thermal image due to temperature or emissivity disparities. Consequently, the initiation and expansion of macroscopic cracks can be distinctly observed. Figure 12.6(a) illustrates a thermal image collected at 91 seconds after deposition started, during which no macroscopic cracks were detected. However, a prominent crack emerged at 115 seconds (Figure 12.6(b)). With continued deposition, the crack propagated upwards as deposition continued (Figure 12.6(c)), and potentially leading to new crack formations (Figure 12.6(d)). In summary, cracks form during the deposition process and are closely associated with the height of deposited metal. Lower deposition heights result in lower stress levels that hinder crack formation while higher deposition heights lead to accumulating stress and significantly increased cracking risks. Conversely, crack formation and expansion partially alleviate stress accumulation.

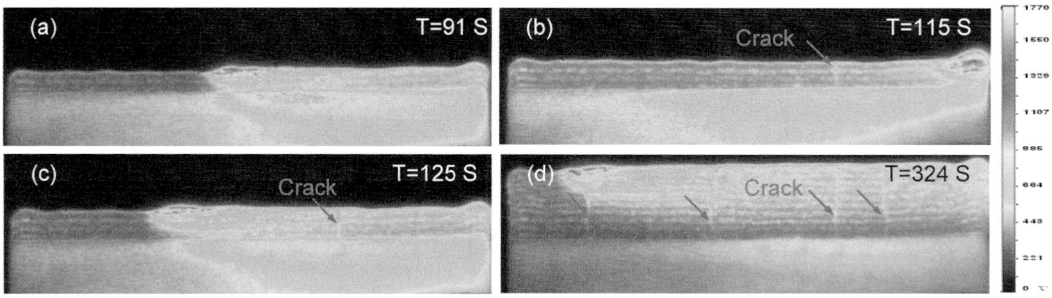

Figure 12.6. The initial and propagation of macro-crack observed by infrared thermal camera, (a), (b), (c) and (d) are the thermal images at different time.

12.2.3 Effect of laser power on macroscopic stress-strain

Similar to welding, the stress and deformation in DED can be categorized into two types: transient stress and strain occurring during the DED process, and residual stress and strain retained in the workpiece after completion of DED. Currently, accurate measurement of residual stress is achievable through drilling method, X-ray diffraction, and neutron diffraction techniques, which have been extensively studied. However, direct detection of transient stress and strain during the DED process is challenging with limited research available. Existing studies primarily rely on finite element simulation technology for analysis. As depicted in Figure 12.6, crack formation and propagation occur during the additive process due to dynamic changes in temperature and deposited layer size. These dynamic effects influence the transient stress-strain state. Therefore, crack generation is inevitably associated with dynamic stress-strain behavior throughout the additive process. Since DIC technology enables real-time tracking of deformation during additive manufacturing processes, it can be combined with relevant knowledge from solid mechanics and elastic-plastic theory to facilitate a more profound understanding of cracking issues.

Figure 12.7 shows the displacement in the X-axis (scanning direction) and the corresponding temperature distribution when the first four layers are deposited with 320 W laser power. Figure 12.7(a) shows the displacement of point A (the top center point of the DIC observation area) along the X-axis direction; Figure 12.7(b) shows the temperature change curve of point A (measured by the infrared thermal imager on the back); Figure 12.7(c) and (d) are the spatial distribution of displacement and temperature corresponding to T1 in (a). Similarly, Figure 12.7(e) and (f) respectively show the corresponding displacement and temperature distribution at time T2 in (a). Among them, the temperature distribution maps (d) and (f) are obtained by measuring the temperature in the symmetrical area of the back of the DEDed thin wall. In order to better correspond to the displacement distribution map, the temperature distribution map is rotated 180° along the z axis.

As can be seen from Figure 12.7(a), before time T1, point A deforms in the opposite direction of the X-axis, and the displacement gradually increases, reaching the extreme value at time T1. It can be seen from this that when the first layer is deposited, the laser heat source moves in the opposite direction of the X-axis to point A, and the basement behind the molten pool is rapidly heated, resulting in thermal expansion effect, which acts as A pushing and extrusion effect on point A at the front of the molten pool, making point A move in the negative direction of the X-axis, resulting in negative displacement. This negative displacement continues to accumulate and reaches a maximum value at T1. Figure 12.7(c) and (d) show the displacement field and temperature field of T1. It can be found that the molten pool is not just above point A where the displacement reaching its a maximum, but located in front of point A. It can be inferred that when the molten pool reaches the top of point A, point A is still subject to the expansion thrust behind the molten pool. After the molten pool passes over point A, the temperature of point A reaches its peak, and the displacement

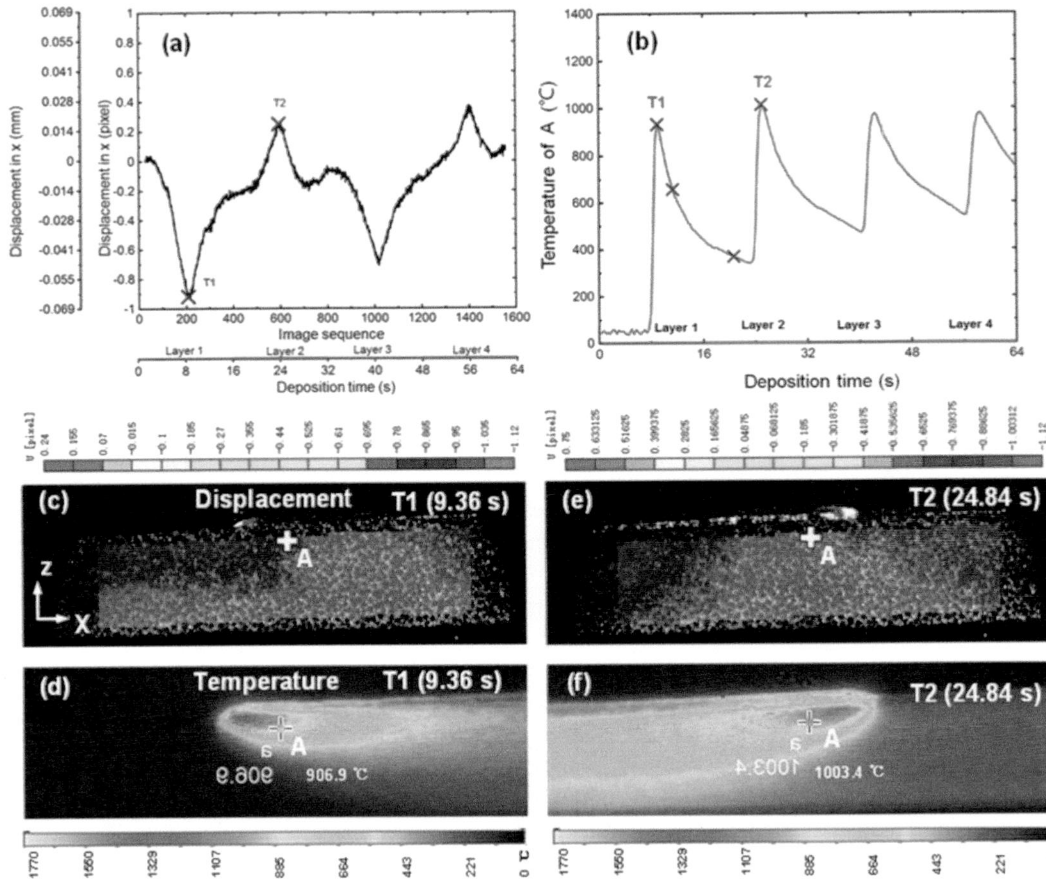

Figure 12.7. Displacement along X-axis and the related temperatures of the substrate deposited with 320 W power, (a) the displacement curve of the top-center point A in the DIC measured area, (c) and (d) are the displacement and temperature distribution at T1 time of (a), (e) and (f) are the displacement and temperature distribution at T2 time in respective.

of point A also reaches its peak. As can be seen from the temperature change curve shown in Figure 12.7(b), the corresponding peak temperature at time T1 is about 907°C. As the high-temperature heating zone of the molten pool passes over point A, the region in front of A generates positive expansion thrust point A, which gradually reduces the negative displacement of point A.

When employing reciprocating bidirectional scanning strategy, the second layer's scanning direction is opposite to that of the first layer, resulting in a rightward movement of the melt pool as shown in Figure 12.7(e). The thermal expansion of the high temperature region starts to push forward point A. This further diminishes and gradually transforms the initial negative displacement of point A into a positive displacement along the x-axis. The positive displacement reaches its peak value (1003°C) at T2. It can be observed from Figure 12.7(b) that at this moment, point A attains its temperature maximum corresponding to the second layer. By examining both the displacement field (Figure 12.7(e)) and temperature field (Figure 12.7(f)) at time T2, it becomes apparent that by then, a certain distance has been covered by the melt pool away from point A as well. Upon comparison, it is evident that while at T1, there exists an absolute value of negative displacement approximately equal to 0.063 mm (0.928 pixels), which surpasses that of T2's absolute value of displacement (0.015 mm or 0.227 pixels). The reason is that during deposition of the first layer, room temperature serves as its initial temperature, however, during deposition of second layer, the initial temperature is about 340°C as shown in Figure 12.6(b). According to thermal expansion laws depicted by Equation 12.1,

$$U^T = e^T L = \alpha \Delta T L = \alpha (T - T_0) L \tag{12.1}$$

Where U^T is the displacement caused by thermal expansion, e^T is the thermal strain L is the initial length of the heated body, α is the coefficient of thermal expansion, T is the heating temperature, and T_0 is the initial temperature. The ΔT corresponding to the first layer is about 877°C which is greater than that of the second layer (about 663°C), so the corresponding thermal strain and the related displacement are larger for the second layer.

Similarly, Figure 12.8 shows the displacement and temperature distribution in BD at 320 W laser power. Where, Figure 12.8(a) is the displacement curve of point A in the Z-axis direction, Figure 12.8(b) is the temperature curve, Figure 12.8(c) and (d) are the displacement distribution and temperature distribution corresponding to T3 moment in Figure 12.8(a), respectively. Figure 12.8(e) and (f) are the displacement distribution and temperature distribution corresponding to T4 moment. As can be seen from Figure 12.8(a), point A displaces in the positive direction of z axis during the first four layers of deposition, and there is a peak and a valley for each layer. This indicates that in this process, point A is always in a state of outward expansion due to heating. In addition, it can also be seen that the peak value of displacement shows a decreasing trend layer by

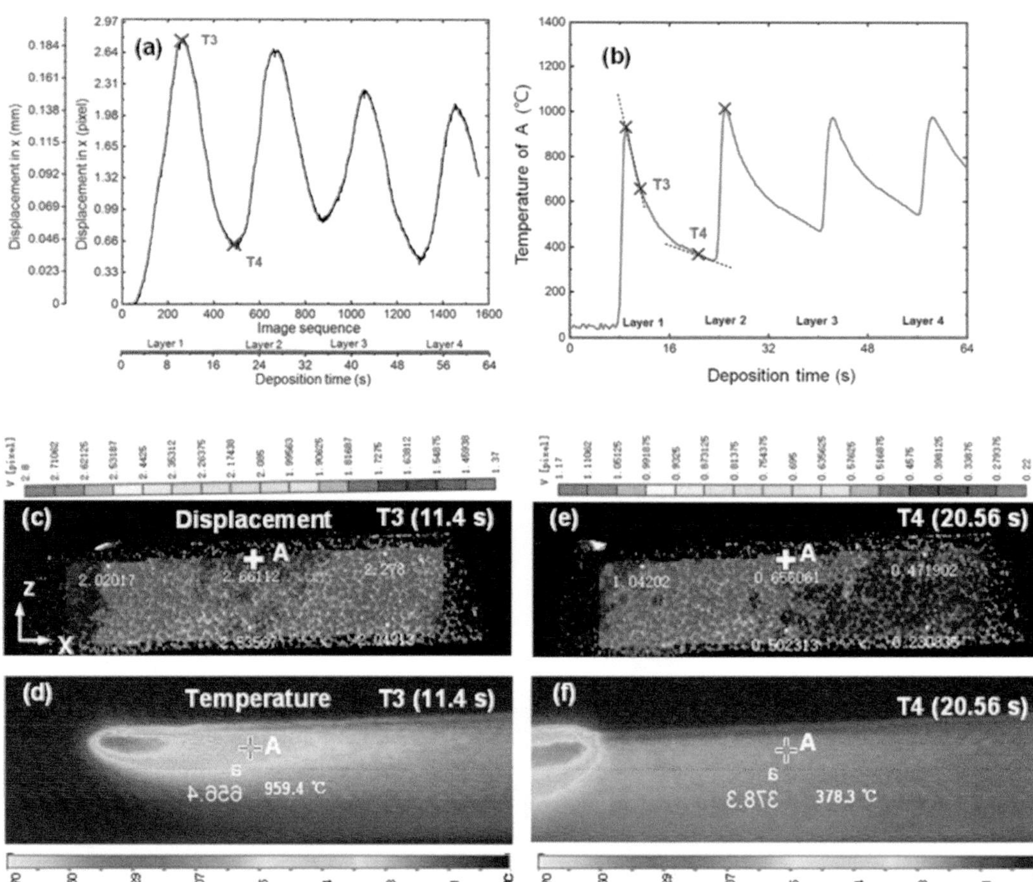

Figure 12.8. Displacement along Z-axis and the related temperatures of the substrate deposited with 320 W power, (a) the displacement curve of the top-center point A in the DIC measured area, (c) and (d) are the displacement and temperature distribution at T3 time of (a), (e) and (f) are the displacement and temperature distribution at T4 time in respective.

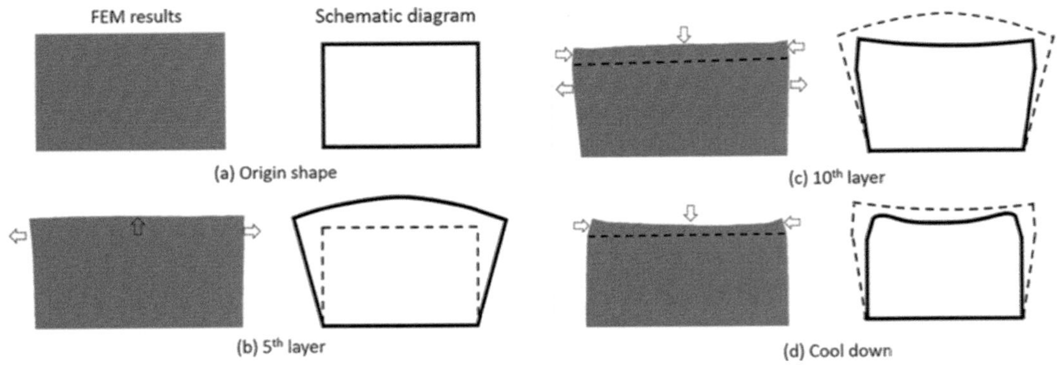

Figure 12.9. Deformation diagram during the whole process of DED [151].

layer. From the temperature change curve shown in Figure 12.8(b), it can be seen that for the peak of the Z-displacement (T3), the corresponding temperature is 656°C and for the valley value (T4), the corresponding temperature is 378°C. Before the T3 moment, the temperature drops sharply, but after the T3 moment, the temperature drops slowly. The temperature drops most slowly after T4 moment.

Based on the final obtained DED samples, it can be observed that the central part of the substrate exhibits a concave shape. Consequently, it can be inferred that during the DED, point A's displacement along the Z-axis changes from positive to negative, which aligns with Biegler et al.'s findings regarding Z-direction displacement [4]. Furthermore, Xie Ruishan et al. [5] also employed DIC technology to investigate deformation in DED processes. The corresponding finite element simulation results depicted in Figure 12.9 demonstrate consistent deformation patterns with our study.

Based on the displacement field obtained by DIC, the corresponding strain field can be calculated. In the two-dimensional case, the strain obtained by DIC is usually Largangian strain, and its definition is shown in formulas 12.2 to 12.4.

$$e_{xx} = \frac{\partial u}{\partial x} + \frac{\left(\frac{\partial u}{\partial x}\right)^2 + \left(\frac{\partial v}{\partial x}\right)^2}{2} \tag{12.2}$$

$$e_{yy} = \frac{\partial v}{\partial y} + \frac{\left(\frac{\partial u}{\partial y}\right)^2 + \left(\frac{\partial v}{\partial y}\right)^2}{2} \tag{12.3}$$

$$e_{xy} = \frac{\frac{\partial u}{\partial y} + \frac{\partial v}{\partial x} + \frac{\partial u}{\partial x}\frac{\partial u}{\partial y} + \frac{\partial v \partial v}{\partial x \partial y}}{2} \tag{12.4}$$

It can be seen from its definition that Largangian strain is a finite strain, including second-order displacement, which essentially reflects the gradient of displacement. Generally, the strain collected in the tensile test is mostly Cauchy strain, that is, engineering strain, which only includes the first-order displacement. Therefore, for small deformation, Largangian strain is close to Cauchy strain, while for large deformation, Largangian strain is often larger than Cauchy strain. Based on the obtained Largangian strain, the principal strain can be calculated according to formulas 12.5:

$$e_{1,2} = \frac{e_{xx} + e_{yy}}{2} \pm \sqrt{\left(\frac{e_{xx} - e_{yy}}{2}\right)^2 + \left(\frac{e_{xy}}{2}\right)^2} \tag{12.5}$$

Where $e_1 > e_2$, and the direction of the principal strain is

$$\tan 2\theta = -\frac{e_{xy}}{e_{xx} - e_{yy}} \tag{12.6}$$

We know that in a complex stress state, the strain distribution has a strong directivity, so the equivalent stress and equivalent strain can be used to describe the deformation resistance. The equivalent stress $\bar{\sigma}$ and equivalent strain $\bar{e}$ can be obtained according to the formulas 12.7 and 12.8, and the $\bar{\sigma}$ and $\bar{e}$ can reveal the train stress and strain.

$$\bar{\sigma} = \frac{1}{\sqrt{2}}\sqrt{(\sigma_1 - \sigma_2)^2 + (\sigma_2 - \sigma_3)^2 + (\sigma_3 - \sigma_1)^2} \tag{12.7}$$

$$\bar{e} = \frac{\sqrt{2}}{3}\sqrt{(e_1 - e_2)^2 + (e_2 - e_3)^2 + (e_3 - e_1)^2} \tag{12.8}$$

Further, according to the obtained DIC strain during DED of IN 738LC, the elastic-plastic strain of point A which is located at the center of the DEDed thin wall can be obtained. Figure 12.10 shows the elastic-plastic strain under different DED power. It can be seen that with the increase of laser power, the maximum of elastoplastic strain increases from 0.6% to 0.68% and 0.84% respectively. It shows that the peak value of stress increases with the increase of power during DED. The increase of stress peak value means the increase of cracking risk. Combined with

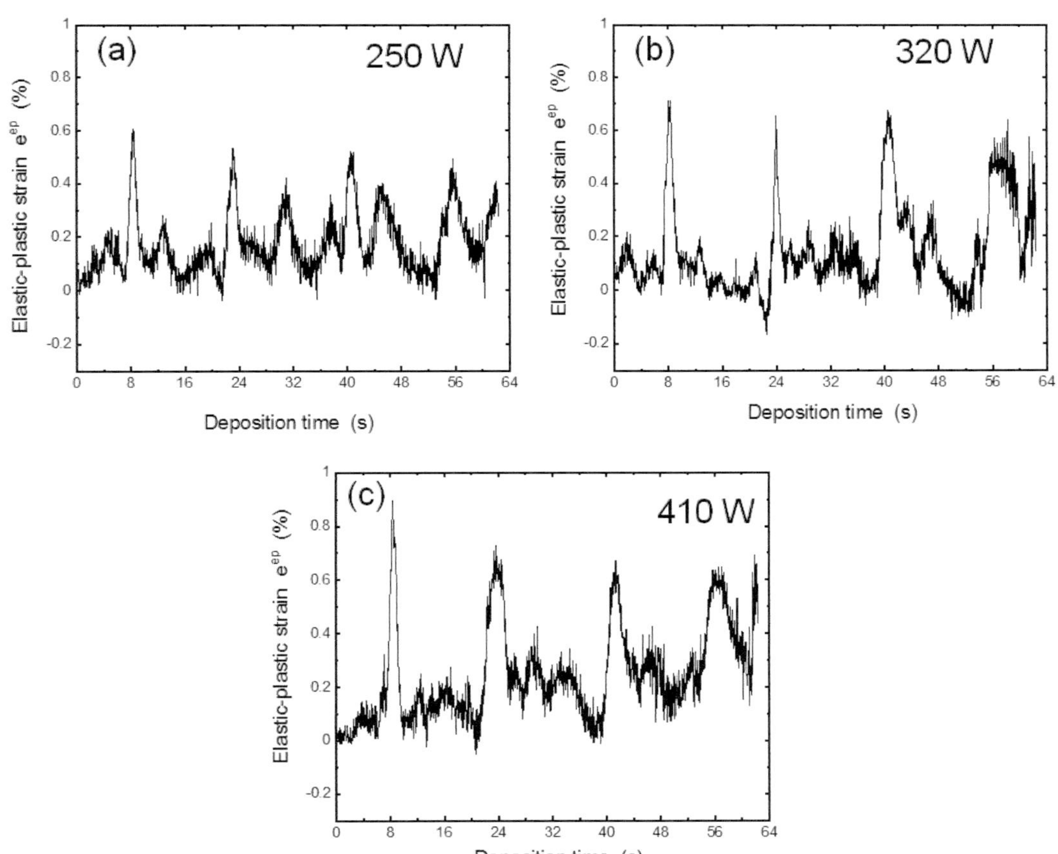

Figure 12.10. The elastic-plastic strain of point A under different DED power, (a) 250 W, (b) 320 W, (c) 410 W.

the actual cracking condition shown in Figure 12.1, we found when the laser power increased to 320 W, the maximum of elastoplastic strain increases to 0.68%, and macroscopic cracks can be found. These results demonstrated that with the increase of DED power, the instantaneous stress maximum during the DED process will increase, thus increasing the risk of cracking. It is worth mentioning that the generation of cracks is not only related to the peak of stress, but also to the loading rate of stress, microstructure and many other factors. Therefore, it is not scientific to quantitatively predict cracking solely based on the stress peak.

12.2.4 Effect of laser power on grain characteristics

The cracks observed by the infrared thermal imager or macroscopic photographs are macroscopic cracks. These macroscopic cracks result from the expanded microscopic cracks. At the grain scale, the formation of microscopic cracks is related to the component phases, such as the $\gamma+\gamma'$ eutectic, MC, borides and zirconides. In addition, it is also related to the grain orientation relationship, grain boundary morphology and micro-strain characteristics. These characteristics not only affect liquation cracking [1, 6], but also solid-state cracking [7–9].

Figure 12.11 presents the EBSD results of samples DEDed under varying laser power. Specifically, Figure 12.11(a), (b), and (c) illustrate the Band Contrast (BC) distribution maps, while Figure 12.11(d), (e), and (f) depict the inverse pole figure (IPF) oriented perpendicular to the deposition height direction (BD). Additionally, Figure 12.11(g), (h), and (i) display the pole figure

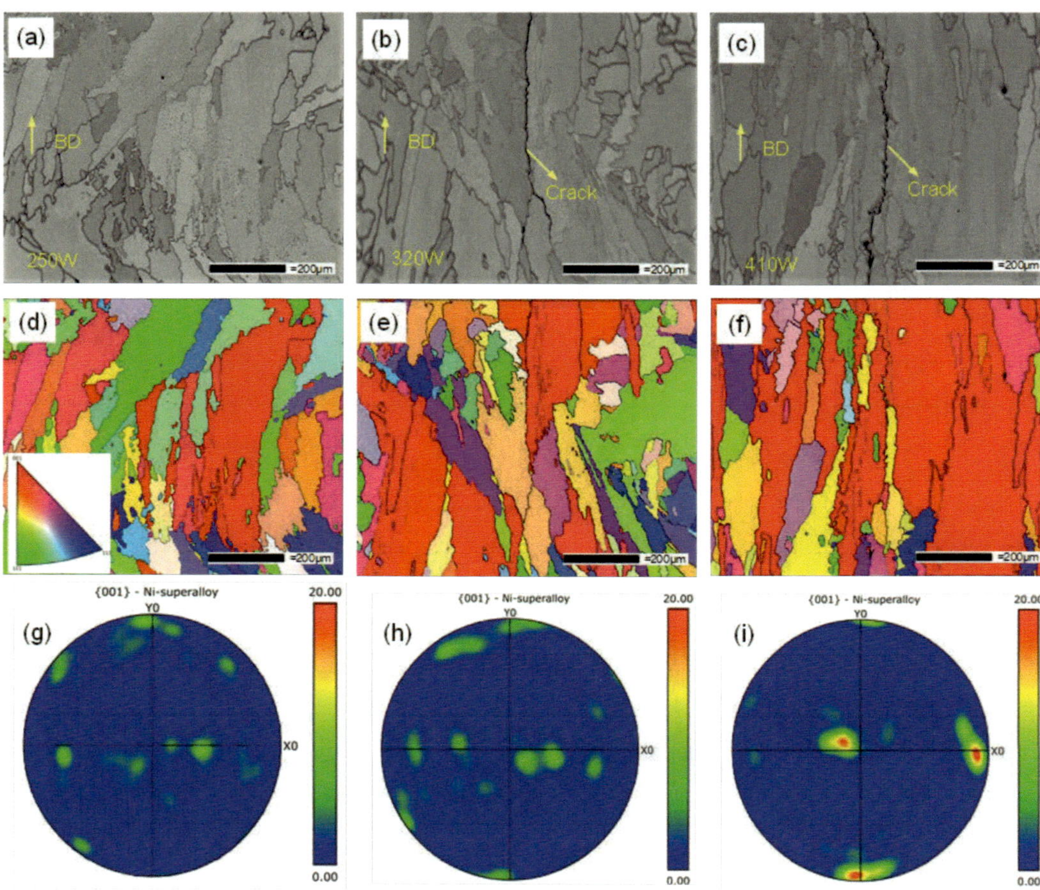

Figure 12.11. EBSD orientation distribution under different laser power, (a), (b) and (c) are the BC Figures, (d), (e) and (f) are the IPF Figures reference to BD, (g), (h) and (i) are the pole Figures of {001}.

for {001}. As indicated in Figure 12.11(a)~(c), an increase in laser power correlates with a more pronounced microcracks, which occurred along the grain boundaries. Figure 12.11(d)~(f) reveals that at lower laser power levels, the orientation within the deposited layer is relatively disordered, specifically, two distinct orientation relationships, namely <101>//BD and <001>//BD, are present within the microstructure, but without significant texture development. Conversely, as laser power increases, a marked enhancement in <001> texture becomes evident. The positive pole figure presented in Figure 12.11(g)~(i) further corroborate this observation.

For the face-centered cubic crystal, the preferred growth orientation is <001>, so the dendrite growth direction is usually in a certain <001> crystal direction that is close to the direction of heat flow. When low heat input is used for bidirectional scanning, the dendrite growth direction (i.e., <001> crystal direction) is approximately at a 45° angle with the BD direction, so <101>//BD texture will appear. As the heat input increases or unidirectional scanning is used, the dendrite growth direction tends to be parallel to the BD direction, so <001>//BD texture will appear. In addition, when the laser power is too small, it cannot completely melt the top of the previous layer's misoriented region, and most grains cannot maintain epitaxial growth based on the previous layer's columnar grain, but rather re-nucleate and grow under certain undercooling conditions, so the overall orientation is not obvious (Figure 12.11(d)). On the other hand, when the laser power is larger, it can fully melt the top of the previous layer's misoriented region, and the grains can maintain epitaxial growth very well and cross the fusion lines between layers, thus maintaining good orientation (Figure 12.11(f)).

Lakshmi L. et al. [10] and Pinkerton et al. [11] also observed a similar phenomenon in their studies on the orientation of DED structures in nickel-based alloys. However, Yuan et al. [1] found that with an increase in laser power and heat input during the additive manufacturing process of Inconel 718, the crystal orientation deteriorated. Gaumann et al. [12] suggested that higher laser power would lead to an increased value of $G^{4.3}/R$ (where G represents temperature gradient and R represents solidification rate), thereby promoting the growth of equiaxial crystals and potentially weakening crystal orientation. Similarly, Pinkerton et al. [11] believed that while the transition between columnar crystals and equiaxial crystals depends on $G^{4.3}/R$, increasing laser power can intensify Marangoni flow in the molten pool, causing fluctuations in temperature gradient (G). It is worth noting that Pinkerton et al.'s study used a range of laser powers from 350 to 560 W, corresponding to line energies ranging from 70 to 112 J/mm; whereas Lakshmi L's study used a line energy of 45~150 J/mm which is comparable to this paper's linear energies (50~82 J/mm). However, Yuan et al.'s study employed much higher line energies ranging from 300 to 600 J/mm compared to this paper as well as the aforementioned studies [11, 12]. Thus, it can be inferred that the influence of heat input on crystal orientation may vary across different process ranges. Although our findings regarding the relationship between power and texture strength do not align with those reported by Yuan et al. [1], both studies indicate an increase in cracking phenomena with higher laser power levels.

Further statistical analysis of grain, grain boundary orientation differences, and grains circumferences were conducted using EBSD. The results are presented in Figure 12.12. Figure 12.12(a), (b), and (c) depict the disorientation angle distribution. In this Figure, the M-Index represents the texture strength, the theoretical value indicates the random inter-grain orientation difference which determined by crystal symmetry itself. Neighbor pairs represent the actual orientation difference between adjacent data points on both sides of the grain boundary. A Random Pair represents the actual orientation difference between two random data points. As observed from Figure 12.12, at laser powers of 250 W, 320 W, and 410 W, corresponding M-Indexes were found to be 0.141, 0.169, and 0.42 respectively, achieving coincidence degrees of random pair with theoretical as high as 93%, 93%, and 62% respectively. This quantitatively character suggests that at low power levels, the distribution of orientation differences between grain boundaries is random and primarily depending on crystal symmetry itself. However, increase in power significantly amplifies deviation between adjacent grains' orientations compared to their theoretical orientations.

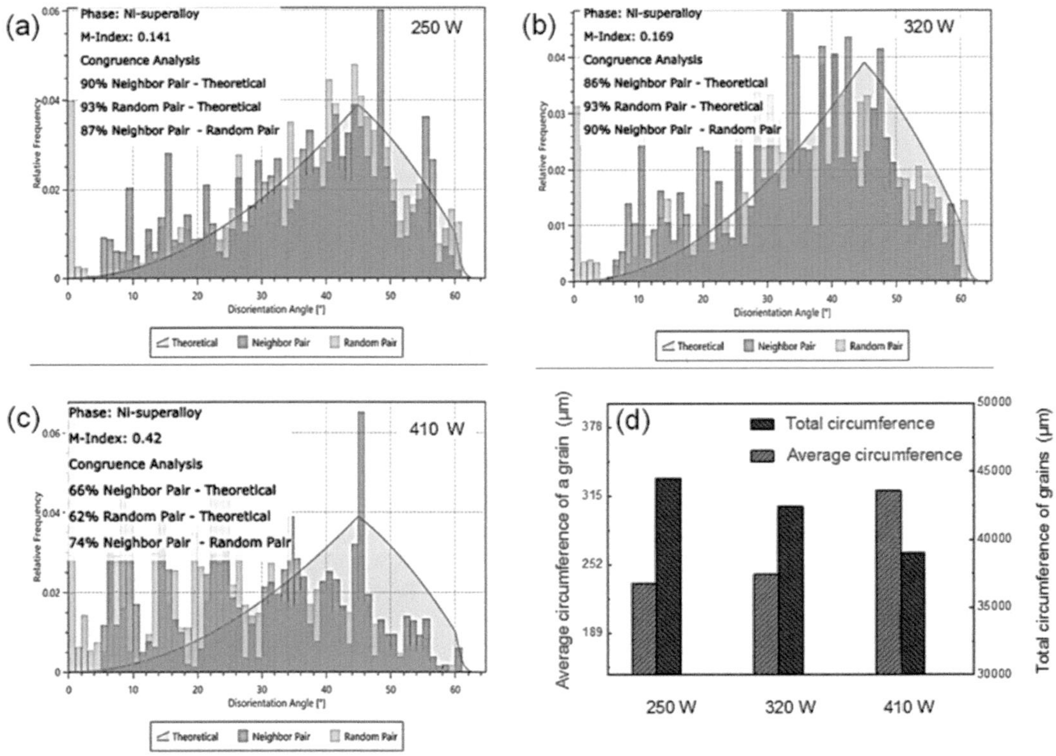

Figure 12.12. EBSD statistical of grain characteristics, (a), (b) and (c) are the disorientation distributions with 250 W, 320 W and 410 W, (d) is the grain circumference statistics.

Additionally, the average grain circumference can serve as a measure for determining grain size while the total length of grain boundary within a certain area can reflect curvature degree. As depicted in Figure 12.12(d), with the increase of laser power, the average grain circumference increases, that is, the average grain size increases. While the total grain boundary length decreases, that is, the degree of zigzag of the grain boundary weakens. Combined with Figure 12.12(a)~(f), it is obvious that cracks tend to occur in some long straight grain boundaries. From the perspective of liquation crack, the eutectics with low melting are more likely to segregate at long straight grain boundaries, and the possibility of forming a continuous liquid film is greater, so it is more conducive to the generation of liquation crack. In terms of the sensitivity of solid-state cracks, for example DDC, the straight grain boundaries are more conducive to the sliding of grain boundaries, thus more readily to the solid-state cracking. From the perspective of stress-strain distribution, the decrease of grain size and the increase of grain boundary length can evenly disperse the stress and strain, so as to prevent the local concentration of stress and strain, which is conducive to the suppression of all kinds of cracks including liquation cracks and solid-state cracks.

Figure 12.13 shows the deformation characteristics of grain under different power. Figure 12.13(a), (b) and (c) show the Kernel Average Misorientation (KAM) of the samples DEDed under 250 W, 320 W and 410 W respectively. Figure 12.13(d), (e) and (f) are the corresponding Grain Orientation Spread (GOS) maps, and Figure 12.13(g), (h) and (i) are the Schmid Factor distributions maps. The distribution of KAM indicates that significant plastic deformation occurred on both sides of the crack, which is consistent with the grain boundary slip mechanism of DDC. In particular, as laser power increases, there is an increase in microscopic stress on the grain, and under this stress, grains undergo rotation and sliding. If the relative sliding between grains exceeds the tolerated limit of grain boundaries, intergranular cracking occurs, leaving behind plastic deformation traces on both sides of the crack and at its front zone.

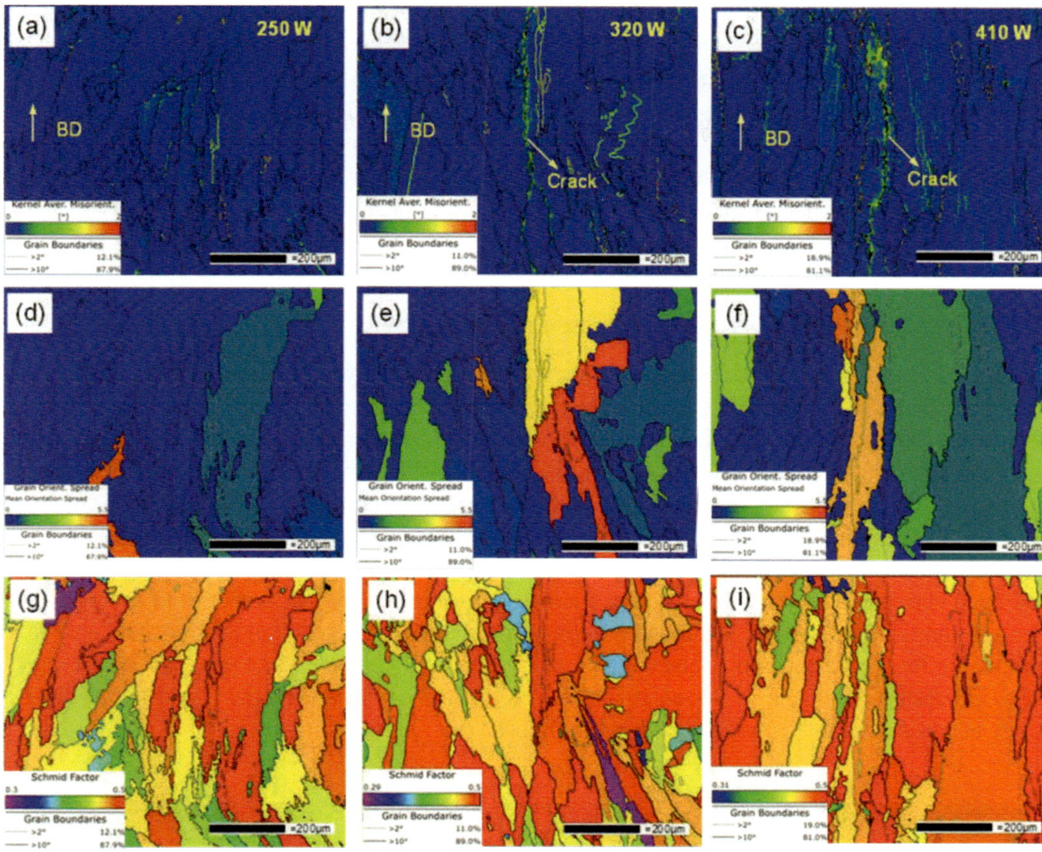

Figure 12.13. Micro deformation under different power, (a), (b) and (c) are the KAM maps of 250 W, 320 W and 410 W respectively, (d), (e) and (f) are the corresponding GOS maps, (g), (h) and (I) are the corresponding Schmid factor maps.

The GOS maps shown in Figure 12.13(d), (e), and (f) describe the maximum or average orientation difference within the entire grain. It can be seen that with the increase of laser power, the number of grains with large deformation gradually increases. The average orientation difference within the grains is in the range of 0~12.5°. Besides, GOS maps demonstrate that cracks generally occur between the severely deformed grains and the weakly deformed grains. The relative sliding of grains caused by uneven deformation between grains is the main inducement factor leading to cracking. We know that crystal slip is mainly carried out under the action of shearing stress. According to Schmid's law (Equation (12.9)), in a single crystal, when the resolving shear stress is greater than the critical resolved shear stress, the slip starts.

$$\tau \geq \tau_c = \sigma_s \cos \emptyset \cos \lambda \qquad (12.9)$$

Where, τ is the resolving shear stress, τ_c is the critical resolved shear stress at which the slip system starts, ∅ is the angle between the slip plane and the external force axis, λ is the angle between the slip direction and the external force axis, and σ_s is the initial yield strength. cos ∅ cos λ is named as Schmid factor, it varies from 0 to 0.12. When the Schmid factor is large, the slip is easy to initiate and is called soft orientation. On the contrary, when the Schmid factor is small, the slip is not easy to start, and it is called hard orientation. According to the Schmid factor distribution diagram shown in Figure 12.13(g), (h) and (i), the number of grains with soft orientation increases significantly with the increase of laser power, so the sliding of grains is more likely to occur. In addition, it can be found that the Schmid factor on both sides of the crack is quite different, that is, the crack occurs mostly between soft-oriented grains and hard-oriented grains, which is also reported in fatigue cracking and tensile cracking [158, 159]. The reason is that the hard-oriented grains are difficult to deform, while the adjacent soft-oriented grains are easy to slip, so the dislocation movement at its junction is difficult to transmit, accumulating a large number of dislocations, resulting in large local plastic deformation. In addition, the junction of dislocation accumulation also often acts as a fast channel for impurity atom segregation and oxidation, which will cause the junction itself to decline in plasticity.

12.3 Effects of scanning speed on crack sensitivity

12.3.1 Cracking conditions at different scanning speeds

Experimental results demonstrate that scanning speed significantly influences cracking during the DED process as well. Figure 12.14 illustrates the macroscopic cracking conditions at scanning speeds of 4 mm/s, 5 mm/s, and 6 mm/s respectively (with other process parameters held constant: laser power of 320 W and powder feed speed of 7.6 g/min). As depicted in the Figure, no apparent macroscopic cracks are observed when the scanning speed is 4 mm/s. However, noticeable cracking occurs at both 5 mm/s and 6 mm/s scanning speed. It should be noted that under a constant laser power condition, changes in scanning speed alter line energy magnitude accordingly. It can be seen from the above that the laser power increases, the line energy increases, and the cracking phenomenon is more serious. The crack patterns shown in Figure 12.14 can thus be attributed to a combined effect resulting from increased scanning speed and decreased line energy. On one hand, solely increasing the scanning speed enhances crack sensitivity, on the other hand, reduced line energy exerts some inhibitory effects on cracking behavior.

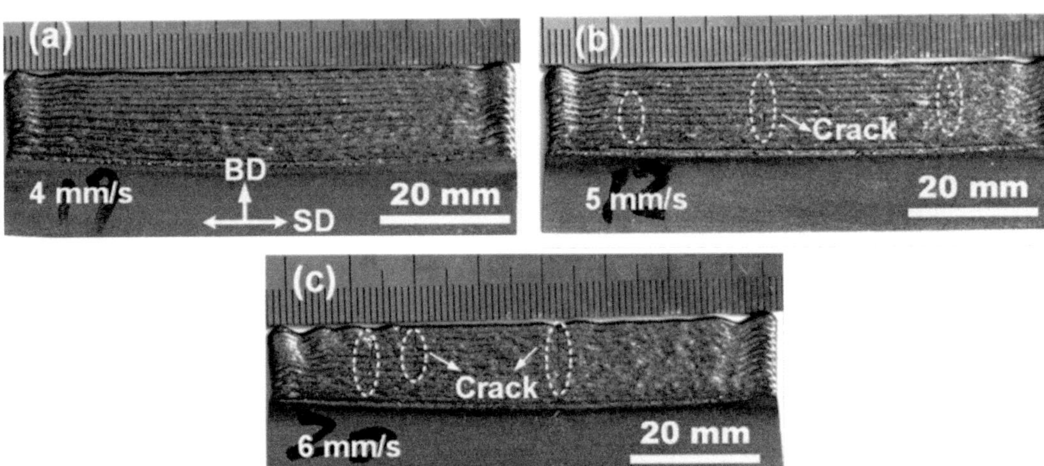

Figure 12.14. Macro-crack morphology in the deposited parts with different scanning speed, (a) 4 mm/s, (b) 5 mm/s, (c) 6 mm/s.

12.3.2 Effect of scanning speed on temperature gradient

Infrared thermal imager was used to measure the temperature field in the deposition process with different scanning speeds. Figure 12.15 shows the temperature gradient at the front of the molten pool along the SD and BD.

It can be seen that for the first layer, when the scanning speed increases from 4 mm/s to 6 mm/s, the temperature gradient in the SD increases from ~400°C/mm to ~500°C/mm, and the temperature

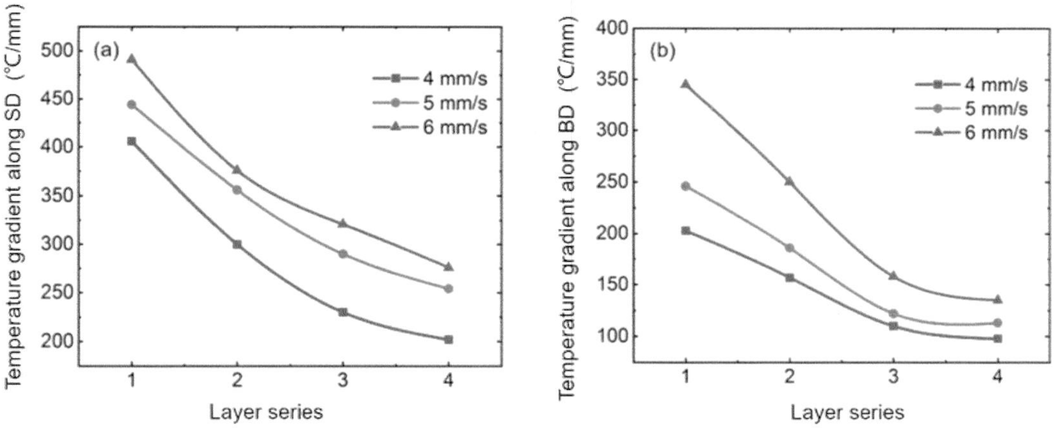

Figure 12.15. Effects of laser scanning speed on the spatial temperature gradient, (a) the SD directional gradient in front of molten pool, (b) the BD directional gradient below the molten pool.

gradient in the BD also increases from ~200°C/mm to ~350°C/mm. In addition, the general trend of the temperature gradient decreases with the increase of the number of deposited layers.

12.3.3 Effect of scanning speed on macroscopic stress-strain

To explore the effect of scanning speed on macroscopic stress and strain, DIC technology was also used to measure deformation conditions at different scanning speeds. In this section, the same method as in Section 12.2.2 is used to calculate the elastic-plastic strain at the top center point A of the DIC observation area, the results are shown in Figure 12.16.

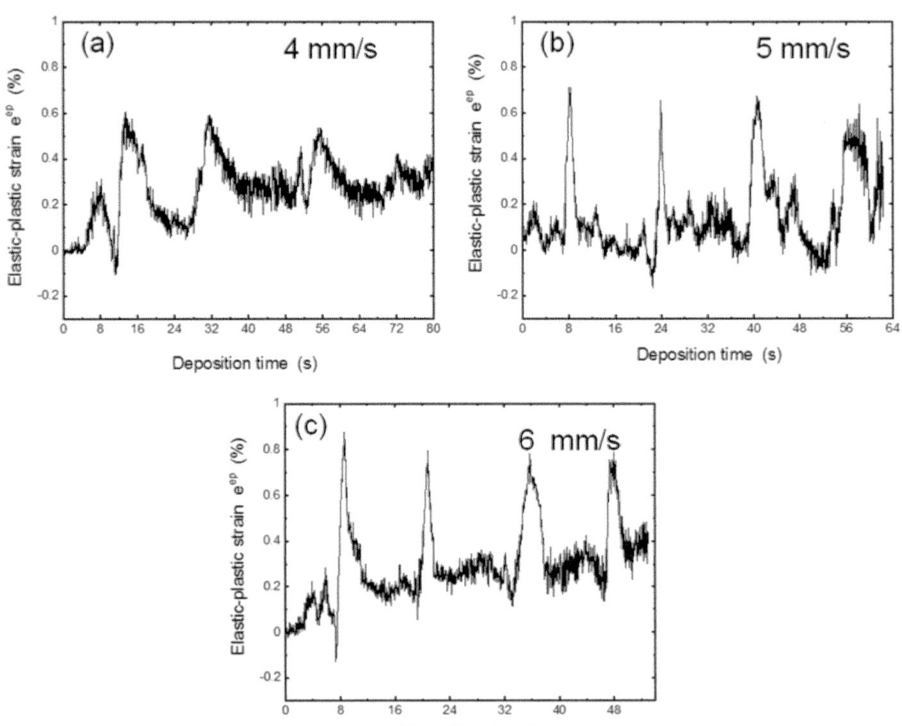

Figure 12.16. Effects of laser scanning speed on the elastic-plastic strain of point A, (a) 4 mm/s W, (b) 5 mm/s, (c) 6 mm/s.

It can be seen from Figure 12.15 that when the scanning velocity is 4 mm/s, 5 mm/s and 6 mm/s, the corresponding peak elastic-plastic strain is ~ 0.6%, ~0.68% and ~0.84%, respectively. This indicates that the peak value of elastoplastic strain tends to increase with the increase of scanning speed, that is, the peak value of dynamic stress tends to increase. It can be inferred that the increase of the peak stress will increase the risk of cracking. Therefore, reducing the laser scanning speed is an effective means to reduce the stress peak, but also an effective means to inhibit cracking. But from the perspective of DED forming efficiency, reducing the scanning speed will undoubtedly reduce the forming efficiency, and reduce the utilization rate of metal powder, and thus increase the production cost.

12.3.4 Effect of scanning speed on grain characteristics

Microscopic deformation characteristics, as depicted in Figure 12.17, can be obtained by EBSD analysis. Figure 12.17(a), (b), and (c) represent the distribution map of Kikuchi Band Contrast (BC) at different scanning speeds, while Figure 12.17(d)–(e), and Figure 12.17(g)–(f) depict the corresponding distribution map of Kernel Average Misorientation (KAM) and Grain Orientation Spread (GOS). The BC distribution map reveals that an increasing trend in crack width is observed with higher scanning speeds, indicating an increase in crack severity. It is noteworthy that, although no visible macroscopic cracks were detected at a scanning speed of 4 mm/s (as shown in Figure 12.17(a)), micro-cracks were still observed under SEM examination. According to the KAM

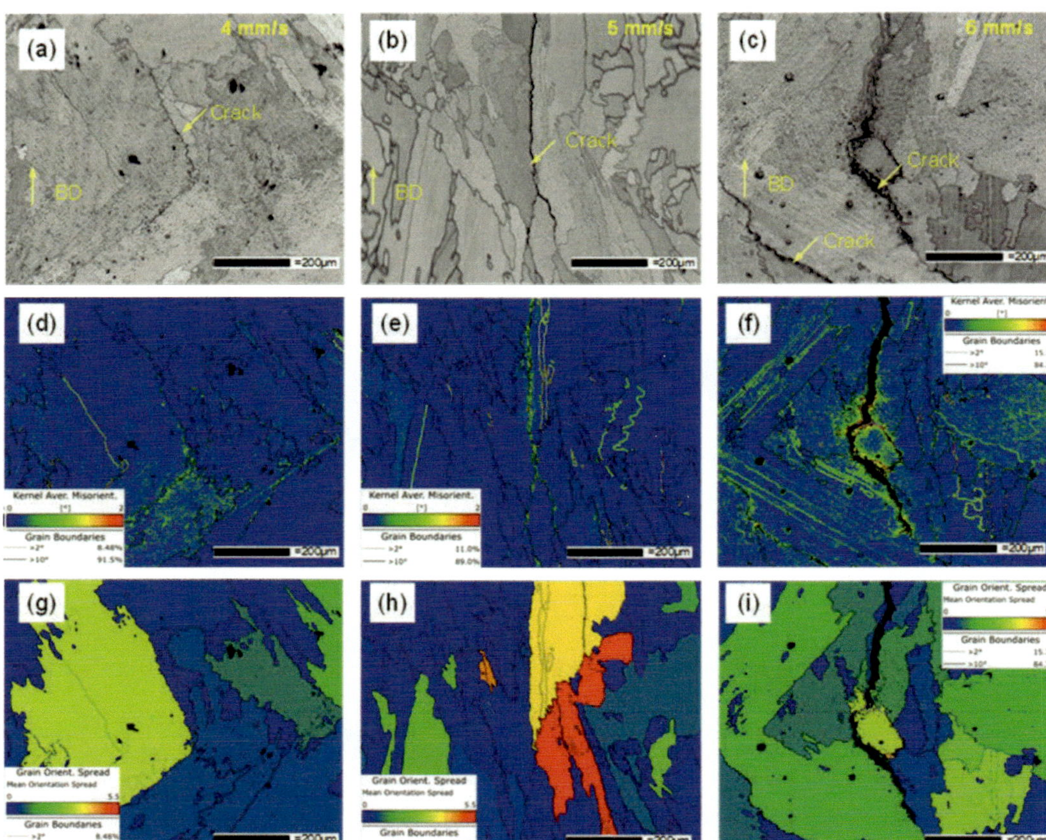

Figure 12.17. Cracking and Micro deformation under different scanning speed, (a), (b) and (c) are the BC maps of 4 mm/s, 5 mm/s and 6 mm/s respectively, (d), (e) and (f) are the corresponding KAM maps, (g), (h) and (I) are the corresponding GOS maps.

distribution diagram, when the scanning speed is set at 4 mm/s, although larger KAM values are found at the tip of microcracks, there is no significant increase in KAM on both sides of the crack. This phenomenon may be attributed to some microcracks being liquefaction-induced, liquefaction cracks formation involves a liquid film which exhibits poor stress and strain transfer ability, thus requiring less severe strain for cracking near the solid phase adjacent to this liquid film region. Once cracks formed, these microcracks generate stress concentration zones around their tips which create favorable conditions for subsequent solid-state cracking. When increasing the scanning speed to 6 mm/s, more pronounced increases in KAM are observed near wider cracks indicating more significant local plastic deformation occurring within those regions. According to the GOS distribution map as shown in Figure 12.17(g)–(f), an increase in scanning speed leads to a corresponding escalation of microscopic deformation in the grains. The cracks still manifest between grains exhibiting significant deformation disparities, and along certain elongated grain boundaries.

12.4 Effects of scanning path on crack sensitivity

12.4.1 Cracking conditions under different scanning paths

The scanning path has a significant impact on cracking behavior. For thin-walled structures with multiple layers, the scanning path is primarily categorized into bidirectional and unidirectional modes. Figure 12.18 illustrates the macroscopic cracking observed under two different scanning methods (with other process parameters held constant: laser power of 320 W, scanning speed of 5 mm/s, powder feed speed of 7.6 g/min). It can be observed that when employing unidirectional scanning, both the number and total length of cracks are significantly increased, with crack propagation occurring at an inclined angle along SD.

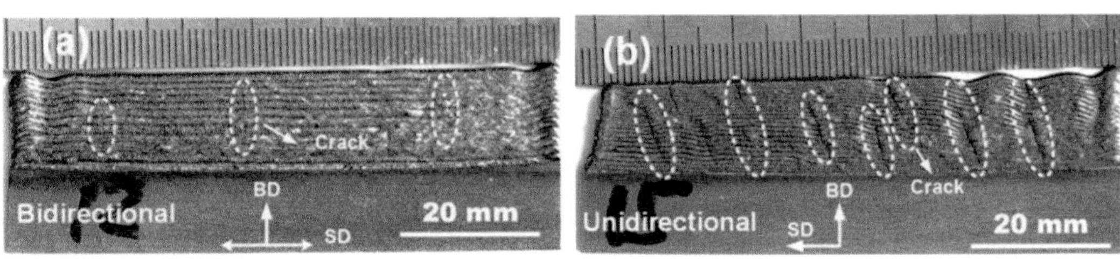

Figure 12.18. Effects of scanning strategy on cracking, (a) bidirectional, (b) unidirectional.

12.4.2 Effect of scanning path on temperature

The thermal cycle curve of the center point A during the DED in unidirectional scanning and bidirectional scanning is illustrated in Figure 12.19. It can be observed that after the initial thermal cycle, the valley temperature at point A for bidirectional scanning reaches approximately 400°C, whereas for unidirectional scanning, it is even lower at around 200°C. This discrepancy arises due to continuous scanning without pause during bidirectional scanning. Conversely, in unidirectional scanning, after each layer scan concludes, the laser is off in the return journey, resulting in a half-empty travel path. The metal deposited along this empty path undergoes better cooling. As thermal cycles progress, there is a rapid reduction in temperature fluctuation range for the sample under bidirectional scanning, however, significant temperature fluctuations still persist for the sample under unidirectional scanning. According to principles governing thermal expansion behavior, these larger temperature differences experienced during unidirectional scanning lead to more pronounced phenomena of thermal expansion and contraction which can induce greater levels of thermal stress and increase susceptibility to cracking.

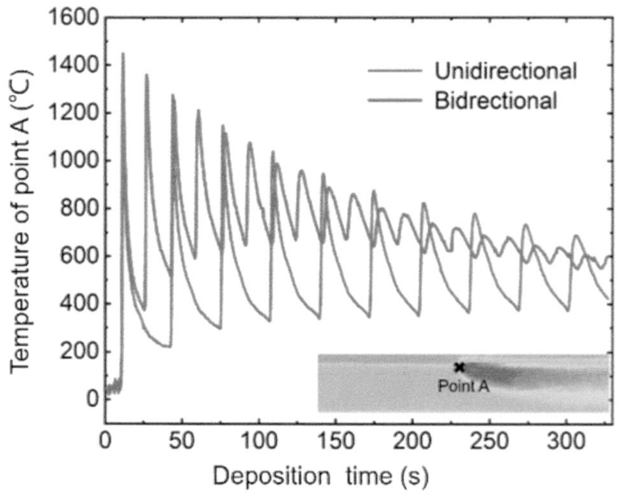

Figure 12.19. The thermal cycles of the top-center point A of the substrate with different scanning strategies.

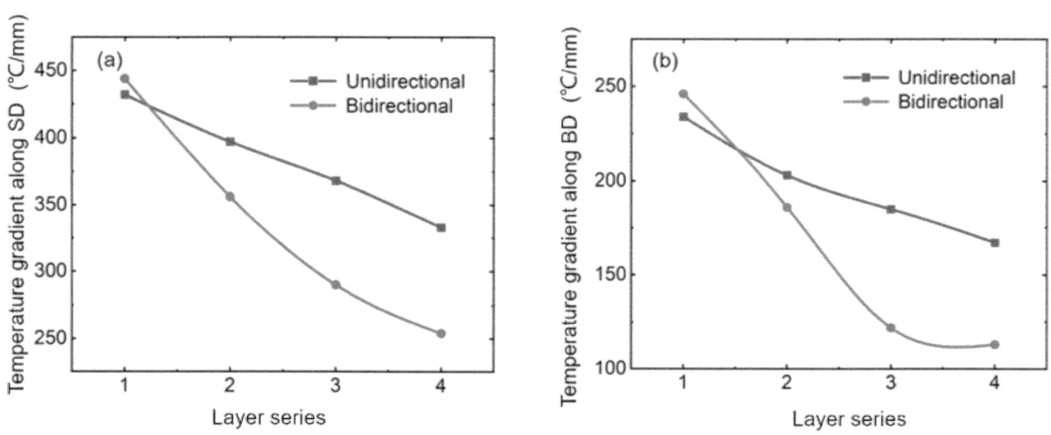

Figure 12.20. Effects of scanning strategy on the spatial temperature gradient, (a) the SD directional gradient in front of molten pool, (b) the BD directional gradient below the molten pool.

The temperature gradient under the two scanning paths was further analyzed, and the results are presented in Figure 12.20. Specifically, Figure 12.20(a) represents the temperature gradient of the molten pool front along the SD, while Figure 12.20(b) illustrates the temperature gradient below the molten pool in BD. It is evident that during initial layer deposition, there is no discernible disparity between unidirectional and bidirectional scanning methods, as both exhibit similar temperature gradients. However, for subsequent layers' deposition, it becomes apparent that unidirectional scanning yields significantly higher temperature gradients compared to bidirectional scanning techniques, regardless of whether it pertains to BD or SD.

12.4.3 Effect of scanning path on macroscopic stress and strain

The DIC method was employed to measure the macroscopic deformation under two scanning paths, and combined with changes in the temperature field obtained through infrared temperature measurement, the elastic-plastic strain corresponding to different scanning paths could be determined (Figure 12.21). It is evident that the peak value of elastoplastic strain during bidirectional scanning

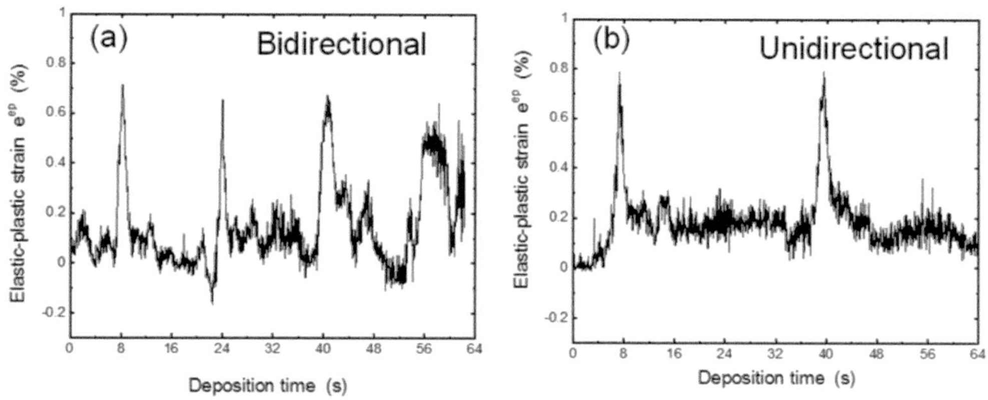

Figure 12.21. Effects of scanning strategy on the elastic-plastic strain of point A, (a) bidirectional, (b) unidirectional.

is approximately 0.79%, which exceeds that was observed during unidirectional scanning (0.68%). Unidirectional scanning results in a higher peak of internal stress, thereby increasing the risk of crack formation. Considering both crack sensitivity and forming efficiency, it can be inferred that reducing laser power and decreasing scanning speed can mitigate cracking sensitivity while also diminishing DED forming efficiency. By employing bidirectional scanning, not only can DED forming efficiency be significantly improved but also cracks can be substantially reduced.

12.4.4 Effect of scanning path on grain characteristics

EBSD analysis was performed on the samples obtained under two different scanning paths, and the corresponding results are presented in Figure 12.22. Among them, (a), (b), and (c) represent the distribution of IPF (BD), KAM, and GOS under bidirectional scanning, respectively, while (d), (e), and (f) depict the IPF map, KAM map, and GOS map under unidirectional scanning. It can be seen from the IPF maps that the sample using the unidirectional scanning path with a stronger <001> orientation and more pronounced epitaxial growth characteristics of columnar crystals compared to those using bidirectional scanning. Additionally, from a grain boundary morphology perspective, enhanced epitaxial growth leads to an increased number of long straight grain boundaries that facilitates sliding between grains and thus elevate the risk of solid-state cracking. Further observation of the IPF diagram shows that some equiaxed grains appear in some regions of the crack side. These equiaxed grains are obviously smaller than other grains and have the characteristics of recrystallized grains. It can be inferred that a large amount of lattice distortion energy is stored near the crack, which may induce recrystallization during DED. Collins et al. [160] also found recrystallization when they studied DDC cracks in welding of nickel-based alloy materials. The existence of recrystallization phenomenon further explains the existence of DDC cracks. The KAM distribution diagrams show that the local strain concentration is more significant in the unidirectional scanning method. In addition to both sides of the crack, there is also strain concentration near some uncracked grain boundaries. Moreover, the subgrain boundaries within the grains also increased significantly (21.4%). Combined with the distribution of the fine recrystallized grains, it can be inferred that the strain concentration region shown in KAM is insufficient to cause recrystallization. When the strain concentration degree and temperature are higher than a certain level, recrystallization phenomenon may occur. The occurrence of recrystallization will consume deformation energy storage, so the strain concentration disappears in the recrystallized grains, and the corresponding KAM value decreases. It can be seen from the GOS distribution diagram that the deformation degree of grains on both sides of the crack presents a large difference, which is similar to the GOS characteristics described in Sections 12.2.4 and 12.3.4. In addition, it can be found that the number of deformed grains in the sample DED-ed by the unidirectional scanning method is more, which indicates that the internal

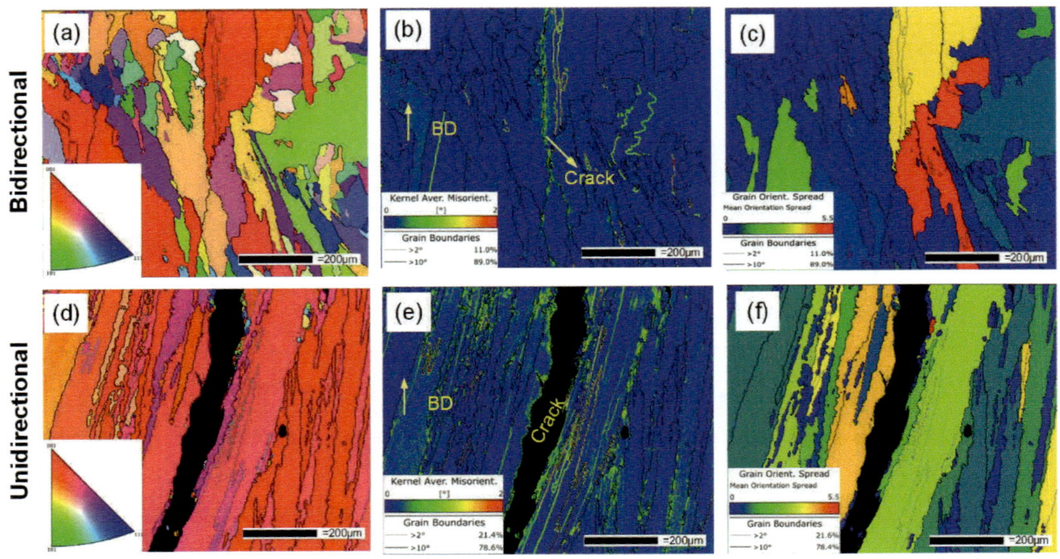

Figure 12.22. Orientation and micro deformation under different scanning strategies, (a), (b) and (c) are the IPF map (//BD), KAM map and GOS map under bidirectional scanning respectively, (d), (e) and (f) are the IPF map, KAM map and GOS map under unidirectional scanning respectively

stress level is higher. In summary, the bidirectional scanning method can not only help to interrupt the epitaxial growth of columnar grains, reduce the number of long straight grain boundaries, but also reduces the internal stress level in the structure and reduce local strain concentration. The above two aspects are conducive to inhibiting the generation of print cracks.

12.5 Effects of substrate preheating on cracks

To suppress the cracking, substrate preheating was used before the DED, the preheated temperature is about 400°C. The sample obtained without preheating and under preheating conditions is shown in Figure 12.23 (laser power of 250 W, scanning speed of 5 mm/s, powder feed speed of 7.6 g/min). As depicted in Figure 12.23, both preheated and un-preheated samples exhibited no macroscopic cracks. This can be attributed to the selection of a smaller laser power Although Figure 12.23 does not fully capture the impact of preheating on crack suppression, it can be deduced by Figure 12.23 that by employing preheating, a smaller laser power can be chosen while ensuring sufficient heat input to significantly enhance porosity and suppressing cracking at the same time. However, solely reducing laser power for cracking suppression may lead to an abundance of porosity defects, thereby hindering the attainment of high-quality DED parts.

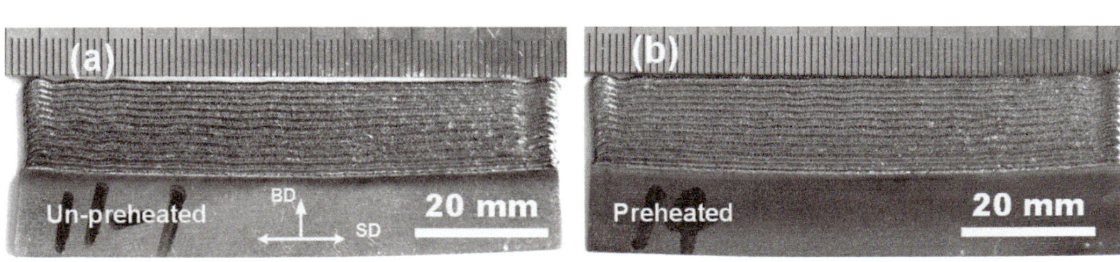

Figure 12.23. The DED parts deposited without (a) and with (b) the preheating of substrates.

12.5.1 Effect of substrate preheating on temperature gradient

Although no significant cracking was observed in the comparison samples depicted in Figure 12.24, it is imperative to discuss the corresponding crack susceptibility as well. Figure 12.24 illustrates the temperature profile at point A, located at the top of the substrate, under preheating and non-preheating conditions. The peak temperature for the preheated sample is higher than that of the un-preheated sample. Furthermore, within each heat cycle, the cooling rate of the preheated sample preheating is much lower which aids in mitigating internal stress levels.

The temperature gradients are shown in Figure 12.25. It can be seen that the preheating can significantly reduce the temperature gradient. The decrease of temperature gradient means that the difference of thermal deformation between adjacent zones is reduced, thus decreasing the binding force between each other and reducing the level of internal stress.

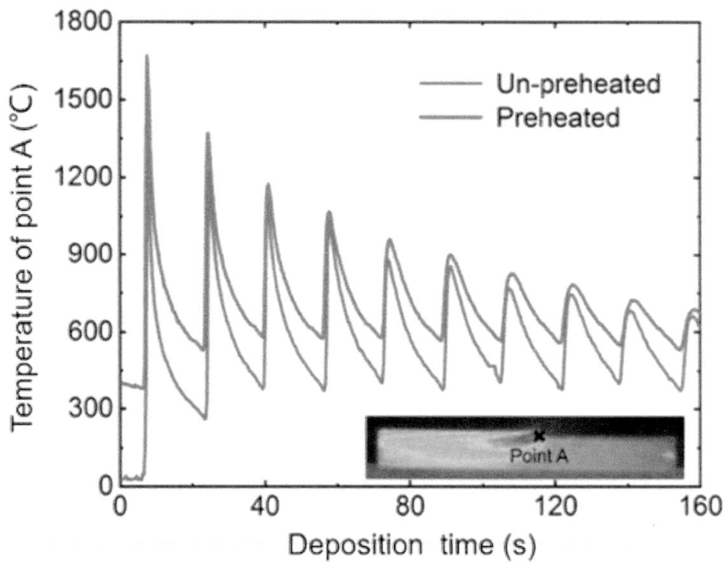

Figure 12.24. The thermal cycles of the top-center point A of the substrate with and without substrate preheating.

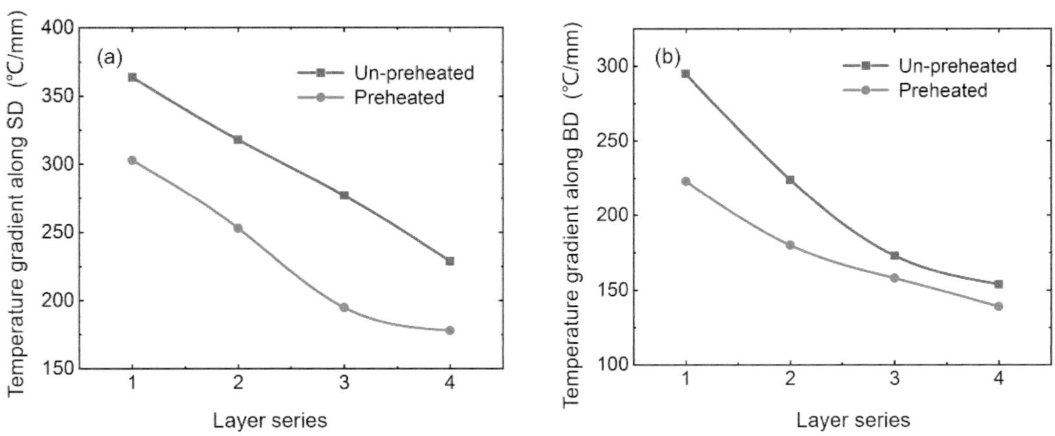

Figure 12.25. Effects of preheating on the spatial temperature gradient, (a) the SD directional gradient in front of molten pool, (b) the BD directional gradient below the molten pool.

12.5.2 Effect of substrate preheating on macroscopic strain

Combined with DIC strain measurement and infrared temperature measurement, the dynamic elastic-plastic strain under preheating and non-preheating conditions can be obtained. Figure 12.26 shows the elastic-plastic strain of point A at the top of the DIC observation area. It can be found that preheating can significantly reduce the peak value of dynamic elastoplastic strain, making the change of elastoplastic strain gentler during the whole deposition process. As mentioned above, a decrease in the peak elastic-plastic strain means a decrease in the peak of the dynamic internal stress, thus helping to reduce the risk of cracking.

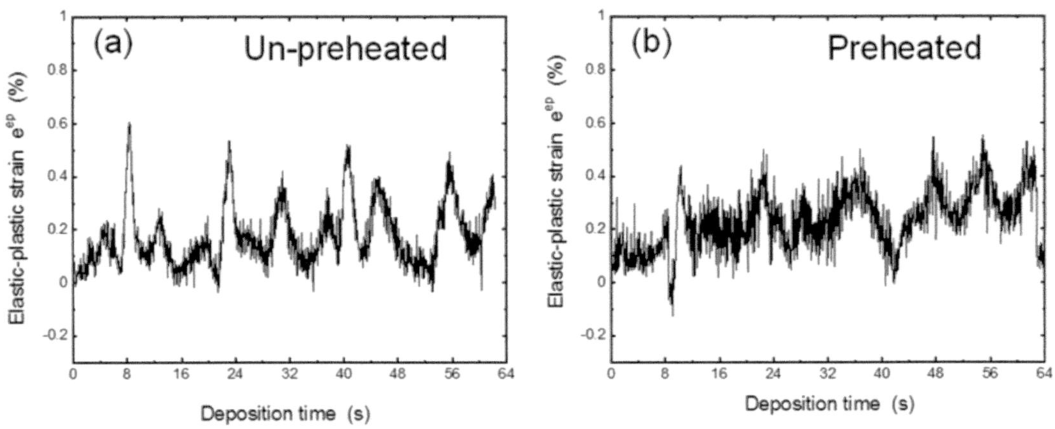

Figure 12.26. The elastic-plastic strain of point A, (a) un-preheated, (b) preheated.

12.5.3 Effect of substrate preheating on grain characteristics

The grain characteristics were analyzed using EBSD for the samples under preheating and un-preheating conditions, and the corresponding results are presented in Figure 12.27. Figure 12.27(a) and (d) depict the BC diagram, (b) and (e) illustrate the IPF diagram parallel to the deposition height, (c) and (f) display the KAM distribution diagram. The BC diagrams reveal that, under the un-preheated condition, there is a significant size disparity between the dendritic grains in the fusion zone and the columnar grains within the deposition layer. These columnar grains generally exhibit specific orientations along the <001> and <101> directions. Conversely, under preheated conditions, grain sizes across all regions are more uniform, with a more random orientation leading to an isotropic grain structure. Given that columnar grains facilitate local stress concentration more effectively than equiaxed grains, they demonstrate heightened crack sensitivity. Examination of grain boundary morphology indicates that in the preheated condition, grain boundaries are more tortuous, this enhances interlocking interactions among boundaries which impedes sliding between grains and results in reduced DDC crack sensitivity. Furthermore, KAM diagrams indicate that the KAM with some grains of the un-preheated sample is lower, which means the preheating can diminish the strain concentration at the grain scale and consequently can lower crack sensitivity.

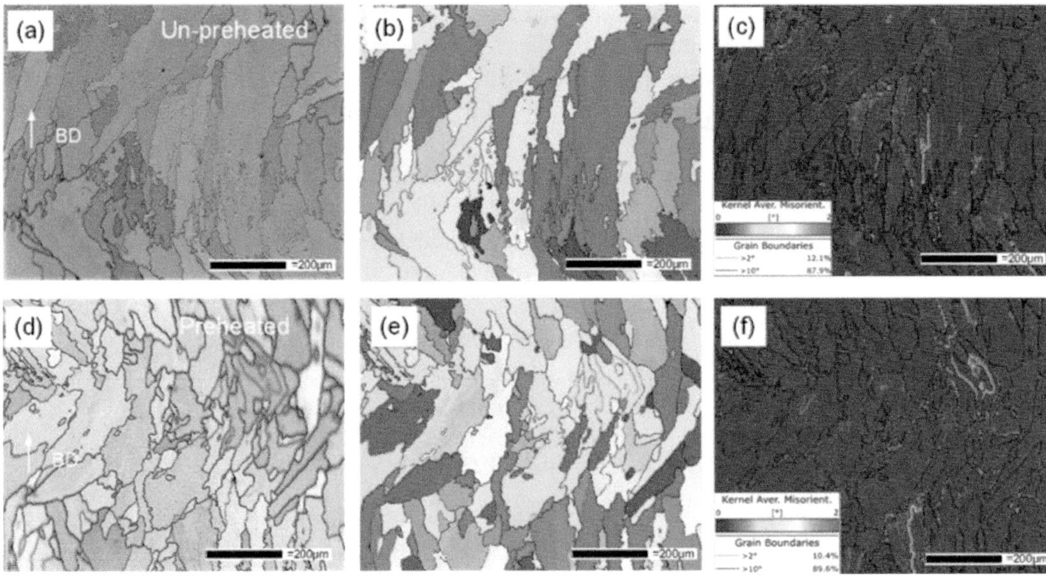

Figure 12.27. Grains features with and without preheating, (a), (b) and (c) are the BC map, IPF map (//BD) and KAM map without preheating, (d), (e) and (f) are the BC map, IPF map (//BD) and KAM map with preheating.

12.6 Effects of DED atmosphere environment on crack sensitivity

12.6.1 Cracking conditions under different atmosphere environments

When DED is carried out in the air, the deposited metal with high temperatures is exposed to the air for a long time, resulting in oxidation. To prevent this phenomenon, global atmosphere protection containing the whole DED part may be required. In our study, the experimental atmosphere protection chamber provided a global Ar gas protection during DED, and the oxygen concentration within the atmosphere chamber was less than 30 ppm. Figure 12.28 shows the samples DED-ed in the air and

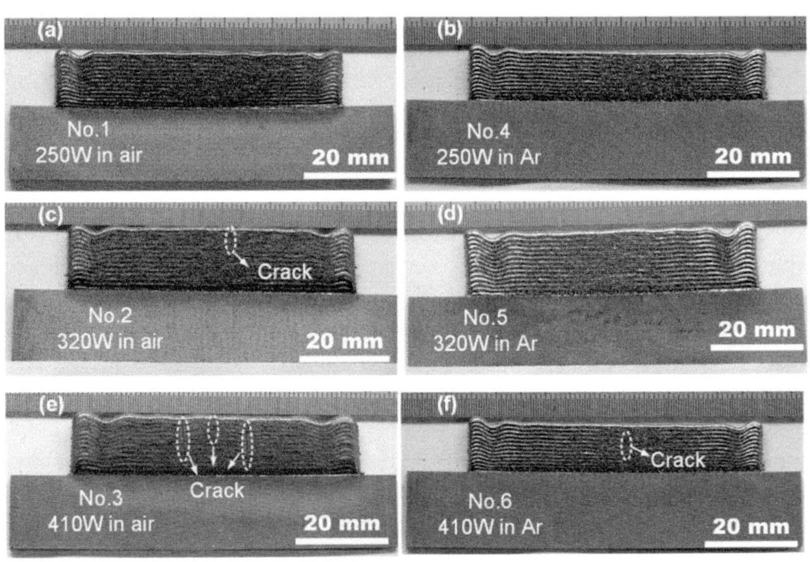

Figure 12.28. Cracks occurred during the DED in air and in Ar, (a), (c) and (e) are the samples deposited in air with 250 W, 320 W and 410 W laser power respectively, (b), (d) and (f) are the samples deposited in Ar with 250 W, 320 W and 410 W laser power respectively.

in the Ar. The corresponding laser power is shown in the figure, and the other process parameters are the same, the scanning speed is 5 mm/s, the powder feeding speed is 7.6 g/min, the scanning mode is bidirectional. As can be seen from Figure 12.28, the cracking degree in the Ar is significantly lower than that in the air atmosphere. Specifically, at a laser power of 250 W, no obvious cracks were observed in either atmosphere. However, at 320 W, macroscopic cracks were evident when DED was performed in the air atmosphere, while no cracking occurred when it was placed in the argon chamber. Increasing the laser power to 410 W resulted in serious cracking phenomena for both atmospheres. The reason is that when DED was conducted in air, stress-assisted diffusion of oxygen can occur and lead to oxide formation at crack fronts, thereby expanding cracks through separation or crushing of oxides from the substrate material. Additionally, short-range diffusion of oxygen may only occur near crack tips and embrittle the substrate material there, thus accelerating crack propagation. In general, these cracks associated with oxygen concentration are the solid-state cracks.

12.6.2 Effect of atmosphere environment on tensile properties

Tensile tests were carried out on specimens without obvious macroscopic cracking, the loaded direction was along the BD. The results are presented in Figure 12.29. It can be observed that, at a power of 250 W, the specimen formed in the Ar gas chamber reached the yield state, with a yield strength of approximately 1,050 MPa, a tensile strength of approximately 1,100 MPa, and a total elongation of merely about 1.2%. However, the other tensile specimens fractured before yielding, and the tensile strength decreased as the laser power increased. Ramakrishnan et al. [13] also studied the DED of IN738, the results indicated that in the as-deposited state, the optimal tensile strength was approximately 1,390 MPa and the elongation was approximately 1.1%, after heat treatment, the elongation increased to 2.76%, but the tensile strength decreased to 1,117 MPa. The heat treatment process employed was as follows: solution treatment at 1,120°C for 2 hours, air cooling to 850°C for aging treatment for 24 hours, and finally air cooling to room temperature. Comparatively, the strength reported by Ramakrishnan et al. was slightly higher than the results in this paper, while the corresponding plasticity was slightly lower. Rickenbacher et al. [14] reported that the room-temperature tensile strength of the cast IN738 alloy was 945 MPa and the elongation was 7.5%. It can be seen that the plasticity of the DED as-deposited state was significantly lower than that of the cast IN738. The aforementioned findings allow for the following deductions. Firstly, although no macroscopic cracks were found in the tensile specimens, there might still exist a large number of microcracks and pores. These defects significantly reduced the strength and plasticity of the material. Secondly, the as-deposited microstructure inherently stored relatively high residual stress, further reducing the ductility and toughness of the material. Thirdly, under the same DED process, the specimens formed in the argon gas chamber exhibited better strength, indicating that the number or the scale of cracks in them was smaller than that of the specimens formed in the air. Therefore, global atmosphere protection can effectively inhibit the generation of cracks.

The morphology of the tensile fracture surfaces under different conditions are shown in Figure 12.30. Specifically, Figure 12.30(a) and (b) are the fracture surfaces of the specimens formed in air and under a laser power of 410 W. As the specimen itself had severe macroscopic cracking before the tensile test and had already fractured before the test, its fracture surface of crack which occurred during DED is retained. It can be observed from Figure 12.30(a) that the cracking occurred between the columnar crystals, and with a large number of liquation ball between the dendrites. These facts confirm the occurrence of hot-cracking. Figure 12.30(c) and (d) show the fracture surfaces of the sample deposited in the air and at a laser power of 320 W. Under these process conditions, the specimen exhibits a tensile strength of approximately 700 MPa, fractures still occur along columnar crystals, but no liquation balls are observed in this section. Judging from the remaining crack on the fracture surfaces (Figure 12.30(d)), a large number of small particles suspected to be oxides were attached to the crack surface. That further confirms the influence of oxidation on crack sensitivity.

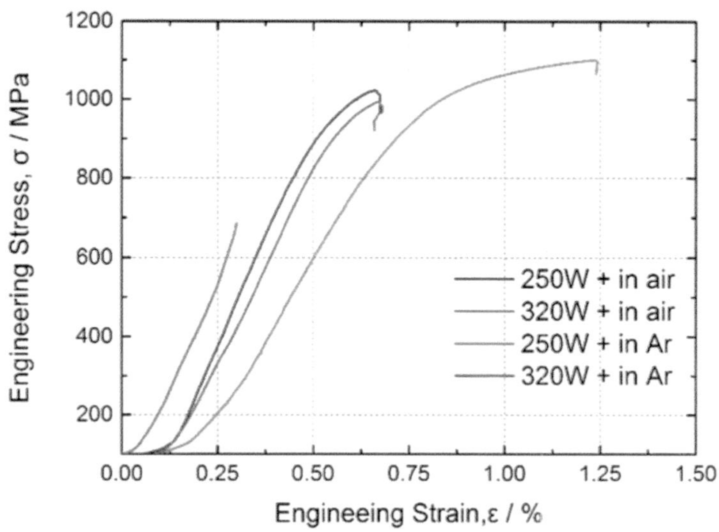

Figure 12.29. Tensile curves at room temperature for as-deposited samples.

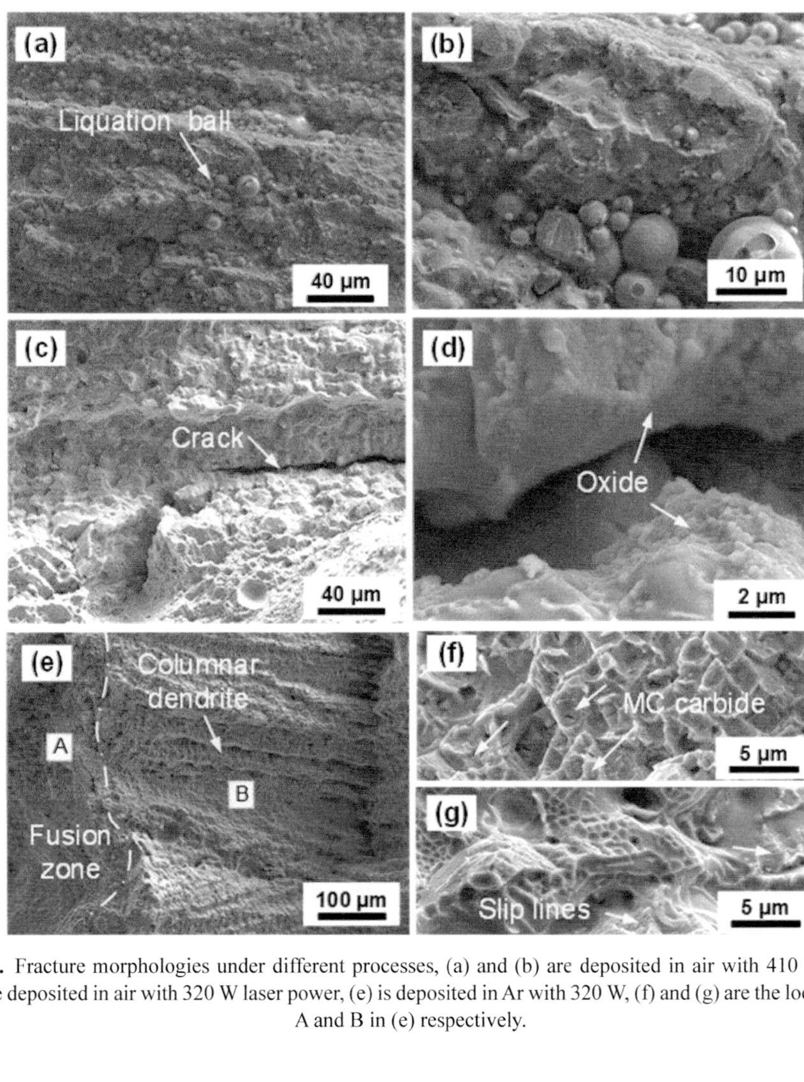

Figure 12.30. Fracture morphologies under different processes, (a) and (b) are deposited in air with 410 W laser power, (c) and (d) are deposited in air with 320 W laser power, (e) is deposited in Ar with 320 W, (f) and (g) are the local amplificated A and B in (e) respectively.

However, for the samples deposited in Ar at 320 W laser power, the tensile fracture morphology is very different (Figure 12.30(e)–(g)). In particular, the fusion zone and the columnar zone can be easily distinguished in the fracture diagram. In the columnar zone, there are some shallow dimples and cleavage planes, which demonstrate that this area has poor plasticity and toughness. In the fusion zone, a multitude of profound dimples, characterized by the presence of MC type carbide particles at their base, can be observed, these characteristics indicate that the fracture here exhibits good plasticity and toughness.

12.7 Summary

This chapter mainly discusses the effects of laser deposition processes on crack sensitivity during the IN738LC DED. The main conclusions are as follows:

1) With the increase of laser power, the crack sensitivity increases. The main mechanism is that the increase in laser power leads to a significant increase in temperature gradient at the front end of the molten pool, which causes uneven thermal deformation inside the material during deposition, thereby increasing internal stress levels. According to dynamic elastoplastic strain analysis during deposition, it can be observed that there is a strain peak within each thermal cycle. With the increase of laser power, both dynamic elastoplastic strain peak and corresponding dynamic tensile stress peak increase, thus crack sensitivity is enhanced. In addition, it was observed that higher laser power promotes stronger epitaxial growth ability and strengthens <001> crystal texture, resulting in more straight grain boundaries, that makes grain boundary sliding action easier and enhances crack sensitivity. Microscopic examination revealed that with increasing laser power, localized strain concentration phenomenon becomes more pronounced. This local strain concentration usually occurs between grains that are prone to deformation and grains that are less deformable. When local strains reach a certain threshold value, adjacent grain boundaries separate and cracks form.
2) With the increase of the laser scanning speed, the crack sensitivity also tends to rise. The increase in scanning speed elevates the temperature gradients near the molten pool, enhances the non-uniformity of deformation in different parts. Additionally, the scanning speed can also raise the peak of the dynamic elastic-plastic strain, thus enable an increase in the peak of tensile stress, thereby enhancing the crack sensitivity. From a microscopic perspective, an increase in the scanning speed will also heighten the deformation degree of grains and the concentration of microscopic strain, thus increasing the crack sensitivity.
3) The scanning path exerts a significant influence on the crack sensitivity. Regarding the DED formation of single-pass multi-layer thin-walled walls, adopting bi-directional reciprocating scanning can notably reduce the cracking risk. The mechanism lies in that unidirectional scanning prolongs the cooling stage of the deposited layer, increases the temperature gradients and raises the peak of tensile stress. Furthermore, unidirectional scanning can evidently increase the orientation of the <001> crystal direction, augment the number of long straight grain boundaries, intensify the concentration of microscopic internal strain, and augment the deformation degree of grains.
4) The preheating of the substrate can effectively decrease the susceptibility to cracks and significantly reduce the density and size of pores simultaneously. The reason lies in that the preheating of the substrate increases the heat input, enhances the melting rate of the metal powder, and prolongs the time available for the gas in the molten pool to escape, thereby reducing the porosity. On the other hand, the preheating of the substrate reduces the temperature gradient and simultaneously lowers the peak of the dynamic tensile stress, thus reducing the susceptibility to cracks. Additionally, the preheating makes the grains in the deposited layer tend to be equiaxed, reduces the number of straight and long grain boundaries, and weakens

the microscopic strain concentration phenomenon, which is also conducive to reducing the susceptibility to cracks.

5) The increase in the oxygen concentration in the environment during the DED will raise the susceptibility to cracks. For IN738LC, it is necessary to provide overall atmosphere protection for the DED component. In an atmosphere with an oxygen concentration lower than 30 ppm, a deposited sample with a tensile strength at room temperature of approximately 1100 MPa can be obtained, but its plasticity is still rather limited, with a total elongation of only about 1.2%. The presence of defects such as micro-cracks and pores as well as the relatively high level of residual stress are the main factors restricting plasticity.

References

[1] Chen, Y., Zhang, K., Huang J., Hosseini, S. R. E. and Li, Z. (2016). Characterization of heat affected zone liquation cracking in laser additive manufacturing of Inconel 718 [J]. Materials & Design 90(JAN.): 586–594.

[2] Yang, J., Li, F., Wang, Z. and Zeng, X. (2015). Cracking behavior and control of Rene 104 superalloy produced by direct laser fabrication [J]. Journal of Materials Processing Tech 225: 229–239.

[3] Idowu, O. A., Ojo, O. A. and Chaturvedi, M. C. (2007). Effect of heat input on heat affected zone cracking in laser welded ATI Allvac 718Plus superalloy [J]. Materials Science and Engineering: A 454-455: 389–397.

[4] Biegler, M., Graf, B. and Rethmeier, M. (2018). *In-situ* distortions in LMD additive manufacturing walls can be measured with digital image correlation and predicted using numerical simulations [J]. Additive Manufacturing 20: 101–110.

[5] Xie, R., Chen, G., Zhao, Y., Zhang, S., Yan, W., Lin, X. et al. (2019). *In-situ* observation and numerical simulation on the transient strain and distortion prediction during additive manufacturing [J]. Journal of Manufacturing Processes 38(FEB.): 494–501.

[6] Wang, N., Mokadem, S., Rappaz, M. and Kurz, W. (2004). Solidification cracking of superalloy single- and bi-crystals [J]. Acta Materialia 52(11): 3173–3182.

[7] Wei, X., Wang, Q. Z., Zhang, M. L., Yu, C., Chen, J. M., Chen, J. S. et al. (2019). Effect of microtextures on the intermediate temperature fracture behavior of additive-manufactured Ni-based overlay [J]. Journal of Alloys and Compounds 795: 103–113.

[8] Kadoi, K., Hiraoka, M., Shinozaki, K. and Obana, T. (2019). Ductility-dip cracking susceptibility in dissimilar weld metals of alloy 690 filler metal and low alloy steel [J]. Materials Science and Engineering: A 756: 92–97.

[9] Wei, X., Xu, M., Wang, Q., Zhang, M., Liu, W., Xu, J. et al. (2016). Effect of local texture and precipitation on the ductility dip cracking of ERNiCrFe-7A Ni-based overlay [J]. Materials & Design 110: 90–98.

[10] Parimi, L. L., Ravi, G. A., Clark, D. and Attallah, M. M. (2014). Microstructural and texture development in direct laser fabricated IN718 [J]. Materials Characterization 89: 102–111.

[11] Pinkerton, A. J., Karadge, M., Syed, W. U. H. and Li, L. (2006). Thermal and microstructural aspects of the laser direct metal deposition of waspaloy [J]. Journal of Laser Applications 18(3): 216–226.

[12] Gäumann, M., Bezençon, C., Canalis, P. and Kurz, W. (2001). Single-crystal laser deposition of superalloys: processing–microstructure maps [J]. Acta Materialia 49(6): 1051–1062.

[13] Ramakrishnan, A. and Dinda, G. P. (2019). Direct laser metal deposition of Inconel 738 [J]. Materials ence and Engineering: A s 740–741(JAN.7): 1–13.

[14] Rickenbacher, L., Etter, T., Hovel, S. and Wegener, K. (2013). High temperature material properties of IN738LC processed by selective laser melting (SLM) technology [J]. Rapid Prototyping Journal 19(4): 282–290.

Chapter 13

L-DED of Hetero-Structured IN738+IN718 Alloys with Excellent Strength and Ductility

13.1 Introduction

Depending on the specific application scenario, different nickel-based alloys are typically chosen for additive manufacturing of components. For examples, IN718 is primarily utilized in the aerospace industry for aircraft engines, high-speed airframe components, cryogenic tankage, petrochemical applications, nuclear power plants, and automotive turbochargers. CMSX-4 represents a second-generation Nickel-based superalloy that is suitable for various high-pressure applications such as jet engine blades, helicopter engine components, high-pressure turbine shrouds, and gas turbines used in power generation. Udimet 720 is commonly employed in gas turbines for hot section components. Waspaloy finds extensive application in jet engines where it is used for cast turbine disks, shafts, low-pressure rings, and seal rings. Rene N6 is an advanced single crystal Nickel-based superalloy specifically designed for military aerospace engines' high pressure blades. IN625 serves as a highly corrosion-resistant Nickel-based superalloy suitable for use in nuclear reactors and oceanographic equipment, it also finds application in structures within thermal power plants along with armoring and plugs within aeroengines. Haynes 242 proves to be a versatile Nickel-based superalloy catering to various aircraft and industrial gas turbine applications. IN738 predominantly finds its usage in gas turbine blade and vane applications while maintaining exceptional performance properties [1]. The hot-end components for extreme heat applications necessitate the nickel-based superalloys with elevated γ' content and exceptional strength in high temperature. However, these superalloys often exhibit susceptibility to cracking during additive manufacturing processes, particularly in laser directed energy deposition, thereby presenting a contradiction between their superior thermal strength and limited printability. Developing superalloys that suitable for additive manufacturing and have excellent mechanical properties at high temperature is the key to solving this problem. Such alloys can be homogeneous or heterogeneous materials, but they often have one character in common, that is, possessing superior strength and ductility synergy [2].

In recent years, inspired by the biological structural materials of shells and pearls in nature, researchers have found that the construction of heterogeneous structures is a new way to enhance the synergy of strength and ductility [3–5]. Heterostructured (HS) materials can be very diverse, including, but not limited to, heterogeneous lamella structure [6], bi-modal structure [7, 8], gradient structure [9–11], dual phase structure [12–14], harmonic structure [15–17], layered structure [18, 19], metal matrix composites [20–24], etc. As an important branch of HS, heterogeneous lamella structure generally constructed by cumulative rolling, hot pressure diffusion, explosive composite welding or electrochemical deposition [25–27]. However, most of these methods have complicated process and poor controllability. As a flexible layered manufacturing technology, laser directed

DOI: 10.1201/9781003406327-13

energy deposition has natural technical advantages in the configuration design and controllable construction of layered heterostructures, and is a promising new method for the construction of layered heterostructures [28, 29]. Chaolin Tan et al. alternately feeded C300 MS and 420 SS poweders during L-DED, and obatined C300 MS/420 SS heterostructures whose tensile strength reached 1.6 GPa while still maintaining 8.1% fracture elongation [30, 31]. Yonggang Sun et al. report a laser-directed energy deposited medium entropy alloy (MEA) with a heterostructure comprising alternate layers of solid-solution and intermetallic-dispersed MEA that possess ultimate tensile strength of 1132.8 MPa and elongation of 50.6%, corresponding to strength-ductility synergy higher than other medium- or high-entropy alloys reported [32]. In this chapter, layer-heterostructured IN738/IN718 was fabricated by DED to suppress cracks and improve the synergy of strength and ductility. This method intends to alternately deposit a small amount of IN718 layers during the DED of IN738, and finally obtain a layered composite gradient structure with IN738 as the main material and IN718 as the auxiliary material. The microstructure evolution and mechanical properties of the composite specimens will be systematically studied, furthermore, the crack suppression mechanism and strengthening and toughening mechanism of will be explored. The study to provide a new idea of crack suppression for the additive manufacturing of difficult welded nickel-based superalloys.

13.2 L-DED of the layer-heterostructured IN738+IN718

Figure 13.1 is the schematic diagram of the L-DED of layer-heterostructured IN738+IN718 (IN738+IN718 LHS). The two kinds of powders are added to two different powder buckets respectively. The deposition is carried out in a bidirectional scanning strategy. The deposition speciments are the single-pass but multi-layered thin walls. They are 80 mm in length and 20 mm in height, the wall thickness is about 1.6–1.8 mm. The layer height is set at 0.42 mm, the laser power is 280 W, the scanning speed is 6 mm/s, and the powder feeding speed is 1 mL/min. The specific DED process is as follows. Firstly, the IN738 powder bucket is enabled and two layers (about 0.84 mm thickness) of IN738 alloy are deposited continuously. Then, the supply of IN738 powder bucket is interrupted, and the IN718 powder bucket is switched on and enabled to deposit one layer (about 0.42 mm thickness) of IN718 powder. After that, the IN718 powder bucket is interrupted,

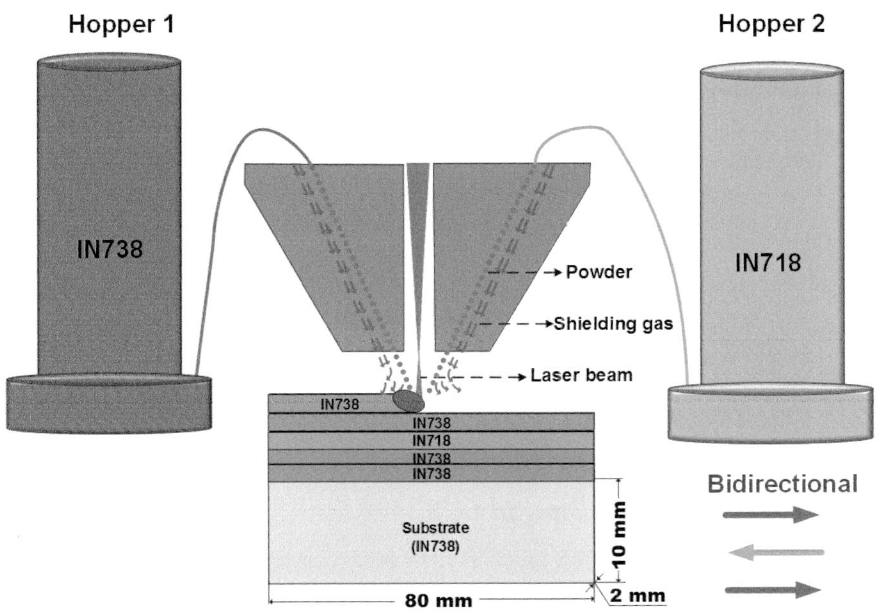

Figure 13.1. The schematic diagram of the layer-heterostructured IN738/IN718.

Table 13.1. Chemical compositions of IN738 and IN718 powder (wt%).

Elements	Cr	Co	W	Fe	Nb	Mo	Ta	Al	Ti	Mn	Si	C	Ni
IN738	15.82	8.62	2.62	0.19	0.92	1.94	1.72	3.7	3.5	0.11	0.16	0.13	Bal
IN718	19.18	--	--	17.65	5.07	3.10	--	0.63	0.94	0.04	0.2	0.04	Bal

and the IN738 powder bucket is enabled again to deposit two layers of IN738 alloy continuously. This cycle goes on and on until the entire CAD model is printed. It should be noted that when the powder is not switched, continuous deposition can be maintained. But when the powder is switched, the laser needs to be turned off and the deposition is paused for 5 seconds. Wait for the upcoming powder to be stably delivered before continuing the deposition.

In this study, the composition of the powders used for the DED is shown in Table 13.1. The deposition ratio of IN738 to IN718 is designed to be 2:1. In fact, the ideal ratio need to be further optimized according to the actual DED condition. The optimization criterion of the ratio is that select a higher deposition ratio of IN738 as possible to inherit the its excellent properties in high temperature but free from cracking. In the preliminary experiments, it was found that under our process condition, if more than 2 layers ($\approx$ 0.8 mm) of IN738 alloy were deposited continuously, micro-cracks may still appear in the deposition, indicating that it is difficult to completely inhibit the generation of cracks when a higher deposition ratio is used.

In addition, pure IN738 and IN718 are deposited respectively with the same process conditions and single powder. Figure 13.2 shows the DEDed samples corresponding to different materials. As can be seen from Figure 13.2(a), under this process condition, the IN738 sample has serious macroscopic cracks, while the IN718 sample and IN738+IN718 LHS sample have no macroscopic cracking phenomenon. The above experimental results preliminarily prove that IN738+IN718 LHS has a good crack inhibition effect.

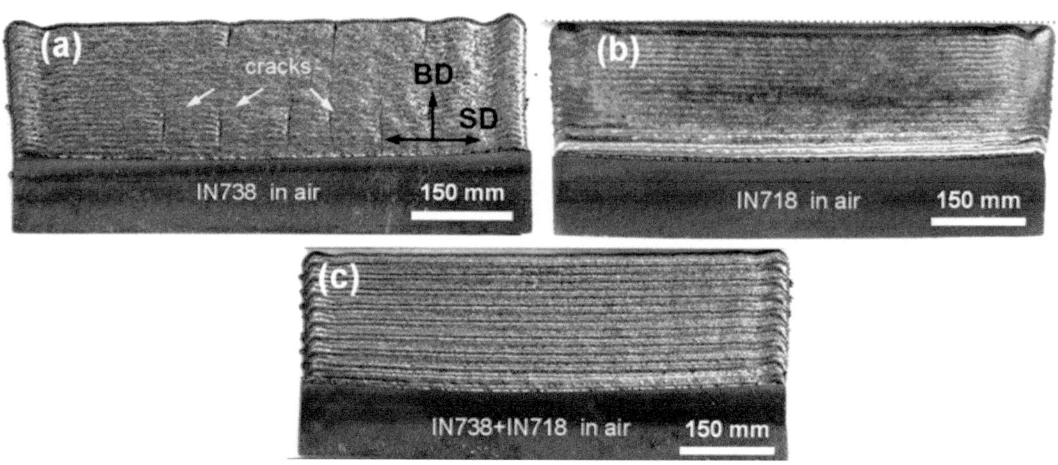

Figure 13.2. As-deposited samples, (a) single-powder-deposited IN738 sample; (b) single-powder-deposited IN718 sample; (c) composite-deposited IN738+ IN718 LHS sample.

13.3 Microstructure of the layer-heterostructured IN738/IN718

13.3.1 Microstructure of IN718 layer in LHS

The Microstructure of the inner IN718 layer in LHS is shown in Figure 13.3. It can be seen that there is a wide interdendrite region between the dendrite trunks. In these interdendrite zones, a large number of particles are scattered (Figure 13.3(a)). Zooming in on these particles, as seen in

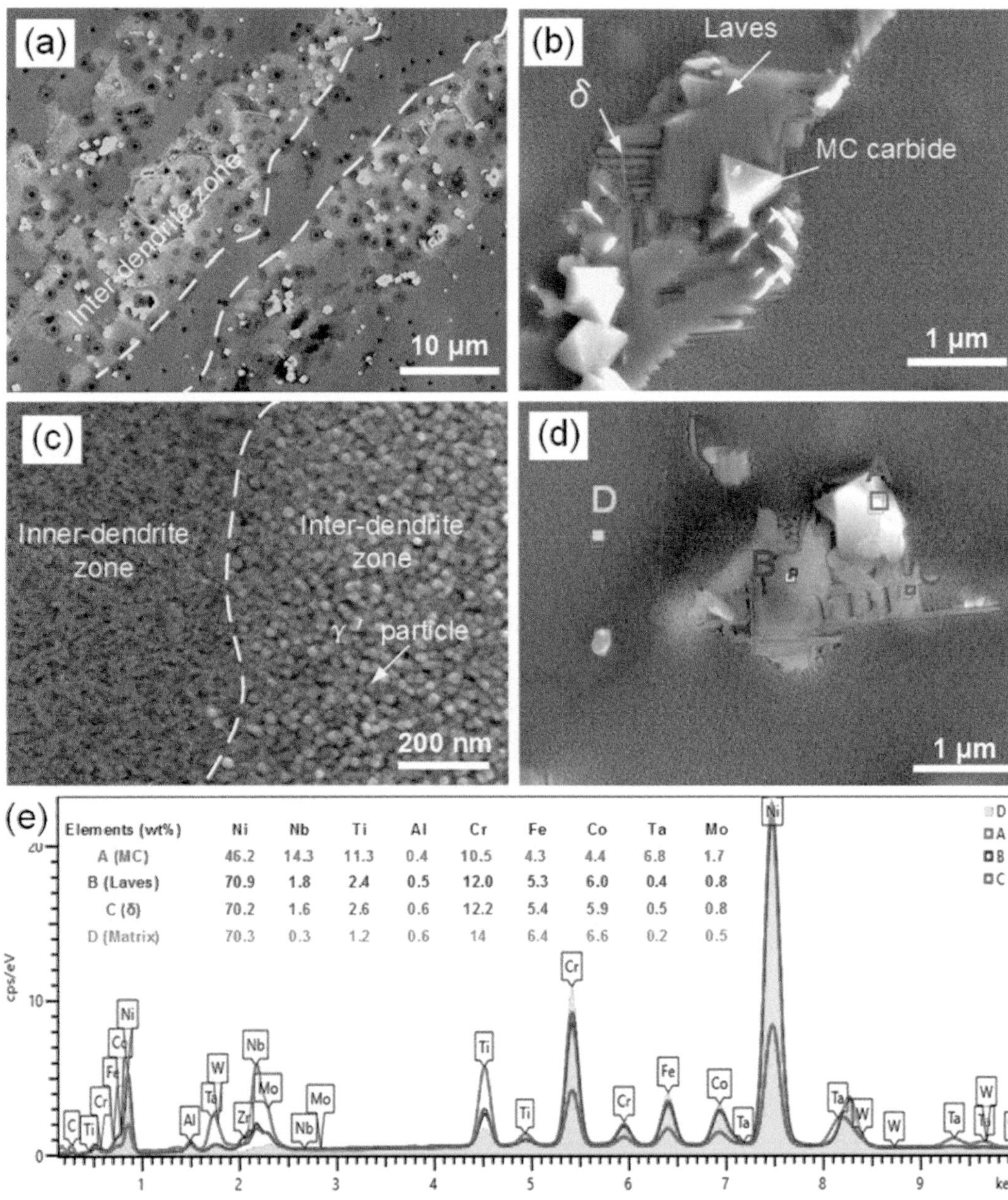

Elements (wt%)	Ni	Nb	Ti	Al	Cr	Fe	Co	Ta	Mo
A (MC)	46.2	14.3	11.3	0.4	10.5	4.3	4.4	6.8	1.7
B (Laves)	70.9	1.8	2.4	0.5	12.0	5.3	6.0	0.4	0.8
C (δ)	70.2	1.6	2.6	0.6	12.2	5.4	5.9	0.5	0.8
D (Matrix)	70.3	0.3	1.2	0.6	14	6.4	6.6	0.2	0.5

Figure 13.3. Microstructure of the inner IN718 layers in the as-deposited IN738+IN718 LHS, (a) shows the columnar dendrites at lower magnification, (b) shows the large particles precipitated in inter-dendrite zone, (c) shows the γ′ precipitation in the inter-dendrite zone and inner-dendrite zone, (d) and (e) show the locations for EDS analysis and the obtained components.

Figure 13.3(b), three different shapes of particulates are found, they are block, island and needle around island. Referring to the reports on the additive manufacturing structure of IN718 alloy [33–35], it can be preliminarily determined that the block-shaped particulates are MC type carbides dominated by NbC, the island-shaped particulates are Laves phase, and the needles around Laves phase are δ phase. Enlarge the matrix at the junction of interdendrite region and dendrite trunk, as shown in Figure 13.3(c), γ′ nanoparticles are found in the matrix of interdendrite region but free in the dendrite trunk region. Approaching the dendrite trunk, the size of these γ′ nanoparticles gradually decreases and the density is gradually reduced. EDS spot scanning was used to analyze the large particles and matrix in the interdendrite zone, the results are shown in Figure 13.3(d, e). It can be found that the interdendrite matrix (D point) is dominated by Ni, Cr, Fe and Co. The content of Nb, Ti, Ta and Mo of MC carbide (point A) is significantly higher than that of matrix D. Compared with matrix D, the contents of Nb, Ti, Ta and Mo in Laves phase (point B) are also increased, but their contents are much barren than that of MC carbides. The composition of δ phase (point C) is very similar to that of Laves phase, but the content of Nb in δ phase is slightly lower than that of Laves, while the content of Ti and Cr is slightly higher than that of Laves phase. Stevens et al. [36] and Parimi et al. [37] also reported the precipitation of acicular δ phase when studying the DED of large-size IN718 component. They believed that δ phase was the product of local dissolution of Laves phase. This result is consistent with the microstructure of IN718 layer in the IN738+IN718 LHS in this study. Generally, the needle δ phase is considered to be a phase that is bad for strength due to its flat edges are often the act as the initial locations of cracks.

Besides the δ phase, there are many other phases in nickel-based alloys that may exhibit needle-like or plate-like morphology, such as η phase, μ phase and σ phase, so it is necessary to use TEM to further confirm the needle-like phase. Figure 13.4(a) shows the TEM bright field images

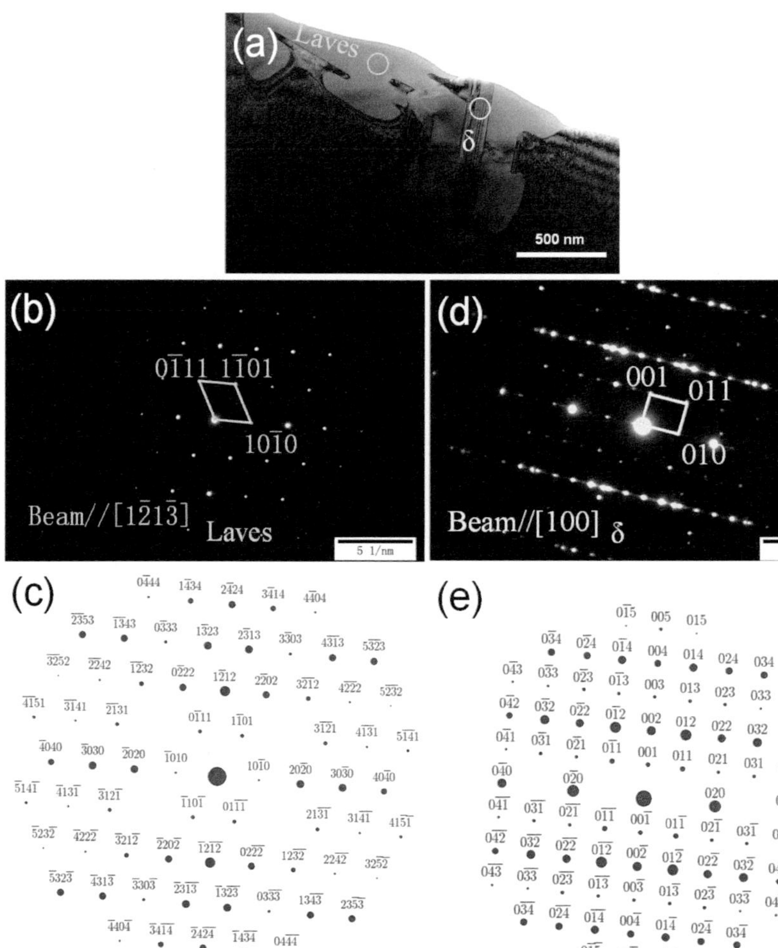

Figure 13.4. TEM analysis for Laves and δ: (a) is the bright field image; (b), (d) is the experimental SADPs of Laves and δ respectively; (c), (e) is the software simulated SADPs of Laves and δ respectively.

Table 13.2. Structural parameters of Laves phase: P63/mmc space group with a = b = 4.659Å, c = 7.743Å, α = β = 90°, γ = 120° [38].

Element	Wyckoff	x	y	z
Mo (1)	4f	0.3333	0.6667	0.5640
Fe (1)	6h	0.17	0.34	0.25
Fe (2)	2a	0	0	0

Table 13.3. Structural parameters of δ phase: Pmmn space group with a = 5.114Å, b = 4.244Å, c = 4.538Å, α = β = γ = 90° [38].

Element	Wyckoff	x	y	z
Ni (1)	2a	0	0	0.3152
Ni (2)	4f	0.7494	0	0.8414
Nb (1)	2b	0	0.5	0.6513

of the needle phase and island-shaped phase in IN718 layer. The corresponding electron diffraction spots are shown in Figure 13.4(b) and (d) respectively. The diffraction spots shown in Figure 13.4(d) are nested with multiple sets of δ bars according to the diffraction spots. As the prediction, the island phase is the Laves phase, while the needle phase is the δ phase, and the corresponding crystal structure parameters are shown in Tables 13.2 and 13.3 [38]. These crystal structure parameters are used to confirm the electron diffraction spots obtained in the experiment. The results show that the crystal structures reflected by the diffraction spots are consistent with those shown in Tables 13.2 and 13.3. Further, PTCLab (V1.14), an auxiliary software for phase transformation crystallography, was used to simulate the electron diffraction spots of Laves and δ phases, and the results were shown in Figure 13.4(c) and (e) respectively. It can be seen that the simulated diffraction spots correspond to the measured diffraction spots. These results above confirm that the acicular phase is indeed the δ phase with orthogonal structure, while the island phase attached to it is the Laves with close-packed hexagonal structure.

To reflect the microstructure differences between IN718-layer in IN738+ IN718 LHS and the purely deposited IN718, the microstructure of purely deposited IN718 are also observed and exhibited in Figure 13.5. It can be found that in the purely deposited IN718, there are a large number of Laves long chains located in the inter-dendrite zones, The Laves morphology is significantly different from the that in IN718-layer of IN738+ IN718 LHS (Figure 13.3(a)). In addition, as shown in Figure 13.5(c), no significant γ' were found in the interdendrite region of the purely deposited IN718, while a large number of γ' nanoparticles were found in the IN718-layer of IN738+ IN718 LHS as shown in Figure 13.3(c). These differences demonstrate that, in the IN738+ IN718 LHS, significant elements exchange occurred between IN718 and IN738 layers. On the one hand, elements such as Fe and Nb which are riched in the In718 layer intruded into the IN738 layer, resulting in a decrease in the content of Nb and Fe in the IN718 layer, thus the Laves phase decomposed from a long chain shape into an island shape. On the other hand, Al, Ti and Co elements with higher content in the IN738 layer also entered the IN718 layer, leading to the enhancing of Al and Ti content in layer In718, these elements are easy to segregate into the dendrites, thereby triggering the precipitation of γ' nanoparticles here.

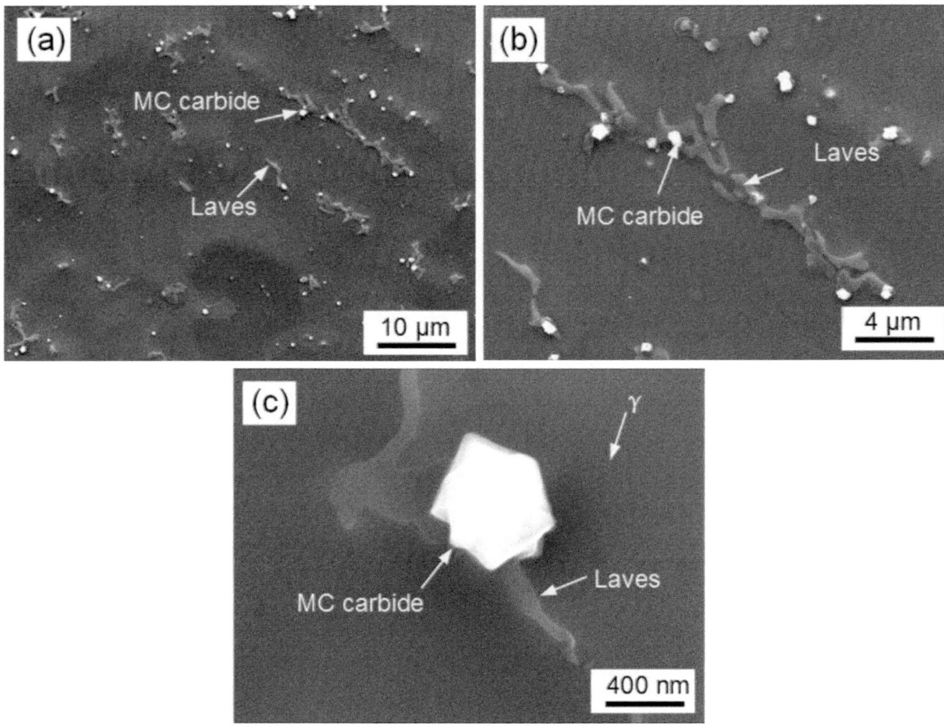

Figure 13.5. Columnar dendrites zone of the purely deposited IN718, (a), (b), (c) are the SEM images with different amplifications.

13.3.2 Microstructure of IN738 layer in LHS

The microstructure of IN718-layer in IN738+ IN718 LHS is shown in Figure 13.6. At low magnification, scattered MC carbides and γ+γ′ eutectic can be found (Figure 13.6(a)). And the enlarged view of these γ+γ′ eutectes is shown in Figure 13.6(b). It can be seen that in the interior of the γ+γ′ eutectic, there are occasionally large particle precipitates which are determined to be MC carbides preliminarily according to the morphology and the solidification of IN738. In particular, MC carbides in IN738 may have different morphologies such as rod shape or block shape, which are related to the growth rate of MC during solidification, the content of elements required by MC, and the surrounding solid phase morphologies. During the solidification with rapid cooling condition, MC+γ+γ′ three-phase eutectic may occur at the end of solidification, thus forming the eutectic microstructure shown in Figure 13.6(b). In addition to the eutectic, densely arranged γ′ particles are found in the matrix. Figure 13.6(c) shows the morphology of γ′ particles in the magnified matrix, in which the γ′ particles exhibit two different sizes. The interdendrite γ′ particles exhibit a square or block-shaped morphology, with a diagonal length of approximately 70 nm. Conversely, the inner-dendrite γ′ particles are smaller in size, measuring around 25 nm, and tends to undergo a transition from cubic to spherical shape. It should be noted that in the compositional transition region from IN738 to IN718 as shown in Figure 13.6(d), there are also some needle-like structures present, while no distinct Laves structures are detected. Considering the location proximity to the IN718 layer where the similar needle-like structures have been identified as δ phase, it can be inferred that the needle-like structures also correspond to δ phase.

Compared to the microstructure of the pure IN738 samples, there are no significant changes in the precipitation characteristics of γ′, MC, and γ+γ′ phases within the IN738 layer of LHS, except for the appearance of the δ needles. From a chemical composition perspective, the elements that undergo component exchange from the In718 layer to the In738 layer primarily consist of Fe, Nb,

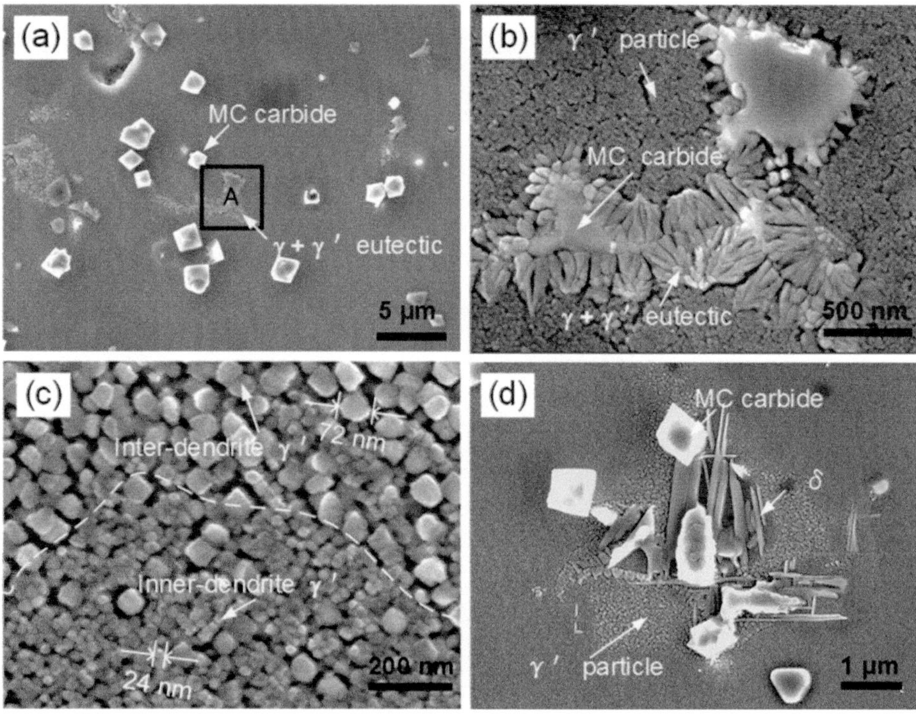

Figure 13.6. Microstructure of IN738 layer within the as-deposited IN738+IN718 LHS: (a) is the microstructure at low magnification, (b) is the local zoom A of (a); (c) shows the magnified matrix; (d) shows the inner-dendritic precipitates nearby the IN738 to IN718 intersection.

and other solid solution strengthening elements. However, their presence is insufficient to induce significant alterations for the primary phases in the IN738 layer. Moreover, it should be noted that the thickness of the In738 layer in our designed LHS is twice that of the IN718 layer. Consequently, there is a greater distance between the central region of the IN738 layer and the IN738/IN718, thereby leading to a weaker degree of element exchange and microstructure evolution in the core of IN738 layer.

13.3.3 Microstructure of IN738/IN718 interface in LHS

Generally, due to the chemical disparities between the component parts constituting the LHS, a chemical transition region inevitably emerges when these two components are metallurgically combined. If the compatibility of the component parts is poor or there is an obvious chemical reaction between them, the composition gradient of this composition transition zone would be very large, and even lead to a sharp interface. However, for the IN738+IN718 LHS reported here, the component parts, i.e., IN738 and IN718, although there are great differences in the content of some elements, they both belong to the nickel-based superalloy with an similar γ phase as their matrix. Therefore, the IN738/IN718 interface in LHS is essentially more like a gradient transition region. Figure 13.7 exhibits the microstructure and composition in this gradient transition region.

The SEM microstructure in the interface of IN718 and IN738 layers is depicted in Figure 13.8. It can be found that this interface exhibits a microstructure transition rather than a distinct boundary, allowing dendrites to grow through it. The particles of precipitated phase between dendrites exhibit gradual changes without significant morphological alterations. Notably, the dendrite stem structure undergoes noticeable modifications. Region A within the IN718-layer dendrite stem and region B within the IN738-layer dendrite stem are magnified at high resolution, as shown in Figure 13.8(b) and (c). It can be observed that there is no apparent particle precipitation in the stem of the IN718

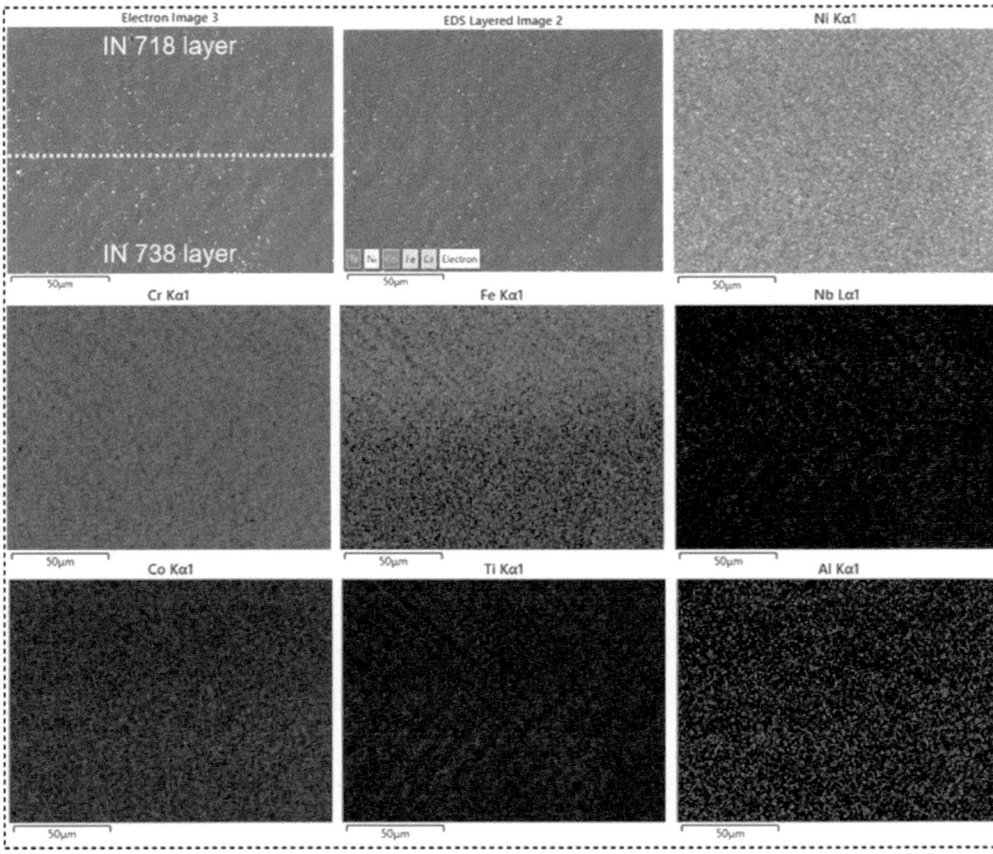

Figure 13.7. EDS maps show both IN718 layer and IN738 layer.

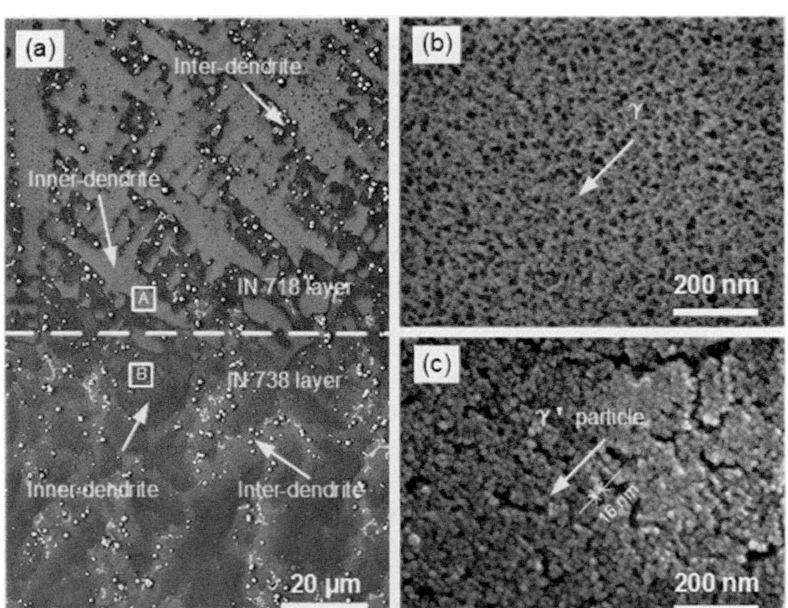

Figure 13.8. Microstructure at the intersection zone of IN718 layer and IN738 layer (a); (b) and (c) are the amplified local zoom of A and B in (a) respectively.

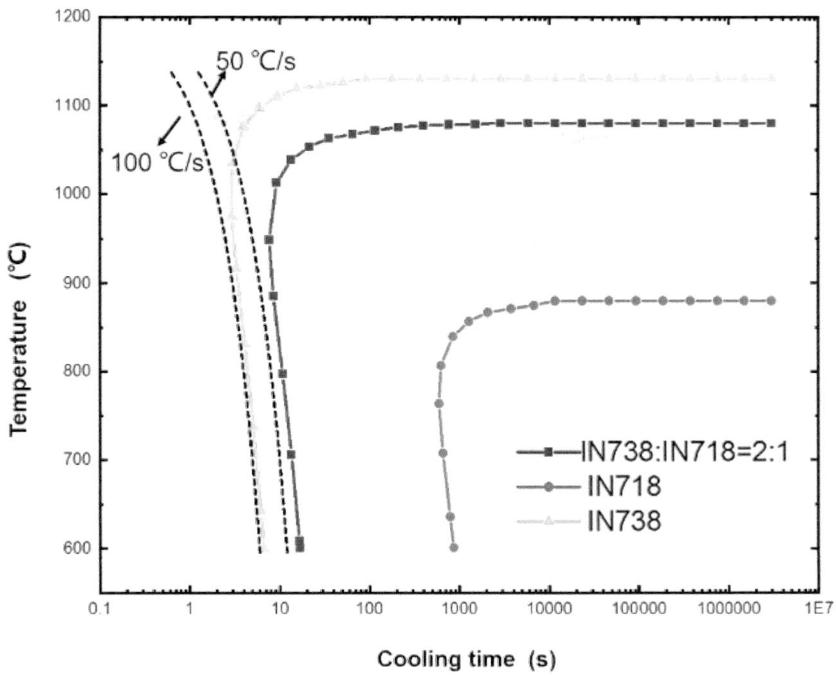

Figure 13.9. Continuous cooling transformation curves of γ′ with different compositions.

layer's dendrites, only the γ′ phase was observed, while densely distributed small particles with diameters ranging from approximately 10 to 20 nm precipitate within the stem of the IN738 layer's dendrites. By considering previous investigations on γ′ phase morphology characteristics discussed, it can be concluded that these tiny particles are the γ′ particles.

Generally, the increase of the content of Al, Ti, Nb, Ta promotes the precipitation of γ′ phase. In order to better understand the precipitation rule of γ′ phase, Jmat-pro software is used here to simulate the Continuous Cooling Transformation (CCT) curve of γ′ phase corresponding to IN738, IN718 and the uniform mixed IN738+IN718. Jmat-pro Software (Sente Software Ltd, Surrey Tech. Centre, Guildford, UK) is a professional calculation software for metal materials that integrates thermodynamics, dynamics calculation and material properties database. The accuracy of its calculation has been highly recognized in the field of metal materials. The CCT curve calculated by Jmat-pro is shown in Figure 13.9. It can be seen that the content of γ′ phase element in IN738 is high, so the CCT curve of γ′ phase is relatively advanced, indicating that γ′ phase is easy to precipitate during cooling. In order to inhibit the precipitation of γ′ phase, it is necessary to maintain a very high cooling rate (above 100°C/s) in the temperature range of γ′ phase precipitation. For the mixed component and IN718, the content of γ′ phase forming elements in the component decreased, corresponding to the backward and downward movement of the CCT curve, indicating that the precipitation of γ′ phase could be ensured only at a slower cooling rate. According to the temperature measurement results of infrared thermal imager, the cooling rate in the deposition thermal cycle is between 50 and 100°C/s within the precipitation temperature range of γ′ phase. Figure 13.9 shows this cooling interval as a dashed line. It can be seen that the CCT curve corresponding to IN738 intersects with the cooling curve in this region, indicating that γ′ phase can be precipitated during the thermal cycle of DED. However, the cooling rate in the range of 50 ~100°C/s is not intersected with the CCT curve corresponding to IN718 and the mixed component, indicating that the cooling rate in the deposition thermal cycle is too fast to precipitate the γ′. From the IN738 and IN718 structures deposited by the pure powder, the observed γ′ precipitation behavior is basically consistent with that reflected by CCT curve. However, in actual deposition, there are obvious composition differences between dendrites and dendrite stems due to the existence of solute redistribution. Generally speaking, higher positive segregation elements such as Ti, Nb, Ta, Mo, Zr [39, 40] are gathered in the dendrites, which happen to be the forming elements of γ' phase, thus the dendrites are usually the preferred precipitation locations of γ' phase. Therefore the interdendrite γ' particle size of IN738 layer is larger than that of the composite deposition, and in the LHS of IN718+IN738 the γ' phase is only found in the interdendrite of IN718 layer.

13.4 Mechanical properties of the LHS

13.4.1 Hardness of heterogeneous layers

Microhardness is often closely correlated with strength, plasticity, and fatigue resistance, making it a rapid, cost-effective, and convenient method for characterizing material properties. In LHS of IN738/IN718, the variation in mechanical properties between different layers plays a crucial role in strengthening and toughening. Therefore, this study employed the micron indentation method to test the top four layers of the LHS. Figure 13.10(a) illustrates the load response curve obtained from the micron indentation test. According to Meyer's law [41], during the loading process, the load response curve can be expressed by equation 13.1.

$$F = Kh^m \quad (13.1)$$

Where F represents the applied load, h denotes the corresponding indentation depth or pressing depth value, K is a proportionality constant, and m signifies the Meyer Index. The unloading phase of this curve can be described as follows.

$$F = \beta(h - h_f)^n \quad (13.2)$$

Where F and h still represent their respective values of load and pressing depth; h_f indicates the remaining indentation depth after unloading; β and n are fitting parameters associated with this behavior. By analyzing these load response curves using equation 13.3.

$$H = \frac{F_{\max}}{A_{pml}} \quad (13.3)$$

Microhardness values can be determined accordingly. Herein, F_{max} refers to maximum load while A_{pml} represents projected area of indentation along loading direction calculated based on Oliver et al.'s work.

In addition to hardness, the load curve also reflects the energy loss during the indentation process, including elastic energy (We) and plastic energy (Wp). Figure 13.10(b) presents the measured load response curves of the top four layers. It can be observed that 1st IN718 is solely influenced by IN738 on one side, resulting in a significantly distinct response curve compared to that of IN738. The 4th IN718 layer can represent the inner IN718 layer of LHS, exhibiting a response curve similar to that of the IN738 layer. Based on these response curves, dimensional microhardness corresponding to each layer can be determined, as shown in Figure 13.10(c). It is evident that the microhardness of 1st IN718 is approximately 270 HV, which aligns with reported literature values for IN718 (273±11.9 HV) [42]. The average microhardness of the 4th IN718 layer measures around 360 HV, considerably higher than that of purely deposited IN718. This indicates a significant increase in hardness for the innermost IN718 layer due to its interaction with IN738. The enhanced hardness in this case is closely associated with intensive precipitation of nanometer γ' particles between dendrites and the enhanced effect of solid solution strengthen. For the IN738 layer, both second and third layers exhibit an average microhardness measuring about 460 HV and 440 HV respectively. However, there exists a maximum deviation range of test samples at approximately ±50 HV. Compared to reports on the purely deposited IN738, whose microhardness is approximately about 480 HV [43], slight reduction in hardness is observed here. Figure 13.10(d)

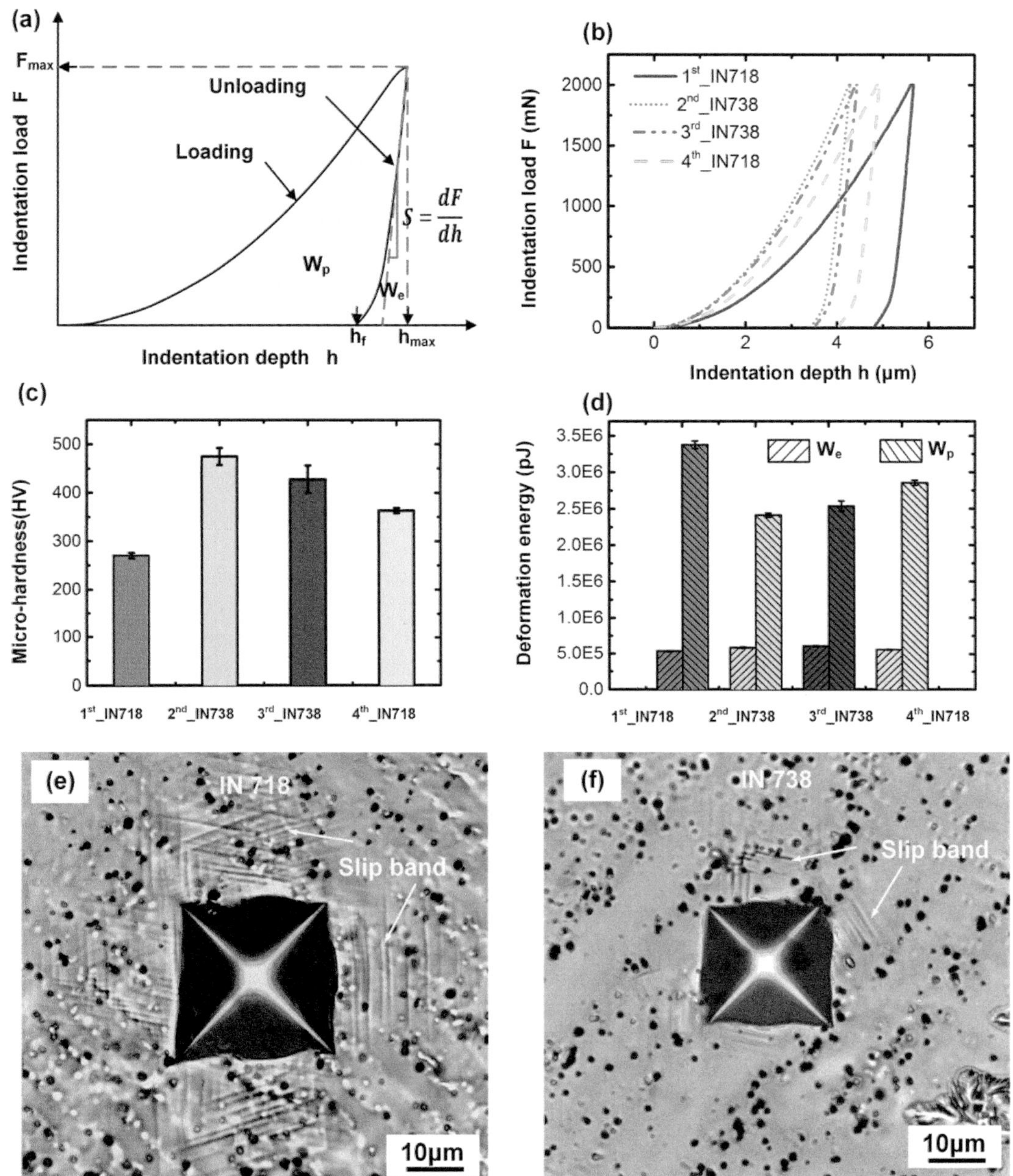

Figure 13.10. Micro-indentation tests of as-deposited IN738+IN718 LHS, (a) schematic curves of response, (b) response curves of the top 4 layers, (c) statistics of micro-hardness, (d) statistics of deformation energy, (e) and (f) residual indentations of the 1st IN718 layer and the 2nd IN738 layer respectively.

shows the elastic and plastic energy dissipation reflected in the response curve. It can be seen that the plastic energy consumption accounts for the vast majority of the entire energy consumption. The difference of plastic energy consumption is large, while the difference of elastic energy consumption is little affected by the difference of composition. The plastic energy consumption of 1st IN718 is the highest, followed by 4th IN718. It can be seen that, compared with the IN738 layer, the IN718 layer has better plasticity and can consume more energy under load conditions. The indentation metallography shown in Figure 13.10(e) and (f) also once again proves the good plasticity of the

IN718 layer. It can be seen from the figure that the indentation area of layer IN718 is larger and there are more slip bands around the indentation. From the above microhardness information, it can be seen that IN738 layer has higher hardness, but poor plasticity, so the deformation resistance is stronger, which also means that its internal stress is difficult to be effectively released through deformation. As a result, internal stresses are preferentially concentrated in certain weak locations, such as $\gamma+\gamma'$ eutectic, or pores formed by constituent liquation. The cavity area and the three intersection points of the grain boundary lead to a higher risk of cracking. On the contrary, IN718 layer has lower hardness, better plasticity and toughness, poor resistance to deformation, and easier to alleviate stress concentration through plastic deformation. It can be inferred that when the IN738 layer and the IN718 layer are combined by the composite deposition method, a micron-level layered gradient material can be formed. Among them, the IN718 layer can provide the necessary plasticity and toughness in order to timely release the huge internal stress transmitted by the IN738 layer, helping to alleviate the stress concentration during the deposition process. In this way, various cracking, including liquation cracks and solid-state cracks, can be effectively inhibited during DED. Moreover the strength of the deformed IN718 layer can be enhanced due to the effect of strain strengthening, which helps to guarantee the strength of the heterogeneous structure.

13.4.2 Tensile properties of the as-deposited LHS

In this section, the tensile properties of the as-deposited structures are evaluated at both room temperature (25°C) and elevated temperatures (600°C, 800°C). Purely deposited IN718 and IN738 are also subjected to tensile testing to make comparison. It is worth mentioning that the purely deposited IN738 in air have severe cracks, rendering the tensile tests impossible. Therefore, the purely IN738 used for tensile tests were deposited in the Ar condition with the oxygen concentration less than 30 ppm. The other tensile samples were deposited in the air condition.

The tensile test results are presented in Figure 13.11. As depicted in 13.11(a), the IN738+IN718 LHS exhibits excellent strength and plasticity at room temperature. While the purely deposited IN738 demonstrates high yield strength but very limited plasticity, it experiences immediate fracture upon reaching yield. In contrast, the purely deposited IN718 possesses lowest strength compared to the LHS and IN738 but displays best plasticity. Notably, both the LHS and purely depostied IN718 exhibit a stable strain strengthening stage after yielding, indicating good toughness which can be quantified by measuring the envelope area beneath the tensile curve. Regarding the ultimate tensile strength (UTS) (Figure 13.11(d)), the LHS attains the highest value with an average of 1131 MPa. The average UTS for purely deposited IN738 is about 1045 MPa while that for purely deposited IN718 is significantly lower at an average of 839 MPa. With respect to yield strength ($\sigma_{0.2}$) (Figure 13.11(e)), on average, the IN738 + IN718 LHS yields at approximately 920 MPa which slightly falls below that of purely deposited IN738 (1010 MPa), yet remains substantially higher than that of purely deposited IN718 (450 MPa). In terms of the total elongation to fracture (TFE) (Figure 13.11(e)), the mean value for TFE in case of LHS reaches up to around 16%, significantly surpassing that of purely deposited IN738 (1.4%). The TFE of purely deposited IN718 is about 35%. It can be found that the purely deposited IN718 or IN738 exhibits imbalanced strength and plasticity, while the LHS has superior synergy of strength and plasticity.

The tensile curve at 600°C, as depicted in Figure 13.11(b), demonstrates that the as deposited LHS exhibits sustained strength and plasticity, while the purely deposited IN738 samples fractures immediately after reaching yield and the purely deposited IN718 has very limited strength. Moreover, both the LHS and IN718 samples exhibit obvious strain strengthening effects even after yielding and there are noticeable zigzag fluctuations are observed on their respective tensile curves. These zigzag fluctuations during tensile at elevated temperature are typically attributed to dynamic strain aging r twin deformation or the austenitic alloys. According to the UTS statistics presented in Figure 13.11(d), the mean value of the LHS reaches a significant level of 930 MPa, surpassing that of purely deposited IN738 (887 MPa) and IN718 (712 MPa). Correspondingly, Figure 13.11(e)

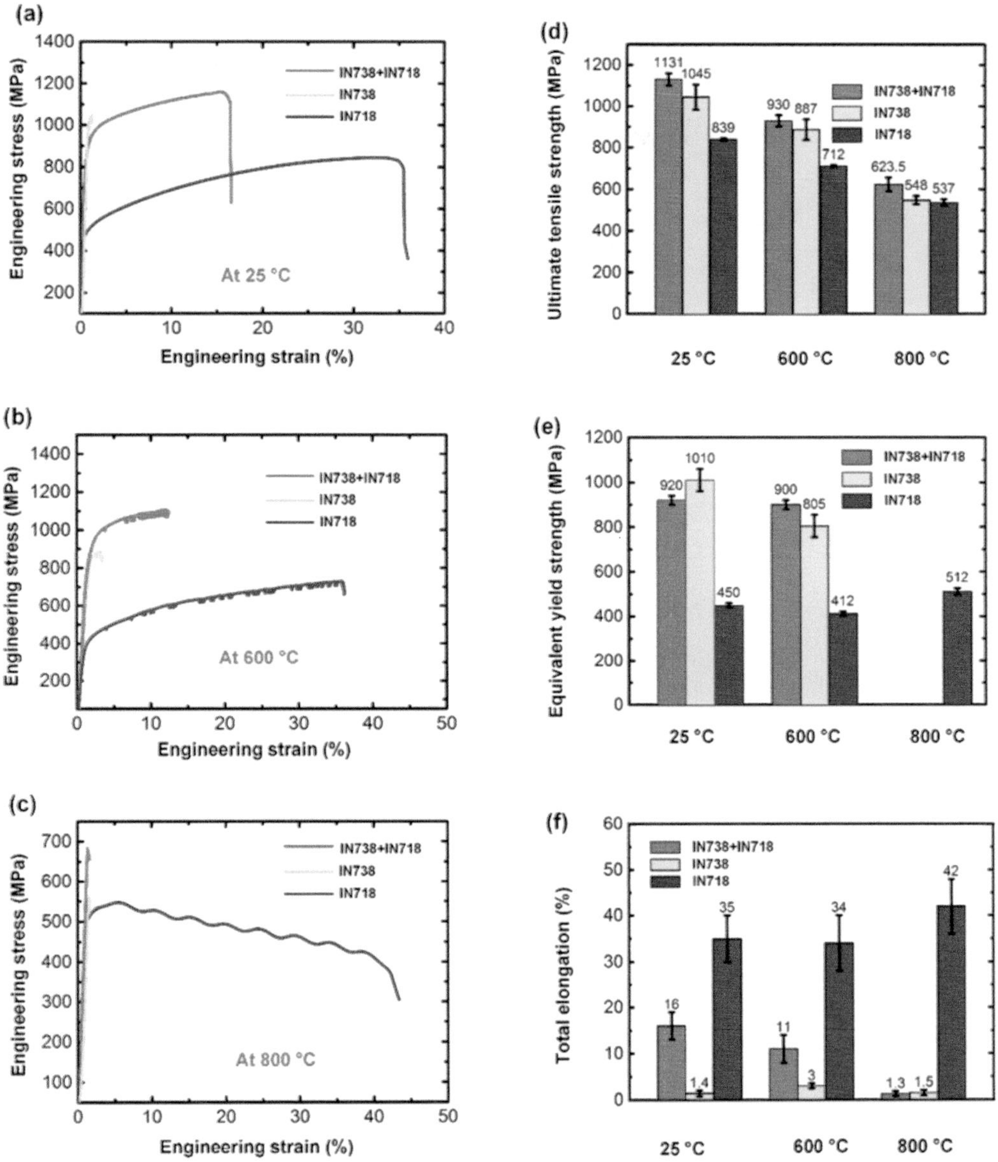

Figure 13.11. The results of tensile tests, (a), (b), (c) are the tensile curves at 25°C, 600°C, and 800°C respectively, (d), (e), (f) show the UTS, σ0.2 and TEF respectively.

illustrates the $\sigma_{0.2}$ of the LHS is about 900 MPa, which is higher than those for the purely deposited IN738 (805 MPa) and IN718 (412 MPa). In addition, the TFE of the LHS stands at an impressive level of 11%, significantly exceeding that of IN738 (3%). These result demonstrate that the superior synergy of strength and plasticity LHS still sustainable at 600°C.

Figure 13.11(c) shows the tensile curve at 800°C, which demonstrates although the as-deposited LHS has excellent strength that reaches to about 623 MPa, their plasticity is very limited just as the purely deposited IN738. On the contrary, the purely deposited IN718 exhibits high TFE but poor strength. Besides, the tensile curve of IN718 has a wavy downward trend after the yielding, implying the strain strengthening effect is failure at 800°C. The poor TEF of LHS possibly deduced to the existence of needle δ in the IN718 layer of LHS. Figure 13.12 illustrates the fracture signs of the component layer in LHS at 600°C. It can be found that the Laves and the needle δ around the Laves

Figure 13.12. The microstructures of the LHS tensile specimen fractured at 600°C, (a) IN718 layer, (b) IN738 layer.

exhibit obvious fracture signs in the IN718 layer of LHS, and there are some more subtle cracks around the γ+γ′ eutectic of IN738 layer. That primarily demonstrates the brittleness of LHS at high temperature is related to the needle δ. The brittleness is expected to be improved by heat treatment.

13.4.3 Microstructure and tensile properties of the heat-treated LHS

Referring to the heat treatment of IN718, the heat treatment of IN738+IN718 LHS was carried out using the process shown in Table 13.4. The first step is the homogenization treatment, the main purpose of which is to eliminate the element segregation between the dendrites, release the residual stress, and provide the basis for the subsequent heat treatment. The main purpose of the second step is to dissolve Laves phase and δ phase. The purpose of the third two-stage aging treatment is to fully precipitate γ′ or γ″ phase to obtain the strong and stabile microstructure.

No cracking occurred during heat treatment of the LHS, the microstructure after heat treatment is shown in Figure 13.13. Among them, the core structure of the IN718 layer is shown in Figure 6.19(a) and (b). It can be found that after heat treatment, the Laves of the IN718 layer decreased significantly, they were dispersed as small particles at the grain boundaries. There is no needle δ phase in the structure, but numerous MC-type carbide particles are still dispersed. Upon further magnification of the matrix, it can be observed that γ′ particles with a particle size ranging from 50 to 70 nm precipitate within the matrix. Figure 13.13(c) and (d) depict the microstructure near the interface of IN718 and IN738, dense precipitation of γ′ particles is also observed in this region. However, these γ′ particles exhibit a bimodal distribution in terms of size. Particularly, the larger square-shaped γ′ particles with a size around 200 nm coexist with the smaller densely packed γ′ particles measuring approximately 20 nm. Figure 13.13(e) and (f) illustrate the microstructure of the core layer of IN738 where it can be seen that after heat treatment the γ+γ′ eutectic phase disappeared and maintaining a bimodal distribution for γ′ particles. Compared to those found near the interface, the contents of the larger γ′ particles is higher. The microstructure characteristics suggest that the heat treatment not only dissolves original Laves phase, δ phase and γ+γ′ eutectic

Table 13.4. Heat treatment of IN738+IN718 LHS.

Steps	Temperature	Holding time	Cooling
1: Homogenizing	1150°C	1 h	Air cooling
2: Solution	980°C	2 h	Air cooling
3: Two-stage aging	720°C	8 h	Furnace cooling to 620°C at 50°C/h rate
	620°C	8 h	Air cooling

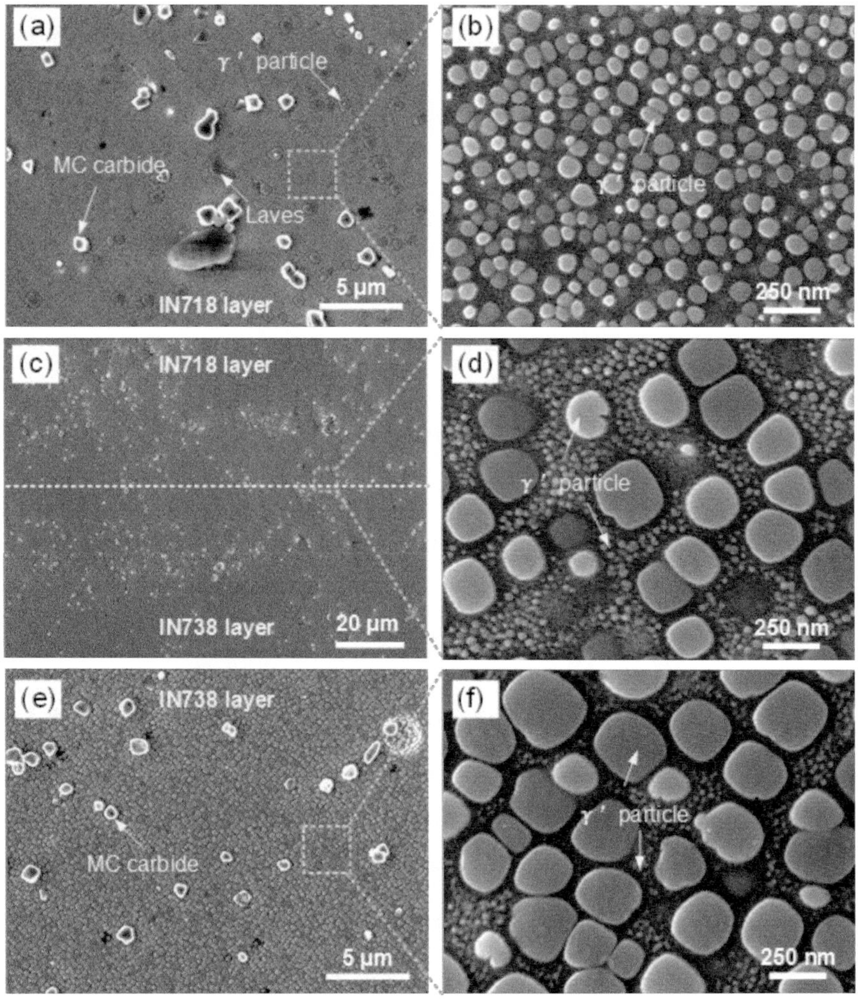

Figure 13.13. SEM images of heat-treated IN738+IN718 LHS: (a) and (b) in the core of IN718 layer, (b) and (d) at the interface of IN718 layer and IN738, (e) and (f) in the core of IN738.

phase but also ensures more even distribution of easily separated elements such as Nb, Ti, Al among dendrites thereby preventing regeneration of Laves phases δ or γ+γ′ eutectics. Moreover, both the IN718 and IN738 layers in LHS are full of γ′ which promises excellent strength and the mechanical properties at high temperature.

The mechanical properties of IN718+IN738 LHS after heat treatment are presented in Figure 13.14. Figure 13.14(a) illustrates the tensile curves at room temperature, 600°C, and 800°C. It can be found that the tensile strength of the heat-treated LHS is significantly enhanced at both 25°C and 600°C, as well as after heat treatment at 800°C. This attributes to the complete precipitation γ′ and the dissolution of harmful phases such as needle δ and γ+γ′. Regarding the plasticity, the heat-treated LHS shows a slight decrease in TEF at room temperature. The possible reason is that the full precipitation of γ′ hinders dislocation slip, resulting in increased strength but reduced plasticity. However, at elevated temperatures such as 600°C and 800°C, heat-treated specimens demonstrate higher TEF compared to the as-deposited specimens due to the dissolution of brittle phases such as needle δ and γ+γ′. Besides, during the tensile tests at high temperatures, γ′ can be precipitated for the as-deposited LHS, this precipitation increased the strengthen and decreased the plasticity, leading

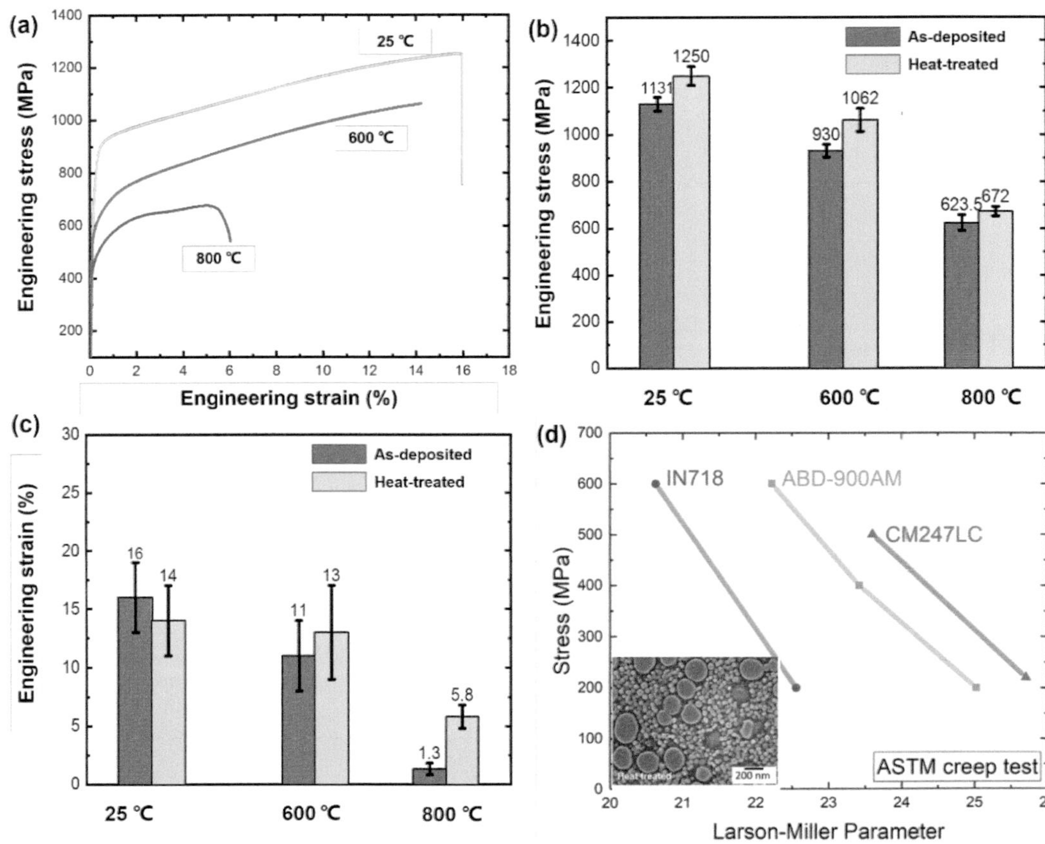

Figure 13.14. Mechanical properties of heat treated IN738+IN718 LHS, (a) tensile curves at 25°C, 600°C and 800°C, (b) and (c) are the tensile strength and elongation before and after heat treatment, (d) is the creep property of ABD-900AM (the inset shows the microstructure of heat treated ABD-900AM) [44].

to the lower plasticity. Conversely, once the γ' has been fully precipitated during heat-treatment, the precipitation of γ' can be very limited during tensile tests at high temperatures.

In addition to tensile properties, creep properties at high temperatures are also crucial mechanical characteristic for superalloys. Previous studies have demonstrated that the γ' phase with a bimodal distribution is more favorable for enhancing the creep resistance at high temperatures. Generally, a smaller spacing between γ' particles leads to increased resistance to dislocation. Under low stress conditions, due to the pinning effect, dislocations cannot slip around the lager γ' particles thus moves through the γ' particles in the form of climbing. However, the climbing requires a higher driving force, thereby leading to the lower creep rates. As stress increases, although dislocations can bypass the large particle γ', forming a dislocation ring around it, they still cannot bypass the small particle γ', and continue climbing through it. When the stress further increases, dislocations can bypass both large massive particles γ' and small spherical particles γ' finally. Figure 13.14(d) illustrates that apart from the IN718 layer core, the LHS primarily consists of bimodally distributed γ', therefore the excellent creep resistance can be ensured. Tang et al. [44] reported a new nickel-based superalloy designed specifically for additive manufacture, namely ABD-900AM, in which the γ' size are bimodally distributed as well, and the γ' content is slightly less than that of the LHS. Therefore, it can be inferred that LHS has comparable or even higher creep resistance than ABD-900AM.

In addition to tensile properties, high temperature creep properties are also important mechanical properties of superalloys. Previous studies have shown that γ' phase with bimodal distribution is more conducive to the improvement of creep resistance at high temperature; in addition, the smaller the spacing between γ′ particles, the greater the resistance of γ′ to dislocation. Under low stress, the dislocation cannot slip around the strengthening phase due to the pinning action of the strengthening relative dislocation, so it mainly passes through the massive large particle γ′ and the spherical small particle γ′ in the way of climbing. However, the climbing of the dislocation requires a higher driving force, so the creep rate is lower. As the stress increases, the dislocation can bypass the large particle γ′ and form a dislocation ring around it, but it still cannot bypass the small particle γ′, and the dislocation still climbs through the small particles γ′. When the stress is further increased, the dislocation can bypass both large massive particles γ′ and small spherical particles γ′. As can be seen from Figure 13.13, except IN718 layer core, the composite sedimentary structure mainly consists of γ′ with bimodal distribution, which can ensure good creep resistance. In addition, the IN718 layer itself is small in number, although the core is mainly spherical γ′ particles, but densely arranged, there is a tendency to shift to γ′ under the action of high temperature for a long time, so the preliminary conclusion is that its damage to the overall creep resistance is small. Recently, Tang et al. [44] reported a new γ-strengthened nickel-based superalloy ABD-900AM designed for additive manufacturing and described its creep resistance after heat treatment in the form of Larson Miller Parameter (LMP).

13.5 Synergistic strengthening and toughening mechanism

The primary reason for the superior combination of strength and ductility in LHS materials is hetero-deformation induced (HDI) strengthening and HDI strain hardening. Many researches have reported the primary mechanism of HDI. Specifically, as depicted in Figure 13.15, geometrically necessary dislocations (GNDs) accumulate and pile up near the boundaries of the soft zones in LHS, thereby resulting in their reinforcement. Additionally, the accumulation of GNDs generates back stress within the soft zones and forward stress within the hard zones, collectively giving rise to HDI stress. Furthermore, this HDI stress induces numerous dispersed microscopic strain bands in LHS materials, facilitating the distribution of plastic strain across the entire gauge length thus enhancing uniform elongation.

The plastic deformation of metals is related to the generation, storage, and movement of dislocations. According to the Nye theory [45], dislocations in metals have two storage forms, namely statistically stored dislocations (SSD) and geometrically necessary dislocations (GND). Among them, SSD refers to the discrete distribution of dislocations in the metal due to random entanglement. GND refers to the stored dislocations that exist to match the uneven plastic deformation of different parts of the material and maintain the material's continuity. The plastic deformation in the material at the microscopic scale is often non-uniform, resulting in misfit phenomena in different

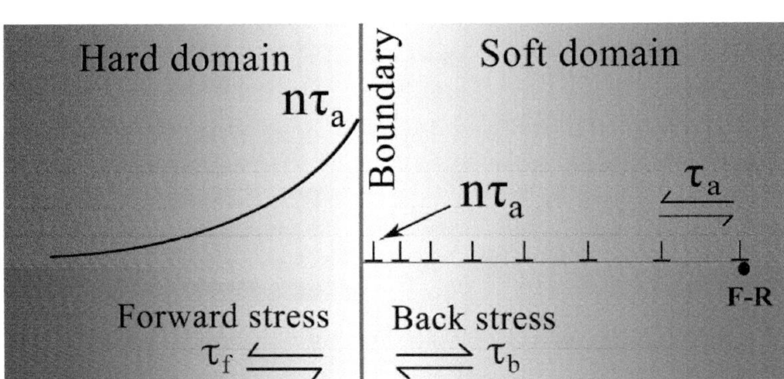

Figure 13.15. Schematic of the pile up of geometrically necessary dislocations, which produces back stress in the soft zone and forward stress in the hardzone.

regions. The misfit phenomena then cause the strain gradient, which needs to be adapted by GND. Therefore, GND is related to lattice bending and can be quantitatively analyzed by EBSD data. The quantification of GND involves various methods, according to the study by Moussa et al. [46], the density of GND ρ_{GND} can be obtained through the following equation:

$$\rho_{GND} = \frac{\alpha\theta}{|b|x} \tag{13.4}$$

Where α is a parameter related to the characteristics of small-angle grain boundaries, and for strongly oriented misfit axes, α can take the value of 3. θ represents the orientation deviation angle, which can be represented by the local average orientation deviation KAM. b is the Burgers vector, and x is the step size. In calculating ρ_{GND}, the kernel size of KAM is set to a 3 × 3 voxel for the purpose of excluding large-angle grain boundaries, and the cut-off angle is selected as 10°. Since the matrix phase γ and γ′ are both face-centered cubic structures, b = 0.5<110>. The EBSD test results of the first three layers of the as-deposited IN738+IN718 LHS are shown in Figure 13.16. From Figure 13.16(a), it can be seen that there is a certain thickness of polycrystalline region in the top IN718 layer, below the polycrystalline region is a well-oriented columnar grain, and the <101> direction of the columnar grain is mainly parallel to the BD direction. It can be concluded from Figure 13.16(b) that there is a significant difference in the GND distribution between the IN738 layer and the IN718 layer. GNDs in the IN718 layer and the interface of IN718/IN738 is clearly higher than that in IN738 layer. This indicates that the IN718 layer undergoes more severe plastic deformation during the DED. The distribution of GND conforming to the principles of HDI reinforcement. Thus provides excellent synergy of strengthening and toughening. For the as-deposited LHS, the differences of GND distribution are caused by the un-uniform deformation under the inner-stress occurred during DED, the resultant HDI strengthening enhances the synergy of strengthening and toughening, further, this synergy implies the LHS posses the strong cracking resistance during DED. For the heat-treated LHS, due to the microstructure differences between IN718 layer and IN738 layer, HDI reinforcement can exist as well during the deformation under external load, which synergistically enhances the strength and toughness of the DEDed component.

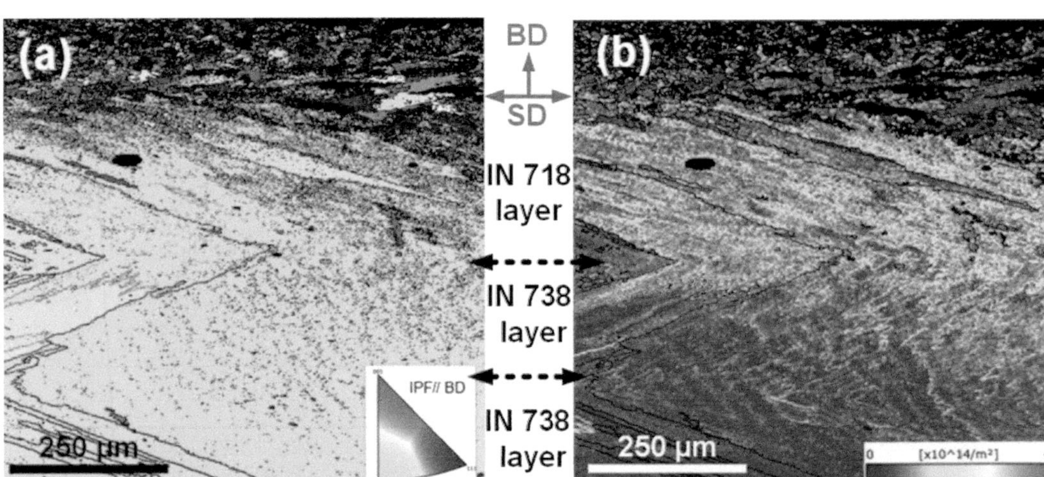

Figure 13.16. EBSD results of as-deposited IN738+IN718 LHS, (a) IPF-BD map, (b) distribution of GND.

13.6 Summary

In this chapter, L-DED of hetero-structured IN738+IN718 alloys with excellent strength and ductility as studied, the microstructure characteristics and the mechanical properties before and after heat treatment were systematically revealed. The main conclusions are as follows.

1) The IN738+IN718 LHS can be obtained through Laser Directed Energy Deposition (L-DED) utilizing an alternative powder feeding method. The obtained LHS is crack-free and have synergistically enhanced strength and toughness.
2) In the IN718 layers of as-deposited LHS, γ' nanoparticles were deposited in the interdendrites, while γ phase remained in the inner-dendrite zone. The interdendritic long chain Laves phase dissolved into islands, accompanied by the precipitation of acicular δ phase around the Laves phase.
3) In the IN718 layers of as-deposited LHS, the precipitation of γ' did not change significantly, the dense precipitation of γ' nanoparticles was maintained between the dendrites and the dendrite stem, and the size of γ' particles between the dendrites was larger, in addition, the γ+γ' eutectic structure still appeared between the dendrites. The difference is that there may be acicular δ phase near the IN718/IN738 interface.
4) The as-deposited LHS has good strength and plasticity below 600°C, and the tensile strength at room temperature and 600°C can reach 1131 MPa and 930 MPa, respectively, and the total elongation is 16% and 11%, respectively. At 800°C, the tensile strength is about 623 MPa, but the plasticity is significantly reduced due to the existence of needle δ.
5) After homogenization at 1150°C for 1 h, solid solution at 980°C for 2 h, and dual aging at 720°C and 620°C for 8 h, Laves phase, δ phase and γ+γ' eutectic phase in the LHS can significantly reduced or even disappeared. Most of the matrix consisted of γ' particles with bimodal sizes. The strength and plasticity at 600°C and 800°C can be synchronously improved due to the HDI strengthening.

References

[1] Mostafaei, A., Ghiaasiaan, R., Ho, I. T., Strayer, S., Chang, K. C., Shamsaei, N. et al. (2023). Additive manufacturing of nickel-based superalloys: A state-of-the-art review on process-structure-defect-property relationship[J]. Progress in Materials Science 136: 101108.

[2] Chaolin Tan, Fei Weng, Shang Sui, Youxiang Chew and Guijun Bi. (2021). Progress and perspectives in laser additive manufacturing of key aeroengine materials [J]. International Journal of Machine Tools and Manufacture 170: 103804.

[3] u, X. L. and Zhu, Y. T. (2017). Heterogeneous materials: a new class of materials with unprecedented mechanical properties. Mater Res. Lett. 5: 527–32.

[4] hu, Y. T. and Wu, X. L. (2019). Perspective on heterogeneous deformation induced (HDI) hardening and back stress. Mater Res. Lett. 7: 393–8.

[5] Li, J. G., Zhang, Q., Huang, R. R., Li, X. Y. and Gao, H. J. (2020). Towards understanding the structure-property relationships of heterogeneous-structured materials. Scr Mater. 186: 304–11.

[6] Wu, X. L., Yang, M. X., Yuan, F. P., Wu, G. L., Wei, Y. J., Huang, X. X. et al. (2015). Heterogeneous lamella structure unites ultrafine-grain strength with coarse-grain ductility. Proc Natl Acad Sci USA 112: 14501–5.

[7] Han, B. Q., Huang, J. Y., Zhu, Y. T. and Lavernia, E. J. (2006). Strain rate dependence of properties of cryomilled bimodal 5083 Al alloys. Acta Mater 54: 3015–24.

[8] Han, B. Q., Lee, Z., Witkin, D., Nutt, S. and Lavernia, E. J. (2005). Deformation behavior of bimodal nanostructured 5083 Al alloys. Metall Mater Trans A 36a: 957–65.

[9] Lu, K. (2014). Making strong nanomaterials ductile with gradients. Science 345: 1455–6.

[10] Chen, A. Y., Liu, J. B., Wang, H. T., Lu, J. and Wang, Y. M. (2016). Gradient twinned 304 stainless steels for high strength and high ductility. Mater Sci. Eng. A 667: 179–88.

[11] Wei, Y. J., Li, Y. Q., Zhu, L. C., Liu, Y., Lei, X. Q., Wang, G. et al. (2014). Evading the strength- ductility trade-off dilemma in steel through gradient hierarchical nanotwins. Nature Comm. 5: 1–8.

[12] Calcagnotto, M., Adachi, Y., Ponge, D. and Raabe, D. (2011). Deformation and fracture mechanisms in fine- and ultrafine-grained ferrite/martensite dual-phase steels and the effect of aging. Acta Mater 59: 658–70.

[13] Li, Z. M., Pradeep, K. G., Deng, Y., Raabe, D. and Tasan, C. C. (2016). Metastable high-entropy dual-phase alloys overcome the strength-ductility trade-off. Nature 534: 227–30.

[14] Park, K., Nishiyama, M., Nakada, N., Tsuchiyama, T. and Takaki, S. (2014). Effect of the martensite distribution on the strain hardening and ductile fracture behaviors in dualphase steel. Mater Sci. Eng. A 604: 135–41.

[15] Sawangrat, C., Kato, S., Orlov, D. and Ameyama, K. (2014). Harmonic-structured copper: performance and proof of fabrication concept based on severe plastic deformation of powders. J. Mater Sci. 49: 6579–85.

[16] Vajpai, S. K., Ota, M., Watanabe, T., Maeda, R., Sekiguchi, T., Kusaka, T. et al. (2015). The development of high performance Ti-6Al-4V alloy via a unique microstructural design with bimodal grain size distribution. Metall Mater Trans A 46: 903–14.

[17] Zhang, Z., Vajpai, S. K., Orlov, D. and Ameyama, K. (2014). Improvement of mechanical properties in SUS304L steel through the control of bimodal microstructure characteristics. Mater Sci Eng A 598: 106–13.

[18] Ma, X. L., Huang, C. X., Moering, J., Ruppert, M., Hoppel, H. W., Goken, M. et al. (2016). Mechanical properties in copper/bronze laminates: role of interfaces. Acta Mater 116: 43–52.

[19] Ma, X. L., Huang, C. X., Xu, W. Z., Zhou, H., Wu, X. L. and Zhu, Y. T. (2015). Strain hardening and ductility in a coarse-grain/nanostructure laminate material. Scripta Mater 103: 57–60.

[20] Beyerlein, I. J., Mayeur, J. R., Zheng, S. J., Mara, N. A., Wang, J. and Misra, A. (2014). Emergence of stable interfaces under extreme plastic deformation. Proc Natl. Acad. Sci. USA 111: 4386–90.

[21] Llorca, J., Needleman, A. and Suresh, S. (1990). The Bauschinger effect in whisker-reinforced metal-matrix composites. Scripta Metall Mater 24: 1203–8.

[22] Thilly, L., Van Petegem, S., Renault, P. O., Lecouturier, F., Vidal, V., Schmitt, B. et al. (2009). A new criterion for elasto-plastic transition in nanomaterials: Application to size and composite effects on Cu-Nb nanocomposite wires. Acta Mater 57: 3157–69.

[23] Tanaka, K. and Mori, T. (1970). Hardening of crystals by non-deforming particles and fibres. Acta Metall 18: 931–1000.

[24] Nie, J. F., Chen, Y. Y., Chen, X., Liu, X. F., Liu, G. L., Zhao, Y. H. et al. (2020). Stiff, strong and ductile heterostructured aluminum composites reinforced with oriented nanoplatelets. Scr. Mater 189: 140–4.

[25] Toshihiko Koseki, Junya Inoue and Shoichi Nambu. (2014). Development of multilayer steels for improved combinations of high strength and high ductility [J]. Materials Transactions M2013382.

[26] uang, M., Xu, C., Fan, G., Maawad, E., Gan, W., Geng, L. et al. (2018). Role of layered structure in ductility improvement of layered Ti-Al metal composite [J]. Acta Materialia S1359645418303574.

[27] Liu, H. S., Zhang, B. and Zhang, G. P. (2011). Enhanced toughness and fatigue strength of cold roll bonded Cu/Cu laminated composites with mechanical contrast[J]. Scripta Materialia 65(10): 891–894.

[28] Carpenter, K. and Tabei, A. (2020). On residual stress development, prevention, and compensation in metal additive manufacturing[J]. Materials 13(2): 255.

[29] Guo, C., Li, G., Li, S., Hu, X., Lu, H., Li, X. et al. (2022). Additive manufacturing of Ni-based superalloys: Residual stress, mechanisms of crack formation and strategies for crack inhibition[J]. Nano Materials Science.

[30] Chaolin Tan, Youxiang Chew, Ranxi Duan, Fei Weng, Shang Sui, Fern Lan Ng et al. (2021). Additive manufacturing of multi-scale heterostructured high-strength steels [J]. Materials Research Letters 9(7): 291–9.

[31] Chaolin Tan, Youxiang Chew, Fei Weng, Shang Sui, Fern Lan Ng, Tong Liu et al. (2022). Laser aided additive manufacturing of spatially heterostructured steels [J]. International Journal of Machine Tools and Manufacture 172: 103817.

[32] Sun, Y., Zhang, C., Ning, Z., Sun, J., Ngan, A. H. and Huang, Y. (2024). Additively manufactured low-gradient interfacial heterostructured medium-entropy alloy multilayers with superior strength and ductility synergy[J]. Composites, Part B. Engineering 2024(Jul.1): 280.

[33] Ferreri, N. C., Vogel, S. C. and Knezevic, M. (2020). Determining volume fractions of γ, γ', γ'', δ, and MC-carbide phases in Inconel 718 as a function of its processing history using an advanced neutron diffraction procedure [J]. Materials Science and Engineering: A 781(Apr.20): 139228.139221–139228.139212.

[34] Kim, S., Choi, H., Lee, J. and Kim, S. (2020). Room and elevated temperature fatigue crack propagation behavior of Inconel 718 alloy fabricated by laser powder bed fusion [J]. Int. J. Fatigue 140: 105802.

[35] Pant, P., Proper, S., Luzin, V., Sjöström, S., Simonsson, K., Moverare, J. et al. (2020). Mapping of residual stresses in as-built Inconel 718 fabricated by laser powder bed fusion: A neutron diffraction study of build orientation influence on residual stresses [J]. Additive Manufacturing 36: 101501.

[36] Parimi, L. L., Ravi, G. A., Clark, D. and Attallah, M. M. (2014). Microstructural and texture development in direct laser fabricated IN718 [J]. Materials Characterization 89: 102–111.

[37] Stevens, E. L., Toman, J., To, A. C. and Chmielus, M. (2017). Variation of hardness, microstructure, and Laves phase distribution in direct laser deposited alloy 718 cuboids [J]. Materials & Design 119(APR.): 188–198.

[38] Dehmas, M., Lacaze, J., Niang, A. and Viguier, B. (2011). TEM study of high-temperature precipitation of delta phase in Inconel 718 Alloy [J]. Advances in Materials Science and Engineering.

[39] Rosenthal, R. and West, D. R. F. (1999). Continuous γ' precipitation in directionally solidified IN738 LC alloy [J]. Metal Ence Journal 15(12): 1387–1394.

[40] Sidhu, R. K., Ojo, O. A. and Chaturvedi, M. C. (2007). Microstructural analysis of laser-beam-welded directionally solidified INCONEL 738 [J]. Metallurgical & Materials Transactions A 38(4): 858–870.
[41] Attaf, M. T. (2004). Connection between the loading curve models in elastoplastic indentation [J]. Materials Letters 58(27): 3491–3498.
[42] Tillmann, W., Schaak, C., Nellesen, J., Schaper, M., Aydinöz, M. U. and Hoyer, K. P. (2017). Hot isostatic pressing of IN718 components manufactured by selective laser melting [J]. Additive Manufacturing 13: 93–102.
[43] Xu, J., Lin, X., Guo, P., Hu, Y., Wen, X., Xue, L. et al. (2017). The effect of preheating on microstructure and mechanical properties of laser solid forming IN-738LC alloy [J]. Materials Science and Engineering: A 691: 71–80.
[44] Tang, Y. T., Panwisawas, C., Ghoussoub, J. N., Gong, Y., Clark, J. W., Németh, A. A. et al. (2021). Alloys-by-design: Application to new superalloys for additive manufacturing [J]. Acta Materialia 202: 417–436.
[45] Ashby, M. F. (1970). The deformation of plastically non-homogeneous materials [J]. The Philosophical Magazine: A Journal of Theoretical Experimental and Applied Physics 21(170): 399–424.
[46] Moussa, C., Bernacki, M., Besnard, R. and Bozzolo, N. (2015). About quantitative EBSD analysis of deformation and recovery substructures in pure Tantalum [J]. Iop Conf Ser-Mat Sci. 89(1): 012038.

Index

I

L

M

N

O

P

R

S

For Product Safety Concerns and Information please contact our
EU representative GPSR@taylorandfrancis.com Taylor & Francis
Verlag GmbH, Kaufingerstraße 24, 80331 München, Germany